COSPAR COLLOQUIA SERIES

VOLUME 15

DUST IN THE SOLAR SYSTEM AND OTHER PLANETARY SYSTEMS

DUST IN THE SOLAR SYSTEM

AND OTHER PLANETARY SYSTEMS

Proceedings of the IAU Colloquium 181
held at the University of Kent, Canterbury, U.K.
4-10 April 2000

Edited by

S.F. Green
Planetary and Space Sciences Research Institute
The Open University
Milton Keynes, U.K.

I.P. Williams
Astronomy Unit, School of Mathematical Sciences
University of London
London, U.K.

J.A.M. McDonnell
Planetary and Space Sciences Research Institute
The Open University
Milton Keynes, U.K.

N. McBride
Planetary and Space Sciences Research Institute
The Open University
Milton Keynes, U.K.

2002

PERGAMON
An imprint of Elsevier Science

Amsterdam – Boston – London – New York – Paris – San Diego – San Francisco – Singapore – Sydney – Tokyo

ELSEVIER SCIENCE Ltd
The Boulevard, Langford Lane
Kidlington, Oxford OX5 1GB, UK

First edition 2002

Library of Congress Cataloging in Publication Data
A catalog record from the Library of Congress has been applied for.

British Library Cataloguing in Publication Data
A catalogue record from the British Library has been applied for.

ISBN: 0 08 044194 7

♾ The paper used in this publication meets the requirements of ANSI/NISO Z39.48-1992 (Permanence of Paper).
Printed in The Netherlands.

PREFACE

This joint IAU and COSPAR Colloquium, held at the campus of The University of Kent at Canterbury from April 10 to 14, 2000 brought together 129 scientists from 18 countries. It was a continuation of the tradition of holding meetings at regular intervals of a few years in order to review the progress in a broad range of disciplines that are relevant to the study of interplanetary dust and to help to unify progress made through observations, both in situ and from the ground, theory and experimentation. The series of meetings started in Honolulu, Hawaii (USA) in 1967, followed by Heidelberg (Germany) in 1975, then Ottowa (Canada) in 1979, Marseilles (France) in 1984, Kyoto (Japan) in 1990 with the last being in Gainesville, Florida (USA) in 1995.

Since the Gainesville meeting, there have been dramatic changes in the field resulting from in-situ space experiments, Earth orbiting satellites and ground based observations. The brightest comet since the early years of the twentieth century, comet Hale-Bopp, appeared, giving an invaluable opportunity to see in action one great source of interplanetary dust. Similarly, the Leonid meteor shower has been at its most active since 1966, producing spectacular displays of meteors and allowing for an array of observational techniques, not available in 1966 to be used, while theory has also been refined to a level where very accurate predictions of the timing of meteor storms has become possible. Prior to the meeting we observed a total eclipse of the Sun in SW England and Northern Europe, traditionally a good opportunity to observe the Zodiacal cloud. Our knowledge of the Near-Earth Asteroid population has also increased dramatically, with the increased study arising from the heightened awareness of the danger to Earth from such bodies. Extrasolar planets have been discovered since the last meeting and it is recognised that we can now study interplanetary dust in other Planetary Systems. Since much of the dust observed in such systems is at a distance of order 100 AU from the star, this brings into focus the production of dust in the Edgeworth-Kuiper Belt of our own system. Recent years have seen a recognition of the importance of dust originating *outside* our own system, that is now present in the near-Earth environment. As is always the case when great strides take place observationally, much theoretical work follows, and the same is true in this instance.

While data about the interplanetary medium from Venus to Jupiter was beginning to be available at the last meeting, the data from both Galileo and Ulysses have now been more fully analysed, with a corresponding increase in our knowledge. Since then however information from SOHO and MSX have become available, giving new insight into the dust population close to the Sun. In addition, ISO allowed us to study the radiation emitted from dust (as opposed to its more normal obscuring properties). There are also new space missions in various stages of planning, Particularly STARDUST and ROSETTA, that will produce a whole new dimension to our knowledge of dust production in the Solar system.

The scientific Organizing Committee was responsible for defining the scientific content and selecting the invited reviews. These proceedings contain 13 invited reviews and invited contributions, and 46 contributed papers. The papers reflect the thematic approach adopted at the meeting, with a flow outwards (from meteors in the atmosphere, through zodiacal dust observation and interplanetary dust, to extra solar planetary systems) and returning (via the Edgeworth Kuiper belt and comets) to the Earth, with laboratory studies of physical and chemical processes and the study of extra-terrestrial samples.

Simon Green, Iwan Williams, Tony McDonnell, Neil McBride.

SCIENTIFIC ORGANISING COMMITTEE

I.P. Williams (UK, Chair)
J.A.M. McDonnell (UK, Co-chair)
W.J. Baggaley (New Zealand)
E. Grün (Germany)
M.S. Hanner (USA)
P. Lamy (France)
A.C. Levasseur Regourd (France)
T. Mukai (Japan)
V. Porubcan (Slovak Republic)
H. Rickman (Sweden)
E. Tedesco (USA)
N. Thomas (Germany)

LOCAL ORGANISING COMMITTEE

J.C. Zarnecki (Chair)
M.J. Burchell
B.J. Goldsworthy
S.F. Green
N. McBride
J.A.M. McDonnell
M.L. Watts

ACKNOWLEDGEMENTS

The Colloquium was sponsored by IAU Commission 22 (Meteors, Meteorites and Interplanetary Dust) and supported by Commission 15 (Physical Study of Comets and Minor Planets), Commission 20 (Positions and motions of Minor Planets, Comets and Satellites), Commission 21 (Light of the Night Sky) and Commission 51 (Bioastronomy: search for Extraterrestrial Life) and also by COSPAR.

We are indebted to several organisations for financial support: The International Astronomical Union, COSPAR, The Royal Astronomical Society, The University of Kent at Canterbury and Unispace Kent. This support allowed us to provide travel grants for students and key speakers who would otherwise have been unable to attend.

It is a pleasure to thank all the members of the Local Organising Committee, as well as many individuals who worked so hard behind the scenes to make the meeting a success: Esther Aguti, Margaret Fowler, James Galloway, Nadeem Ghafoor, Jon Hillier, Michael Müller, Jo Mann, Naveed Moeed, Manish Patel, Tim Ringrose, and especially Jane Goldsworthy and Mary Watts; Andrew Thompson and his team for flawless organisation of the local tours and Sir Harry Kroto for entertaining us as guest of honour at the conference dinner. Finally, we thank Louise Hobbs, Michael Mucklow, James Garry, Mary Watts and Michael Willis for assistance with preparation of these proceedings.

33 YEARS OF COSMIC DUST RESEARCH

"Welcome to Canterbury 2000", extended to the Interplanetary Dust community, was phased to mark progress in research over 33 years at Kent. The group, founded by Roger Jennison and myself in 1967, commenced research with space dust experiments involving collaboration with Otto Berg of NASA GSFC, later taking a big stride forward with the NASA and USSR Lunar Sample analyses. Deep space experiments on Pioneers 8 and 9, developed by Merle Alexander and Otto Berg showed the potential, and high reliability, needed for measurements in sparsely populated interplanetary space. With dust accelerators then at Kent and at Heidelberg, experiments such as those on Ulysses and Galileo were able to be proposed and, vitally, calibrated; impact detectors for the Giotto Halley Mission, for Cassini and now for Stardust followed. Results, which will be flowing for many years, provide that vital in-situ link between distant regions and observations at planet Earth. Equally vital to this "ground truth", albeit in space, are the fields of modelling, laboratory measurements, radar studies and extended astronomical measurements such as those of the Zodiacal Light. Without these different approaches and the different data acquired, each would be the weaker. These proceedings underscore the breadth and strength which has developed since that first coherence was created in "Cosmic Dust" (1978).

The Canterbury welcome coincided with farewells from the majority of space academics who, with their equipment, expertise and experience, joined the well established lines of success developed by Colin Pillinger at the Open University, Milton Keynes.

Success for a research group is very much due to the efforts and response of each individual; the essential contributions are not confined to academics. I thank therefore all of the group members throughout my time at Kent and all of the UK and International colleagues who have been both a stimulus and pleasure in sharing a career at Canterbury.

From The Open University ….. where even greener pastures may unfold!

Tony!

LIST OF ATTENDEES

S. Abe
P. Abraham
E. Aguti
D.J. Asher
P.B. Babadzhanov
D.E. Backman
W.J. Baggaley
L.R. Bellot Rubio
S. Benzvi
D.E. Brownlee
M.J. Burchell
A. Bursey
M. Burton
B.C. Clark
L. Colangeli
M.J. Cole
J. Crovisier
S.F. Dermott
V. Dikarev
C. Dominik
J.R. Donnison
G. Drolshagen
E. Epifani
F. Esposito
G.J. Flynn
S. Fonti
M. Fulle
D.P. Galligan
J. Galloway
M.J. Genge
N.A.L. Ghafoor
F. Giovane
B.J. Goldsworthy
M.M. Grady
G.A. Graham
A.L. Graps
S.F. Green
I.D.S. Grey
K. Grogan
E. Grün
B.Å.S. Gustafson
E. Hadamcik
Y. Hamabe
M.S. Hanner
V. Haudebourg
R.L. Hawkes
S. Helfert
M.K. Herbert
J.K. Hillier
T.-M. Ho
E.K. Holmes
S.S. Hong
J.E. Howard
S.I. Ipatov
M. Ishiguro
D. Janches
S. Jayaraman
P. Jenniskens
E.K. Jessberger
T.J.J. Kehoe
H.U. Keller
S. Kempf
K.V. Kholshevnikov
H. Kimura
D. Koschny
A.V. Krivov
N.A. Krivova
H. Krüger
J. Kuitunen
S.M. Kwon
P.L. Lamy
M. Landgraf
M.R. Leese
A.-C. Levasseur-Regourd
G. Linkert
J-C. Liou
C.M. Lisse
K. Lumme
J.C. Lyra
Y. Ma
J. Mann
M. Matney
N. McBride
J.A.M. McDonnell
N.S. Moeed
M. Müller
K. Muinonen
T. Mukai
I.S. Murray
H. Nübold
H. Ohashi
R. Ohgaito
E. Palomba
C. Park
M.R. Patel
A. Pellinen-Wannberg
S.B. Peschke
T. Poppe
H. Rickman
F.J.M. Rietmeijer
T.J. Ringrose
S. Sasaki
G. Schwehm
H. Sdunnus
Z. Sekanina
H. Shibata
N.R.G. Shrine
A.A. Sickafoose
M.B. Simakov
R. Srama
D.I. Steel
M. Stübig
H. Svedhem
S. Takahashi
H. Tanabe
E.A. Taylor
S.P, Thompson
K. Torkar
P. Tsou
R. Vasundhara
R. VickramSingh
K.W.T. Waldermarsson
M.K. Wallis
I.P. Williams
M.J. Willis
J.-C. Worms
H. Yano
S. Yokogawa
J.C. Zarnecki

CONTENTS

I Meteors and Meteoroid Streams

II Observations of the Zodiacal Light

III Interplanetary Dust

IV Dust in the Outer Solar System and Other Planetary Systems

V Cometary Dust

VI Laboratory Studies

VII The Near-Earth Environment

VIII Evidence from Meteorites

Index

I Meteors and Meteoroid Streams

Meteoroid streams and meteor showers

I.P.Williams[a]

[a]Astronomy Unit, Queen Mary and Westfield College,
Mile End Rd, London E1 4NS, UK

The generally accepted evolution of meteoroids following ejection from a comet is first spreading about the orbit due to the cumulative effects of a slightly different orbital period, second a spread in the orbital parameters due to gravitational perturbations, third a decrease in size due to collisions and sputtering, all in due course leading to a loss of identity as a meteor stream and thus becoming part of the general sporadic background. Finally Poynting-Robertson drag causes reduction in both semi-major axis and eccentricity producing particles of the interplanetary dust complex. The aim of this presentation is to review the stages involved in this evolution.

1. HISTORICAL BACKGROUND

This meeting is about dust in our Solar System and Other Planetary Systems. Planets have been discovered in about 30 nearby systems, but in these we have not as yet observed dust. On the other hand, a number of young stars are known to have a dust disk about them, but in these direct detection of planets is absent. At present, our system is the only one where dust and planets, as well as comets and asteroids to provide a source for the dust is present. Many phenomenon show the presence of the interplanetary dust complex, the zodiacal light, grains captured in the near-Earth environment as well as a number of in-situ measurements from spacecraft both in Earth orbit and in transit to other regions of the Solar System. We start the discussion with proof that must have been visible to humans since pre-history, namely the streaks of light crossing the sky from time to time, popularly called *shooting stars*, but more correctly known as *meteors.* Indeed, many of the ancient Chinese, Japanese and Korean records, talk of *stars falling like rain*, or *many falling stars.* A detailed account of these early reports can be found in the work of Hasegawa [1]. The same general thought probably gave rise to the English colloquial name for meteors, namely *Shooting Stars.* In paintings of other events, meteors were often shown in the background (see for example [2]). These historical recordings are very valuable, for they show that the Perseids for example have been appearing for at least two millenia. Recording and understanding are however two different things so that the interpretation of these streaks of light as interplanetary dust particles burning in the upper atmosphere is somewhat more recent. The reason probably lies in the belief that the Solar System was perfect with each planet moving on its own well determined orbit. Such beliefs left no room for random particles colliding with the planets, especially the Earth. Meteors were thus regarded as some effect in the atmosphere akin to lightning, hence the name. About two centuries ago the situation changed. First, there were a number of well observed meteorite falls where fragments were actually recovered. This at least proved that rocks could fall out of the sky though it did not by itself prove that they had originated from interplanetary space. however, as more observations of meteors took place, so thoughts changed. The measurement of the height of meteors as about $90km$ by Benzenberg & Brandes in 1800 [3] in essence spelt the end of the lightning hypothesis. When Herrick (1837, 1838) [4,5] demonstrated that showers were periodic on a sidereal rather than a tropical year, the inter-planetary rather than terrestrial in origin was proved.

2. OVERVIEW OF METEOR SHOWERS

Meteors can be seen at any time of the year, appearing on any part of the sky and moving in any direction. Such meteors are called sporadic and the mean sporadic rate is very low, no more than about ten per hour. Nevertheless, the flux of sporadics, averaged over a reasonable time span, is greater than the flux from any major stream averaged over the same time span. The major streams appear at well-determined times each year with the meteor rate climbing by two or three orders of magnitude. For example around 12 August meteors are seen at a rate of one or two per minute all apparently radiating from a fixed well determined point on the sky, called the *radiant.* This is the Perseid meteor shower, so named because the radiant of this shower lies in the Constellation of Persius. This behaviour is generally interpreted in terms of the Earth passing through a stream of meteoroids at the same siderial time each year. Olmstead [6] and Twining [7] are credited with first recognizing the existence of a radiant. Many of the well-known showers are rather consistent from year to year, but other are not. The best-known of these latter is the Leonids, where truly awesome displays are sometimes seen. For example, in 1966, tens of meteors per second were seen. Records show that such displays may be seen at intervals of about 33 years, with the displays of 1799, 1833 and 1966 being truly awesome, but good displays were also seen for example in 1866 and 1999. These early spectacular displays helped Adams [8], LeVerrier [9] and Schiaparelli [10], all in 1867, to conclude that the mean orbit of the Leonid stream was very similar to that of comet $55P/Tempel-Tuttle$ and that 33 years were very close to the orbital period of this comet. Since then comet-meteor stream pairs have been identified for virtually all recognizable significant stream.

These simple facts allow a model of meteor showers and associated meteoroid streams to be constructed. Solid particles (*meteoroids*) are lost from a comet as part of the normal dust ejection process. Small particles are driven outwards by radiation pressure but the larger grains have small relative speed, much less than the orbital speed. Hence these meteoroids will move on orbits that are only slightly perturbed from the cometary orbit, hence in effect generating a meteoroid cloud about the comet which is very close to co-moving with the comet. As the semi-major axis of each meteoroid will be slightly different, each will have a slightly different orbital period, resulting in a drift in the epoch of return to perihelion. After many orbits this results in meteoroids effectively being located at all points around the orbit. With each perihelion passage a new family of meteoroids is generated, but, unless the parent comet is heavily perturbed, the new set of meteoroids will be moving on orbits that are almost indistinguishable from the pre-existing set. Various effects, drag, collisions, sputtering, will remove meteoroids from the stream, changing them to be part of the general interplanetery dust complex and seen on Earth as Sporadic meteors.

An annual stream is thus middle-ages, with meteoroids having spread all around the orbit so that a shower is seen every year. In a very old stream where the parent comet may not still be very active, the stream is never very noticeable, but again constant each year. A very young stream on the other hand will only show high activity in certain years only since the cloud of meteoroids has had insufficient time to spread about the orbit.

3. THE LIFE OF A METEOROID STREAM

The basic physics behind the process of ejecting meteoroids from a cometary nucleus became straightforward as soon as a reasonably correct model for the cometary nucleus became available. Such a model for the nucleus was proposed in 1950 by Whipple [11], the so called *dirty snowball* model, in which dust grains were embedded in an icy matrix.

As the comet approaches the Sun, the nucleus heats up until some of the ices sublime and become gaseous. The heliocentric distance at which this occurs will depend on a number of parameters, the composition, the albedo and the rotation rate for example, but the process which follows this is independent of these details. When sublimation occurs, the gaseous material flows outwards away from the nucleus at a speed which is comparable to the mean thermal velocity of the gas molecules.

Any grains, or meteoroids not still embedded in the matrix will experience drag by the outflowing gas. The outward motion of the meteoroid will be opposed by the gravitational field of the comet nucleus and a meteoroid will escape from the cometary nucleus into inter-planetary space only if the drag force exceeds the gravitational force. Now, drag is roughly proportional to surface area while gravity depends on mass, thus smaller grains might experience a greater acceleration while gravity will win for grains over a given size. Hence there is a maximum size of meteoroid that can escape, though this size might vary from comet to comet depending on the size and activity level of the comet. The final speed achieved by any meteoroid that does escape will similarly depend on these factors as well on the grain properties. These considerations were first quantified by Whipple [12]. He obtained

$$\dot{R}^2 = 4.3 \times 10^5 R_C \left(\frac{1}{b\sigma r^{2.25}} - 0.013 R_C \right), \tag{1}$$

where σ is the bulk density of the meteoroid of radius b and r the heliocentric distance in astronomical units. R_C is in kilometers and all other quantities in cgs units.

A number of authors have suggested modifications to this general formula, for example Gustafson [13] pointed out that the drag formula was incorrect if the meteoroids were non-spherical while Harris and Hughes [14] suggest that the gas outflow down a tube or cone is slightly faster than is suggested by considering the mean thermal velocities. Both these points are undoubtedly correct but the end result leads to only a slight increase in the ejection velocity. Finson and Probstein [15] produced a model for dust outflow that related the observed brightness variations along the cometary tail to the dust flow rate. The dust that causes light scattering in the tail is somewhat smaller than dust that evolves into meteors, but nevertheless, there is no major difference between the dust velocities given by this approach and that given for example by Whipple's formula. The main conclusion, in terms of meteoroid stream formation, is that the ejection velocity is in all cases considerably less than the orbital velocity of the parent comet.

As an illustration, consider comet 1P/Halley. Grains of up to a few centimeters will escape, while at 1AU, a one millimeter meteoroid would have an ejection speed of about $70ms^{-1}$. The orbital speed at 1AU is of the order of $30kms^{-1}$.

The effect of the meteoroid being ejected with a speed given by the mechanism above relative to the comet will be to produce differences between the orbit of the meteoroid and that of the comet. These changes will of course depend on the direction at which the meteoroid is ejected and the point on the cometary orbit at which the ejection takes place. There will always be a change in the specific energy E. Now, standard theory of Keplerian motion tells us that

$$E = \frac{-GM_{\odot}}{2a}, \tag{2}$$

and that

$$P^2 = a^3. \tag{3}$$

where a is the semi-major axis of the orbit in Astronomical Units and P the orbital period in years. Hence we can obtain

$$\frac{\Delta E}{E} = \frac{-\Delta a}{a} = \frac{-2\Delta P}{3P}. \tag{4}$$

a change in semi-major axis and period thus is an inevitable consequence of the ejection process, but since $\frac{\Delta E}{E}$ is likely to be small in view of the fact that the ejection velocity is small compared to the orbital velocity, changes in a and P are also likely to be small. Observationally, it will be very difficult to detect such changes in the semi- major axis. However, changes in the orbital period are different in that their effect is cumulative. After n completed orbits, the time difference between a meteoroid and the comet passing perihelion will be $n\Delta P$. For a typical situation, in about 50 orbits meteoroids will be found at all points of the orbit that is an annual stream is formed.

If there is a component of the ejection velocity in the transverse direction, then the specific angular momentum h will also be changed. we have

$$h^2 \; = \; GM_{\odot}p, \tag{5}$$

where p is the semi-parameter of the orbit, that is $p \; = \; a(1-e^2)$.
This yields

$$\frac{\Delta h}{h} = \frac{\Delta p}{2p} = \frac{\Delta a}{2a} - \frac{e\Delta e}{(1-e^2)}. \tag{6}$$

This implies that in general there is a change in eccentricity as well. Detecting changes in the eccentricity from observations of meteors will also be very difficult.

Unless the ejection took place exactly at perihelion, the changes in a and e, together with the requirement that the ejection point is on both the comet and meteoroid orbit, implies that a change in the argument of perihelion ω must also take place. Since the orbit is assumed to be Keplerian, its equation is known, and from this we can obtain

$$e\cos(f_0+\Delta\omega)-e\cos f_0=\frac{(2e+e^2\cos f_0+\cos f_0)}{2e}\frac{\Delta p}{p}-\frac{1-e^2}{2e}\cos f_0\frac{\Delta a}{a} \tag{7}$$

where f_0 is the true anomaly of the ejection point. Though the changes in a, e and ω may each be small and indeed undetectable without very accurate observations, a combination of them can cause a change that is of fundamental importance in the observability of a meteor shower, namely the nodal distance, r_N.

The nodal distances are derived from the standard equation for an ellipse with the true anomaly being taken as $-\omega$ or $\pi-\omega$, that is

$(1-e\cos\omega)r_N \;=\; p$ and $(1+e\cos\omega)r_N \;=\; p.$

Hence, we can obtain

$$(1-e\cos\omega)\frac{\Delta r_N}{r_N}=\frac{1-e^2}{2e}\cos\omega\frac{\Delta a}{a}-\frac{(e^2\cos\omega+\cos\omega-2e)}{2e}\frac{\Delta p}{p}-e\sin\omega\Delta\omega \tag{8}$$

for the first node with a similar equation for the other node.

Again, the changes in the nodal distance may appear to be small but whereas a 1% change in a, e or ω is fairly hard to detect a one percent change in r_N is $0.01AU$, or about 4 times the Earth-Moon distance. This is rather a large distance when the meteoroid stream has to hit the Earth in order to produce a meteor shower.

The ejection velocity will generally also have a component perpendicular to the comet orbital plane. In consequence, the meteoroid orbital plane will be different from that of the comet. Since the line of intersection of the orbital plane with the ecliptic is defined as the line of nodes and the displacement of this from first point of Aries is defined as the longitude of the ascending node, Ω, any such a velocity component will induce a change in Ω.

Deriving the expression for $\Delta\Omega$ is mathematically rather tedious and will not be repeated here . The derived expression is

$$\Delta\Omega=\frac{r_0\sin(\omega+f_0)}{h\sin i}v\sin\phi \tag{9}$$

where r_0 and f_0 are the heliocentric distance and the true anomaly of the ejection point, i is the inclination of the orbit, and ϕ is angle between the direction of ejection and the orbital plane so that $v\sin\phi$ is the component of the ejection velocity perpendicular to the orbital plane.

Since Ω measures the time at which a shower is seen, then this is also sensitive to small changes and is important in the study of meteor showers.

Hence, the effect of the initial ejection velocity is to change all the orbital parameters by a small amount, but these small changes can also produce a change in the nodal distance which is a very sensitive parameter for the production of a meteor shower. For a very young stream, perhaps one which generates a meteor storm such as the Leonids, these effects may be the dominant ones, but, as soon as the meteoroid is ejected from the immediate vicinity of the comet, it becomes an independent moving body in the Solar System and subject to all the evolutionary effects that any body is subject to.

Solar radiation falling directly on a body generates a force which is radial and depends on the strength of the incident radiation and so is proportional to the inverse square of heliocentric distance, like gravity. It can thus be regarded as weakening gravity and is usually represented by writing the effective force acting on the body as

$$F = -\frac{GM_{\odot}(1-\beta)}{r^2}, \tag{10}$$

and, when numerical values for standard constants are inserted, β is given by (eg [16]

$$\beta = \frac{5.75 \times 10^{-5}}{b\sigma}, \tag{11}$$

where as before b is the meteoroid radius in centimeters and σ the relative bulk density in gcm^{-3}. It is self-evident that meteoroids will be lost from the Solar System if $\beta \geq 1$, since the net force is then outwards. However, as Kresák [17] first pointed out, meteoroids will be lost whenever their total energy is positive. A meteoroid moving with the parent comet will have a specific energy E' given by

$$2E' = V^2 - \frac{2GM_{\odot}(1-\beta)}{r}. \tag{12}$$

But,

$$V^2 = GM_{\odot}\left(\frac{2}{r} - \frac{1}{a}\right), \tag{13}$$

so that E' is positive provided

$$2\beta \geq r/a \tag{14}$$

At perihelion, $r = a(1-e)$, and here, meteoroids for which

$$\beta \geq (1-e)/2 \tag{15}$$

will be lost. This is much more restrictive limit than $\beta = 1$, so that larger grains are lost than is implied by the $\beta = 1$ limit. Taking our numerical example again, for comet 1P/Halley, $e = 0.964$, so that meteoroids for which $\beta \geq 0.018$ will be lost. Taking a bulk density of $0.5gcm^{-3}$, meteoroids smaller than about $6 \times 10^{-3}cm$ will be lost from the stream.

Since the radiation may be absorbed and then re-emitted from a moving body, there can be a loss of angular momentum from the body, affecting its orbit. This effect was first mentioned by Poynting [18] and but in a relativistic frame by Robertson [19] and is now generally known as the Poynting-Robertson effect. This effect has been studied by many authors. The first to apply this to meteoroid streams was probably Wyatt and Whipple [20]. More recent accounts of this effect can be found in Hughes *et al.* [21] and Arter and Williams [22]. In discussing changes caused to the orbital parameters a and e, it is more convenient to use a parameter η, rather than β to characterize the effects of radiation. The relationship between the two parameters is

$$c\eta = GM_{\odot}\beta, \tag{16}$$

where c is the speed of light. η has a numerical value 4.4×10^{15} that of β in *cgs* units. Note that while β is dimensionless, η is not. Using this notation, all the authors mentioned give the following two equations, (using the same units as those used to express η)

$$\frac{da}{dt} = \frac{-\eta(2+3e^2)}{a(1-e^2)^{3/2}}, \tag{17}$$

and

$$\frac{de}{dt} = \frac{-5\eta e}{2a^2(1-e^2)^{1/2}}. \tag{18}$$

In order to obtain the change in a given orbit, it is necessary to specify the dimensions of the meteoroid so that the value of η can be obtained and then numerically integrate these equations, the latter task not being particularly difficult. However, some insight into the effect of this can be obtained without performing numerical integrations. Using the chain rule on the two above equations gives,

$$\frac{da}{de} = \frac{2a(2+3e^2)}{5e(1-e^2)}, \tag{19}$$

an equation which can be integrated to give

$$a(1-e^2) = Ce^{4/5}, \tag{20}$$

where C is a constant of integration.

Since time has been eliminated, this equation gives no indication of how long it takes for an orbit to evolve to any given state. However an estimate of the time required to significantly change orbits can be obtained by substituting the value of a from equation (20) into equation (18), giving

$$\frac{de}{dt} = \frac{-5\eta(1-e^2)^{1/2}}{2C^2e^{3/5}}. \tag{21}$$

Apart from factors of general order unity, the typical time-scale of this equation is given by C^2/η. For the case we have so far used as an example, namely a meteoroid of $1mm$ radius and density $0.5gcm^{-3}$ associated with comet 1P/Halley, this time-scale is of order $3 \times 10^5 years$. Though this is short by the standards of evolution generally in the solar system, it is a long time compared to our time-span of observation of meteor showers and is towards the top end of estimates for stream life-times. The time to significantly change the orbital parameters will also vary from stream to stream, so that the above value should be regarded as only an indication of the time scale for the Poynting-Robertson drag to be important.

Like other bodies in the Solar System, the motion of the meteoroid will be affected by the gravitational fields of all the other bodies in the system, with all the accompanying problems of accurately dealing with these perturbations that are familiar to all that have worked on orbital evolution in the Solar System. It is known since the work of Poincaré in 1892, (see [23]) that no analytical solution exists to the general problem of following the orbital evolution of more than two bodies under their mutual gravitational attraction exists. Hence, following the motion of meteoroids implies some form of numerical integration of the equations of motion.

The concepts involved in considering planetary perturbations are very easy to understand though following through the consequences is somewhat harder. Each planet produces a known gravitational field. Hence, if the position and velocity of each body in the system is known at any given instant, then the force due to each body and hence the resulting acceleration can be calculated which allows a determination of the position and velocity of the body at a later time. Of course, this is only strictly true for an infinitesimal time interval and so the problem in reality is to chose a time step that is short enough to maintain a desired level of accuracy while at the same time making progress in following the evolution. The methodology described above was known and used in the mid-nineteenth century by the astronomers that calculated the orbits of comet, though, the 'computers' they used had a rather different meaning then from now. In those days it meant a low paid assistant who computed myriads of positions using hand calculators. Some of the earliest calculations on the evolution of meteoroid streams which included planetary perturbations were carried out by Newton between 1863 and 1865 ([24–26]), where he investigated the generation of Leonid meteor storms. A number of other early calculations are described in Lovell's classical text book on the subject [27]. Though some useful early results were obtained by these early calculations, it is clear that no real progress in following the evolution of a large number of meteoroids can be made by such labour intensive means and further development had to wait until the human computers were replaced by electronic ones.

The early electronic computers were also to small and slow to be able to follow a realistic number of meteoroids over realistic time-scales. In order to overcome these shortcomings, effort was spent on refining the mathematical modelling, in particular on the idea of averaging the perturbations over an orbit so that only secular effects remained. The real gain with such methods is that the whole assembly of meteoroids are replaced by one mean orbit with a consequential huge gain in effort. At first, such 'secular perturbation' methods only worked for nearly circular orbits, good for following the evolution of satellite systems and main-belt asteroids, but of little value in following the evolution of meteoroids on highly eccentric (and possibly also highly inclined) orbits. In 1947, Brouwer [28] generated a secular perturbation method that worked well even for orbits of high eccentricity (though not for values very close to unity) and this method was used by Whipple and Hamid [29] in 1950 to integrate back in time the orbit of comet 2/P Encke and the mean orbit of the Taurid meteoroid stream. They showed that 4700 years ago, both the orbits were very similar and suggested that the two were related. This was the first time that a link between a comet and a stream had been suggested based on a past similarity in orbits rather than a current similarity. This also established an age of 4700 year for the Taurid stream. Other secular schemes were also used, for example, Plavec [30] used the Gauss-Hill method to investigate the changes with time in the nodal distance of the Geminid stream.

One of the more popular (in terms of general usage) secular perturbation methods that were developed is the Gauss-Halphen-Goryachev method, described in detail in Hagihara [31]. This was used for example by Galibina [32] to investigate the lifetime of a number of meteoroid streams and by Babadzhanov and Obrubov [33] to investigate the changes in the longitude of the ascending node (rather than nodal distance as investigated by Plavec) of the Geminid stream. The same authors also used this method extensively during the

1980's to investigate the evolution of a number of streams (for example, [34]).

The disadvantage of the secular perturbation methods is that the averaging process, by its very nature, removes the dependence of the evolution on the true anomaly of the meteoroid. It is thus impossible to answer questions regarding any difference in behaviour between a clump of meteoroids close to the parent comet and a typical meteoroid in the stream. As computer hardware improved, the use of direct numerical integration methods became more widespread. By direct methods, we mean where the evolution of individual meteoroids, real or hypothetical, is followed rather than the evolution of an orbit. The first such investigation was probably by Hamid and Youssef [35] who in 1963 integrated the orbits of six actual Quadrantid meteoroids. In 1970, Sherbaum [36] generated a computer programme to numerically integrated the equations of motion using Cowell's method which was used by Levin *et al.* [37] to show that Jovian perturbations caused an increase in the width of meteoroid streams. In the same year, Kazimirchak-Polonskaya *et al.* [38] integrated the motion of 10 α Virginid and 5 α Capriconid meteoroids over a 100 year interval. Seven years later, the number of meteoroids integrated was still small and the time interval over which the integration was performed remained short, with Hughes *et al.* [39] in 1979 following the motion of 10 Quadrantid meteoroids over an interval of 200 years, using the self adjusting step-length Runge-Kutta method. This however marked the start of significant increases in both the number of meteoroids integrated and the time interval, and by 1983, Fox *et al.* [40] were using 500 000 meteoroids, indicating that in five years computer technology had advanced from allowing only a handful of meteoroids to be integrated to the situation where numbers to be used did not present a problem.

The direct integration methods used in meteoroid stream studies fall into two broad categories, the single step methods of which the best known is the Runge-Kutta method, (see Dormand *et al.* [41] for a fast version of this method) and the 'predictor-corrector' methods following Gauss (see Bulirsch and Stoer [42] for the methodology)

By the mid eighties, complex dynamical evolution was being investigated, Froeschlé and Scholl [43], Wu and Williams [44] were showing that the Quadrantid stream behaved chaotically because of close encounters with Jupiter, and the proximity of mean motion resonances. A new peak in the activity profile of the Perseids, roughly coincident in time with the perihelion return of the parent comet 109P/Swift-Tuttle caused interest with models being generated for example by Williams and Wu [45] . Babadzhanov *et al.* [46] investigated the possibility that the break-up of comet 3D/Biela was caused when it passed through the densest part of the Leonid stream. By now, numerical integrations of models for all the major streams have been carried out. In addition to those mentioned earlier, examples of streams for which such numerical modelling exists are : the Geminids, (Gustafson [47], Williams and Wu [48]), April Lyrids (Arter and Williams [49]), η Aquarids (Jones and McIntosh [50]), Taurids (Steel and Asher [51]), α Monocerotids (Jenniskens and van Leeuwen [52]), 9the Giacobinids (Wu and Williams [53]) and the Leonids Asher *et al.* [54]). From the point of view of the discussion here, it is sufficient to say that numerical modelling has now reached a stage where it is possible to follow the evolution of given meteoroids from their formation over any time scale that appears to be of interest.

4. THE END OF A STREAM

A stream will stop being a stream when one can no longer recognize that a family of meteoroids are moving on similar orbits. There are two distinct possibilities here. Either individual meteoroids experience some catastrophic event so that they cease to be able to produce observable meteor trails, or the individual orbits have changed, so that, though the individual meteoroid still exists, the resulting meteor is no longer recognizable as being part of a known shower.

All the mechanisms discussed above lead to changes in the orbital parameters, but they lead to a dispersal of the stream only if they produce different changes to the orbital elements of different meteors. They are also different in their effect, the Poynting-Robertson effect may be quite efficient at changing the orbital parameters over a short time period, but it moves similar sized meteoroids by the same amount. Hence, all large (or visible) meteors say will be affected by the same amount which will be smaller than the changes experienced but radio meteors. Nevertheless, though the visible meteors may now be on a different orbit, they will be on a recognizable orbit and so have not merged into the sporadic background.

Gravitational perturbations depend on the exact distance of the meteoroid from each of the planets. Hence every meteoroid experiences a different perturbation and can in theory evolve differently. Unfortunately over many orbits, these perturbations average out and most experience the average perturbation with only a small variance about this. The stream may move and become wider but the meteoroids in general still appear to belong to a stream. Other effects must thus play their part in dispersing a stream.

The most obvious loss mechanism from a meteoroid stream is the production of a meteor shower. Every dust grain that is seen as a meteor has burnt up in the Earth's atmosphere and so has been lost from the stream. But this mechanism is simply a meteoroid removal mechanism which leaves the surviving stream unaffected. However, for every meteoroid that hits the Earth, many more have a near miss and they will be scattered by the gravitational field of the Earth. Those affected however represent only a fraction of the stream, a few Earth radii is a tiny part of the circumference of a typical stream.

Other mechanisms that have been proposed are inter-meteoroid collisions, in particular high velocity collisions as discussed by Williams *et al.* [55]. Again unlikely to be important to the stream as a whole. Fragmentation following collisions with solar wind electrons, which leads to an increased efficiency of radiation forces also leads to meteoroid loss. A mechanism that has not received much attention is the sublimation of residual ices which again leads to fragmentation. A much less dramatic effect is the combined perturbation of the planets that slowly change the orbital parameters so that coherence is gradually lost and the stream appears to get weaker and weaker and of longer and longer duration. From the point of view of a stream none of these effects may appear dramatic, but they all do the same thing, they feed the inter-planetary dust complex with small grains. All streams do this and so the cumulative effect is significant.

5. CONCLUSIONS

In its broadest sense, the evolution of meteoroid streams and the generation of meteor showers has been understood for some considerable time. However, it is only in recent years that the computational capabilities have been available to allow realistic models of meteoroid streams to be developed and much success has been obtained in doing this. The aim of this review was to discuss the underlying principles of meteor stream evolution, including formation. Many of the aspects touched upon here will be revisited in following chapters.

REFERENCES

1. I. Hasegawa, in Meteoroids and their parent bodies, (eds. J. Stohl and I.P. Williams) Slovak Academy of Sciences, Bratislava (1993) 209.
2. R.J.M. Olson and J.M. Pasachoff, Fire in the Sky, Cambridge University Press, Cambridge, UK, 1998.
3. J.F. Benzenberg and H.W. Brandes, Annalen der Phys 6 (1800) 224.
4. E.C. Herrick, American Jl. Sci. 33 (1837) 176.
5. E.C. Herrick, American Jl. Sci. 33 (1838) 354.
6. D. Olmstead, American Jl. Sci. 25 (1834) 354.
7. A.C. Twining, American Jl. Sci. 2 (1834) 320.
8. J.C. Adams, Mon. Not. R. astr. Soc. 27 (1867) 247.
9. U.J.J. Le Verrier, Comptes Rendus 64 (1867) 94.
10. G.V. Schiaparelli, Astronomische Nachrichten 68 (1867) 331.
11. F.L. Whipple, Astrophys. J. 111 (1950) 375.
12. F.L. Whipple, Astrophys. J. 113 (1951) 464.
13. B.Å.S. Gustafson, Astrophys. J. 337 (1989) 945.
14. N.W. Harris and D.W. Hughes, Mon. Not. R. astr. Soc. 273 (1995) 992.
15. M.L. Finson and R.F. Probstein, Astrophys. J. 154 (1968) 327.
16. I.P. Williams, in Dynamical trapping and evolution in the Solar System (eds. V.V. Markellos and Y. Kozai) D. Reidel (1983) 83.
17. L. Kresák, Bull. Astron. Inst. Czechos. 13 (1974) 176.
18. J.H. Poynting, Proc. R. Soc. London 72 (1903) 205.
19. H.P. Robertson, Mon. Not. R. astr. Soc. 97 (1937) 423.
20. S.P. Wyatt and F.L. Whipple, Astrophys. J. 111 (1950) 134.
21. D.W. Hughes, I.P. Williams and K. Fox, Mon. Not. R. astr. Soc. 195 (1981) 625.
22. T.R. Arter and I.P. Williams, Mon. Not. R. astr. Soc. 286 (1997) 163.
23. H. Poincaré, Les Méthodes nouvelles de la Mécanique Céleste, Dover Publications, New York, 1957.
24. H.A. Newton, American J. of Sci. and Arts 36 (1863) 145.
25. H.A. Newton, American J. of Sci. and Arts 37 (1864) 377
26. H.A. Newton, American J. of Sci. and Arts 39 (1865) 193.
27. A.C.B. Lovell, Meteor Astronomy, Oxford University Press, New York, 1954.
28. D. Brouwer, Astron. J. 52 (1947) 190.
29. F.L. Whipple and S.E. Hamid, Sky and Telescope 9 (1950) 248.
30. M. Plavec, Nature 165 (1950) 362.
31. Y. Hagihara, Celestial Mechanics MIT Press, London, 1972.
32. I.V. Galibina, in The Motion, Evolution of orbits, and Origin of Comets (eds. G.A. Chebotarev, H.I. Kazimirchak-Polonskaya and B.G. Marsden) D. Reidel, Dordrecht

(1972) 440.
33. P.B. Babadzhanov and Y.Y. Obrubov, in Solid Particles in the Solar System (eds. I. Halliday and B.A.McIntosh) D. Reidel, Dordrecht (1980) 157.
34. P.B. Babadzhanov and Y.Y. Obrubov, in Highlights in Astronomy (ed. R.M. West) D.Reidel, Dordrecht (1983) 411.
35. S.E. Hamid & M.N. Youssef, Smithson. Cont. Astrophys. 7 (1963) 309.
36. L.M. Sherbaum, Vestun. Kiev. Un-ta Ser. Astron. 12 (1970) 42.
37. B.Y. Levin, A.N. Simonenko and L.M. Sherbaum, in The Motion Evolution of Orbits and Origin of Comets (eds G.A. Chebotarev, H.I. Kazimirchak-Polonskaya and B.G. Marsden), D. Reidel, Dordrecht (1972) 454.
38. H.I. Kazimirchak-Polonskaya, A. Beljaev, I.S. Astapovic and A.K. Terenteva, in The Motion, Evolution of Orbits and Origin of Comets, (eds. G.A. Chebotarev, H.I. Kazimirchak-Polonskaya and B.G. Marsden) D. Reidel, Dordrecht (1972) 462.
39. D.W. Hughes, I.P. Williams and C.D. Murray, Mon. Not. R. astr. Soc. 189 (1979) 493.
40. K. Fox, I.P. Williams and D.W. Hughes, Mon. Not. R. astr. Soc. 205 (1983) 1155.
41. R.J. Dormand, M.E.A. El-Mikkaway and P.J. Prince, IMA J. Numer. Anal. 7 (1987) 423.
42. R. Bulirsch and J. Stoer, Numer. Math. 8 (1966) 1.
43. C. Froeschlé and H. Scholl, in Asteroids Comets Meteors II, (eds. C.-I. Lagerkvist, B.A. Lindblad, H. Ludsted & H. Rickman) Uppsala Universitet Reprocentralen, Uppsala (1986) 523.
44. Z. Wu and I.P. Williams, in Chaos, Resonance and Collective Dynamical Phenomena in the Solar System (ed. S. Ferraz-Mello) Kluwer, Dordrecht (1992) 329.
45. I.P. Williams and Z. Wu, Mon. Not. R. astr. Soc. 269 (1994) 524.
46. P.B. Babadzhanov, Z. Wu, I.P. Williams and D.W. Hughes, Mon. Not. R. astr. Soc. 253 (1991) 69.
47. B.Å.S. Gustafson, Astron. Astrophys. 225 (1989) 533.
48. I.P. Williams and Z. Wu, Mon. Not. R. astr. Soc. 262 (1993) 231.
49. T.R. Arter and I.P. Williams, Mon. Not. R. astr. Soc. 298 (1997) 721.
50. J. Jones and B.A. McIntosh, in Exploration of Comet Halley, ESA-SP 250, Paris (1986) 233.
51. D.S. Steel and D.J. Asher, in Physics, Chemistry and Dynamics of Interplanetary Dust (eds. B.Å.S. Gustafson and M.S. Hanner, Pub. Astron. Soc. Pacific Conference Series (1996) 125.
52. P. Jenniskens and G.D. van Leeuwen, Planet. Space Sci. 45 (1997) 1649.
53. Z. Wu and I.P. Williams, Planet. Space Sci. 43 (1995) 723.
54. D.J. Asher, M.E. Bailey and V.V. Emel'yanenko, Mon. Not. R. astr. Soc. 304 (1999) L53.
55. I.P. Williams, D.W. Hughes, N. McBride and Z. Wu, Mon. Not. R. astr. Soc. 260 (1993) 43.

Thermal gradients in micrometeoroids during atmospheric entry.

M. J. Genge and M. M. Grady

Department of Mineralogy, The Natural History Museum, Cromwell Road, London SW7 5BD, UK.

Melted rims found on micrometeorites recovered from Antarctic ice indicate that micrometeoroids as small as 50 μm in diameter can maintain temperature differences of at least 600 K between their surfaces and cores. We present the results of finite element simulations of the thermal evolution of micrometeoroids during entry heating that indicate that large thermal gradients cannot arise simply as a result of the non-steady state heating of particles. The generation of thermal gradients resulting in melted rims may occur in fine-grained micrometeorites due to energy losses at the melt-core boundary due to the endothermic decomposition of volatile-bearing phases. However, the occurrence of melted rims on many coarse-grained particles that lack such low-temperature phases suggests this is not the primary cause of the temperature differences. Large mass losses due to vaporisation and energy losses due to fusion may therefore be involved in the generation of melted rims. The presence of thermal gradients in micrometeoroids during atmospheric entry increases the likelihood that low-temperature primary phases such as abiotic carbonaceous compounds will survive atmospheric entry heating.

1. INTRODUCTION

The thermal behavior of micrometeoroids determines their survival of atmospheric entry and their state of alteration and thus strongly influences the sample of the interplanetary dust population that can be collected on the Earth. Models of the atmospheric entry of micrometeoroids specifically assume that particles are thermally homogeneous during heating [1]. This simplification significantly reduces the complexity of simulations and is based on a formulisation of the Biot number adapted to radiative heat loss under steady state heating and thus may not be appropriate under non-steady state transient heating by the hypervelocity collisions with air-molecules during atmospheric entry.

Micrometeorites larger than 50 μm collected on the Earth's surface, however, exhibit clear evidence for thermal gradients developed during entry heating [2]. Cored micrometeorites have vesicular melted rims consisting of Fe-rich olivine microphenocrysts in glassy mesostases and unmelted cores some of which retain phyllosilicates (Fig. 1). These particles suggest that temperature differences between the surface and core of the micrometeoroid can exceed 600°C [2].

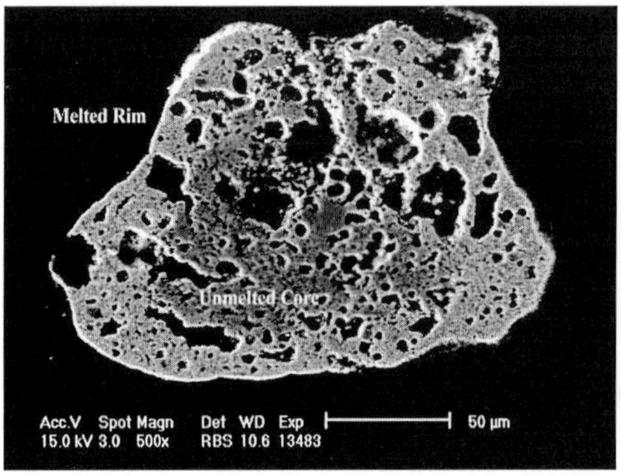

Figure 1 A melted rim (light coloured outer layer) on an otherwise unmelted fine-grained micrometeorite.

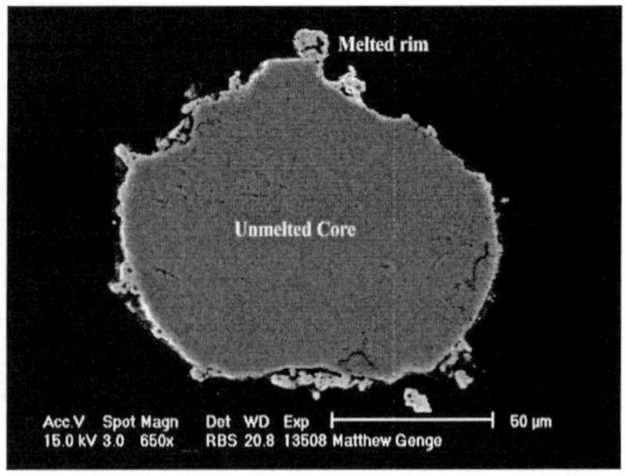

Figure 2 A backscattered electron image of a coarse-grained micrometeorite with a thin melted rim.

The origin of large temperature gradients in micrometeoroids is problematic because only a small fraction of the incident energy flux provided by the collision of air molecules is required to heat the particle to peak temperature [1]. Low effective thermal conductivities, due to high porosity, and energy losses due to the vaporisation of low temperature phases are possible explanations for the development of large thermal gradients in small micrometeoroids. The occurrence of melted rims on compact coarse-grained micrometeorites (Fig. 2; [3]) that lack low temperature, volatile components, indicate that neither decreases in thermal conductivity or energy sinks due to devolatilisation are the primary cause of thermal heterogeneity.

On the basis of the thermal evolution of micrometeoroids predicted by 'homogeneous' particle entry heating models we have suggested that thermal gradients might be supported due to the rapid increase in the surface temperature of particles during deceleration [4]. To determine whether thermal gradients develop simply in response to non-steady state, single-pulse heating we have conducted two- and three-dimensional finite element simulations of the thermal evolution of micrometeoroids during entry heating.

2. FINITE ELEMENT MODEL

The thermal model adopted for the simulation of heat flow during entry heating estimates the temperature profile across a model elliptical micrometeoroid consisting of up to 4000 cubic finite elements by approximating a solution to the partial differential equations controlling heat transfer. Because we are specifically interested in whether the increase in surface temperature of micrometeoroids support the temperature profile through the particle a constant surface heating rate was used. Thermal diffusivity was taken as 1.45×10^{-6} m^2 s^{-1} equivalent to well compacted sandy soil to model the porous nature of many fine-grained micrometeorites.

3. RESULTS

The finite element simulations suggest that thermal gradients are an unavoidable consequence of non-steady state heating of homogeneous particles irrespective of size due to the thermal lag in the equilibration of the core of the particle relative to the surface. The rate of increase of temperature of the core of the particle reaches that applied to surface only after a specific equilibration time which is dependent on particle size. Equilibration times are ~5 ms for a 100 μm diameter and ~0.1 s for a 500 μm diameter particle and are independent of the heating rate. The temperature difference maintained across a particle is thus determined by particle size, which controls equilibration time, and the heating rate with smaller temperature differences at higher surface heating rates.

The temperature differences calculated for 100 μm and 500 μm diameter particles are much lower than observed in micrometeorites recovered from the Earth's surface. The calculations indicate that although non-steady state heating does maintain temperature gradients across micrometeoroids these are only ~30 K for particles 500 μm in diameter and ~3 K for particles 100 μm in diameter at heating rates of 500 K s^{-1}.

Simulations were also performed to model the equilibration of thermal gradients at peak temperature using the temperature profiles generated in the heating calculations and a constant surface temperature. The results of these simulations indicate that the small temperature differences generated during heating disappear rapidly (i.e. ~0.1 s for a 500 μm particle).

4. DISCUSSION

Typical heating rates for asteroidal particles (entry velocities 12 km s^{-1}) suggested by entry heating models are ~500 K s^{-1} [1]. The finite element simulations therefore suggest that such micrometeoroids could only support thermal gradients of ~30 K (for a 500 μm diameter particle) if these result only from non-steady state heating and that thermal gradients will quickly equilibrate at peak temperature. Core-rim temperature differences of 30 K would be sufficient to generate the melted rims observed on micrometeorites recovered from the Earth's surface, however, only those particles whose surfaces reached temperatures close to the melting point would be expected to preserve melted rims. This is contrary to the large number of fine grained micrometeorites that have melted rims and unmelted cores. The observation that cored particles vary from those with rims a few microns in sizes to those which contain one or more small areas of unmelted fine-grained matrix suggests that melted rims are a general feature of the melting process of micrometeorites. The simulations also indicate that temperature differences of up to 600 K in particles as small as 100 μm in size do not result from non-steady state heating.

Previous steady-state calculations on the thermal evolution of phyllosilicate-bearing micrometeoroids by Flynn et al., [5] that included the contribution of the latent heat required for endothermic decomposition of water-bearing phyllosilicate minerals produce temperature discontinuities similar to those observed in micrometeorites. A dehydration/melting front thus probably exists in fine-grained micrometeorites that migrates into the particle during heating with the thermal decomposition acting as a sink for energy that maintains the lower temperature of the micrometeoroid core. Other devolatilisation and decomposition reactions such as the pyrolysis of carbonaceous materials and the breakdown of sulphide minerals may also contribute significantly to this affect and enable temperature differences of the magnitude observed in some micrometeorites to be maintained.

Melted-rims are, however, also frequently observed on coarse-grained micrometeorites that consist mainly of anhydrous silicates and glass. These particles contain no volatile-bearing minerals to maintain the temperature differences and yet particles with melted rims are abundant. The melted rims on these coarse-grained micrometeorites might arise through the melting of small amounts of fine-grained matrix material, which has a lower melting temperature, attached to the exterior of the particle. However, the observation that unmelted coarse-grained particles with fine-grained matrix are rare amongst micrometeorites is contrary to the high abundance of particles melted rims. Potentially melted rims on coarse-grained micrometeorites could be generated by temperature differences of only a few degrees since there is no means of identifying what the peak temperature the cores of these particles attained. The abundance of particles with unmelted rims is, however, not consistent with such an origin since only a small fraction of coarse-grained micrometeorites should have peak temperatures in close to the melting point of their constituent minerals.

One final possibility is that the temperature differences are in part maintained by energy losses to vaporisation at the surface of particles. If the vaporisation rate is high enough that mass losses cause significant decreases in particle size then significant energy losses could occur due to the latent heat of fusion at the melt-core boundary and the latent heat of vaporisation at the particle surface. If this process is an important factor in the development and survival of the temperature differences observed in micrometeorites then particles with melted rims have probably experienced significant mass loss and care must be taken when considering the particle-size distribution of the different micrometeorite types.

The development of melted rims on micrometeoroids during entry heating will enhance the survival of unmelted primitive extraterrestrial materials as the cores of heated particles. Micrometeoroids with relatively high geocentric velocities may therefore be more likely to be preserved to reach the Earth's surface with at least a proportion of the original nature of their refractory components intact by virtue of surface melting. Similarly low-temperature volatile materials such as abiotic hydrocarbons may also survive atmospheric entry without complete decomposition in particles with low geocentric velocities. These materials would have been a potentially important source of pre-biotic carbon on the early Earth and may have played a role in the origin of life on our planet.

REFERENCES

1. S. G. Love and D. E. Brownlee, Icarus 89 (1991) 26.
2. M. J. Genge, M. M. Grady and R. Hutchison, Geochim. Cosmochim. Acta 61 (1997) 5149.
3. M. J. Genge, R. Hutchison and M. M. Grady, Meteoritics Planet. Sci. 31 (1996) A49.
4. M. J. Genge and M. M. Grady, Lunar Planet. Sci. XXX (1999) 1578.
5. G. J. Flynn, Lunar Planet. Sci. XXVI (1995) 405.

Direct determination of the micrometeoric mass flux into the upper atmosphere

J.D. Mathews[a],D. Janches[a] and D. D. Meisel[b]

[a]Communications and Space Sciences Laboratory, Department of Electrical Engineering, Penn State University, University Park, PA, USA

[b]Department of Physics and Astronomy, SUNY-Geneseo, Geneseo, NY, USA

The advent of radar micrometeor observations at Arecibo Observatory (AO) has enabled direct estimates of the meteoric mass flux into the upper atmosphere. These observations yield on average ~3200 events per day in the 300 m diameter Arecibo beam. Doppler velocity estimates are found for approximately 50% of all events and of these, approximately 55% (26.5% of the total) also yield measurable (linear) decelerations. Assuming spherical particles of canonical density 3 gm/cc, the meteoric masses obtained range from a few micrograms to a small fraction of a nanogram. This approach yields an average mass of 0.31 microgram/particle for the 26.5% of all particles that manifest observable deceleration. The 45% with velocities, but not decelerations, correspond to particle masses larger than a few micrograms. However if we assume that all observed particles average 0.31 micrograms each, we find a mass flux of about 1.4×10^{-5} kg/km^2-day over the whole Earth. Detailed annual whole-Earth mass flux per decade of particle mass is calculated and compared with those of Ceplecha et al. [1]. Our results fall below those of Ceplecha et al. for observed mass fluxes however inclusion of those particles for which we cannot explicitly determine mass yield similar fluxes.

Many of the particles we observe show evidence of catastrophically disintegrating in the meteor zone. We thus suggest that the majority of micrometeoroid mass is deposited in the 80-115 km altitude region where ionospheric and atmospheric manifestations such as sporadic E and neutral atomic metal layers are well documented. We further suggest that the "background"diurnal micrometeor mass flux is sufficient to dominate the average lower atmosphere mass influx from the annual meteor showers.

1. Introduction

The meteor classical momentum equation [2] can be written in terms of the meteor ballistic parameter (BP) [3] — ratio of the meteor mass to cross-sectional area — as:

$$\frac{dV}{dt} = -\frac{\Gamma \rho_{atm} V^2}{BP} \tag{1}$$

where dV/dt is the meteor deceleration, V the velocity, ρ_{atm} is the atmospheric density and Γ is the drag coefficient assumed to be 1 for the remainder of this paper. In this scenario the BP is based only on observed velocity and deceleration while the atmospheric density is based on the MSIS-E-90 model atmosphere. As it has been discussed in Janches et al. [3], the micrometeor deceleration observed at AO appears to be linear, at least during the time they are observed by the radar. Furthermore, if we assume the meteoroids to have a spherical shape and a canonical mass density equal to 3 gm/cc then the particle masses can be derived. This approach permits the direct determination of meteoric mass flux in the upper atmosphere utilizing ground based techniques.

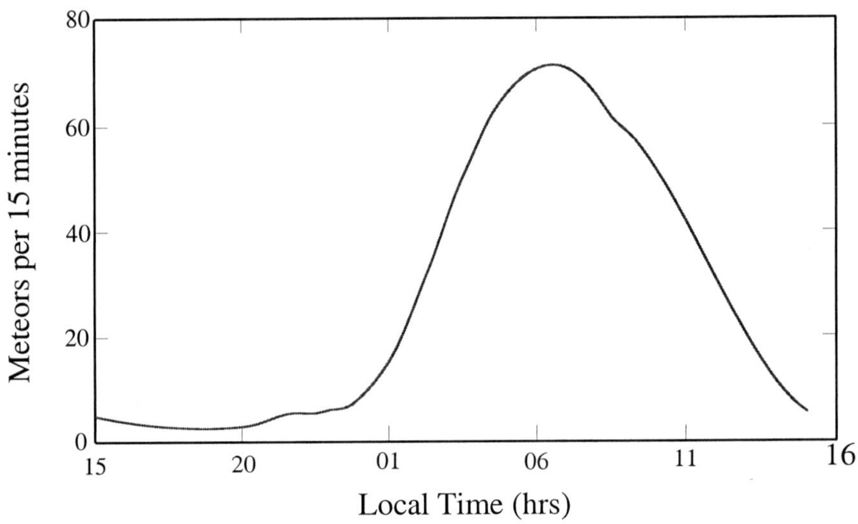

Figure 1. Typical diurnal average of $\sim$ 3200 meteors are observed in the 300 meter AO beam

2. Results

The typical diurnal count rate observed in the 300 m diameter 430 MHz beam results in an average of ~3200 events per day (Figure 1). This result combine with our very good meteor observed time, altitude and velocity distribution allows us to calculate the upper atmospheric meteoric mass influx and compare with past results. Preliminary results of this method are display in Figure 2. Curve a in Figure 2 represents the results reported in Ceplecha et al. [1] where the authors gathered data from several sources of observational flux. Curve b shows the mass flux measured at AO based on the $\sim$ 26 % of events that showed linear deceleration allowing the determination of the meteor BP. As it can be observed, these results fall below those reported by Ceplecha et al. [1]. However if the events for which velocity but no deceleration (i.e. no BP) was determined are included

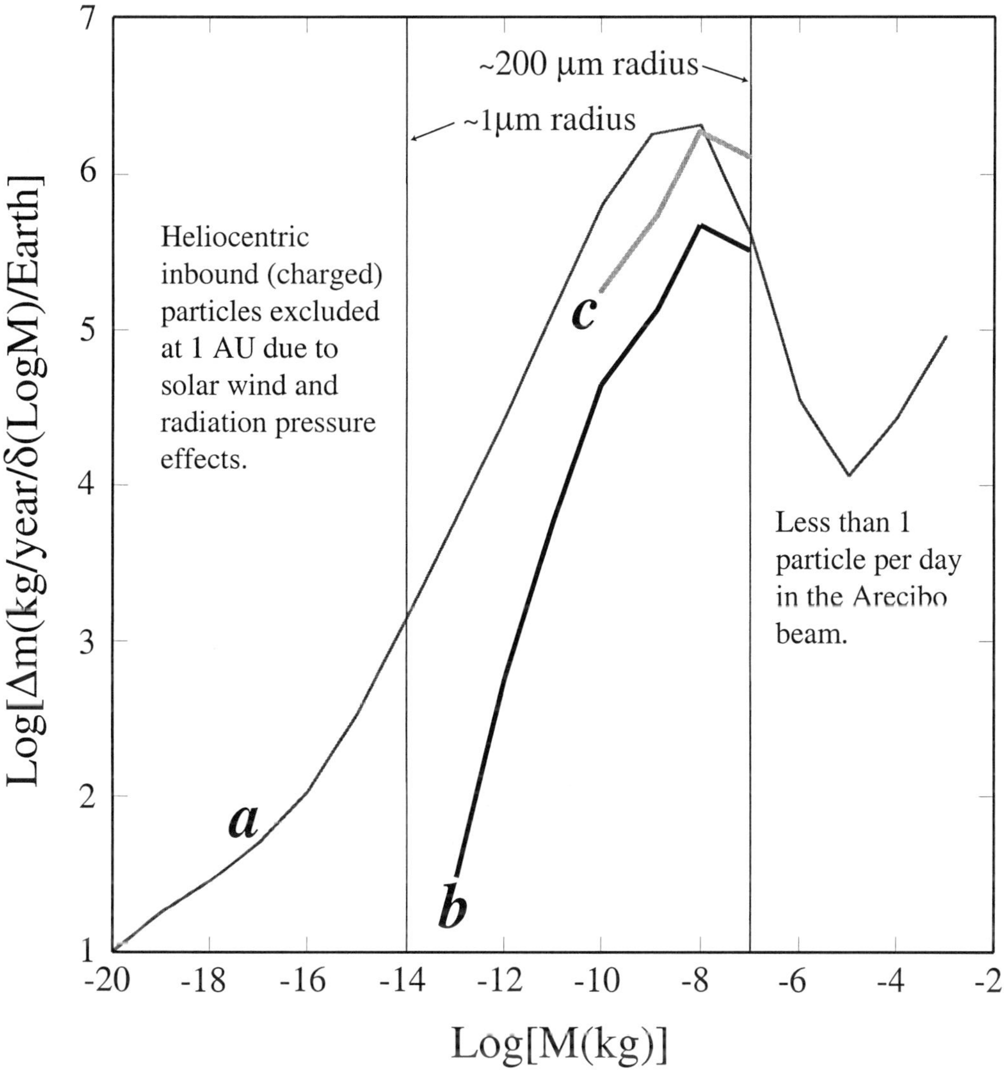

Figure 2. Yearly whole-Earth mass flux per decade of particle mass. The different lines are described in the text.

by evenly distributing them into the 3 mass decades below the maximum, our numbers (curve c) in Figure 2 are in better agreement with those of Ceplecha et al. For lack of a better approximation we distributed all these events in the top three decades. The reason why these events do not show deceleration remains unclear. This question along with approaches to better determine deceleration and thus BP/mass is under current investigation.

In Figure 2 we note two mass limits of considerable interest. The upper limit is simply determined by the small area of the 300 m diameter AO radar beam [4] for which incidence of particles larger than 10^{-7} kg is quiet improbable. The lower limit, is that of so-called β-meteoroids [5] that should not reach Earth from outside Earth's orbit. Interestingly, the flux observed falls off much more quickly than the Ceplecha et al. results as this limit is approached. It remains to be seen if this trend will be sustained as we continue to add to our database.

3. Conclusions

In this paper we have presented preliminary results obtained using the 430 MHz AO radar of the determination of the micrometeoric mass flux into the Earth's upper atmosphere. The Doppler-based velocity/deceleration results provide a direct method to determine this flux. We compared our results with those reported by Ceplecha et al. [1] and find reasonably good agreement if we include those events that no deceleration is observed. We will greatly enhance our meteor database in the next year as well as refine our deceleration determinations. This should yield firmer flux estimates.

REFERENCES

1. Z.J. Ceplecha, J. Borovicka, W.G. Elford, D.O. Revelle, R.L. Hawkes, V. Porubcan and M. Simek, Space Sci. Rev. 84 (1998) 327.
2. F.L. Whipple, Proc. N. A. S. 36(12) (1950) 687.
3. D. Janches, J.D. Mathews, D.D. Meisel, and Q.H. Zhou, Icarus 145 (2000) 53.
4. J.D. Mathews, D.D. Meisel, K.P. Hunter, V.S. Getman and Q.H. Zhou, Icarus 126 (1997) 157.
5. E. Grün, P. Staubach, M. Baguhl, D.P. Hamilton, H.A. Zook, S. Dermott, B.A. Gustafson , H. Fechtig, J. Kissel, D. Linkert, G. Linkert, R. Srama, M.S. Hanner, C. Palnskey, M.Horanyi, B.A. Lindblad, I. Mann, J.A.M. McDonnell, G.E. Morfill, and G. Schwehm, Icarus 129 (1997) 270.

The size of meteoroid constituent grains: Implications for interstellar meteoroids

R.L. Hawkes[a], M.D. Campbell[a,b], A.G. LeBlanc[a,c], L. Parker[a], P. Brown[b], J. Jones[b], S.P. Worden[d], R.R. Correll[e], S.C. Woodworth[a,f], A.A. Fisher[a,g], P. Gural[h], I.S. Murray[a,i], M. Connors[j], T. Montague[k], D. Jewell[l] and D.D. Babcock[m]

[a]Physics Department, Mount Allison University, Sackville, NB Canada.

[b]Physics and Astronomy Department, University of Western Ontario, London, ON Canada.

[c]Astronomy and Physics Department, Saint Marys University, Halifax, NS Canada

[d]United States Air Force, Pentagon, Washington, DC USA

[e]Headquarters United States Air Force Space Command and NASA, Washington, DC USA

[f]Engineering Physics Department, McMaster University, Hamilton, ON Canada.

[g]Department of Physics and Astronomy, University of Calgary, Calgary, AB Canada.

[h]Science Applications International Corporation, Arlington, VA USA.

[I]Department of Physics, University of Regina, Regina, SK Canada.

[j]Department of Physics, Athabasca University, Athabasca, AB Canada.

[k]Air Force Research Lab, Kirtland AFB, NM USA.

[l]United States Space Command, Colorado Springs, CO USA.

[m]Centre for Research in Earth and Space Science, York University, Toronto, ON Canada.

The most widely accepted model for the structure of cometary meteoroids is a dustball with grains bound together by a more volatile substance [1]. In this paper we estimate the size distribution of dustball grains from meteor flare duration, using image intensified CCD records of 1998 Leonid meteors. Upon the assumption of simultaneous release of dustball grains at the beginning of the flare, numerical atmospheric ablation models suggest that the dustball grains in these Leonids are of the order of 10^{-5} to 10^{-4} kg, which is somewhat larger than estimates obtained by other methods. If the dustball grain sizes determined here are representative of cometary meteoroid structure in general, only the most massive (O and B0) type stars could eject these grains into interstellar space by radiation pressure forces.

1. INTRODUCTION

There is now clear proof for the influx into our solar system of meteoroids of interstellar origin in the size ranges covered by radar [2], image intensified video [3], micrometeor radar, spacecraft dust detectors [4] and as meteorite inclusions [5]. It is not clear that interstellar meteors of the size range covered by photographic methods are present in detectable numbers [6], and the flux of interstellar meteoroids seems to be sharply mass dependent [7].

Most cometary meteoroids are a conglomeration of grains, a "dustball" [8,9,1]. The two component dustball model [1] views these grains as being bound by a more volatile substance, and this model has been successful in matching a number of meteor observational features [10,11,12,13]. This paper addresses the question of whether fragmentation of dustball meteoroids, coupled by subsequent ejection from a planetary system by radiation pressure forces, is an important mechanism in the production of interstellar meteoroids.

2. OBSERVATIONAL DATA AND NUMERICAL MODELLING

The 1998 Leonid shower was rich in bright fireballs, some of which produced intense flares. We use observational data collected in Mongolia for 316 Leonid meteors observed with microchannel plate image intensified CCD detectors (see [14] for more details on the equipment and observations). Four of these meteors had intense flares – see Figure 1. The duration of meteor flares can be used to estimate the size of the constituent grains if one assumes that a rapid commencement flare is the result of simultaneous detachment of many grains [15].

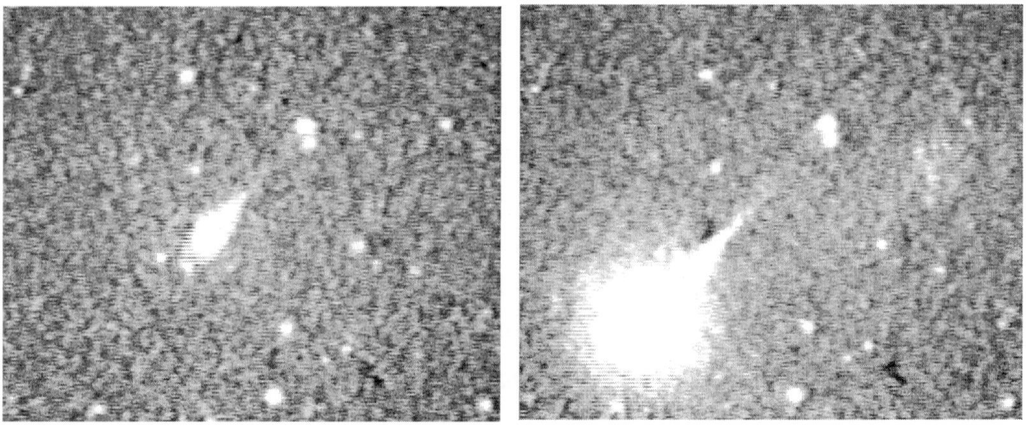

Figure 1. Leonid meteor recorded at 22:37:48 UT on Nov. 16, 1998. These two images are only 5 video frames (0.167 s) apart. This meteor displayed a single intense flare.

These flares were so bright that precise absolute photometry is impossible. Although the CCD auto-gain circuitry was turned off during observations, several of these events were bright enough to enable the protection circuitry in the microchannel plate image intensifiers (which then reduced the intensifier gain momentarily). If we extrapolate techniques used for image intensified CCD meteor photometry [11,13,16,17] we can determine light curves for these events. We demonstrate in Figure 2 the light curve for the early part of the 22:37:48 UT Nov 16 1998 event. It is clear that there was a well defined meteor light curve which suddenly brightened to produce an intense flare. A single station technique which utilizes the known radiant and velocity and the apparent angular velocity from the video data [18] can be used to estimate the heights of these meteors to a precision of about 2.0 km. The data is shown in Table 1.

If we assume that the flares are a consequence of simultaneous detachment of a large number of meteoroid grains we can match the observed flare duration with predictions based on numerical modeling of the atmospheric ablation of these grains [12]. We assume that the grains are spherical, with an average bulk density of 1000 kg m^{-3}, and with a sum of latent heat of vaporization plus fusion of $6x10^6$ J kg^{-1}. The grain mass which best matches the height of maximum luminosity of the flare is given in the final column of Table 1.

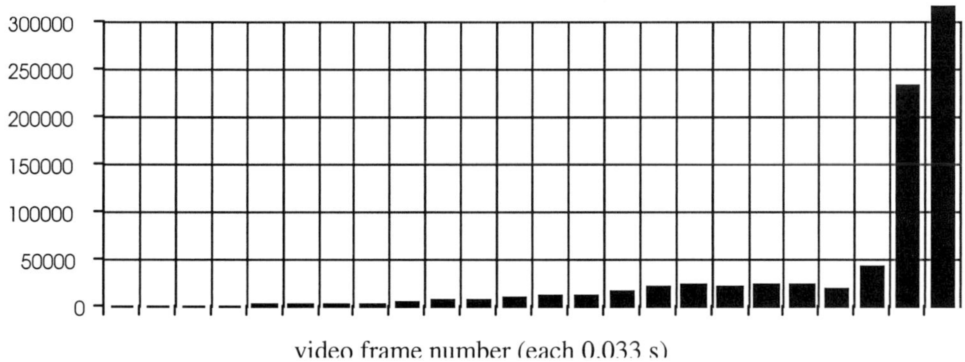

Figure 2. Early part of light curve for the meteor of Figure 1. Luminous intensity (arbitrary units) is plotted versus time (each bar represents 33.3 ms). The flare began in the last two time units displayed here.

Table 1
Heights (in km) for the four Leonids with intense flares.

UT Nov 16	first ht.	last ht.	zenith	flare beg.	flare peak	flare end	best fit
19:35:00	115.8	90.8	74.9	115.8	101.3	93.4	$1x10^{-4}$
20:02:15	135.8	95.1	47.2	112.5	97.5	90.5	$4x10^{-5}$
20:15:00	150.5	93.7	44.8	124.5	101.8	92.1	$5x10^{-5}$
22:37:48	150.0	88.8	27.1	109.8	97.5	88.8	$2x10^{-5}$

First and last heights are the heights of the first and last observed points, zenith is the zenith angle in degrees, and the last three columns give the heights (in km) when the flare began, displayed peak intensity and ended. The last column gives the grain mass (in kg) which best matches the flare maximum.

3. DISCUSSION

The grain sizes determined here are considerably larger than those determined by overall light curve shape modeling [11]. Radiation pressure forces from main sequence stars can only eject grains of this size from the most massive O and possibly B0 stars [7,19]. However, by our flare duration technique we cannot rule out the presence of smaller grains in addition to the larger ones needed to model the flare duration. Some authors [20] have assumed that the grains within each dustball meteoroid may follow the same mass distribution law as meteoroids themselves. An interesting question is whether dustball meteoroids may fragment in space, with their grains being subsequently ejected from the planetary system by radiation pressure forces. While this must occasionally occur, a consideration of the solar wind energy flux suggests that hundreds to thousands of Leonid orbital passages would be needed for a typical Leonid to remove the volatile component by solar wind sputtering. This is supported by the fact that obviously separated clusters appear relatively rare [21,22] although the transverse spread Leonids [13,23] may be less strongly separated clusters. In any case we conclude that it is likely that ejection from the early stages of planetary system formation [24] is probably a more significant source of interstellar meteoroids.

REFERENCES

1. R.L. Hawkes and J. Jones, Mon. Not. R. Astron. Soc. 173 (1975) 339.
2. A.D. Taylor, W.J. Baggaley and D.I. Steel, Nature 380 (1996) 323.
3. R.L. Hawkes and S.C. Woodworth, J. Roy. Astron. Soc. Can. 91 (1997) 218.
4. E. Grün, B. Gustafson, I. Mann, G.E. Morfill, P. Staubach, A. Taylor, and H.A. Zook, Astron. Astrophys. 286 (1994) 915.
5. E. Anders and E. Zinner, Meteoritics (1993) 490.
6. M. Hajdukova, Astron. Astrophys. 288 (1994) 330.
7. R.L. Hawkes, T. Close and S.C. Woodworth, in Meteoroids 1998 (eds W.J. Baggaley and V. Porubcan) Slovak Academy of Sciences, Bratislava (1999) 257.
8. L.G. Jacchia, Astrophys. J. 121 (1955) 521.
9. F. Verniani, Space Sci. Rev. 10 (1969) 230.
10. M. Beech, Mon. Not. R. Astron. Soc. 211 (1984) 617.
11. M.D. Campbell, R.L. Hawkes and D.D. Babcock, in Meteoroids 1998 (ed. W.J. Baggaley and V. Porubcan) Slovak Academy of Sciences, Bratislava (1999) 363.
12. A.A. Fisher, R.L. Hawkes, I.S. Murray, M.D. Campbell and A.G. LeBlanc, Planet. Space Sci. 48 (2000) 911.
13. I.S. Murray, R.L. Hawkes and P. Jenniskens. Meteoritics Planet. Sci. 34 (1999) 949.
14. M.D. Campbell, P. Brown, A.G. LeBlanc, R.L. Hawkes, J. Jones, S.P. Worden and R.R. Correll, Meteoritics Planet. Sci. 35 (2000) 1259.
15. A.N. Simonenko, Soviet Astronomy – A.J. 12 (1968) 341.
16. D.E.B. Fleming, R.L. Hawkes and J. Jones, in Meteoroids and Their Parent Bodies (ed. J. Stohl and I.P. Williams) Slovak Academy of Sciences, Bratislava (1993) 261.
17. R.L. Hawkes, K.I. Mason, D.E.B. Fleming and C.T. Stult,. in International Meteor Conference 1992 (eds. D. Ocenas and D. Zimnikoval) Int. Meteor Org, Antwerp (1993) 28.
18. P. Brown, M.D. Campbell, K. Ellis, R.L. Hawkes, J. Jones, P. Gural, D. Babcock, C. Barnbaum, R.K. Bartlett and M. Bedard, Earth Moon Planets 83 (2000) 167.
19. R.L. Hawkes and S.C. Woodworth, J. Roy. Astron. Soc. Can. 91 (1997) 91.
20. I.S. Murray, M. Beech, M. Taylor, P. Jenniskens and R.L. Hawkes, Earth Moon Planets 82 (2000) 351.
21. P.A. Piers and R.L. Hawkes, WGN J. Inter. Meteor Org. 21 (1993) 168.
22. M. Kinoshita, T. Maruyama and S. Sagayama, Geophys. Res. Lett. 26 (1999) 41.
23. A.G. LeBlanc, I.S. Murray, R.L. Hawkes, S.P. Worden, M.D. Campbell, P. Brown, P. Jenniskens, R.R. Correll, T. Montague and D.D. Babcock, Mon. Not. R. Astron. Soc. 313 (2000) L9.
24. T.G. Brophy, Icarus 94 (1991) 250.

Radar meteoroids: advances and opportunities.

W. J. Baggaley[a]

[a]Department of Physics and Astronomy, University of Canterbury, Private Bag 4800, New Zealand.

Radar sensing of meteoric plasma is a powerful tool for probing the spatial structure of meteor streams, the mass distribution of their member particles, and the dynamics of individual meteoroids. With their enhanced sensitivity, radars are able to provide information that complements photographic, TV, and video techniques and also to probe areas inaccessible to other methods. An outline will be given of presently operating radar systems and current programmes that contribute to our knowledge of inner Solar System dust.

1. INTRODUCTION

Radars probe the plasma irregularities generated by ablating interplanetary dust grains in the upper atmosphere generally heights 80-120 km. From observational programmes we ultimately want to know about the physical and dynamical characteristics of the dust. There are certain properties of interplanetary dust for which radars are an especially valuable probing agent. Radar surveys, sampling individual meteoroids, can provide us with information about the space environment determining especially:

- the influx rate for a given mass hence spatial density and mass distribution;
- for discrete streams the time variations of rates and mass distributions with any associated fine structure are valuable signatures of processes like comet ejection mechanisms and dynamical history of streams where sampling in longitude is valuable;
- determining the velocity vector of a meteor's atmospheric trajectory provides the heliocentric orbit;
- measurements of meteoroid atmospheric decelerations or recording of body fragmentation are valuable in providing evidence of their physical characteristics and cohesive structure.

As in other dust observational techniques there are important biases that must be taken into account: for example to derive the dust heliocentric orbital distribution severe correction factors must be recognised: the impact probability with the Earth; Earth focusing; atmospheric effects and the radar detection function. The size of Earth-impacting dust that can be sampled by radar systems depends on radar transmitted powers available, operating frequency and antenna system used but has a lower useful size limit of some tens of μm (set by the radar transmitter power available and antenna gain) while the ultimate lower limit is set by the fact that very small grains ($\lesssim 10\,\mu$m) suffer incomplete ablation. The upper size is set by the area of the atmosphere (acting as a detector) illuminated, and statistical sampling: for a single radar the meteoroid population of sizes $\gtrsim$ cms is sparsely sampled.

2. RADAR GEOMETRIES

The type of echo recorded–and therefore the quality of information to be gained– depends on the geometrical relation between the plasma train created by the ablating meteoroid and the radar: additionally, radars may employ multi-station, monostatic or bistatic arrangements.

2.1. Transverse reflection

Here the trajectory of the meteor is orthogonal to the (mono-static) radar. The scattering of radio waves by the ionization created by the meteor can be analysed in terms of Fresnel diffraction and the analysis has a convenient analogue in optical diffraction at a straight edge. For meteor scattering the Fresnel zone length is about 1 km for HF radars and as ionization is progressively deposited more Fresnel zones contribute with different phases and in summation most of the reflected energy is produced from a region on the meteor train of length $\sim$ 1 km centred at the geometrically orthogonal point. The instant in time when the meteoroid reaches that orthogonal position is termed the t_0 point and the received radar signal is termed the 'body echo'.

The ionization column (cylindrical in the absence of an external magnetic field) is created with a finite diameter: additionally ambipolar diffusion of the plasma will lead to an increasing column diameter with time: if the column size is comparable to the operating wavelength phase differences in the scattering from individual electrons in a train cross-section will result in destructive interference and a reduction in the reflected energy.

The time-history of the reflected energy to produce a radar echo can be conveniently analysed with the aid of the Cornu spiral (depicting phase behaviour) with the presence of ambipolar diffusion (leading to an exponential decay of the meteor echo) introducing a modification of the classical behaviour. In the absence of meteoroid fragmentation or irregular plasma the frequency of post t_0 amplitude oscillations give a measure of the meteor's scalar speed. Conversely the post t_0 phase oscillations are too small ($< 30°$) to be useful speed indicators whereas the large pre-t_0 phase changes are valuable for meteoroid speed measurements. Radars with phase capability can employ the pre-t_0 rapid phase changes to secure accurate speed measurements because the ionization train in its initial formation has no adverse effects arising from train diffusion, no ionization irregularities and no disruption by grain fragmentation and for small times atmospheric wind shear has not sufficient time to operate. Good examples of echo behaviour are well illustrated in Elford [1] Figures 1 and 2. A contributing factor to the suppression of post t_0 amplitude oscillations is the presence of continuous fragmentation along its trajectory of the ablating grain. If, on plasma train creation, the orthogonality condition does not hold so that the central Fresnel interval is outside the main radiation pattern of the radar, then the classical meteor echo is not formed so that the rapid leading edge is absent: however, the phase changes are still present and speed measurements can be made on such echoes (see Elford [1] Figure 3.)

2.2. Radial reflection

An ablating meteoroid not only deposits ionization along its path (and that ionization quickly attains dynamic equilibrium with the ambient atmosphere so is stationary unless transported by the atmospheric neutral wind) but also creates a plasma spheroid surrounding the meteoroid itself. This plasma ball shares the meteoroid's motion. The scattering from such a plasma ball produces what is termed a 'head echo': the scattering cross-section depends on the radar wavelength but the reflection coefficient is very small compared to that for transverse reflection (the body echo) so that the echo is not not discernible for orthogonal geometry. However, if the geometry is radial so that the meteoroid is moving in the line-of-sight then the body echo is absent and the head echo dominates. The radar-approaching plasma ball acts as a moving target that directly represents the meteoroid atmospheric speed: the echo will rapidly decrease in range traversing successive range bins and also with a phase-sensitive radar system rapid phase changes will occur. Notice that for radar sampling pulse rates even as high as 1 kHz the plasma target will move through several wavelengths between samples and results in phase aliassing: however, the range shift and phase changes can be combined to produce an accurate (uncertainty $\sim 0.3\%$) radial speed.

With a single station radar the trajectory aspect angle is unknown so that there is an uncertainty in the radial speed and direction. For accurate results therefore, a narrow pencil beam $\sim 1^\circ$ is required and provision for measuring the across-beam angle. Using such, both the meteor trajectory (the upstream direction of which is termed the 'radiant') and speed can be deduced and hence, after appropriate transformations and corrections, the heliocentric orbit.

2.3. Oblique reflection

In this geometry the radar transmitter and receiver have a ground separation large compared to the meteor target height so that the specular condition results in a large scattering angle (the angle between the normal to the meteor train and the incident wave propagation direction, ϕ, where $\phi = 0$ for transverse, backscattering case). In effect the Fresnel zone length for such a forward scatter configuration is increased by a factor $(\cos\phi)^{-2}$ and the radar wavelength is effectively increased by a factor $(\cos\phi)^{-1}$. Two valuable consequences compared to the strict transverse reflection result: the scattering cross-section is larger and the echo decay due to ambipolar diffusion is less rapid with consequential benefits for detecting high altitude rapidly diffusing meteors.

3. CURRENT PROGRAMMES

It's useful to list those radars currently operational with on-going programmes. Some radar facilities are able to operate with different geometries but here we list them according to their major operating role.

3.1. Transverse reflection

3.1.1. Measuring individual orbits

The Advanced Meteor Orbit radar (AMOR) operates at 26.2 MHz radiating 100 kW peak pulse power. The facility uses three $\sim$ 8 km spaced stations to provide time-of-flight measurements of echoes to give velocity components while elevation is secured via a dual baseline interferometer. The antenna system is specifically designed [2] to have narrow (1.6°) azimuthal beams and broad in elevation. FM UHF data channels provide links between stations. The facility is in continuous operation in programmes devoted to: the distribution of solar system dust from heliocentric orbit surveys; the identification of interstellar dust in the inner solar system; the dynamical structuring of cometary and asteroidal streams; and the measurement of atmospheric winds and turbulence.

The 45.6 MHz MU radar at Shigaraki near Kyoto Japan has a programme mainly devoted to middle atmospheric dynamical work but the system can sense individual meteor radiants by rapid beam switching with meteoroid speeds determined from Doppler pulse compression characteristics. An array of 475 crossed Yagi antennas is used for transmitting and receiving with each being driven by individual transmitter units. The system antenna beam has a half-power width of 3.7° and target zenith angles of up to 30° can be accessed. Astronomical projects concentrate on the times of major streams [3].

3.1.2. Echo directions but no individual orbits

The Chung-Li radar in Taiwan operating at 52 MHz employs a transmitter array providing a $\sim$ 10° width vertical pointing beam with echo direction determined by relative phases measured using a 0.86λ spacing triple Yagi array The meteor programme has focused principally on the Leonid shower influx [4].

In Canada stream parameters have been measured using a 40.68 MHz 10 kW facility. This system (CLOVAR) consists of single transmitter Yagi combined with five Yagis as a multi-spacing interferometer of spacing 2.0 and 2.5 λ to determine echo directions to $\sim$ 2°. Stream meteors are identified according to the directions with respect to the expected shower radiant [5].

The Adelaide Buckland Park facility in Australia operates at 54.1 MHz using a TX/RX square antenna filled array sides 16 λ giving a full width half power radiation beam of 3.2°. Antenna element phasing can tilt the beam 30 east or west of zenith and accurate ($\sim$ 0.8 %) meteor speeds can be determined. The programme has been devoted to stream flux characteristics and the probing the velocity distribution within stream population (e.g. [6]).

3.1.3. Fluxes

One of the most sustained radar surveys has been that carried out at the Ondrejov facility in the Czech republic. The 37 MHz operation employs a steerable antenna 36° beam and has maintained flux measurements of the major streams for several decades. Range-time plots yield valuable longitude cover for fine structure in streams, long term rates influenced by atmospheric changes and data on head echoes (see e.g. [7]).

In South Africa the 28 MHz Grahamstown radar with echo position determined by 4-antenna phase comparisons and with large angular sky coverage but lacking range and velocity information has been able to provide maps of apparent sporadic sources after subtraction of the major streams [8].

3.2. Radial reflection

The first measurements of speeds and decelerations using radial geometry were those of the Adelaide (Australia) group [9, 10]. The 54 MHz Buckland Park facility employing radial configuration provided accurate speeds (0.2 %) as well as decelerations and fragmentation event measurements. Examples of such down-the-beam-echoes are well presented in Elford [1] Figures 5 and 6.

The radio-astronomy instrument at Arecibo has been operated in meteor mode for limited periods. The 430 MHz facility employs a near-vertical pointing 305 m dish with principle focus steering deployed to scan up to 15° from zenith. Because of the high gain beam width of 0.16° the radiants of incident meteors can be located accurate to a fraction of a degree. The use of triple transmitter pulses yields enhanced precision and good meteoroid decelerations though the sky coverage is restricted: the antenna configuration provides limited viewing direction near zenith [11]. Since the Arecibo instrument has a full astronomical programme dedicated meteor operation is limited.

The European incoherent scatter radar (EISCAT) operating at 930 MHz is an example of a system designed for ionospheric work that has proved valuable as a meteor probe, providing analyses of head echoes [12] and fluxes. A tristatic geometry (radars at Kiruna, Sweden, Tromso, Norway and Sodankyla, Finland) will enable trajectories and hence orbits to be secured [13].

3.3. Oblique scatter

The only dedicated facility known to the author is that operating in Italy over paths of 700 km between Budrio (near Bologna) and south-east to Leece and also 600 km north-west to Modra in Slovakia. The 1 kW continuous wave Budrio transmitter using 42.7 MHz operates to encompass the major shower times. This technique is able to provide standard yearly influx data [14].

Forward scatter links are operated by many groups world-wide and particularly active are those in USA, Japan, Europe and Finland using passive operations employing transmitters such as TV, FM broadcasts and commercial beacons. Providing a wide global coverage, these programmes are valuable in monitoring time changes in flux representing structure in stream spatial density. Such monitoring at the times of e.g. Leonid Storm epoch can sample spatial changes in the dust stream that cannot be sampled by a single radar station.

4. PROGRESS ON AIDS TO INTERPRETATION

To correctly interpret radar data it is important to incorporate realistic physical effects. Here mention is made of three recent aids in the area.

To gain absolute meteoroid mass calibration and flux calibration, account needs to be taken of the attenuating effect of the meteoric plasma column radius at formation. Using simultaneous multiple wavelength records of Leonid echoes, Campbell [15] has measured train formation cross-sections as a function of height: this 'height-ceiling' effect can have gross effects on estimates of meteor fluxes and masses.

At heights in the atmosphere where the electron gyro frequency exceeds the electron-neutral collision frequency, the rate at which a meteor train diffuses depends on the orientation of the train and radar line of sight to the local geomagnetic field. Elford and Elford [16] have provided numerical values showing how the effective diffusion can be inhibited: small high-speed meteoroids inaccessible to many radars because of the rapid diffusion of their plasma column can have extended echo life-times depending in the relevant geometry.

Though radio wave absorption will be negligible at the frequencies utilised by many meteor radars, it is expected that Faraday rotation produced by the day-time lower E-region ionization situated below the reflection point can be significant. Many meteor radars employ linearly polarised antennas so that polarisation rotation can lead to effective signal attenuation [17].

5. FUTURE DIRECTIONS

Several current active radar programmes are dedicated to monitoring both background interplanetary influx and stream spatial densities and structure. There are some areas where valuable insight may be gained about the meteoric process and therefore improvements in our models of radar reflection mechanisms and related processes of the meteoric plasma. There are specific areas where programmes might be valuably directed.

Employing geometrical arrangements to select head echoes to gain information about meteoroid Earth-impacting trajectories needs input about the details of the plasma that surrounds the ablating meteoroid; its production and maintenance. The role of meteoroid fragmentation needs targeting; how structural characteristics of the grains affect the created ionization and the form of the echo: are radars seeing all types of meteoroids or are our samples biased: there is a need to better understand the nature of the fragmentation (gross or minor) if we want unbiased sampling of interplanetary dust. Measurements of ablation coefficients and its effect on meteoroid deceleration needs further examination to fix more firmly the pre-atmospheric orbital speeds of grains sampled by ground-based radars.

REFERENCES

1. W.G. Elford, in Meteoroids 1998, (eds. W.J. Baggaley and V. Porubcan) Astronomical Inst. Slovak Acad. Sci. Bratislava (1999) 21.
2. W.J. Baggaley, (2000) this volume.
3. T. Nakamura et al., Adv. Space Res. 19 (1997) 643.
4. Y.-H. Chu and C.-Y. Wang, Radio Sci. 32 (1997) 817.
5. P. Brown, H.W. Hocking, J. Jones and J. Rendtel, Mon. Not. R. astron. Soc. 295 (1998) 847.
6. D.P. Badger W.G. Elford, in Meteoroids 1998 (eds. W.J. Baggaley and V. Porubcan) Astronomical Inst. Slovak Acad. Sci. Bratislava (1999) 195.
7. M. Simek and P. Pecina, Earth Moon & Planets 68 (1995) 555.
8. L.M.G. Poole, Mon. Not. R. astron. Soc. 290 (1997) 245.
9. A.D. Taylor, M. Cervera and W.G. Elford, in Physics, Chemistry and Dynamics of Interplanetary Dust, Astronom. Series Pac. Conference Series 104 (1996) 75.
10. M. Cervera, W.G. Elford and D.I. Steel, Radio Sci. 32 (1997) 805.
11. J.D. Mathews, D.D. Meisel, K.P. Hunter, V.S. Getman and Q.-H. Zhou, Icarus 126 (1997) 157.

12. G. Wannberg, A. Pellinan-Wannberg and A. Westman, Radio Sci. 31 (1996) 497.
13. A. Pellinan-Wannerg, A. Westman and G. Wannberg in, Meteoroids 1998, (eds. W.J. Baggaley and V. Porubcan) Astronomical Inst. Slovak Acad. Sci. Bratislava (1999) 83.
14. G. Cevolani, M.F. Gabucci, A. Hajduk, M. Hajdukova, V. Porubcan and G. Trivellone, Il Nuovo Cimento C19 (1996) 447.
15. Campbell, M. Personal communication.
16. W.G. Elford and M.T. Elford in, Meteoroids 1998, (eds. W.J. Baggaley and V. Porubcan) Astronomical Inst. Slovak Acad. Sci. Bratislava (1999) 71.
17. W.G. Elford, J. atmos. Sol. Terr. Phys. 59 (1997) 1021.

Dynamical and orbital properties of the Aricebo micrometeors

D. Janches[a], D.D. Meisel[a,b] and J.D. Mathews[a]

[a]Communications and Space Sciences Laboratory, Department of Electrical Engineering, Penn State University, University Park, PA, USA

[b]Department of Physics and Astronomy, SUNY-Geneseo, Geneseo, NY, USA

Using the Arecibo Observatory (AO) 430 MHz Radar we have developed a Doppler technique to measure very precise micrometeor instantaneous velocities directly from the meteor head-echo. In addition, a large number of these observations show deceleration. With the velocity, the deceleration, the assumption of a spherical shape, and a mean micrometeoroid mass density (3 g cm^{-3}) we have obtained estimates of in-atmosphere particle sizes. Therefore we can produce a more realistic orbital analysis than previously obtained for micrometeors. We first use an MSIS standard atmosphere model and the measured deceleration in order to obtain the meteor extra-atmospheric speeds, assuming that sputtering is the only mass-loss mechanism that these particles undergo prior and during the time we detect them. So far, over 7000 detections obtained during the Leonids 1997 (L97) and 1200 during the Leonids 1998 (L98) observation campaigns have been analyzed. Out of these detections, we present elements without correction for perturbations (i.e. radiation pressure, perturbation by the Jovian planets and photoelectric charging effects) of over 500 events from 1997 and 200 from 1998.

1. INTRODUCTION

Current VHF/UHF radar micrometeor studies at Arecibo Observatory (AO), Puerto Rico, utilize the highly resolved (in time and height) meteor "head-echo"(as contrasted with classical meteor radars that detect the "trail-echo") to obtain meteor altitude, velocity and deceleration [1–3]. In addition, very precise beam pointing information provides good radiant information of large numbers of interplanetary dust particles (IDPs) [1] entering the earth atmosphere. The dynamical masses inferred using this approach appear to range from a small fraction of a nanogram to a few micrograms ($<$ 1 - 100 μm diameter). These results combine to make AO uniquely situated and suited for the study and understanding of the IDP near 1 AU using ground-based observations.

Top-of-atmosphere velocity and mass estimates are obtained by integrating upwards the meteor mass and momentum equations. For this, we utilize the MSIS-E-90 model atmosphere (see http://www.wdc.rl.ac.uk/wdcc1/ msis90.html). In addition, to test the effect of possible mass-loss, an empirical sputtering model with a sputtering yield that increases with energy [4,5] was adopted. However the orbital results discussed in the next section do not statistically change if our empirical model is neglected or if the in-atmospheric meteor speeds are taken as the velocity at infinity. The Geocentric and heliocentric velocities are then determined and they include rigorous correction for diurnal aberration, zenith attraction and earth orbit ellipticity. The determination of the immediate (unevolved) meteoroid orbits follows from the determination of the meteor radiant and extra-atmospheric velocity [6].

2. RESULTS

The corresponding orbital elements for L97 and L98 datasets are displayed in Figure 1 for the events which resulted in particle radii greater than 0.5 microns. We excluded smaller particles because the calculated possible magnetospheric forces exerted on them due to dust charging is of the same order as the earth attraction term or larger. Also it is unclear how particles of these sizes have survived such large deceleration, reaching relatively low altitudes, and it is possible that they are the result of larger meteoroids which have an across-the-beam velocity component. For a more detailed description on these results, the reader may refer to Janches et al. [3].

Panels (a) and (b) in Figure 1 give the perihelion times of the entire IDP set showing nearly equal pre- and postperihelion numbers in the case of L97 and a lack of preperihelion numbers, due to the bias introduced by using only 6 hours of observation, in the L98 case. Panels (c) and (d) give the semimajor axis (a) versus the eccentricity (e). In addition, the Whipple K and Pe criteria for particle asteroid/comet origin and the evolutionary path due to drag effects are shown for various values of the constant C [7]. As can be noted, most of the particles detected at AO are concentrated in asteroidal type orbits of a$\leq$2 AU (i.e. above the criteria curves) at the time of earth interception. The comparison between our results and the evolutionary paths or drag contours show that the a/e of the AO particles have evolved (downward along the curves) from orbits within the distance of Saturn, with the majority coming from within the orbit of Jupiter. This evolution is produced under non-gravitation influences such as Poynting-Robertson effect and radial solar corpuscular radiation pressure [8]. These diagrams also suggest that the particle orbit semi-major axes are reduced to ~1 AU by radiation pressure evolution followed by eccentricity increases (at nearly constant a). We interpret this increase as produced in a manner similar to the electromagnetic resonance (with interplanetary magnetic field sector boundary crossings) mechanism predicted by Morfill and Grün [9]. The meteoroid orbital inclinations displayed in panels (e) and (f) show that while both prograde and retrograde ecliptic concentrated particles are present for the preperihelion case, the postperihelion IDPs are dominated by the presence of retrograde orbits with a peak at ~140 degrees. Finally the perihelion distances (q) are displayed in panels (g) and (h). A dramatic reduction of "sungrazers", for the case of L97, in the postperihelion sample can be seen. For the L98 case a weaker lack rather than a pronounced reduction is indicated because of the low preperihelion particle orbits number. This reduction (or lack) is presumed in both years to be due to solar evaporation or other thermal destruction of these particles. Both the L97 and L98 datasets show a concentration of orbits with q between Mercury and Venus. The large remaining fraction of postperihelion orbits with q within the orbit of Mercury provides again a strong indication that the AO micrometeors are particularly durable compared with classical cometary meteoroids.

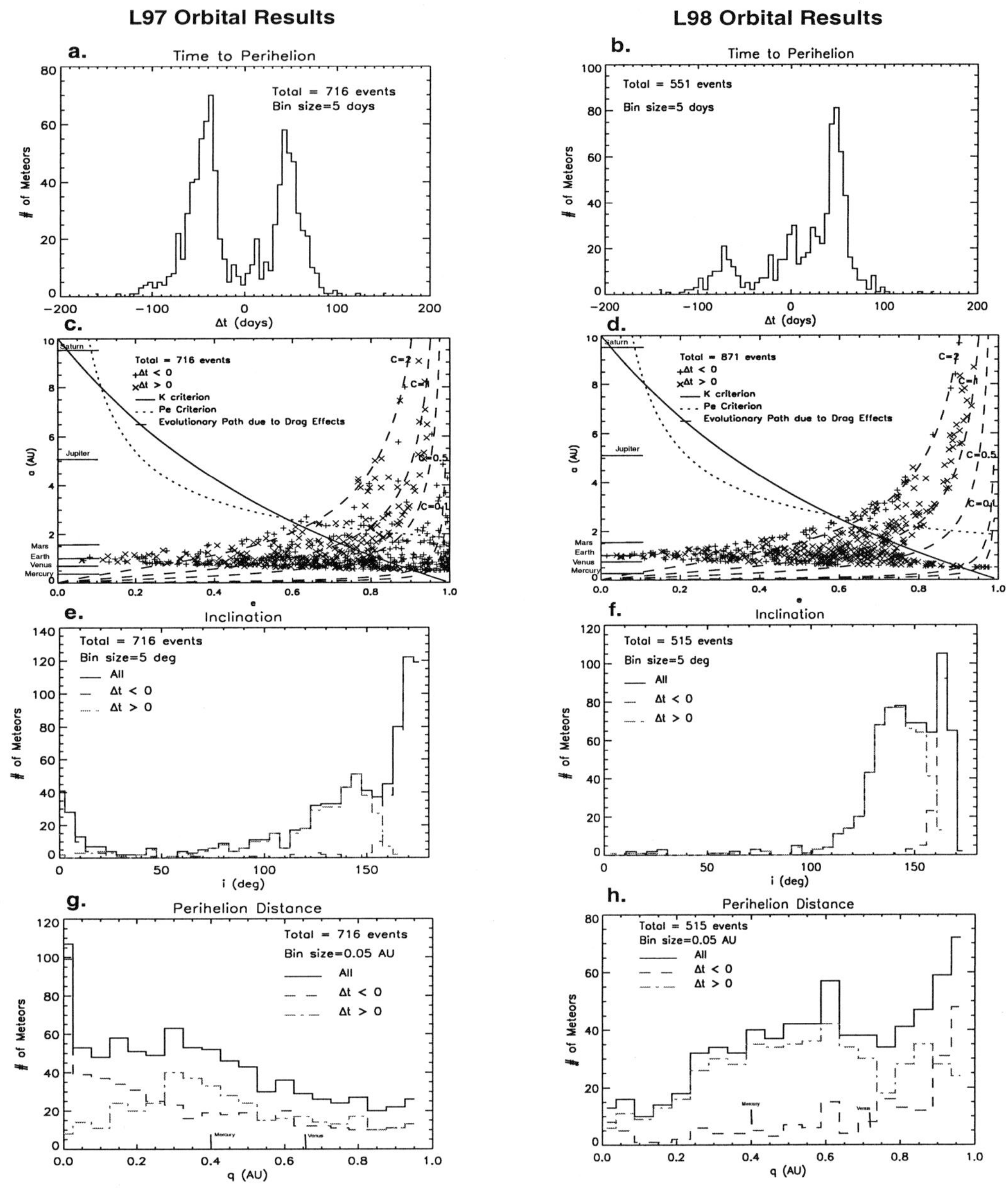

Figure 1. Selected orbital elements for the L97 and L98 samples

3. CONCLUSIONS

In this paper preliminary results from the first orbital study of a large number of AO micrometeors is presented. These events are mainly ecliptic particles down to sub-micron sizes, and a large portion of these meteoroids appear to have orbits lying entirely within the orbit of Mars. However, they appear to have evolved from within 4 AU indicating that older particles, evolved by drag effects from eccentric orbits, are predominant in these samples. Finally, preliminary comparison with the IAU Meteor Database using the Drummond [10] criteria limited to D$\leq$0.05 shows no correspondence between our data and the major November meteor showers (i.e. Leonids, Taurids, etc.). Likewise we find no intrasample groups to the same D limit. The apparent lack of association within our results as well as with the IAU database, seems to indicate that the micrometeoroid orbits are well randomized and decoupled from each other, indicating once again that individual particle histories are largely unsolvable.

REFERENCES

1. D. Janches, J.D. Mathews, D.D. Meisel, V.S. Getman and Q.H. Zhou, Icarus 143 (2000) 347.
2. D. Janches, J.D. Mathews, D.D. Meisel and Q.H. Zhou, Icarus 145 (2000) 53.
3. D. Janches, D.D. Meisel and J.D. Mathews, Icarus 150 (2001) 206.
4. P. Sigmund, Phys. Rev. 184 (1969) 383.
5. K. Wasa and Hayakawa, Handbook of Sputter Deposition Technology: Principles, Technology and Applications, Noyes Publications, 1992.
6. J.M.A. Danby, Fundamentals of Celestial Mechanics, Willmann-Bell, Richmond VA, 1988.
7. L. Kresák, L., in Meteor Orbits and Dust, (ed. G.S. Hawkins) NASA SP-135/SCA-11, Cambridge MA (1965) 9.
8. S.P. Wyatt and F.L. Whipple, Astrophys. J. 111 (1950) 134.
9. G.E. Morfill and E. Grün, Planet. Sp. Sci. 27 (1979) 1269.
10. J.D. Drummond, Icarus 45 (1981) 545.

Update on new developments of the Advanced Meteor Orbit Radar AMOR

W.J. Baggaley[a], R.G.T. Bennett[a], S.H. Marsh[a], G.E. Plank[a] and D.P. Galligan[a]

[a]Department of Physics and Astronomy, University of Canterbury, Private Bag 4800, New Zealand.

AMOR is a continuously operating radar facility for measuring the heliocentric orbits of Earth-impacting grains down to sizes of about 40 μm. Recent developments including additional extensive antenna arrays and multi-antenna independent phase measurements are providing an expanded sky coverage and increased meteoroid velocity resolution yielding an enhanced system capability.

1. THE AMOR FACILITY

The southern hemisphere meteor orbit radar (AMOR) (geographical coördinates 172°39′ E, 43°34′ S) operated by the University of Canterbury, Christchurch, New Zealand designed [1] to provide near-continuous data has been operational since 1990. The facility yields archived orbital information valuable for interplanetary dust studies [2,3]. There are three major innovations that characterise this radar facility.

The system uses narrow antenna beams which implicitly locate the meteor and increase the radar gain making possible the detection of smaller particles.

Raw AMOR data can be used to calculate the time intervals between the occurrence of meteor echo profiles on three spaced receivers to determine the particle velocity components in the atmosphere. Two remote sites are located $\sim$ 8 km west and north of the central site which incorporates the system control, transmitting antenna and echo elevation-finding receiving antennas. UHF FM data links transmit signal data from these sites. Employing Fresnel diffraction patterns as the only method of speed measurements employed by some meteor radar systems suffers from two disadvantages. To secure well-defined echo amplitude fluctuations for fast meteors a high pulse sampling rate is necessary, thus introducing aliasing in range measurement. Additionally, processes such as fragmentation may produce multiple sources destroying the resultant Fresnel pattern, and atmospheric wind-shear may distort the echo profile. Routine speed measurements provided by AMOR do not use diffraction patterns but rather the velocity is determined directly from the time intervals between the recording of the echo profiles on 8 km–spaced receivers. This gives a considerable increase in the proportion of meteor observations for which speeds can be determined compared with other Fresnel diffraction methods. However, the system does make use of the overall velocity calibration provided by the archived Fresnel speed values.

Computerised data acquisition, storage, reduction and display make it possible to reduce observational data rapidly. The association of data acquisition and reduction software with diagnostic display routines provides a powerful environment in which to maintain the system while a graphical display suite provides ready access to the AMOR database.

2. ENHANCED SKY COVERAGE

The technique employs narrow fan antenna radiation patterns confined to the geographic meridian for the elevation-finding dual-spacing interferomEter.

The location of echo targets within the narrow main beam is unnecessary since the main beam azimuthal width of 1.6° (FWHP) is comparable with the elevation accuracy $\sim 0.5°$ and both uncertainties are comparable with the angular errors arising from the uncertainties in the velocity vector determined from the orthogonal velocity components measured from time-of-flight. However this N-S fan of echo coverage does results in a bias in sampling the celestial sphere. Enhanced sky sampling is now (operational since Jan 2000) provided by an additional orthogonal radar sampling direction in the E-W meridian: one result is that those influxing meteoroids with radiants near the local zenith (declinations $\sim -46°$) produce target ionisation trails which are sampled more fully.

A new antenna system providing orthogonal transmitter and receivers (six co-linear arrays in total) have been installed with underground ducted coaxial cables feeding time-switched receivers from the elevation-finding dual interferometer (see Figure 1).

The radar system samples in orthogonal directions by switching at regular (~ 10 min.) intervals with transmitter, receivers and remote site directional switching carried out by GPS time synchronisation.

3. MULTIPLE ANTENNA PHASE MEASUREMENT

A meteor's velocity components are derived from time-of-flight differences between spaced-station echoes with scalar speeds being also provided independentaly by diffraction behaviour. As an impacting meteor deposits ionisation the reflection cross section grows as more Fresnel zones contribute and the instant when the meteoroid has reached the specular geometrical condition it termed the t_0 point. Analysis of post-t_0 echo amplitude fluctuations arising from diffraction provides speed values.

As a meteor traverses those Fresnel zones prior to the t_0 condition large phase changes occur which provide speed values more accuratly than the amplitude oscillations later in the echo. This is because for times early in the creation of a meteoric plasma a radar echo is not subject to the long-term processes present (especially after echo maximum) of plasma diffusion, effects of ionization irregularities, meteoroid fragmentation, and atmospheric wind shear. These processes can contribute to disrupting the smooth character of the plasma column with resulting distortion and often destruction of the classical phase behaviour: many ($\sim 60\%$) meteor radar echoes do not exhibit post-t_0 amplitude oscillations.

In the present upgrade independent signal phases are available at all six of the elevation-finding antennas. This permits echo elevation measurement from the relative phases (three values for each meteor echo) to provide unambiguous elevation and echo height over the height regime 60 to 150 km.

Each of the three independent antenna phases maps the Fresnel diffraction function over

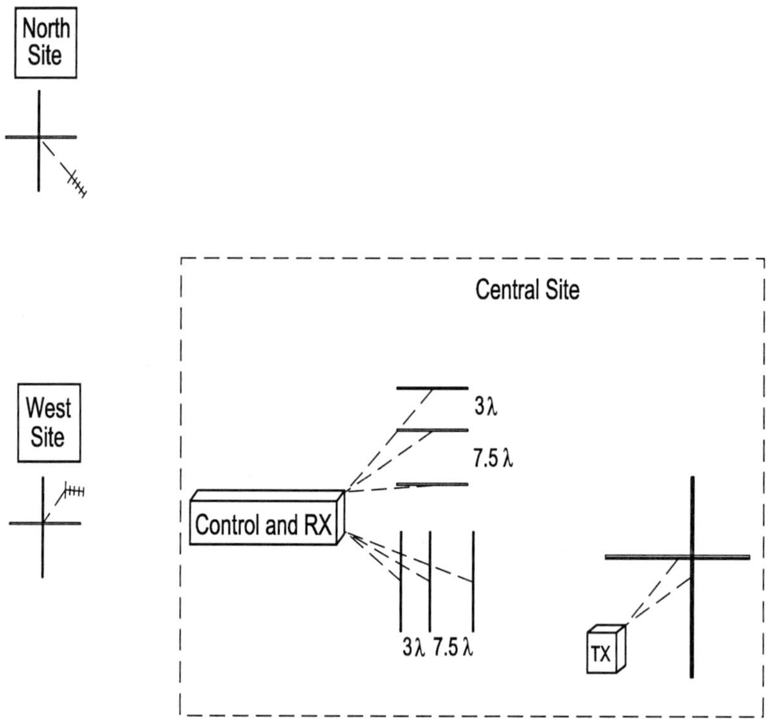

Figure 1. Schematic of the AMOR facility. The central site comprises the transmitters; orthogonal transmitting antenna arrays; elevation finding duel-spacing interferometer receiving antennas for each direction; and operations control.

the pre-t_0 time Fresnel zones during the amplitude rising edge of the echo. In addition to the lack of distortion of the diffraction pattern suffered in the pre-t_0 regime, the technique provides high sensitivity to phase oscillations of the echo signal: echoes of low signal/noise can be provide accurate diffraction speeds resulting in a high data rate of successful speed measurements. An example of an archived echo is shown in Figure 2 which depicts:

- Post-t_0 record phase behaviour (times > 90 radar pulses on the Central site echo) represents a radial train motion of 19 $\mathrm{m\,s^{-1}}$
- The phase detector locks for echoes of signal/noise ratio > 8 dB.
- Independent pre-t_0 phase oscillations are clear for all three height-finding antennas.
- The pre-t_0 phase oscillations indicate a meteor scalar speed of 38.8 ± 0.3 $\mathrm{km\,s^{-1}}$

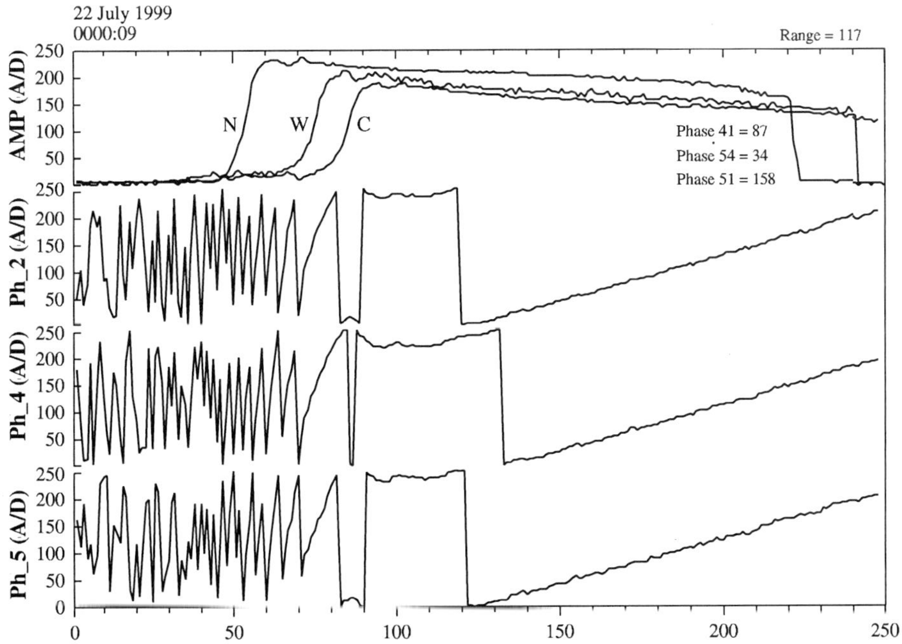

Figure 2. Amplitude and phase record of a 107 km range echo. Upper traces: the three receiver voltages with each station labelled—central site C, north station N, west station W. Lower traces: receiver phases (digitized 0–255 representing phases 0–2π) on the three Central site elevation-finding antennas.

4. MONITORING AND DATA ARCHIVING

The radar facility maintains a high efficiency of orbit collecting operation with $< 5\%$ down-time while overseeing the operation of the system by active telephone provides monitoring of the meteor echo collection process.

Archived data at the radar site is stored on hard disk for transfer via portable hard-drive to the institution: in addition raw data can be daily uploaded using a telephone line and modem transfer.

REFERENCES

1. W.J. Baggaley, R.G.T. Bennett, D.I. Steel and A.D. Taylor, Q. J. R. astr. Soc. 35 (1994) 293.
2. W.J. Baggaley, Earth Moon and Planets 68 (1995) 127.
3. D.P. Galligan and W.J. Baggaley, Wavelet enhancement for detecting shower structure in radar meteoroid orbit data. II Application to the AMOR data set, this volume.

Wavelet enhancement for detecting shower structure in radar meteoroid orbit data.
I. Methodology

D. P. Galligan[a] and W. J. Baggaley[a]

[a]Dept. of Physics and Astronomy, University of Canterbury,
Private Bag 4800, Christchurch, New Zealand.

A new method is introduced for the detection of shower structure in a predominantly sporadic radar meteoroid orbit data set for the case where no *a priori* knowledge of this structure is assumed. Wavelet enhancement within radiant regions, in conjunction with speed partitioning and sliding windows in time, is used to detect these showers. A test is developed to determine the significance of any structure detected. This method is successfully tested on the AMOR data set and is shown, as an example, to clearly detect the Southern δ Aquarids meteor shower.

1. INTRODUCTION

The AMOR meteoroid orbit radar facility [1] has been in operation since 1990. Over the 1995 to mid-1999 period $\sim 5 \times 10^5$ high quality meteoroid orbits were catalogued using this radar. The meteoroids detected ranged in size down to ~ 40 μm. It is expected, given such a small limiting size and the mass distribution index, that the majority of these would be sporadic—detecting shower structure against such a background is difficult. Wavelet enhancement techniques have been used in many previous studies to detect structure in similarly noisy astronomical (but not meteor) environments (e.g. [2,3,6]). A related method which has been applied to the AMOR meteoroid orbit data set is introduced here.

2. DEFINITION OF SHOWER STRUCTURE

A meteoroid orbit is defined by the five classical orbital elements $(q, e, i, \omega, \Omega)$; it is equivalently given by four directly observable parameters: the radiant position, geocentric speed (V_G), and time of detection. In the current study the latter method of orbit definition is adopted; the time of detection is measured in terms of the mean solar longitude ($\lambda_\odot$) and the radiant position is measured in terms of the ecliptic longitude with respect to the Sun ($\lambda - \lambda_\odot$) and ecliptic latitude (β). (All angular parameters are here referred to the J2000.0 equinox.)

A meteor shower is defined as an over-density in the $(\lambda - \lambda_\odot, \beta, V_G, \lambda_\odot)$ parameter space. Visual inspection of the non-shower orbit characteristics reveals the most obvious large-scale features present in this space to be the antihelion, helion and apex sporadic sources (e.g. [4])—any shower present within the data set must necessarily stand out against this background. The 2-D radiant position sub-space is searched in the current study using wavelet enhancement techniques to extract structure. While the search is inherently a 4-D problem, it is found that geocentric speed is strongly dependent on radiant position: only large-scale partitioning in V_G is applied. The overlapping partitions used are defined by 20 $\mathrm{km\,s^{-1}}$ wide bins about centres of 20, 30 and 40 $\mathrm{km\,s^{-1}}$ for prograde meteoroid orbits; these centres are changed to 50, 60 and 70 $\mathrm{km\,s^{-1}}$ for retrograde orbits. In each case, in addition to this partitioning, a search of unpartitioned data is also carried out. The time dimension is dealt with by sliding windows in $\lambda_\odot$; both 2° and 6° widths are used to step in 1° increments between vernal equinoxes.

3. VISIBILITY OF METEOR SHOWERS IN LARGE-SCALE ACTIVITY CURVES

Figure 1 shows the activity profiles for all meteors detected by AMOR over four years of operation. Detections during night-time (1800–0600 NZST) and day-time (0600–1800 NZST) are shown separately; the expected activity times of the three strongest showers observed by AMOR are highlighted. Note the lack of visible shower activity in these realms, apart from the tentative increase in the night-time profiles over the period of the Southern δ Aquarids (the strongest shower in the AMOR data set).

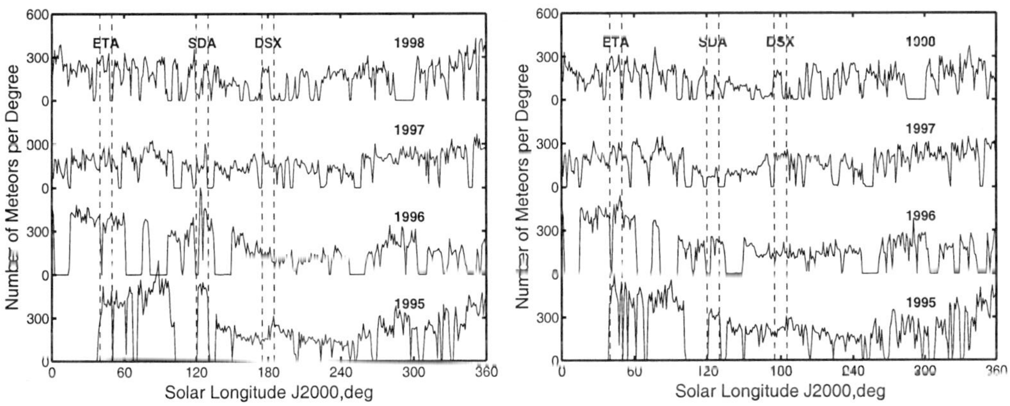

Figure 1. Daily meteor activity over four years of AMOR operation. Night-time profiles are shown on the left and day-time on the right. Years stretch between vernal equinoxes with mean solar longitude as a time unit. The three strongest shower periods are marked for the η Aquarids (ETA), Southern δ Aquarids (SDA) and the Daytime Sextantids (DSX) respectively.

The time periods selected in Figure 1 are expected to distinguish the antihelion and helion source activities with the retrograde apex region meteors being split evenly between the two regimes. Selecting the prograde and retrograde meteors separately is found to slightly improve the relative strength in the Southern δ Aquarids period, but with no noticeable effect in the other two shower periods.

It is clearly that even the major showers are a second order effect on the sporadic background. A method of probing this background more deeply is therefore desirable in order to enhance the underlying shower structure for detection.

4. WAVELET STRUCTURE ENHANCEMENT

The wavelet enhancement method highlights features in an image at the size-scale of the wavelet. There are a large set of possible wavelets—that generally chosen for structure enhancement is the so-called Mexican Hat (Maar) wavelet (e.g. [5]). This is defined as the second derivative of the classical Gaussian function:

$$\psi(x, y, a) = \left(2 - \frac{x^2 + y^2}{a^2}\right) \exp\left(-\frac{x^2 + y^2}{2a^2}\right), \tag{1}$$

where a determines the size of the wavelet and, therefore, the scale which is probed. The 2-D wavelet transform, obtained by correlating $\psi(x, y, a)$ with an image $f(x, y)$, is given by

$$W(x, y, a) = \int_{-\infty}^{+\infty} \int_{-\infty}^{+\infty} \psi(\xi, \eta, a)\, f(x, y)\, d\xi\, d\eta. \tag{2}$$

In practice, for speed of execution, this correlation is performed by a convolution in discrete Fourier space.

A notable feature of $W(x, y, a)$ are the regions of negative amplitude surrounding positive cores—the latter correspond to under-densities while the former correspond to over-densities in the original image at the size-scale under study.

4.1. Determination of Significant Structure

Studies such as Bendjoya [2] determine the significance of structure in a given area of an image by finding the maximum wavelet transform coefficient expected in that area from a set of pseudo-random images which are based on the large-scale structure of the original. In the current study only the strongest feature in each wavelet transform is noted owing to the number of dimensions over which the structure is spread and the strength of the sporadic sources. Two wavelet size-scales are used in order to enhance features with different spatial extents: a 3° probe size is assumed to probe at the scale expected from a shower source embedded in the sporadic background while a 6° probe is used to detect showers which occur closer to the size-scale of the background itself. Because the sporadic source regions specified above differ markedly, their regimes are examined separately.

All orbital data from the vernal equinox of 1995 through to the end of July 1999 were combined to produce a virtual equinoctial year for study—the assumption here is that coherent showers must occur year after year to some degree in order to be detected and verified within the AMOR data set. In total 64 different wavelet transform sets are created for all combinations of the two time window sizes, two wavelet probe sizes, four speed partitions and four sporadic source regions. Each of these sets contains 360 wavelet transforms of selected radiant region images; the data extracted from each transformed image are the maximum wavelet transform coefficient (W_M), the position of the latter in

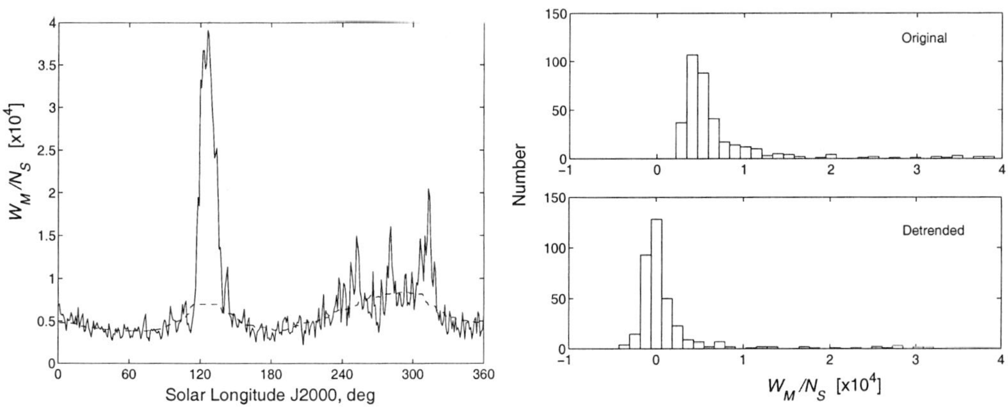

Figure 2. A normalised amplitude profile containing the Southern δ Aquarids peak about the $\lambda_\odot \sim 125°$ point is shown on the left, with the background, determined by a 60° sliding median window overlaid (dashed line). On the right the distribution of normalised amplitude values measured throughout the virtual equinoctial year is shown.

radiant space $((\lambda - \lambda_\odot)_M, \beta_M)$ and the total number of meteors recorded in the sporadic source region under study, with no speed constraints, during the corresponding time window (N_S). N_S is used to normalise W_M for fluctuations in the background meteor rate based on the assumption that a meteor shower constitutes only a small part of the overall population in a sporadic source region at a given time: this is found to be reasonable for the AMOR data set where, as shown in Figure 1, only the Southern δ Aquarids show a slight hint of existence in the night-time activity profiles.

A typical profile obtained from the current antihelion region data is shown in Figure 2 where the antihelion region is probed in the 40 ± 10 km s^{-1} speed partition, using the 3° wavelet probe and a 2° wide time window. The normalised (maximum) transform coefficient amplitude shows a clear maximum about $\lambda_\odot \sim 125°$; there are a number of less clearly significant peaks in this diagram, however, whose reality must be determined. The assumption made is that the normalised maximum transform coefficient will describe a Gaussian noise distribution for most of the year with occasional unexpectedly strong signal strength corresponding to the presence of a shower. There are some long term background trends in the profile: these are determined by means of a 60° wide sliding median window, producing the background level shown. The latter figure also shows the distributions obtained from both the original set of normalised maximum transform coefficients and from that set corrected for the background trend. It is clear that the detrended distribution is approximately Gaussian but with some very far outlying points which are attributed to shower presence. Points further than 3σ from the mean in the detrended distribution are removed in order to diminish the effect of much of the shower structure and the standard deviation of the truncated distribution resulting is then determined which is assumed to correspond to the noise distribution spread (σ_N). Tests on profiles such as that shown in Figure 2 are applied at the $3\sigma_N$ and $4\sigma_N$ levels in order to determine shower significance at the 99% and 99.99% levels respectively.

4.2. Application Example

An example of the results obtained from a typical search is shown in Figure 3. This is from the same antihelion region search shown in Figure 2. The upper two sub-plots in this figure show the radiant position coordinates at which the maximum transform coefficient occurred in each 1° time step. The lower sub-plots show the maximum coefficient amplitude in both original and background normalised forms. The twin dotted lines in these amplitude profiles correspond to the 99% and 99.99% significance levels: amplitude peaks rising above these are considered significant at the respective levels.

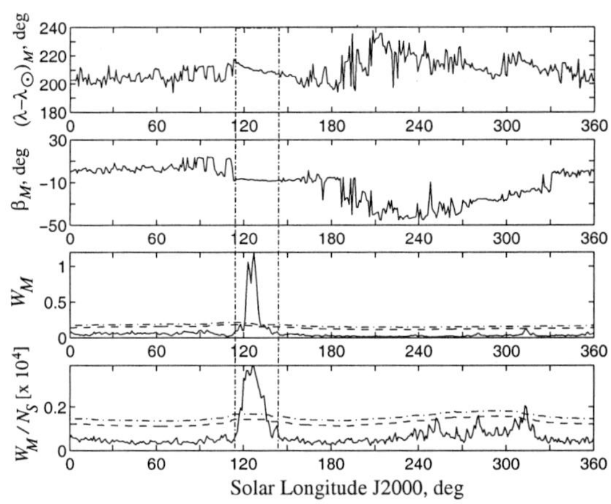

Figure 3. A selected wavelet enhancement search of the antihelion region highlighting the Southern δ Aquarids shower peak.

The Southern δ Aquarids shower is clearly evident, appearing above the 99.99% level in both of the amplitude profiles for a period centred about $\lambda_\odot \sim 125°$. The central radiant position of this shower experiences a "steady-state" occurring both in $(\lambda-\lambda_\odot)_M$ and in β_M over the active period. It is found, however, that the radiant position is not trustworthy as a primary indicator of shower activity. Because the sky area of radar sensitivity changes with season, the sporadic sources themselves move in systematic long-term patterns which may appear shower-like while true showers, such as the Southern δ Aquarids, generally experience some daily motion in their ecliptic longitude with respect to the Sun. This motion, appearing as a non-unity $(\lambda - \lambda_\odot)_M$ gradient in Figure 3, is unpredictable. It is therefore difficult to design a computer algorithm to automatically pick out these "steady-state" radiant position patterns: they are therefore used as secondary indicators of shower activity, lending credibility to a structure which is already considered significant due to a local maximum in the normalised amplitude profile.

An interesting point to note in the Southern δ Aquarids profiles are the differences in the time periods over which the original profile appears to be significant as compared to the normalised profile. The latter is corrected to a longer period due to the lower background activity experienced to each side of $\lambda_\odot \in [120°, 130°]$; this is in better agreement with that expected from the corresponding radiant position "steady-state".

Additionally in Figure 3 a peak significant at the 99.99% level appears about $\lambda_\odot \sim 315°$. The meteors producing this peak are not found to coincide with any known shower. It is an open question as to the reality of this shower candidate—it is possible that this is simply a particularly favourable alignment of the radar-Earth platform with the sporadic background. Further investigations on this candidate's characteristics are carried out in the second paper in this series.

5. CONCLUSION

A new method for the detection of meteor showers against the sporadic background in radar meteoroid orbit data sets has been introduced and tested on the AMOR catalogue. No *a priori* knowledge of the shower structure is assumed by this method. The possibility of determining the statistical significance of a shower against the surrounding background in time has been established. A comprehensive survey of the data set has been performed and the second paper in this series presents the full results of this application.

ACKNOWLEDGEMENTS

Galligan gratefully acknowledges support from a University of Canterbury Doctoral scholarship and a New Zealand Marsden Fund grant (contract: UOC911).

REFERENCES

1. W.J. Baggaley and R.G.T. Bennett, in Physics, Chemistry, and Dynamics of Interplanetary Dust, ASP Conf. Ser. 104; IAU Colloq. 150 (1996) 65.
2. P. Bendjoya, P., Astron. Astrophys. Supp. 102 (1993) 25.
3. F. Damiani, A. Maggio, M. Micela, S. Sciortino, in Astronomical Data Analysis and Systems V. (eds. G.H. Jacoby and J. Barnes) AIP Conf. Proc. 101 (1996) 143.
4. J. Jones and P. Brown, Mon. Not. R. astr. Soc. 265 (1993) 524.
5. R. Murenzi, Wavelets (eds. J.M. Combes, A. Grossman and P. Tchamitchian) Springer Velag, Heidelberg, 1988.
6. E. Slezak, F. Durrect, D. Gerbal, Astron. J. 108 (1994) 1996.

Wavelet enhancement for detecting shower structure in radar meteoroid orbit data.
II. Application to the AMOR data set

D. P. Galligan[a] and W. J. Baggaley[a]

[a]Dept. of Physics and Astronomy, University of Canterbury,
Private Bag 4800, Christchurch, New Zealand.

A wavelet enhancement based method developed by Galligan and Baggaley [4] to search for showers in radar meteoroid orbit data sets has been systematically applied to data obtained from the AMOR radar facility. This class of technique, applied for the first time here, has proved to be powerful. Less than 1% of this data set are found to be within recognisable shower groupings—generally only the major Southern Hemisphere meteor showers (η Aquarids, Southern δ Aquarids, α Capricornids and Daytime Sextantids) are detectable. Further probing with wavelets to select shower members, once the existence of the latter has been established, is found to be convenient. The orbital characteristics of the shower meteors thus selected are explored where it is found that the orbital parameter spreads in most showers do not exceed those expected due to measurement uncertainty.

1. INTRODUCTION

Galligan and Baggaley [4] introduce a method of detecting, and testing the significance of, shower structure in the $\sim 5 \times 10^5$ meteoroid orbit data set provided by the AMOR system [2] from 1995 to mid-1999. AMOR is a particularly sensitive system with a limiting meteoroid size of $\sim 40\,\mu$m; it is expected that many meteoroids having sizes close to this limit will have no detectable shower affiliation owing to the higher mass distribution index expected for showers relative to that of the non-shower background.

The four year data set is formed into a virtual equinoctial year stretching between vernal equinoxes with time measured in terms of the mean solar longitude ($\lambda_{\odot}$). (All angular parameters are here referred to the J2000.0 equinox.) Partitioning based on overlapping geocentric speed (V_G) ranges and sliding windows in time is applied as discussed in Galligan and Baggaley [4]. Wavelet enhancement of the resultant meteor radiant position distributions in the antihelion, helion and apex apparent source regions is found to be useful in probing the sparse structure.

2. SHOWER DETECTIONS

Separate searches have been performed for all 64 combinations of two wavelet probe sizes (3° and 6°), two time window widths (2° and 6°), sporadic source region selection(antihelion, helion, prograde apex and retrograde apex) and geocentric speed (windows 20 km s^{-1} wide centred at 20, 30 and 40 km s^{-1} for prograde and centred at 50, 60 and 70 km s^{-1} for retrograde). While a number of significant peaks have been identified in these wavelet enhancement based searches, few of these survive close scrutiny. Only the well-established major showers, in addition to a very small number of other showers, are worthy of further comment. The searches in each sporadic source region in which these definite showers appear are discussed below—only the optimal search in each case is shown although there are generally several others in which each shower is also found significantly.

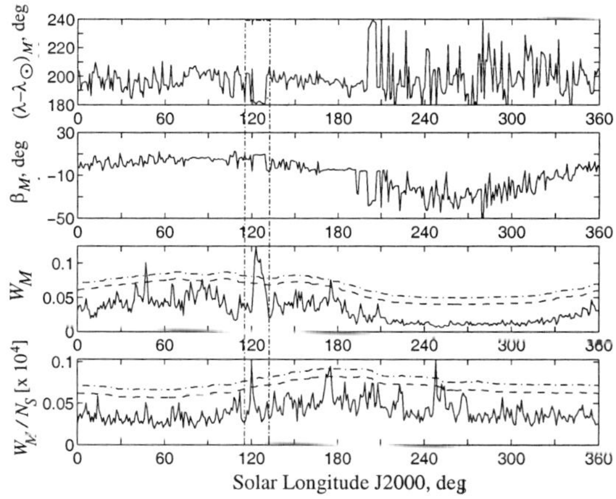

Figure 1. Search in the antihelion region using a 3° wavelet, a 2° sliding $\lambda_\odot$ window and V_G of 20 ± 10 km s^{-1}.

In the antihelion region the Southern δ Aquarids (SDA) and α Capricornids (CAP) appear significantly. Although these showers have radiants occurring in a similar region over the same period of the year, it is possible to resolve them due to the speed partitioning—the average V_G of the meteoroids in these streams is separated by $\sim$ 20 km s^{-1}. Figure 1 shows an example of the CAP shower detection; a similar example of the SDA detection has been already given in Galligan and Baggaley [4] in a profile in which an unidentified shower candidate was also found about $\lambda_\odot = 313°$—the latter candidate is labelled "Peak 1" and defined in Table 1. All of these profiles were obtained using a 3° wavelet probe and a 2° wide time window. In contrast, the 6° wavelet probe is found to remove the significance of the CAP due to its tendency to focus on the sporadic background: similarly, the wider 6° time window tends to minimise the time over which the shower is significant to the point where weak showers such as the CAP are undetectable. Figure 1 presents four profiles: the upper two profiles show the change in radiant position of the maximum $((\lambda - \lambda_\odot)_M, \beta_M)$ over the equinoctial year while the third shows the maximum amplitude (W_M) and the fourth shows the normalised maximum amplitude (W_M/N_S) (N_S is the total number of meteors of all speeds within the whole radiant region under study); for each

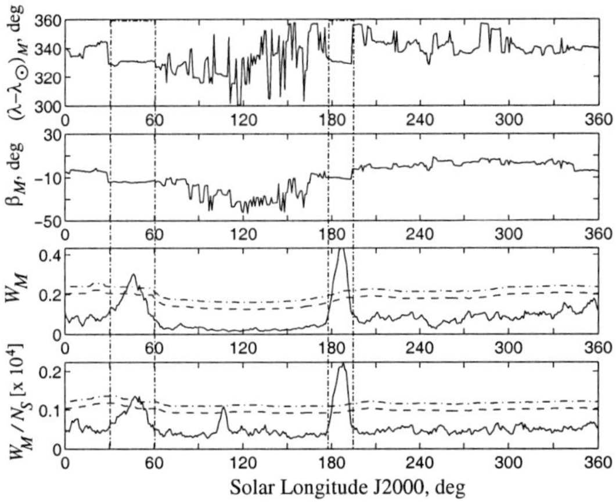

Figure 2. Search in the helion region using a 3° wavelet, a 6° sliding $\lambda_\odot$ window and V_G of 30 ± 10 km s^{-1}.

of the amplitude profiles the lower dotted line represents the 99% confidence level while the upper one corresponds to 99.99%. The CAP is shown to be briefly 99.99% significant in normalised amplitude at $\lambda_\odot \sim 120°$. The SDA has been shown in Galligan and Baggaley [4] to be particularly strong after this point and it is possible that the CAP significance has been contaminated by the perturbation on the background rate used in the normalisation. This possibility appears to be supported by the radiant position profiles over the $\lambda_\odot \in [120°, 130°]$ period where a different steady-state compared with the surrounding time-period appears and also by the clear significance of the non-normalised amplitude over this period. However, removal of the selected SDA meteors from the data set is not found to increase the normalised profile significance of the CAP. The number of meteors detected from the SDA per year is an order of magnitude higher that that detected from the CAP; the CAP is therefore much closer to the "noise" level than is the SDA making their reality more difficult to establish.

In the helion region the Daytime Sextantids (DSX) is found and, as shown in Figure 2, this shower appears well above the 99.99% significance level. Due to some equipment outages and sporadic-E interference at the time of the shower over the years covered, the 6° time window is found to be more appropriate than the 2° window. The DSX is found to be significant for all helion region parameter permutations, apart from the 20 ± 10 km s^{-1} V_G partition which includes some very weak/possible shower peaks. The strongest peaks are found in the 30 ± 10 km s^{-1} partition in good agreement with the ~ 30 km s^{-1} mean shower V_G expected based on other studies. Another significant peak occurring about $\lambda_\odot \sim 45°$ is also identified in Figure 2. There are reasons to doubt the reality of this shower candidate. The normalised amplitude profile peak is only significant for a few degrees of solar longitude whereas the peak structure itself lasts for $\sim 30°$ with a gradual rise to, and fall from, the peak over that time. Such motion is suggestive of an Earth observation effect on the sporadic background; the period of observation is also rather

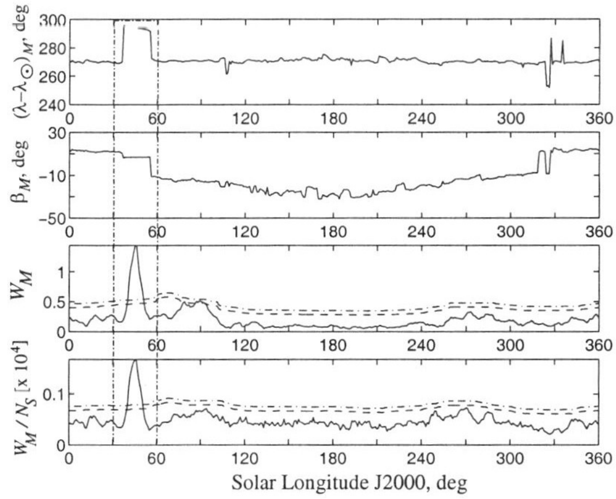

Figure 3. Search in the retrograde apex region using a 3° wavelet, a 6° sliding $\lambda_\odot$ window and meteoroids with geocentric speeds of 60 ± 10 km s^{-1}.

long for a shower (though the Southern δ Aquarids has been shown to be active for a similar period in Galligan and Baggaley [4]). This peak is labelled "Peak 2" for further study later in this paper. A third significant peak in this figure at $\lambda_\odot \sim 105°$ is dismissed as an artifact of the low background rate and shower activity as evidenced in the original amplitude profile.

As expected, the retrograde orbit apex region is found to be based almost completely on the sporadic meteors. The well-known bias towards retrograde meteoroid detection on Earth-based platforms is responsible for this. Figure 3 shows the only shower detected by the wavelet enhancement method in this region to be the η Aquarids (ETA) which is detected at the 99.99% significance level for a period centred about $\lambda_\odot \sim 46°$. This shower is found to be present in all V_G partitions, apart from 50 ± 10 km s^{-1}, owing to the relatively large geocentric speed measurement uncertainty on ETA shower meteors. The motion of the apex region maximum radiant position generally follows a symmetrical curve throughout the year, however, about the time of the ETA there is a pronounced change in the ecliptic longitude and to a lesser extent the ecliptic latitude of the radiant maximum—these distinctive changes help to confirm the reality of this shower.

The prograde apex region shows no significant structure. This region is sparsely populated containing only high inclination or low latitude orbits. The dearth of meteors and the uncertainties in the measurements of each meteoroid's orbit make the recording of only background noise in the search profiles inevitable.

3. MAJOR SHOWER ORBITAL PARAMETERS

For each of the showers detected using the wavelet enhancement method, the properties of their member meteoroid orbits are now discussed. These members are identified using the wavelet transform of the appropriate sporadic source region over the peak activity period of the shower with no speed constraints. The shower is defined as the region of positive wavelet coefficients, bounded by a delineating negative region, in the transform. The zero-boundary between these regions is never perfect—in practice the shower is here defined as the region of wavelet coefficients which occur above $\sim 10\%$ of the maximum wavelet coefficient (W_M) in the particular transform.

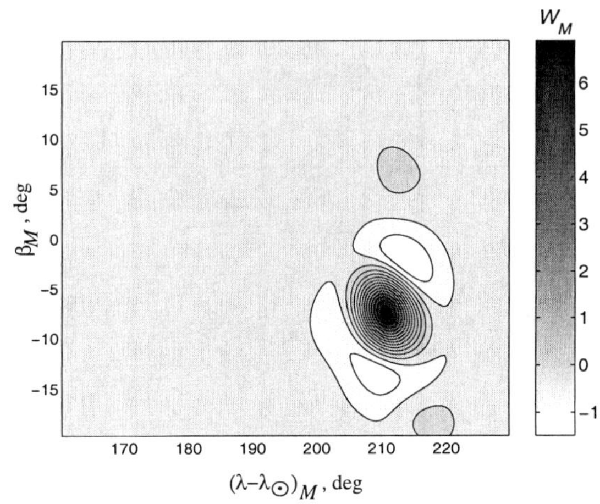

Figure 4. The wavelet transform of the SDA radiant region using a Mexican Hat probe 3° wide. Constraints on the two dimensions, not shown, are defined by V_G of 40.4 ± 4.9 km s^{-1} and $\lambda_\odot$ of $130° \pm 15°$.

The SDA is used to introduce the methodology used for the exploration of each shower region. This shower, the strongest detected by AMOR, shows many interesting features whose visibility is enhanced by the shower's high meteor rate and relatively low measurement uncertainties. The wavelet transform of the SDA radiant region, using a Mexican Hat probe of size $a = 3°$, is shown in Figure 4. Radiant positions are reduced to $\lambda_\odot = 125°$, this leads to a clear (daily motion corrected) maxima at $(\lambda - \lambda_\odot, \beta) = (210.2°, -7.2°)$ which changes very slightly to $(210.2°, -7.8°)$ if only the central activity period ($\lambda_\odot \in [120°, 130°]$) is used, reflecting perhaps a small unaccounted for radiant instability or drift. A cut-off in this image at 5% of the maximum coefficient is found to remove all surrounding background. For all meteors appearing in the selected radiant region over the central shower activity period a 2 km s^{-1} wide sliding window in V_G is moved from 20 km s^{-1} to 60 km s^{-1} to produce the profile shown in Figure 5. This profile, representing the combination of shower and background meteoroid speeds, is fitted well by a Gaussian defined by $\overline{V}_G = 40.4$ km s^{-1} and $\sigma_V = 4.9$ km s^{-1}. Shower meteoroids are arbitrarily selected to have speeds within 1σ of the mean and occurring within the selected radiant region over the period $\lambda_\odot \in [115°, 145°]$. Table 1 lists the definitions used for the other showers found in Section 2 (a 2σ V_G cut-off is used in the case of the η Aquarids due to its relatively large speed measurement uncertainty); the radiant centres in each case are reduced to the central mean solar longitude listed.

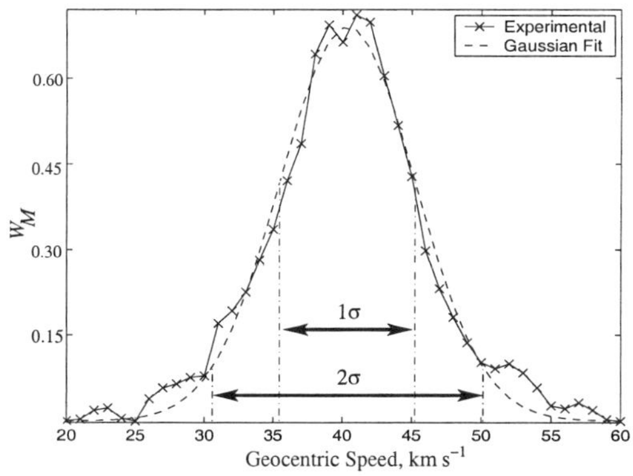

Figure 5. The distribution in speed of the SDA shower meteoroids as determined by moving window scans of the radiant region with a window width of 2 km s^{-1}. The profile is fitted by a Gaussian curve with parameters $\overline{V}_G = 40.4$ km s^{-1} and $\sigma_V = 4.9$ km s^{-1}.

Shower	V_G (km s^{-1})	$\lambda_\odot$	Radiant Centre $(\lambda - \lambda_\odot, \beta)$	Probe Size
CAP	23.4±3.1	$120^\circ \pm 10^\circ$	$(181.2^\circ, 9.8^\circ)$	3°
SDA	40.4 ± 4.9	$130^\circ \pm 15^\circ$	$(210.2^\circ, -7.2^\circ)$	3°
Peak 1	43.0 ± 4.6	$312^\circ \pm 8^\circ$	$(214.8^\circ, -18.2^\circ)$	3°
Peak 2	37.0 ± 4.8	$145^\circ \pm 15^\circ$	$(330.8^\circ, -13.8^\circ)$	3°
DSX	31.3 ± 4.1	$186^\circ \pm 7^\circ$	$(330.2^\circ, -11.2^\circ)$	3°
ETA	65.1 ± 7.6	$46^\circ \pm 9^\circ$	$(294.2^\circ, 6.8^\circ)$	2°

Table 1
Definitions of meteor showers detected.

3.1. Daily Motion of Parameters

As noted in the previous section, the average radiant position of a meteor shower experiences a linear daily motion generally parallel to the ecliptic. In order to remove this motion a central epoch is chosen and all data are reduced to this. Table 2 lists radiant motion measured for each of the five meteor showers under study, using every shower point in an uncertainty weighted linear least squares fit. The large number of shower meteors ($\sim 2 \times 10^2$ to $\sim 2 \times 10^3$) in each case enables a good fit for many of the parameters to be obtained. Note the generally more pronounced daily motions in equatorial as compared with the ecliptic Sun referenced radiant positions, thus justifying the choice of the latter for searches within the radiant space.

	CAP	SDA	Peak 1	Peak 2	DSX	ETA
$\lambda - \lambda_\odot$	-0.03 ± 0.02	-0.20 ± 0.02	-0.23 ± 0.03	-0.08 ± 0.02	-0.08 ± 0.02	-0.15 ± 0.02
β	0.034 ± 0.005	-0.042 ± 0.005	0.11 ± 0.01	0.019 ± 0.005	-0.098 ± 0.005	0.013 ± 0.005
α	0.91 ± 0.02	0.73 ± 0.02	0.68 ± 0.03	0.80 ± 0.02	0.84 ± 0.02	0.73 ± 0.02
δ	0.25 ± 0.01	0.26 ± 0.01	-0.16 ± 0.01	0.38 ± 0.01	-0.43 ± 0.01	0.31 ± 0.01

Table 2
Daily motion of shower radiant parameters. Days are given in the approximately equivalent degrees of mean solar longitude rendering the motion measurements unit-less.

The radiant motion measurements are dependent on the meteors selected on which to perform these measurements; the selection of these shower meteors is made using some assumed daily motion in the radiant position. This intertwined relationship allows the measurements to easily vary even on the same data set. The most important feature of the measurements are their use in correcting the perceived motions in the current data set. However, in order to verify that these are close to those expected from other surveys, bearing in mind the measurement instabilities mentioned above, comparisons are now made. Cook [3] lists a daily motion for the SDA of $0.8° \text{ d}^{-1}$ in right ascension and $0.18° \text{ d}^{-1}$ in declination which is similar to that obtained here. More recently Kronk [5] and Rendtel, Arlt and McBeath [9] list this motion at $0.9° \text{ d}^{-1}$ in right ascension and $0.4° \text{ d}^{-1}$ in declination. As the AMOR values are based on a larger sample of orbits than these, with correspondingly small uncertainties, it is more likely that the AMOR low uncertainty values are closer to the "true" motion. The ETA right ascension measured daily motion is close to Lindblad's [7] $0.76° \text{ d}^{-1}$: however, that in declination differs to a greater extent from Lindblad's $0.422° \text{ d}^{-1}$ motion. Lindblad only covers solar longitudes from 43° to 47° with 23 single data points, while the current study uses $\sim 10^3$ points over a longer period albeit with higher individual uncertainties. The ~ 0.1 longitudinal motion found for the CAP is approximately that expected according to Cook [3]. There appear to be no published daily motion for comparison with the DSX, this shower shows less longitudinal motion with respect to the Sun than do the ETA and SDA.

In addition to the radiant position other orbital parameters also experience measurable daily changes. Sometimes these are very small but in a number of cases they are clearly important. The SDA provide a good example of the latter; the perihelion distance (q) and argument of perihelion (ω) orbital elements of its members exhibit clear daily motion as shown in Figure 6. The fitted lines in the latter do not appear to be correctly centred in some cases, this is due to the increased uncertainties at higher q and lower ω respectively biasing the fitting routine (the median Δq is 0.014 AU for $q_N < 0.07$ AU and 0.019 AU for $q_N > 0.07$ AU; the median $\Delta\omega$ is 3.2° for $\omega_N < 155°$ and 2.2° for $\omega_N > 155°$ (q_N and ω_N are motion corrected values)). The reason for such daily motions can be clearly illustrated; for example, Figure 7 shows that as the component of the meteoroid's pre-impact heliocentric velocity in the direction of the apex of the Earth's way (V_x) increases the range of shower meteoroid q and ω allowed changes in a very constrained fashion. As the Earth moves in its orbit the direction of the apex does change, hence the average V_x measured from a coherent stream source also changes leading to the daily motion in the orbital elements shown.

For any parameters (P) showing a clear daily motion, as measured by linear least-squares ($dP/d\lambda_\odot$), all shower meteors are reduced to a central time ($\lambda_\odot^c$) using

$$P_N = P - \frac{dP}{d\lambda_\odot}(\lambda_\odot - \lambda_\odot^c). \quad (1)$$

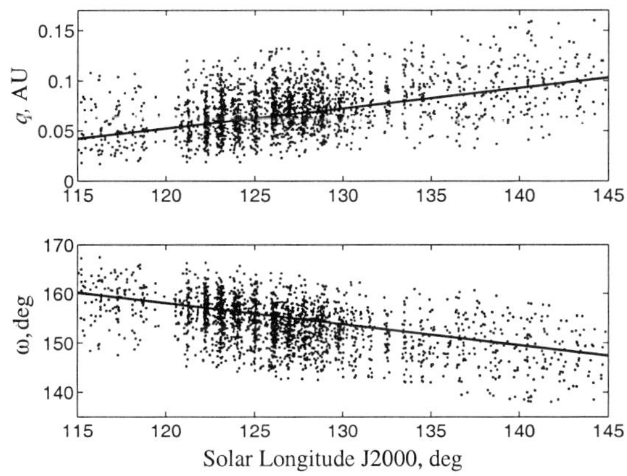

Figure 6. Trends in the SDA q and ω measured value over its active period.

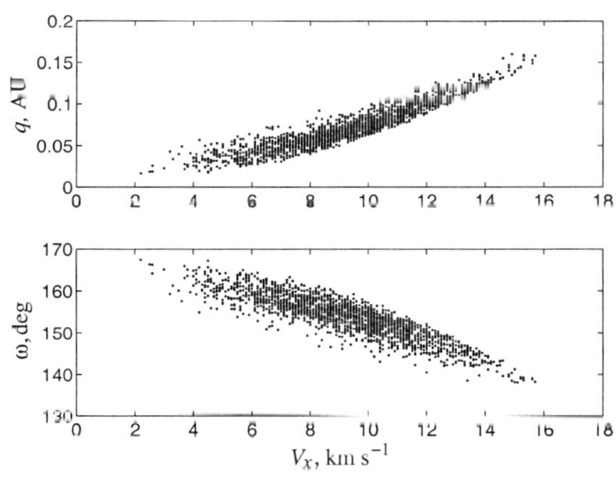

Figure 7. The relationship between the on-ecliptic heliocentric velocity component in the direction of the apex of the Earth's way and q and ω for the SDA orbits.

3.2. Determination of Uncertainties

It is important to provide an estimate of the parameter uncertainties for individual orbits in order to determine the extent to which the observed shower spread may be attributed purely to such uncertainties. Uncertainties in high-level parameters, such as the orbital elements, propagate from uncertainties in the low-level direct meteor observation measurements. An orbit, as determined by AMOR, is derived from four directly measured parameters [1]: two receiving station time-lags (Lag_{12} and Lag_{13}) and the elevation (Ψ) and azimuthal (A) angles of the echo-point. Representative uncertainties in these are 1.4 radar pulses in the time-lags, 0.5° in Ψ and 2° in A. The reduction of these parameters to orbital elements consists of many steps so that a simple classical uncertainty analysis approach is therefore difficult. Two alternative methods have been tested in the current study. The first method is similar to that used by Taylor [13]: it extracts random combinations of the four parameters from Gaussian distributions having mean values based on the original measured parameter values and standard deviations based on the assumed parameter uncertainties; each of these parameter quartets are run through the orbit reduction computer program in order to produce a full set of high-level output parameters—the spread in the distributions in the latter, obtained from 30,000 such randomised simulations, is assigned to be the uncertainty in the respective parameters. The second method is found to provide equivalent results to those of the first method, however it has the advantage of a negligible computer time requirement in contrast to the rather lengthy time required for the first. Classical analytic uncertainty techniques are used; however, where normally the uncertainty in a single function is determined, here the computer reduction program is treated as that function with four input variables and many output variables. The uncertainty δf in a function $f(V_1, V_2, ...V_n)$ is given by

$$(\delta f)^2 = \sum_{k=1}^{n} \left(\frac{\partial f}{\partial V_k} \right)^2, \tag{2}$$

where the partial derivatives follow from

$$\frac{\partial f}{\partial V_k} = \lim_{h \to 0} \frac{f(V_k + h) - f(V_k - h)}{2h}. \tag{3}$$

f is a particular high-level parameter for which the uncertainty is required and $V_1..V_4$ are the low-level input parameters (Lag_{12}, Lag_{13}, Ψ and A); $h = 10^{-3}$ is used in this equation as this is sufficiently small to give reasonable answers but large enough to avoid numerical instabilities found as $h \to 0$.

3.3. Shower Statistics

Calculation of the shower orbital statistics may seem straightforward but there are a number of issues which must be addressed. As noted in Section 3.1, many orbital parameters experience an apparent daily motion as they are detected by the moving Earth observation platform over time; the observed distribution in such parameters is smeared according to the perceived (generally non-uniform) distribution in time of shower activity. In the current study this problem has been removed for affected parameters by using Equation 1 together with linear least squares motion measurements. Another difficulty in dealing with Earth-detected meteors is the necessary relationship between the corresponding meteoroid orbit's q, e and ω elements:

$$q = \frac{1 \pm e \cos \omega}{1 + e}, \tag{4}$$

where the $\pm$ sign signifies detection at either the ascending (+) or descending (−) nodes. In obtaining mean values in these parameters one may determine the mean in e and ω, and then base the mean of q on this; alternatively one can treat these parameters independently, simply noting that Equation 4 may not be exactly obeyed for the means obtained. The latter approach is taken in the current study as the former leads to difficulties in determinations of the statistical spread in the dependent parameter.

Table 3 lists the orbital parameter statistics for the five showers under study. These parameters (apart from $\lambda_\odot$ and Ω) have been corrected where appropriate for measured daily motion. The reduction epochs used in Equation 1 for these corrections are $\lambda_\odot = 125°$, 125°, 313°, 45°, 187° and 46° for the CAP, SDA, Peak 1, Peak 2, DSX and ETA showers respectively. Representative uncertainties in each shower's parameters are obtained from the median measurement uncertainties calculated as discussed in Section 3.2. It is important to note that generally the representative uncertainties are very similar to the measured spread (σ) in the distributions—in most cases this means that little can be learned about the physical parameter spreads apart from that they lie within these bounds. The uncertainties in the declination and ecliptic latitude angles which are directly related to the echo elevation angle are exceptions in all cases: their spreads, however, are influenced by the wavelet probe size choice and hence the "physical spread" data obtained are of limited use. The inclination angle generally has a spread exceeding that expected from uncertainty: this is one of the more stable elements measured by AMOR and plays a significant role in defining a shower. Exceptionally, the SDA has a large $\sim 10°$ uncertainty in i. Of the showers studied here, the ETA has the largest spread in most orbital elements owing to the high geocentric speed of its constituent meteoroids. The spread in ω found in this shower is less than that expected due to the representative measurement uncertainty spread—some ETA meteors have been omitted due to their distance from the shower centre in the defining 4-D space.

Owing to the very small standard errors in the mean parameters of each of the showers, we may have very good confidence in these (daily motion corrected) means. This confidence is in contrast to Lindblad [7], for example, who ignores radar meteors in his survey of the ETA due to the poor mean orbits which result. While AMOR's strength may lie in determining high quality shower mean parameters, measuring the physical spread, as discussed above, is non-trivial and in many cases unrealistic.

The shower mean orbits were found using Southworth and Hawkin's [12] criterion of orbital similarity (D_{SH}) to agree favourably with several previous examples in the literature.

The SDA has dissimilarities of 0.05, 0.08 and 0.06 with the mean orbits of Nilsson [8], Cook [3] and Sekanina [11] respectively. Such D_{SH} values show close agreement, particularly those for radar based means measured by Nilsson and Sekanina which show negligible differences.

The ETA shows more variation in literature comparisons, with dissimilarities of 0.10, 0.16, 0.10, 0.08, 0.07, 0.12 and 0.05 when compared with Cook [3]; Lindblad [6] photo-

		$\lambda_\odot$	q	e	i	ω	Ω	V_H	V_G	α	δ	N
		deg	AU		deg	deg	deg	km s^{-1}		deg	deg	
	$\overline{x}$	122.3	0.550	0.768	7.7	273.3	122.3	37.0	23.4	306 .7	-9.3	
CAP	σ	5.2	0.036	0.059	1.2	5.3	5.2	1.3	1.7	2.7	1.3	269
	$\sigma/\sqrt{N}$	0.3	0.002	0.004	0.1	0.3	0.3	0.1	0.1	0.2	0.1	
	Unc.	0.0	0.026	0.044	0.7	3.1	0.0	1.0	1.3	1.5	0.5	
	$\overline{x}$	127.2	0.067	0.966	30.8	154.5	307.2	36.0	40.2	340 .4	-16.3	
SDA	σ	5.8	0.021	0.018	9.8	4.1	5.8	2.3	2.7	2.9	1.5	2,413
	$\sigma/\sqrt{N}$	0.1	0.000	0.000	0.2	0.1	0.1	0.0	0.1	0.1	0.0	
	Unc.	0.0	0.015	0.023	9.0	2.8	0.0	2.7	3.4	2.5	0.6	
	$\overline{x}$	313.1	0.143	0.920	64.3	141.9	133.1	36.1	42.7	162 .1	-13.3	
Peak 1	σ	3.5	0.032	0.034	7.8	5.3	3.5	2.4	2.6	2.3	3.1	327
	$\sigma/\sqrt{N}$	0.2	0.002	0.002	0.4	0.3	0.2	0.1	0.1	0.1	0.2	
	Unc.	0.0	0.027	0.029	8.3	6.0	0.0	3.0	3.5	3.0	0.6	
	$\overline{x}$	45.0	0.133	0.916	37.1	214.9	225.0	34.4	36.5	19.0	-7.0	
Peak 2	σ	7.9	0.035	0.029	7.5	6.0	7.9	2.3	2.6	3.3	2.1	970
	$\sigma/\sqrt{N}$	0.3	0.001	0.001	0.2	0.2	0.3	0.1	0.1	0.1	0.1	
	Unc.	0.0	0.025	0.024	3.8	4.6	0.0	1.9	2.2	2.4	0.5	
	$\overline{x}$	186.1	0.151	0.855	23.1	212.5	6.1	30.4	31.2	154.5	-1.5	
DSX	σ	3.5	0.025	0.030	5.0	3.5	3.5	1.7	2.2	2.8	1.5	410
	$\sigma/\sqrt{N}$	0.2	0.001	0.001	0.2	0.2	0.2	0.1	0.1	0.1	0.1	
	Unc.	0.0	0.023	0.023	3.9	3.0	0.0	1.5	1.6	2.7	0.5	
	$\overline{x}$	45.6	0.545	0.953	165.1	91.9	45.6	40.5	65.0	339 .0	-1.4	
ETA	σ	3.6	0.080	0.162	2.1	13.5	3.6	3.4	3.5	1.8	0.8	942
	$\sigma/\sqrt{N}$	0.1	0.003	0.005	0.1	0.4	0.1	0.1	0.1	0.1	0.0	
	Unc.	0.0	0.133	0.239	2.3	21.3	0.0	5.4	5.5	2.9	0.5	

Table 3
Statistics of the parameters of four meteor showers found in the AMOR data. Mean ($\overline{x}$), standard deviation (σ), standard error in the mean ($\sigma/\sqrt{N}$) and representative (median) uncertainty (Unc.) are shown for each shower of size N. Parameters experiencing daily motion are corrected to central solar longitudes.

graphic and radar; Lindblad [7] photographic, TV and combined photographic/TV; and Taylor [13] mean orbits respectively. The best agreement is with Taylor's mean and this is fortuitous as his mean orbit derives from meteors detected on the original configuration of the AMOR system (circa 1990). The level of agreement indicates continuity and consistency in the measurement of this shower at AMOR's 26.2 MHz transmission frequency thus inspiring confidence in the quality of the long-term AMOR calibration regime. Lindblad [7] notes that Cook [3] derived his value from a single photographic meteor—it is ignored here. Lindblad's [6] photographic mean shows a particularly high level of dissimilarity: it differs from the AMOR mean orbit determination in the method of detection and also in the number of meteors forming the mean (Lindblad uses 11 orbits while $\sim 10^3$ AMOR orbits are used currently). It is quite possible that the different meteoroid size ranges probed yield different orbital means or that the small sample of Lindblad may have led to an inaccurate mean. In support of the latter, the AMOR shower is found to be much more similar to the radar mean of Lindblad [6] and to the (closer to radar size range) TV mean of Lindblad [7].

In the case of the CAP two photographic means listed by Kronk [5] and the radar mean of Weiss [14] agree best with the AMOR mean orbit at dissimilarities of 0.06, 0.06 and 0.05 respectively. These are followed by Kronk's other mean orbit at $D_{SH} = 0.24$, an unacceptably high value for stream association of near-ecliptic stream comparisons. Finally the radar orbits of Sekanina are even more dissimilar with $D_{SH} = 0.3$ for Sekanina [10] and an extreme $D_{SH} = 1.18$ for Sekanina [11]. These very large dissimilarities are not unexpected given the very low inclination angle (0.9°) in Sekanina [10] and the very hyperbolic eccentricity (1.92) in Sekanina [11]; such parameter values are both outside the range expected for this shower. Kronk's [5] third orbit which also disagrees badly, having a dissimilarity of 0.24, likewise has a very low inclination of 1.1°.

The DSX has not been detected in many major orbit reducing studies—the daytime nature of this shower renders it only visible by radar methods. Nilsson [8] and Sekanina [11] present mean orbits obtained, in both cases from nine orbits, at the Adelaide and Harvard Radio Project radars respectively. Nine meteors is a rather meager sample of these relatively high uncertainty orbits on which to base a mean; Nilsson, however, agrees well with the current result at $D_{SH} = 0.03$ while Sekanina's mean is very dissimilar with $D_{SH} = 0.21$ which is expected given the $\sim$ 10° difference between his inclination angle and that provided by both AMOR and Nilsson. The inclination angle in the latter studies is closer to that generally accepted.

On further examination of the shower candidates (Peaks 1 and 2 of Section 2) no match for Peak 1 is found while Peak 2 on close analysis is found to be similar to the omicron Cetids tabulated by Sekanina [11]. A dissimilarity of $D_{SH} = 0.09$ between the mean orbits of the latter and the former indicates excellent agreement.

4. SUMMARY

Galligan and Baggaley's [4] wavelet enhancement technique for detecting showers in radar data has been systematically applied to the AMOR data set. This is the first time such a technique has been applied in the field of meteor science. It has proven to be robust, efficient and objective further development in this field is recommended.

In the current study, apart from a few possible other showers, one of which has been discussed here, only the major well established showers have been detected over four years. Due to its sensitivity to small meteors this result is expected; the less than 1% of the shower in compact and distinct groupings leads to the particularly useful role of AMOR as a probe of the general meteor background structure.

The showers which have been found in the data set have had their orbital structure analysed with account taken of the daily motions in many of the parameters. The mean resultant orbits have been shown to generally agree well with other studies. Daily motion in radiant position has been measured and removed. Such motion has also been demonstrated in other orbital parameters; it is unusual in meteor observation programs that one has enough shower meteors and low enough uncertainties to make these measurements the removal of this motion, where appropriate, undoubtedly provides a better estimate of the true means of the parameter distributions.

ACKNOWLEDGEMENTS

Galligan gratefully acknowledges support from a University of Canterbury Doctoral scholarship and a New Zealand Marsden Fund grant (contract: UOC911).

REFERENCES

1. W.J. Baggaley, R.G. Bennett, D.I. Steel and A.D. Taylor, Q. J. R. astr. Soc. 35 (1994) 293.
2. W.J. Baggaley and R.G. Bennett, in Physics, Chemistry, and Dynamics of Interplanetary Dust, ASP Conf. Ser. 104: IAU Colloq. 150 (1996) 65.
3. A.F. Cook, in Meteor Research Program by A.F. Cook, M.R. Flannery, H. Levy II, R.E. McCrosky, Z. Sekanina, C.-Y. Shao, R.B. Southworth and J.T. Williams, NASA CR-2109, Smithsonian Astrophysical Observatory, Washington DC, 1972.
4. D.P. Galligan and W.J. Baggaley, part I of this paper, this volume.
5. G. Kronk, Meteor Showers: A Descriptive Catalog, Enslow Astronomy Series, 1988.
6. B.A. Lindblad, in Third International Symposium on Asteroids, Comets, Meteors (eds. C.-I. Larerkvist, P. Magnusson and H. Rickmann) Uppsala, Sweden (1989).
7. B.A. Lindblad, Planet. Space Sci. 42 (1994) 133.
8. J. Rendtel, R. Arlt and A. McBeath (eds.), in IMO Monograph No.2: Handbook for Visual Meteor Observers, International Meteor Organisation, 1995.
9. C.S. Nilsson, Aust. J. Phys. 17 (1964) 205.
10. Z. Sekanina, Icarus 18 (1973) 253.
11. Z. Sekanina, Icarus 27 (1976) 265.
12. R.B. Southworth and G.S. Hawkins, Smithson. Contrib. Astrophys. 7 (1963) 261.
13. harvarditemTaylor1991Taylor1991 A.D. Taylor, PhD Thesis in Dept. of Physics and Astronomy, University of Canterbury, Christchurch, New Zealand, 1991.
14. A.A. Weiss, Mon. Not. R. astr. Soc. 120 (1960) 387.

Predictability in meteoroid stream evolution

D.J. Asher [a]

[a]Armagh Observatory, College Hill, Armagh, BT61 9DG, UK

Each time an active comet returns to perihelion, it releases particles that progressively stretch into a dense, narrow trail of meteoroids and dust. When a section of a trail makes a close approach to a planet, particles are scattered into the meteoroid stream as a whole. Before trails disperse chaotically into the stream (which in turn is before the stream disperses into the zodiacal background) they not only remain narrow, but also undergo a dynamical evolution that is quite predictable. Since meteor storms occur when the Earth passes through dust trails, calculation of gravitational perturbations enables the prediction of meteor storms and outbursts. Although this chaotic scattering limits trail lifetimes, parts of trails that are in mean motion resonances can remain coherent over substantially longer timescales. Approaches to Earth have been the main cause of dispersing trails in the Leonid stream over recent centuries.

1. DUST TRAILS AND METEOR STORMS

Meteor storms or sharp meteor outbursts occur when the Earth passes through dense, narrow dust trails of the type discovered by the Infra-Red Astronomical Satellite IRAS [1,2]. Such structures exist since the dispersion in orbital period among meteoroids that have been ejected from a cometary nucleus leads to particles getting progressively further ahead of or behind the comet, thus stretching into a trail [3]. A particle's orbital period differs from the comet's both because of the velocity relative to the nucleus induced during the ejection process, and because of the radiation pressure acting on the particle during its subsequent motion. The latter is parameterised by β, the ratio of radiation pressure to solar gravity.

The ejection process induces a spread in all the orbital elements, not only the period. This causes a trail to have a nonzero width. For meteor producing streams, it is most relevant to measure this width (two dimensional cross section) in the ecliptic near $r = 1$ AU. Ejection velocities expected for the size of particles that produce visual meteors [4–6] are generally small enough that a trail's width, although nonzero, is rather narrow compared to the cross section of the stream as a whole. For particles having $\beta \neq 0$, ejection away from $r = 1$ AU can also affect the position in the orbit at $r = 1$ AU and thus increase the trail width slightly.

Gravitational perturbations are a very important influence on the evolution of meteor streams, and trails within streams. Even over a single orbital revolution, they can alter the nodal position of particles by an amount that is significant compared to the stream's width, and even by several times the width of a single trail. It follows that perturbations can move a trail away from or towards Earth intersection, and that calculating them is essential for determining the occurrence of meteor storms.

As the node of the comet itself is similarly perturbed, particles released at different returns of the comet to perihelion have significantly different initial orbits. Additionally the configuration of the planets differs between different returns, and so the various sets of particles have different perturbation histories. Therefore the trails embedded in a stream, one corresponding to each previous return of the comet, tend to be separated in space, and the Earth may or may not encounter any of them during its annual passage through the stream. The trails only tend to converge in the vicinity of the comet itself.

Although different particles are perturbed differently, the perturbations on particles at a single point along a single trail are very similar. This is because particles in a single trail have been perturbed over the same interval of time (i.e., since ejection), and moreover in order to be at the given point along that trail, they have been comoving (whereas a different point along the trail may be months or years ahead or behind). Comoving particles are always in nearly the same position relative to each perturbing planet. The orbital period is continuously perturbed, but it is perturbed in the same way for comoving particles, which consequently remain comoving.

If, however, the particles at some point along a trail undergo a close approach to a planet, then the position relative to the planet is not the same for all those particles, even though they had been comoving around their orbits. This can cause that part of the trail to be scattered into the stream as a whole. This effect limits the lifetime of trails. But before a (part of a) trail is dispersed in such a way, it almost exactly retains its original width (this width being due to ejection velocities and radiation pressure), and the dilution of the density of particles in space is due only to the gradual lengthening of the trail in the along orbit dimension. This provides the potential for a sharp, intense meteor outburst if the Earth encounters a trail. While the trail still exists, the perturbations on any part of it can be precisely evaluated, so that the nodal position can be calculated and the occurrence or non-occurrence of a storm predicted.

The main purpose of the remainder of this paper is to exemplify the above principles in the case of the Leonids by means of suitably chosen numerical integrations.

2. EARLY EVOLUTION: PREDICTABILITY

As a first example we consider the Leonid trail generated when 55P/Tempel-Tuttle returned to perihelion in 1800. To produce Figure 1, particles were ejected at perihelion and tangential to the direction of motion, as this is sufficient to generate particles of any orbital period, and therefore any subsequent perturbation history (before the onset of chaotic behaviour). The quantity Δa_0, the difference in semi-major axis from the comet at the time of ejection, can be used to parameterise the period.

In the absence of differential perturbations and radiation pressure and if Δa_0 is small, the difference in mean anomaly M after n revolutions is

$$\Delta M = -360^\circ \times n \times \frac{3}{2}\frac{\Delta a_0}{a_0} \qquad (1)$$

Figure 1 illustrates this, approximately showing a slope of M against Δa_0 that increases proportionally to the number of revolutions. However, there are two further features of

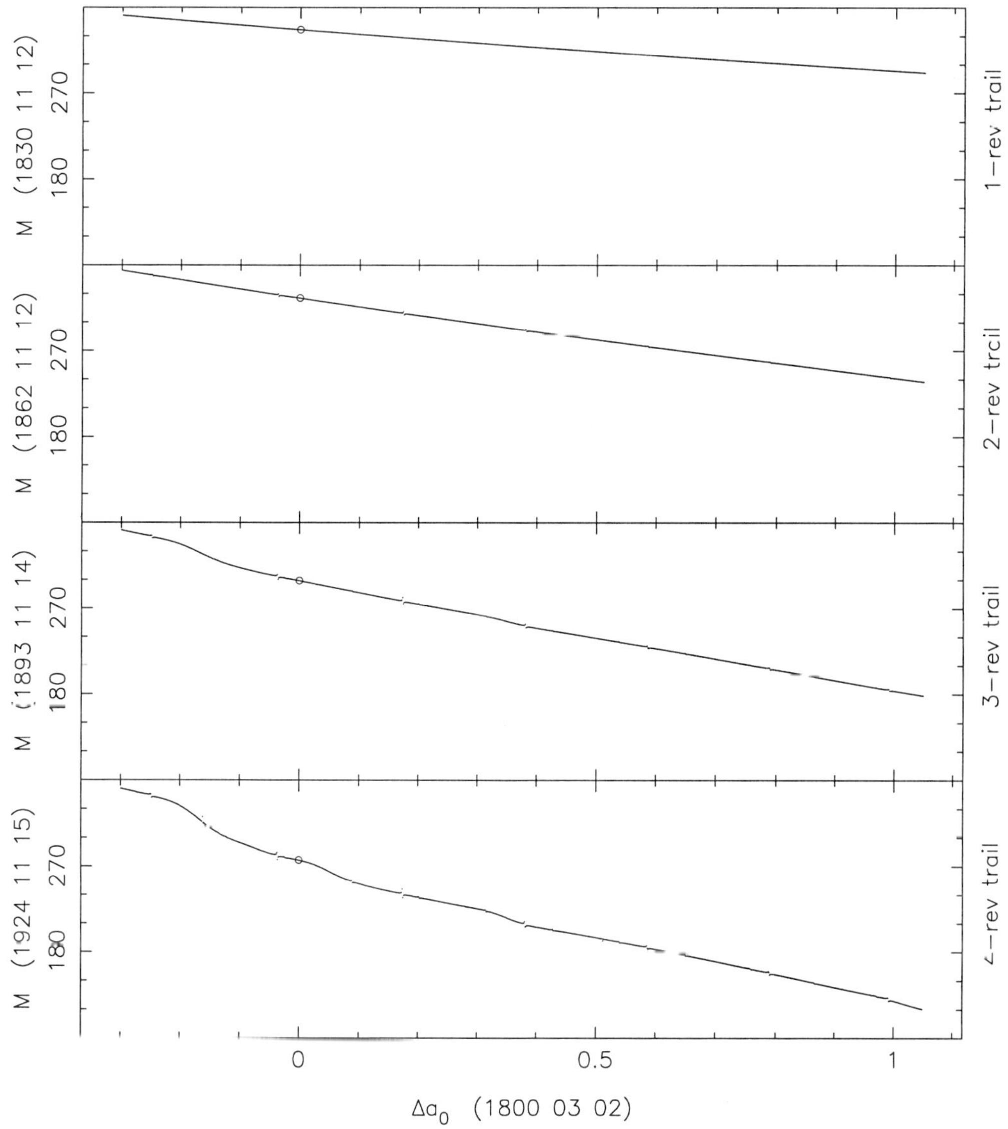

Figure 1. Early evolution of 1800 Leonid trail. Mean anomaly M shown as function of Δa_0 (difference in semi-major axis from comet at ejection), after 1, 2, 3 and 4 revolutions of particles at $\Delta a_0 = -0.3$ AU (after 4 revolutions, particles at $\Delta a_0 = +1.0$ are over half a revolution behind). Comet shown by circle.

Figure 1 to be noted. Firstly, after a few revolutions there are gentle changes in the M vs Δa_0 slope along the trail's length. This is because points at different distances along the trail have become substantially separated in space and therefore the gravitational effect (even moderately distant, i.e. away from close approaches) of Jupiter and Saturn is quite different. In particular, the orbital period is perturbed differently and so the rate at which different parts of the trail stretch (i.e., at which ΔM increases) varies. The second feature is small gaps at various points along the trail, to be discussed in Section 3.

The aim of Figure 1 is to illustrate dynamical evolution, not spatial density variations along the trail. The latter depend on the ejection process from the nucleus, since a distribution in ejection velocities corresponds to a distribution in Δa_0. For reasonable ejection models, the distribution will be centred around $\Delta a_0 = 0$. However, a meteoroid susceptible to radiation pressure and with a given Δa_0 (defining Δa_0 as being calculated from instantaneous position and velocity without adjusting the attraction towards the Sun by β) comoves around its orbit with a $\beta = 0$ particle of higher Δa_0. An typical value of β for meteoroids that produce visual Leonids is $\sim$0.001, corresponding to $\Delta a_0 \approx +0.2$ [7]. Therefore the peak in the particle density as a function of Δa_0 will effectively be at $\sim$0.2 (depending on β) with significant levels at up to a few $\times$ 0.1 AU on either side.

Figure 1 shows that M is largely a well behaved function of Δa_0 for a young trail, suggesting that the trail's evolution is predictable. Section 3 will confirm that the same is true of the nodal position.

3. ONSET OF CHAOTIC BEHAVIOUR

The Minimum Orbit Intersection Distance, or MOID, between two orbits evolves under orbital precession. During epochs when its value is very small, close approaches between objects moving on the two orbits are possible if the objects reach their near-intersection point at a similar time.

The MOID between 55P/Tempel-Tuttle and Earth is moderately small for some millennia about the present. This means that at any given time there are some particles in the Leonid stream with a small enough MOID to permit close enough approaches to be substantially deflected by the Earth's gravity. Of course, the closest approaches of all are impacts, resulting in meteors, and the comet's small MOID is the reason why meteor storms are possible, since just a small perturbation of the nodal position is required to bring orbits to Earth intersection.

Some breaks in the smooth, slowly varying behaviour of the trail are evident in Figure 1 (easiest to see in the last of the 4 plots as the gaps progressively widen). For example, parts of the trail at $\Delta a_0 \approx -0.04$, $+0.17$, $+0.38$, etc. came close to the Earth at one year intervals in the 1830s epoch (-0.04 in 1832, $+0.17$ in 1833, etc.). There is a similar sequence corresponding to the 1860s epoch, e.g., $+0.09$ in 1866, $+0.20$ in 1867, $+0.31$ in 1868, although some gaps are smaller (and may not be noticeable at the resolution plotted in Figure 1) because the approach distance to the Earth was greater.

Therefore after a close approach, a break is present at that point in the trail at subsequent times. Although trails are generally disrupted because of approaches to the giant planets [1], the most noticeable gaps in Figure 1 are identifiable with Earth approaches. This is consistent with the fact that for the past several centuries, the MOID between 55P/Tempel-Tuttle (aphelion near orbit of Uranus) and Jupiter and Saturn has been well above zero. Precession in the argument of perihelion means that the MOID to Saturn approached zero ~0.9 kyr ago. At present we are approaching an epoch when the orbit of Uranus is intersected, although currently a near-commensurability keeps the comet, and parts of trails not too distant from the comet, safe from the closest possible Uranus approaches [7]. Thus Leonid trails that have formed during recent centuries have evolved under circumstances where the only very close approaches have been to Earth.

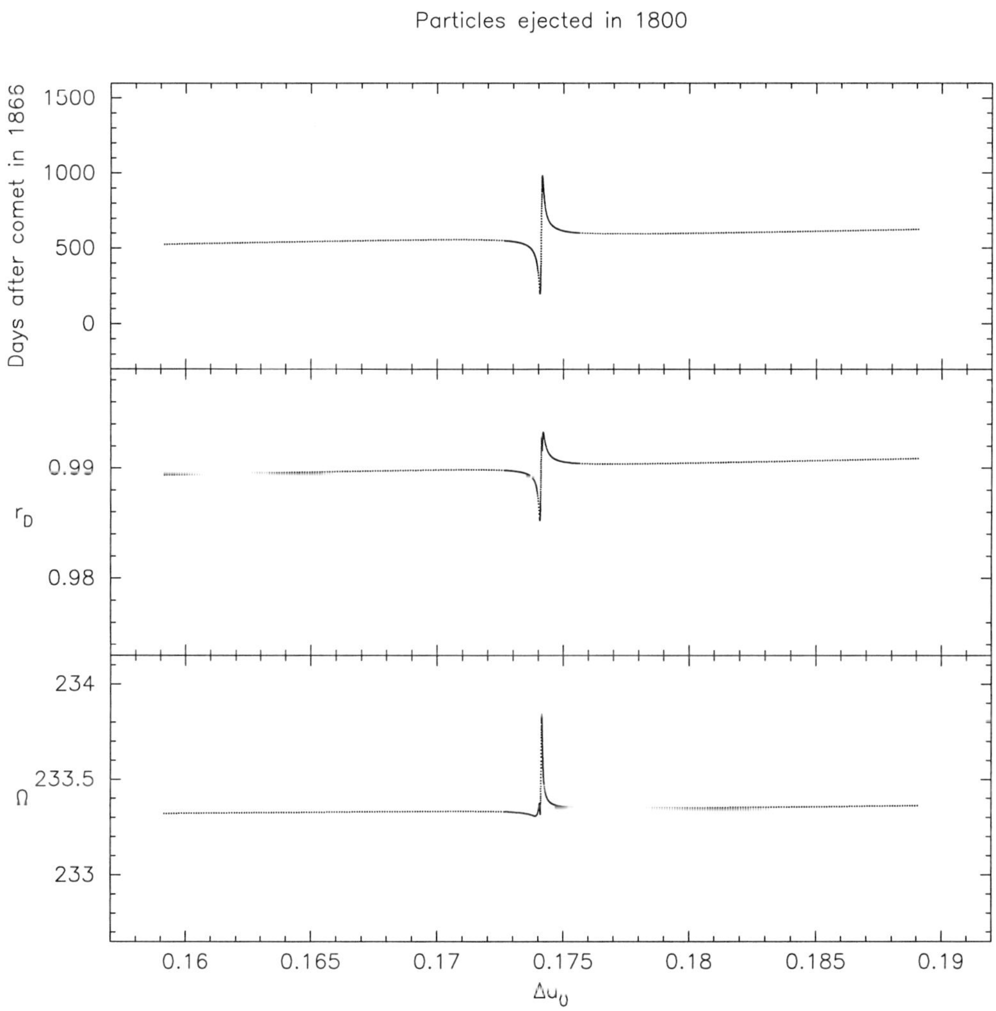

Figure 2. Integrations of particles ejected tangentially at perihelion in 1800. Time particles reach descending node 2 revolutions later, heliocentric distance of descending node r_D, and longitude of ascending node Ω shown as functions of Δa_0. Particles with Δa_0 near +0.174 reach the ecliptic in 1833 November when the Earth is nearby and are gravitationally scattered.

Figure 2 illustrates more details of the orbital evolution near one of these breaks in the 1800 trail, namely around the value of Δa_0 for which the 1833 close approach occurred. The time of nodal passage is plotted (essentially equivalent to M in Figure 1) along with the two coordinates of the node, one revolution after the approach. The Δa_0 for which the closest approach (0.0002 AU) occurred is clear in Figure 2. Particles having Δa_0 within about ± 0.002 of this value approached to within less than 0.01 AU. The further Δa_0 is from that value (ahead or behind in the trail), the greater the miss distance from the Earth and the smaller the effect of the gravitational deflection on the subsequent orbit.

As only one parameter (tangential ejection velocity at perihelion, equivalently Δa_0 with perihelion distance held fixed) is varied in this idealised model, subsequent orbital elements are a function of this parameter. Figure 3 shows that the time of nodal passage and the coordinates of the node are, within narrow limits, unchanged under a more realistic model that includes ejection away from perihelion and in arbitrary directions from the nucleus. No claim is made that the particular model used for Figure 3 is true, but it is representative of such realistic models, and specifically it is useful for demonstrating the how the width of trails is affected by dispersion in orbital elements.

Apart from at the centre of the plots (where the trail has been disrupted by the 1833 Earth approach), the elements in Figure 3 are dispersed by only small amounts about the single valued functions depicted in Figure 2. Thus away from disrupted (widely scattered) sections of trails, orbital elements can be calculated by considering the idealised trail (with tangential ejection at perihelion) only. The idealised model is sufficient for storm prediction, since storms are due to the highest density regions, which have not been widely scattered. This idealised trail has been defined [8] as the 'centre' of the real trail, which has a nonzero width. Although the idealised trail is a convenient representation, avoiding all need for physical models of ejection, it should be noted that the value of r_D at the trail centre may be slightly displaced from the real mean r_D of particles at that point along the trail [8]. The trail width (dependent on Δa_0 and the ejection model used) is given by the ranges in r_D and Ω in Figure 3.

Moreover, comparing Figure 3 with Figure 4 shows that the width does not increase with time during the first few revolutions of trail evolution, but is instead due only to the range in ejection velocities around the arc of the comet orbit during which ejection occurs. Radiation pressure when coupled with ejection over that arc also makes a small contribution to the trail width, but this increased dispersion in nodal position due to radiation pressure (for Leonids with $\beta = 0.001$) is very small compared to typical shifts in nodal position that subsequently occur due to gravitational perturbations. Radiation pressure (constant β on any given particle) and planetary perturbations do not cause trails to widen with time for several revolutions, although it is possible that the Yarkovsky effect on spinning meteoroids can have a widening effect [9].

The fact that all particles in Figure 4 have very similar nodal passage time merely means that a trail has negligible length during the perihelion passage when it is formed. Within a revolution it has elongated greatly, and thereafter is a dense, narrow structure broken only at points where close approaches have occurred. To investigate the effect of approaches, the central part of Figure 3 had extra particles integrated. No significance should be attached to the increased density of points, which are to only provide better number statistics for illustrating the scatter. In reality particles would be expected to

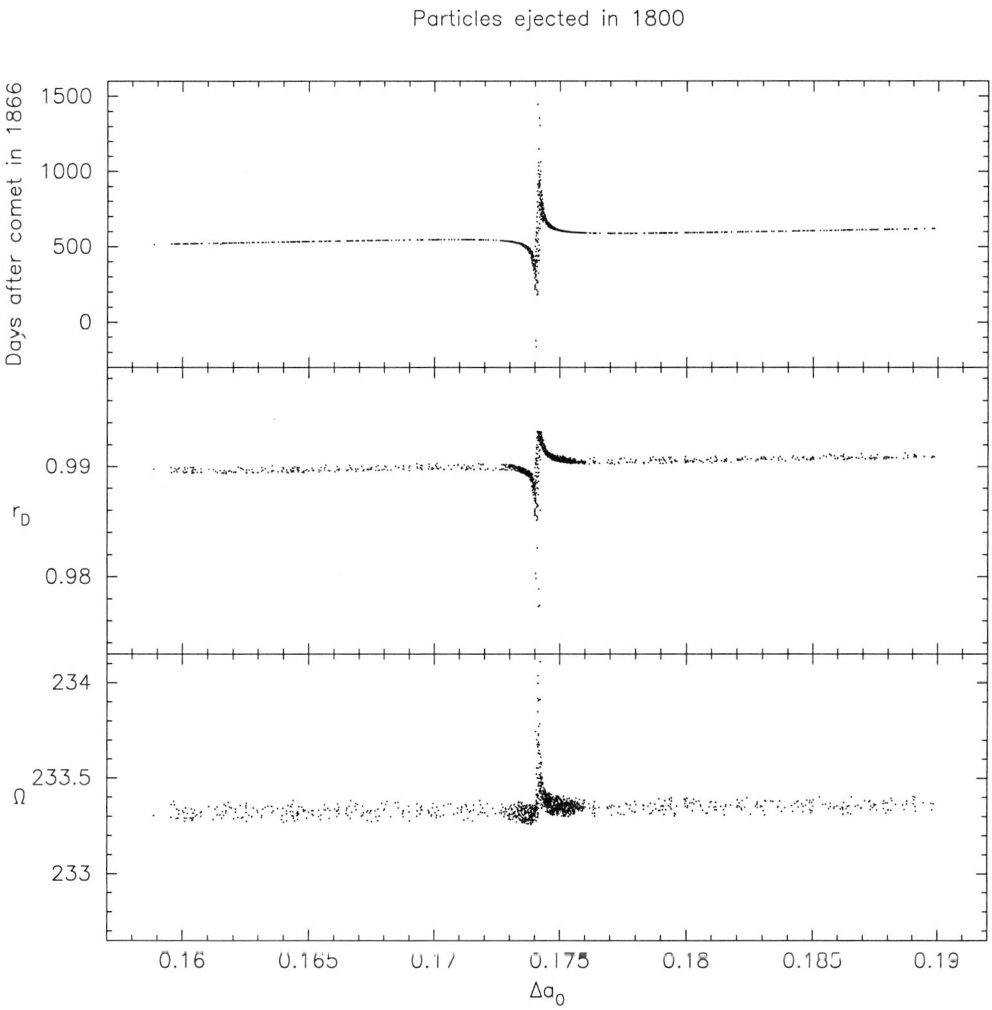

Figure 3. Particles ejected using a Monte Carlo program and only integrated if their Δa_0 was within the range shown here. The program (model) has isotropic ejection, uniformly random in true anomaly for heliocentric distances r below about 3.4 AU, and ejection speed $25/r$ m/s. In this limited range of Δa_0, this ejection speed in fact limits ejection to having $r \lesssim 1.1$ AU. Cf. Figure 2.

be uniformly distributed in Δa_0, over small ranges of Δa_0 (such as the entire range in Figure 3).

Figure 3 shows that away from disrupted trail sections, nodal passage time t_D is very close to being a single valued function of Δa_0, so that trail width, i.e. the ranges in r_D and Ω that are present, can equally be considered as a function of t_D or Δa_0. The parameter

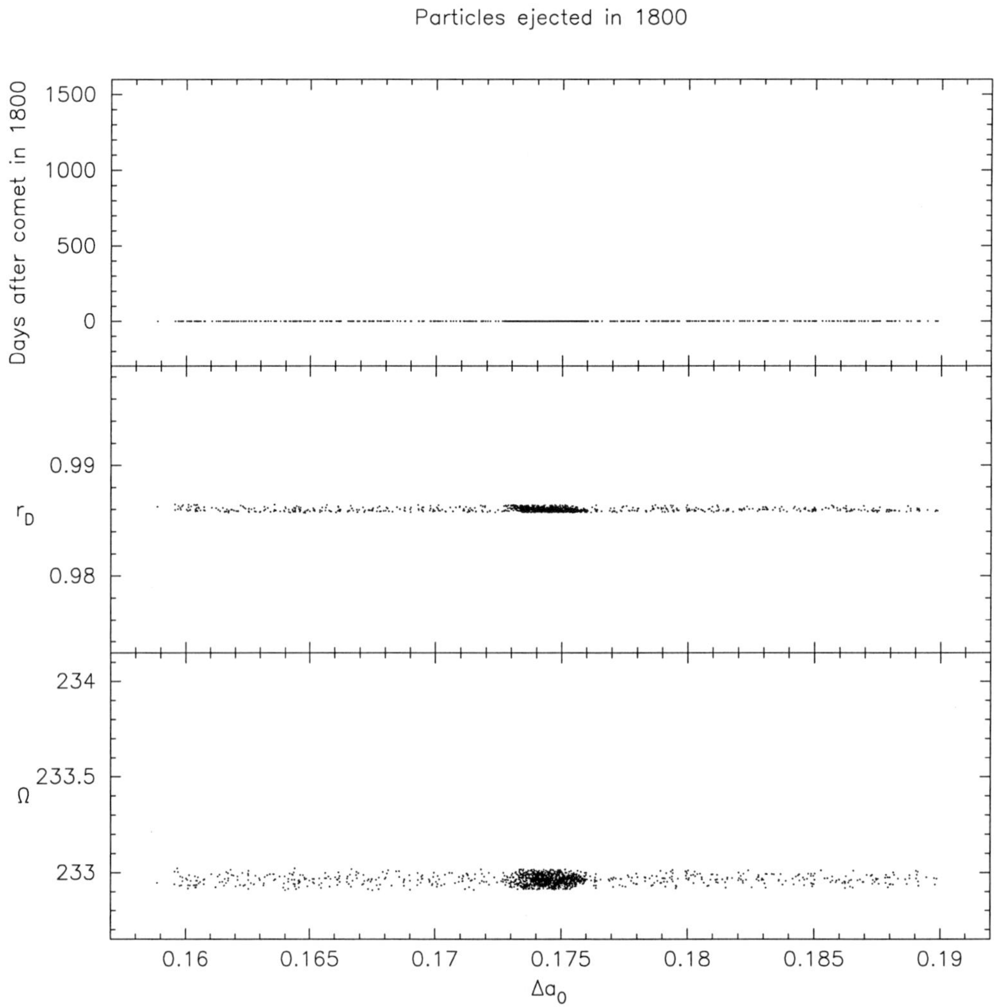

Figure 4. Initial orbital elements (i.e., referred to time of nodal passage during perihelion passage of ejection), of particles integrated in Figure 3. For particles ejected after nodal passage of comet, elements are projected (integrated) back to their ecliptic crossing.

Δa_0 relates more closely to the ejection velocity distribution, and t_D to the physical existence of a trail in space at later times. Near the disrupted section, a large range in all three quantities t_D, r_D, Ω is present and so the latter two are plotted as functions of t_D in Figure 5. This shows the close 1833 encounter to remove particles that by the time of the next return would span about a month along the trail (i.e., there is a very clear gap of this length in the trail), and to scatter these particles over orbital periods that span ~1000 days, by the time of the next return. However, although t_D spans such a large

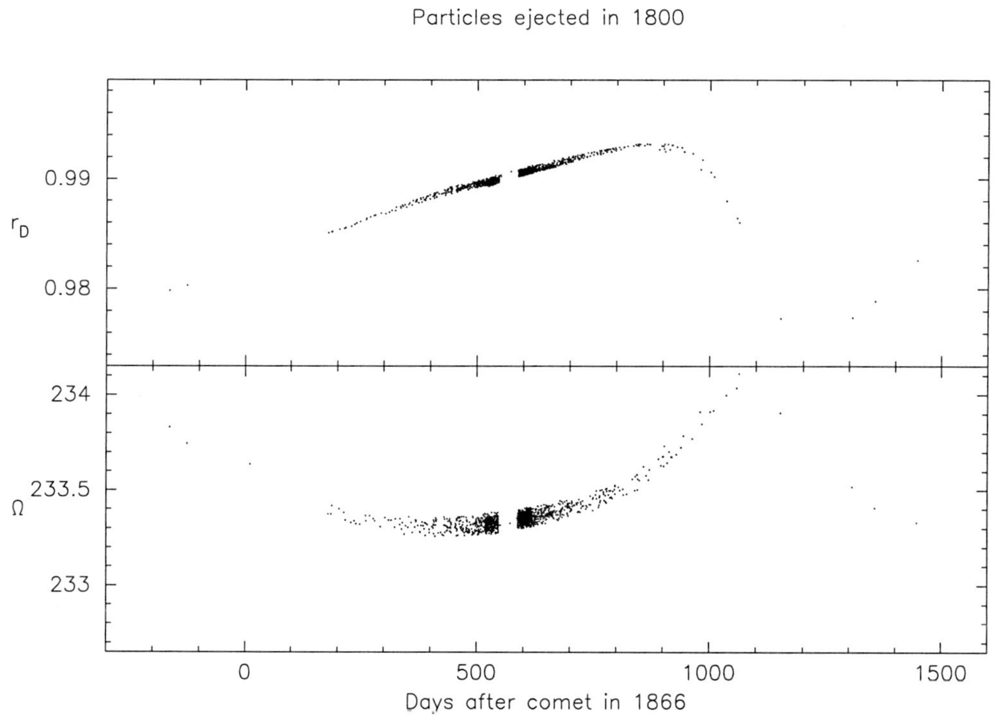

Figure 5. Particles integrated in Figure 3 but with x-axis being nodal crossing time instead of Δa_0.

range, r_D and Ω in Figure 5 are essentially still functions of t_D, apart from the scatter originally generated at ejection time (Figure 4).

The dispersion at subsequent times of particles in a close encounter depends on both the range of elements induced by the encounter and the extent to which subsequent perturbations amplify the scatter. In Figure 5 there was very little such amplification of the scatter in r_D and Ω, although this may be because only one revolution after 1833 was integrated. The particles removed from the trail were dispersed over a huge distance in the along trail dimension but, initially at least, remain in quite a narrow structure, essentially a separate filament of the stream albeit with very low density because of the along trail dispersion.

4. EFFECT OF RESONANCES

The encounter whose effects were shown in Figure 2 was one of the closest possible. It gave rise to the great 1833 storm, and removed particles over a substantial length of the trail. Less close encounters remove smaller amounts of a trail. Overall, an increasing proportion of a trail will be removed with time. However, mean motion resonances can

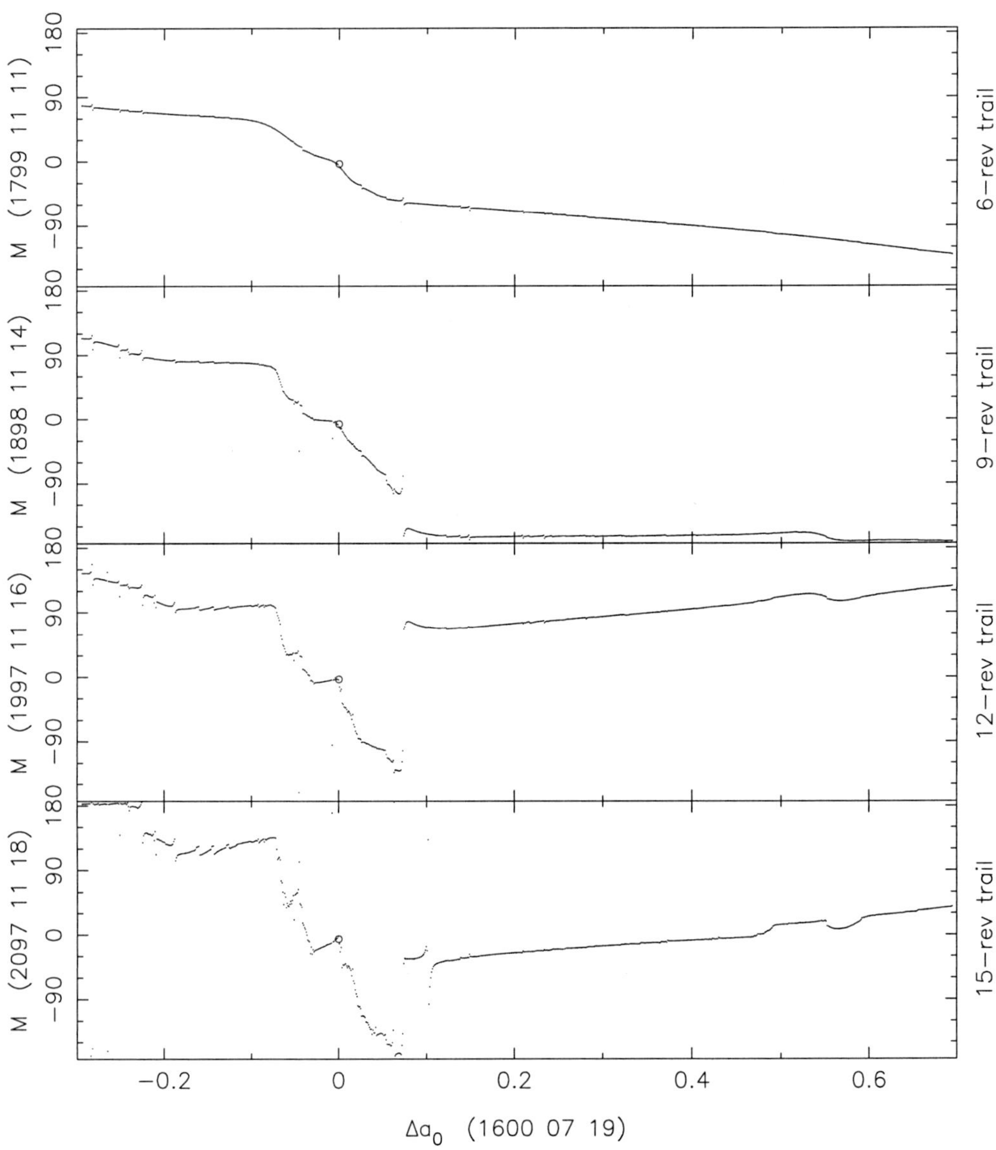

Figure 6. Evolution of 1600 Leonid trail. Mean anomaly M shown as function of Δa_0 (difference in semi-major axis from comet at ejection), after 6, 9, 12 and 15 revolutions of comet, which is shown by a circle.

restrict the process of scattering. Figure 6 shows some integration results, using the tangential ejection at perihelion model, for the trail generated at the 1600 return of the comet. At some values of Δa_0, M is quite scattered after 15 revolutions. However, particles in Figure 6 with e.g. Δa_0 between -0.04 and 0 are in the 5:14 Jovian resonance (55P/Tempel-Tuttle is also in this resonance), and particles in the large range in Δa_0 to the right of $+0.07$ are in the 1:3 resonance. The resonances cause M to be a more coherent function of Δa_0. Moreover, in general the dispersal of nodal positions is also greatly restricted by resonances, owing to resonant motion being near-regular rather than stochastic [10]. Since compact, resonant structures can survive for increased timescales, outbursts can occur due to Leonid meteoroids released from the comet many centuries ago, as was the case in the 1998 fireball display [11–13].

5. LEONID TRAIL ENCOUNTER PARAMETERS

Forecasts of Leonid outbursts and explanations for past displays, based on determining that the Earth encounters specific trails, have been published elsewhere (e.g. [8,14–18]). In Table 1 the nodal positions of trail centres are listed for encounters in years near the present. To generate these data, the point along the trail must be found that reaches the descending node in mid-November; the effect of perturbations on the nodal position of particles at that point can then be determined. As the time of nodal passage is generally quite a smooth function of Δa_0 for the youngest trails, there is usually no problem in finding the desired Δa_0. This technique has been applied to many streams, e.g., the June Bootids [19] and Draconids [20].

It is possible to identify details of the perturbation history of particles in Table 1. For example, Ω of particles in the 4-revolution trail at $\Delta a_0 = +0.14$ increases by $\sim 2°$ between 1866 and 2001, and this happens mainly during two moderately distant approaches (~ 1 AU) to Jupiter in 1898 and 1901, before and after perihelion. At those times, the comet is ~ 2 AU ahead of the particles, so that the geometry of the approaches is different and Ω only changes by half as much. The r_D of those particles changes a few times during the 4 revolutions, sometimes by ~ 0.01 AU during the course of a single, moderately distant, planetary approach. It is therefore rather fortuitous that this part of the trail is brought

Table 1
Encounters with young Leonid trails during years around present return of 55P/Tempel-Tuttle. Listed are age of trail in revolutions, point along trail parameterised by Δa_0, longitude of nodal position given as date when Earth reaches that longitude, and difference in heliocentric distance (AU) between Earth and descending node of trail.

Trail	Δa_0	Date	$r_E - r_D$
3-rev	+0.14	1999 11 18.09	−0.0007
8-rev	+0.06	2000 11 18.16	+0.0008
4-rev	+0.11	2000 11 18.33	+0.0008
7-rev	+0.08	2001 11 18.42	−0.0004
4-rev	+0.14	2001 11 18.76	+0.0002
9-rev	+0.04	2001 11 18.73	+0.0002
7-rev	+0.11	2002 11 19.17	−0.0001
4-rev	+0.17	2002 11 19.44	0.0000
2-rev	+0.96	2006 11 19.20	−0.0001

so close to Earth orbit intersection in 2001 November, although given sufficiently many long trails, it is likely that the Earth will encounter some of them from time to time.

ACKNOWLEDGEMENTS

This work was supported by PPARC and used Starlink facilities. John Chambers kindly made available his MERCURY integration package [21].

REFERENCES

1. L. Kresák, in Meteoroids and their Parent Bodies (eds. J. Štohl and I.P. Willaims), Slovak Acad. Sci., Bratislava (1993) 147.
2. M.V. Sykes and R.G. Walker, Icarus 95 (1992) 180.
3. M. Plavec, Publ. Astron. Inst. Czechosl. Acad. Sci. 30 (1957) 1.
4. P. Brown and J. Jones, Icarus 133 (1998) 36.
5. J. Jones, Mon. Not. R. astr. Soc. 275 (1995) 773.
6. F.L. Whipple, Atrophys. J. 113 (1951) 464.
7. I.P. Williams, Mon. Not. R. astr. Soc. 292 (1997) L37.
8. R.H. McNaught and D.J. Asher, WGN 27 (1999) 85.
9. E.J. Lyytinen and T. Flandern, Earth, Moon and Planets 82 (2000) 149.
10. V & .V. Emel'yanenko and M.E. Bailey, in Physics, Chemistry and Dynamics of Interplanetary Dust (eds. B.Å.S. Gustafson and M.S. Hanner) ASP Conf. Ser. 104, Astron. Soc. Pacific, San Francisco (1996) 121.
11. D.J. Asher, M.E. Bailey and V.V. Emel'yanenko, Mon. Not. R. astr. Soc. 304 (1999) L53.
12. D.J. Asher, M.E. Bailey and V.V. Emel'yanenko, Irish Astron. J. 26 (1999) 91.
13. P. Brown and R. Arlt, Mon. Not. R. astr. Soc. 319 (2000) 419.
14. P. Brown and J. Jones, in Physics, Chemistry and Dynamics of Interplanetary Dust (eds. B.Å.S. Gustafson and M.S. Hanner) ASP Conf. Ser. 104, Astron. Soc. Pacific, San Francisco (1996) 113.
15. P. Brown and B. Cooke, Mon. Not. R. astr. Soc. 326 (2001) L19.
16. E.D. Kondrat'eva and E.A Reznikov, Sol. Syst. Res. (19) (1985) 96.
17. E.D. Kondrat'eva, I.N. Murav'eva and E.A. Reznikov, Sol. Syst. Res. 31 (1997) 489.
18. E. Lyytinen, Meta Res. Bull. 8 (1999) 33.
19. E.A. Reznikov, Trudy Kazan. Gor. Astron. Obs. (in Russian) 47 (1983) 131.
20. E.A. Reznikov, Trudy Kazan. Gor. Astron. Obs. (in Russian) 53 (1993) 80.
21. J.E. Chambers, Mon. Not. R. astr. Soc. 304 (1999) 793.

A dust swarm detected after the main Leonid meteor shower in 1998

Y.-H. Ma[ab*] Y.-W. He[c] and I.P.Williams[b]

[a]Purple Mountain Observatory, Academia Sinica, Nanjing 210008, China;
National Astronomical Observatories, Academia Sinica, China

[b]Astronomy unit, Queen Mary and Westfield College, London E1 4NS, UK

[c]Electric Wave Propagation Institute of China, Xinxiang, 453003, China

The Leonid meteor shower in November 1998 was widely observed by both visible and radio astronomers. A significant outburst, which included several fireballs, was observed on the morning of Nov.17, about 16 hours before the predicted maximum of the main shower. The main shower was also observed as expected on Nov.18 and many observations of both these events have been recorded. About 18 hours after the main shower, an abnormal peak of the ionosphere characteristic value f_bE_s was detected by two separate ionosphere observational stations, Guangzhou and Hainan. The very high f_bE_s value was maintained over one hour. The most likely explanation for this abnormally high activity is that the ionosphere was bombarded by a swarm of dust particles, much smaller than those which produce a visible or radio trail. The near coincidence in time between this event and the Leonid shower strongly suggests that we should look for an explanation for this dust swarm in terms of the dynamics and evolution of the Leonid meteoroid stream and this paper explores these aspects.

1. INTRODUCTION

The Leonid meteor shower is a well-known periodic meteor shower. The Leonid parent comet, 55P/Tempel-Tuttle, has an orbital period of about 33.2 yr, and storms usually occur in years around the perihelion passage of the parent comet. Since comet 55P/Tempel-Tuttle passed perihelion on February 28, 1998, the Leonid meteor shower of 1998 was predicted [1, 2, 3, 4, 5] to produce a strong shower. Such strong displays were seen and details can be found in many published articles. Showers have not always appeared when expected although the general behaviour of meteor streams has been reasonably well modelled. In 1998 a strong component, rich in bright meteors, appeared about 16 hours before the expected maximum of the main shower. An explanation for this has been given by Asher *et al.* [6]. An unexpected new peak in the Perseids occurred in the early nineties slightly separated in time from the traditional main peak [7], and a similar case occurred in the Quadrantids [8]. We found another unusual phenomenon, which was an abnormal level of ionization detected in the ionosphere about 18 hours after the main Leonid shower of 1998. The same phenomena occurred in the ionosphere around the strong showers of the Leonids, Perseids and Draconids

* Supported by National Natural Science Foundation of China (No.19873020 \& No.49474242) and the Royal Society KC Wong fellowship.

direction of ejection and the orbital plane, so $v\sin\phi$ is the component of the ejection velocity perpendicular to the orbital plane.

Suppose the ejection occurred at perihelion, Eq.(1) becomes

$$\Delta\Omega = \frac{q\sin\omega}{h\sin i}v\sin\phi \tag{2}$$

From Eq.(2), substituting the orbital parameters of 55P/Tempel-Tuttle, we can get

$$\Delta(v\sin\phi) = v_1\sin\phi_1 - v_0\sin\phi_0 = 1.67\times10^3\Delta\Omega\ (m\ s^{-1}) \tag{3}$$

v_1 and v_0 are ejection velocity of different sized particles, and $\Delta\Omega$ is in degrees.

Following Whipple [14] and Wu and Williams [15], the ejection velocity can be given as

$$v = \frac{C}{(bc)^{0.5}}r^{-1.125} \tag{4}$$

where C is a constant. If they were ejected simultaneously in the same direction, we have

$$\frac{v_1\sin\phi}{v_0\sin\phi} = \frac{v_1}{v_0} = \sqrt{\frac{b_0c_0}{b_1c_1}} \tag{5}$$

It is well known that the small meteoroids are principally affected by the solar gravity and an anti-gravity force arising from solar radiation pressure after they were ejected from their parent body, and they move basically along the original orbit. Different sized particles will have different orbital periods due to the different influence of solar radiation force on them. They will therefore separate gradually in the process of orbital movement. Since the force from solar radiation pressure is opposite to the direction of solar gravity, its effect is to 'weaken' the gravity so that the quantity GM is replaced by $GM(1-\beta)$, where $\beta=F_r/F_g$ (F_r: the force from solar radiation pressure and F_g: the solar gravity) has the value about $\beta\sim10^{-5}/bc$ [16], b being the radius and c the bulk density of the grain, both in *cgs* units. From the Kepler's third law, and remembering that the meteoroids move basically along the same orbit, we have the orbital period of the meteoroids

$$P = \frac{P_c}{(1-\beta)^{1/2}} \tag{6}$$

where P_c is the orbital period of the parent comet, so in this case P_c =33.2 yr.

In addition to radiation pressure causing the period to vary with particle size, the small energy change by ejection affects the period of ejected particles. The energy change per unit mass by ejection is

$$\Delta E = Vv\cos\theta + \frac{1}{2}v^2 \approx Vv\cos\theta \tag{7}$$

[9,10]. These phenomena were thought to be due to the impact of a swarm of very small particles into the ionosphere. We shall analyse the small particle peak of Leonid shower of 1998 in the following section.

2. OBSERVATIONS

An overview of the optical observations of the 1998 Leonid activity was given by Arlt [11]. An unexpected component rich in bright meteoroids appeared about 16 hours before the predicted maximum of the main shower. The actual "storm" component was observed as expected with a peak at November 17, 20:30UT.

We had a radio observation at Electric Wave Propagation Institute of China in Xinxiang, Henan. The result was also consistent with the prediction, and the peak was over 2500 h^{-1}.

Several ionosphere observational stations in China surveyed the ionization effect of the meteor shower during Nov.14-20. The ionosphere characteristic value f_bE_s was selected as the observational parameter. About 18 hours after the main shower, an abnormal peak of the f_bE_s value was detected simultaneously by two ionosphere observational stations, Guangzhou and Hainan. The very high f_bE_s value was maintained for over one hour. The abnormal phenomenon showed that the ionosphere was bombarded by a swarm dust that could not be observed by optical and radio.

We can understand this phenomenon as follows. According to Hughes' [12] suggestion that typical stream meteoroids have a mass distribution index s=2.27 and the number of meteoroids N, with masses greater than m is proportional to $m^{(1-s)}$, we have the number of meteoroids N, with radii greater than b proportional to $b^{3(1-s)}$. So $N \sim m^{-1.27} \sim b^{-3.81}$. If a region of the stream contains 1 meteoroid with radius greater than 1 centimeter, it should contain $10^{3.81}$ meteoroids with radii greater than 0.1 centimeter, $10^{7.62}$ meteoroids with radii greater than 0.01 centimeter and so on. The large population of tiny grains bombarding the ionosphere can produce enough positive ions to create abnormal ionization effects in the ionosphere [9,10]. There was no abnormal ionization effect during the main shower because the number of the grains was insufficient.

3. THE EFFECTS OF EJECTION VELOCITY AND RADIATION PRESSURE

The ejection velocity of the meteoroid from the comet and the solar radiation pressure upon it are obviously dependent on the mass of the meteoroids, which can cause the meteoroids of different sizes to be separated.

The dust swarm 18 hours after the main shower illustrates that the longitude of the ascending node of the tiny grains in the Leonid stream has a 0.74° difference from that of the visible ones. If they suffer the same planetary perturbation, the node can be changed only by ejection. From the fundamentals of the celestial mechanics, we can get [13]

$$\Delta\Omega = \frac{r\sin(\omega + f)}{h\sin i} v\sin\phi \qquad (1)$$

where $\Delta\Omega$ is the longitude change of the ascending node caused by ejection, r is the heliocentric distance, h is angular momentum, v is ejection velocity, ω is the argument of the perihelion, f is the true anomaly, i is the inclination of the orbit, ϕ is angle between the where V is the orbital velocity of the comet at the position of ejection, v is the ejection velocity, θ is angle between the directions of V and v.

If $\theta = \pi/2$, from Eq.(7), $\Delta E \approx 0$, no energy change is caused by ejection. This is a special case but in general it is possible and it simplifies our discussion here.

Taking b_0=0.1cm, c_0=3gcm^{-3} (b_0c_0=0.3) for visible meteors, from Eq.(6) and the observation (ΔP=18 hours), we can get b_1c_1=0.064 for the tiny grains which caused abnormal ionization effects in the ionosphere.

Substituting the values of b_1c_1 and b_0c_0 into Eq.(5), and combining Eqs.(3) and (5), we get

$$v_1 \sin\phi = 2.24\times10^3 \ m\,s^{-1}, \quad v_0 \sin\phi = 1.00\times10^3 \ m\,s^{-1}.$$

These results suggest that the initial ejection velocities of the streamlet responsible for Leonid meteor shower in 1998 must have a component perpendicular to the orbital plane and are of order $10^3\ m\,s^{-1}$ if the ejection occurred at the last perihelion. This is much different from that of Brown and Jones [17] who derived that the initial ejection velocities are of order $5\ m\,s^{-1}$ and do not exceed $20\ m\,s^{-1}$ for any of the Leonid storms over the last 200 years.

REFERENCES

1. P. Brown and J. Jones, in Physics, Chemistry and Dynamics of Interplanetary Dust (eds B.A.S. Gustafson and M.S. Hanner) ASP Conf. Ser. vol. 104, Astron. Soc. Pac., San Francisco (1996) 113.
2. P. Jenniskens, Meteorit. Planet. Sci. 31 (1996) 177.
3. Z. Wu and I.P. Williams, Mon. Not. Roy. Astr. Soc. 280 (1996) 1210.
4. D.K. Yeomans, K.K. Yau and P.R. Weissman. Icarus 124 (1996) 407.
5. R. Arlt and P. Brown, WGN 26 (1998) 161.
6. D.J. Asher, M.E. Bailey and V.V. Emel'yanenko, Mon. Not. Roy. Astr. Soc. 304 (1999) L53.
7. P. Brown and J. Rendtel, Icarus 124 (1996) 414.
8. D.W. Hughes, I.P. Williams and K. Fox, Mon. Not. Roy. Astr. Soc. 195 (1981) 625.
9. Y.-W. He and P.-X. Xu, Electric Wave Science Acta 12 (1997) 65 (in Chinese).
10 Y.-W. He, Annual of Chinese Geophysics Society (1998) 241 (in Chinese).
11. R. Arlt, WGN 26 (1998) 239.
12. D.W. Hughes, in Cosmic Dust (ed. J.A.M. McDonnell) John Wiley & Sons, Chichester, (1978) 123.
13. I.P. Williams, This volume.
14. F.L. Whipple, Astrophys. J. 113 (1951) 464.
15. Z. Wu and I.P. Williams, Planet. Space Sci. 43 (1995) 723.
16. I.P. Williams, Mon. Not. Roy. Astr. Soc. 292 (1997) L37.
17. P. Brown and J. Jones, in Meteoroids 1998, Astron. Inst., Slovak Acad. Sci., Bratislava, (1999) 159.

Meteor showers associated with Near-Earth Asteroids in the Taurid complex

P.B. Babadzhanov

Institute of Astrophysics, Dushanbe 734042, Tajikistan

The existence of observed meteor showers associated with some of Near-Earth asteroids (NEAs) is one of the few criterion that such asteroids may be considered to be candidates for extinct cometary nuclei. In order to reveal new NEA-meteor shower associations we calculated the secular variations of the orbital elements of 17 Taurid Complex asteroids with allowance for perturbations from six planets (Mercury-Saturn) over one cycle of variations of perihelia arguments. The Earth-crossing class of these NEAs and theoretical geocentric radiants and velocities of their meteor showers were determined and compared with available observational data. It turns out, that each Taurid Complex asteroid is associated with four meteor showers. This is evidence for the cometary origin of these asteroids.

1. INTRODUCTION

According to generally accepted opinion, meteoroid streams are formed as a result of the disintegration of cometary nuclei. The presence of meteor showers associated with some near-Earth asteroids (NEAs) give evidence that such asteroids have a cometary origin, i.e. they are extinct cometary nuclei [1,2,3,4,5]. The existence of asteroids identifiable with extinct or dormant comets (2060 Chiron, 4015 Wilson-Harrington, 1986 TF Parker Hartley) confirm the cometary origin for some NEAs.

Investigation of NEA-meteor showers associations is important not only for confirmation or denial of NEA cometary origins, but also for the receipt of important information about NEA sources – comets from outer regions of the Solar System, and real asteroids from the main belt.

The calculation of theoretical meteor radiants is the first step in revealing the generic relationship between a given near-Earth object (comet or asteroid) and its possible meteor showers. However, methods for the determination of the theoretical radiants of comets and asteroids approaching the Earth's orbit close than 0.1-0.3 AU, which were used by different authors, e.g., [6,7,8] until recently, did not take into account the meteoroid stream evolution and could only roughly predict one or two radiants of the given comet or asteroid.

As follows from basic principles of meteoroid stream formation and evolution [9,10,3,4,11] related meteor showers can be possessed also by those comets whose orbits are located presently at distances more than 0.3 AU from the Earth's orbit (Table 1) but which crossed it in the past. The orbit of the parent body for the moments of its crossings of the Earth's orbit can be determined by studying its evolution under the gravitational perturbing action of the major planets.

2. FORMATION OF METEOROID STREAMS

Ejection velocities of meteoroids from their parent bodies and radiation pressure (for small particles) cause an initial dispersion in orbital elements of ejected meteoroids. Because of differences in the semi-major axes (and orbital periods) between the meteoroids and their parent body, some meteoroids lag behind the parent body, while others, overtaking it, spread along the entire orbit and form a complete loop in a comparatively short time [12,13].

After the meteoroids are distributed along the orbit of the parent body, due to differences in the planetary perturbing action on stream meteoroids of different semi-major axes and eccentricities, the rates and cycles of variations in the angular orbital elements (ω, Ω, i) will be different for different meteoroids. As a result, the orbits of different meteoroids will be at different evolutionary stages as distinguished by their arguments of perihelia, that is the stream meteoroids occupy all evolutionary tracks of their parent body. This process increases considerably the size of the meteoroid stream, first of all, its thickness (the breadth of a stream is determined by the value of the meteoroids' orbital semi-major axes).

If the Earth's orbit is assumed to be circular, then it may be intersected by those stream meteoroids which have the orbital node at r=1 AU, i.e. satisfying the expression:

$$\pm\cos\omega = \frac{a(1-e^2)-1}{e}. \tag{1}$$

As shown earlier [10,3,4] the number of meteor showers produced by a meteoroid stream is determined by the Earth-crossing class of the parent-body orbit. For example, if it is a quadruple crosser of the Earth's orbit (i.e. during one cycle of variations of the perihelion argument of its orbit under the perturbing action of the major planets a parent body crosses the Earth's orbit four times) the meteoroids of the stream that separated from parent can produce four meteor showers: two at the pre-perihelion intersections and two at the post-perihelion intersections with the Earth. Crossing before perihelion gives rise to two nighttime showers, and after the perihelion to two daytime showers. These two pair of showers are formed by the same meteoroid stream, each pair consisting of a northern and a southern branch.

3. METEOR SHOWERS ASSOCIATED WITH TAURID COMPLEX ASTEROIDS

The object of the present paper is to reveal the meteor showers associated with the Taurid Complex asteroids, which has Encke's comet as a member and according to Clube & Napier [14] and Asher *et al.* [15] have a common cometary origin. Possible association of daytime fireballs and some Taurid complex asteroids (4486 Mithra, 1990 SM and 1991 BA) was suggested by Hasegawa [16]. But the existence of observable associated meteor showers is the only substantial index that a given NEA is a candidate for comet origin.

Asher *et al.* [15] assumed that those near-Earth objects (NEOs) belong to the Taurid complex asteroids if their longitudes of perihelion $\pi=\Omega+\omega$ lie within the limits of $100° < \pi < 190°$, and if their orbital parameters (a, e, i) satisfy the criterion $D\leq 0.2$, where

$$D^2 = \left(\frac{a_1-a_2}{3}\right)^2 + (e_1-e_2)^2 + \left(2\sin\frac{i_1-i_2}{2}\right)^2. \tag{2}$$

and a_1=2.1 AU, e_1=0.82, and i_1=4° and the subscript 2 denotes a near-Earth object's orbit. The criterion (2) is a modified Southworth & Hawkins' [17] criterion of orbital similarity as applied to the membership for the Taurid complex.

Among 536 near-Earth asteroids known by August 8, 1998 we find 17 whose orbits satisfy the criterion (2). Data for these NEAs are given in Table 1, where a is semi-major axis, e is the eccentricity, q is the perihelion distance, i is the inclination to the ecliptic, Ω is the longitude of the ascending node, ω is the argument of perihelion, H is the absolute magnitude and d the equivalent diameter, calculated using the expression [18,19]:

$$\log d = 312 - 0.2\ H - 0.5 \log p\,. \tag{3}$$

The candidates for comet origin should be dark asteroids of C,P and D-types of low albedos in the range of 0.02 to 0.08. The values of d in Table 1 are given for an assumed albedo p=0.08. R_a and R_d are the radius-vector of the ascending and descending nodes respectively.

Table 1

Near-Earth Asteroids of the Taurid Complex

Asteroid	q AU	a AU	e	i^0	Ω^0 2000.0	ω^0	π	D	H	d km	R_a AU	R_d AU
1993 KA2	0.502	2.227	0.775	3.2	239.6	261.3	140.9	0.06	29.0	0.01	1.0	0.8
6063 Jason	0.522	2.216	0.764	4.8	170.0	336.5	146.5	0.07	15.1	4.5	0.5	3.1
1996 SK	0.494	2.428	0.796	2.0	198.3	283.4	121.7	0.12	17.0	1.9	0.7	1.1
2201 Oljato	0.630	2.176	0.711	2.5	76.9	96.0	172.9	0.12	15.3	4.1	1.2	1.0
5025 P-L	0.647	2.140	0.697	3.1	341.6	153.8	135.5	0.12	15.9	3.1	2.9	0.7
1991 TB2	0.394	2.397	0.836	8.6	297.2	195.6	132.8	0.13	17.0	1.9	3.7	0.4
5143 Heracles	0.420	1.834	0.771	9.2	310.8	226.4	177.2	0.13	13.9	7.7	1.6	0.5
4197 1982 TA	0.522	2.297	0.773	12.2	10.2	119.2	129.4	0.14	14.5	5.9	1.5	0.7
1995 FF	0.674	2.321	0.710	0.6	175.6	293.1	108.7	0.14	26.5	0.02	0.9	1.6
1991 BA	0.713	2.242	0.682	2.0	118.9	70.7	189.6	0.15	28.5	0.01	1.0	1.6
1991 GO	0.663	1.956	0.661	9.7	25.0	88.6	113.6	0.19	19.0	0.7	1.1	0.9
4183 Cuno	0.718	1.980	0.637	6.8	295.9	235.2	171.1	0.19	14.5	5.9	1.8	0.9
5731 Zeus	0.785	2.262	0.653	11.6	282.8	215.6	138.4	0.19	15.5	3.7	2.8	0.8
4341 Poseidon	0.588	1.835	0.679	11.9	108.2	15.5	123.7	0.20	15.6	3.5	0.6	2.9
8201 1994 AH2	0.730	2.527	0.711	9.6	164.4	24.8	189.2	0.20	16.3	2.6	0.8	3.5
1990 HA	0.782	2.571	0.696	3.9	184.8	308.3	133.1	0.20	16.0	3.0	0.9	2.3
1996 RG3	0.790	2.000	0.605	3.6	158.5	299.9	98.4	0.20	18.5	0.9	1.0	1.8
2P/Encke	0.331	2.209	0.850	11.9	334.7	186.3	161.0				1.0	0.3

Using the Halphen-Goryachev method [20] we calculated the secular variations of the orbital elements of each asteroid mentioned in Table 1 with allowance for perturbations from six planets (Mercury-Saturn) over one cycle of the variations in the arguments of perihelia (4,000-10,000 yrs). The results of calculations are that all these asteroids (exception 1991 BA) are quadruple crossers of the Earth's orbit, and, therefore, their hypothetical meteoroid streams might produce four meteor showers each. The smallest NEA (1991 BA) is twice-crosser and might produce two meteor showers. We calculated the theoretical orbital elements, geocentric radiants (the right ascension α and declination δ) and geocentric velocities, solar longitudes and corresponding dates of activity of all meteor showers associated with 17 Taurid Complex asteroids. A computerized search for the predicted showers was carried out in the catalogues published by: Cook [21], Kashcheev *et al.* [22],

Lebedinets *et al.* [23], and Sekanina [24,25]. This search took into account the closeness in the positions of the theoretical and the observed radiants (the requirement used was $\Delta\alpha = \Delta\delta = \pm 10°$), in velocity values ($\Delta Vg \leq \pm 5$ km/s) and period of activity ($\Delta t \leq +15$ days) for $D_{S\text{-}H} \leq 0.2$, where $D_{S\text{-}H}$ is the criterion of Southworth and Hawkins [17], which, in the case under consideration, serves as a measure of the similarity between the theoretical and the observed orbits. With the use of these catalogues, 58 out of the 66 theoretically predicted showers were identified with observed showers. Probably, the other 8 showers also are active, but they are not distinguished yet amongst the sporadic background. The values of $D_{S\text{-}H}$ – criterion showed good agreement between the parameters of theoretically predicted and the observed showers.

Figure 1 represents the theoretical (dots) and observed (crosses) radiants of the meteor showers associated with Taurid Complex asteroids. The theoretical and observed radiants are linked in pairs. The curve delineates the ecliptic. Fig. 1 shows that in most cases theoretical and observed radiants coincide or are close with each other, but in some cases their differences reach 10-12° due to daily motion of the radiant which was not taken into account because the published catalogues often don't contain corresponding data.

At present about 2000 minor meteor showers and associations are detected from optical and radar observations of meteors, but in the overwhelming majority of cases the parent comets of these showers have not been recognized. As shown by the results of the present paper, the lack of parent comets of many meteoroid streams may be explaned not only by different changes of the stream and parent comet's orbits, but also by extinction and transformation of parent comets into asteroid-like bodies.

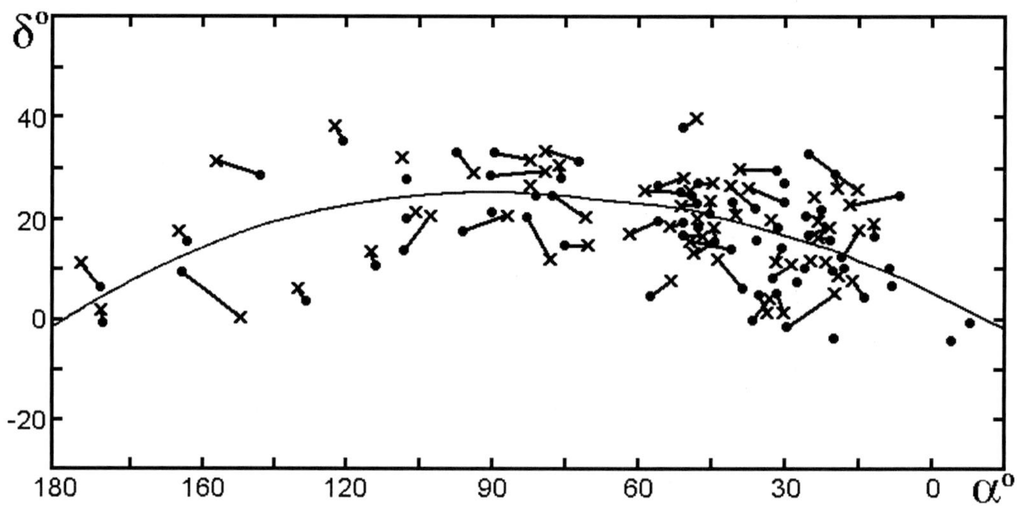

Fig. 1. The predicted (dots) and observed (crosses) radiants of the meteor showers associated with the Taurid Complex asteroids. The predicted and corresponding observed showers are linked in pairs. The curve delineates the ecliptic.

4. CONCLUSIONS

The present investigation shows that each Taurid complex asteroid is associated at least by two, but in most cases with four detected meteor showers. This is evidence for the cometary origin of these asteroids.

For many meteor showers and associations the parent bodies were revealed. This reduces the disproportion between the number of meteor showers and their parent bodies.

The northern and southern branches of meteor showers and their corresponding daytime (or nighttime) twins were determined.

It turns out that a number of meteor showers are common to several members of the Taurid Complex asteroids. So, for example, Northern and Southern Taurids, ζ-Perseids and β-Taurids, Northern and Southern May Aquarids, Northern and Southern Khi Orionids, Northern and Southern Piscids and others associated with 2P/Encke [26], are also associated with the NEAs 2101 Oljato, 4341 Poseidon, 5143 Heracles, 6063 Jason, 1993 KA2, 1997 GL3, 1996 SK, 1991 TB2 and others. These results confirm the assumpton [14,2] that Taurid Complex consist of meteoroids of all possible sizes, including large asteroid-like bodies (extinct cometary nuclei or their fragments), which have produced meteoroid substreams and entered a valuable contribution in the formation of enormous Taurid meteoroid stream producing in the Earth atmosphere more than sixty meteor showers during nearly the whole of the year.

ACKNOWLEDGEMENTS

It is a pleasure to thank members of the Scientific Organizing Committee, (Chairman Prof. I.P. Williams) for the invitation to present this paper. I would like to thank Dr. D.I. Steel for critically reading the manuscript and helpful comments which led to an improvement of this paper.

REFERENCES

1. D.I. Olsson-Steel, Icarus 75 (1988) 64.
2. D.I. Steel, Earth, Moon, and Planets 68 (1995) 13.
3. P.B. Babadzhanov, Solar System Research 30 (1996) 499.
4. P.B. Babadzhanov, Celest. Mech. & Dyn. Astron. 69 (1998) 221
5. D.F. Lupishko, Program and Abstracts of the IAU Colloquium 173 "Evolution and Source Regions of Asteroids and Comets", Aug. 24-28, Tatranska Lomnica, Slovakia (1998) 33
6. E.N. Kramer, Problem. Kosm. Fiz. No 8, Kiev: Izd. Kiev Univ. (1973) 123
7. J.D. Drummond, Icarus 49 (1982) 143.
8. O.D. Artoos, WGN J. of the IMO, 22 (1994) 85 and 189.
9. P.B. Babadzhanov and Yu.V. Obrubov, in Interplanctary Matter (eds. Z. Ceplecha and P. Pecina) Publ. Astr. Inst. Czecho-Sl. Acad. Sci. 2 (1987) 141.
10. P.B. Babadzhanov and Yu.V. Obrubov, Celest. Mech. & Dynam. Astron. 54 (1992) 111.
11. D.I. Steel, in Asteroids, Comets, Meteors 1993 (eds. A. Milani et al.) (1994) 111.
12. D.W. Hughes, in Asteroids, Comets, Meteors II (eds. C.-I. Lagerkvist, B.A. Lindblad, H. Lundstedt and H. Rickman), Uppsala Universitet, Reprocentr., HSC, Uppsala (1986) 503.

13. I.P. Williams, Earth, Moon, and Planets 68 (1995) 1.
14. S.V.M. Clube and W.M. Napier, in The Galaxy and the Solar System (eds. R. Smoluchowski, J.N. Bahcall, and M.S. Matthews), University of Arizona Press, Tucson (1986) 260.
15. D.J. Asher, S.V.M. Clube and D.I. Steel, Mon. Not. R. Astron. Soc. 264 (1993) 95.
16. I. Hasegawa, Earth, Moon, and Planets, 72 (1996) 311.
17. R.B. Southworth and G.S. Hawkins, Smit. Contrib. Astrophys. 7 (1963) 261.
18. E. Bowell and K. Lumme, in Asteroids, (ed. T. Gehrels), Univ. of Ariz. Press, Tucson (1982) 132.
19. B.H. Rowe, Sky & Telescope 85 (1993) 83.
20. N.N. Goryachev, Halphen's Method for Calculation of Planetary Secular Perturbations and its Application to Ceres, Tomsk, Krasnoe Znamya (1937).
21. A.F. Cook, in Evolutionary and Physical Properties of Meteoroids, NASA SP-319 (1973) 183.
22. B.L. Kashcheev, V.N. Lebedinets and M.F. Lagutin, Meteoric Phenomena in the Earth's Atmosphere, Nauka, Moscow (1967).
23. V.N. Lebedinets, V.N. Korpusov and A.K. Sosnova, Trudy Inst. Eksperim. Meteorol. No 1(34), (1973) 88.
24. Z. Sekanina, Icarus 18 (1973) 253.
25. Z. Sekanina, Icarus 27 (1976) 265.
26. P.B. Babadzhanov, Yu.V. Obrubov and N. Makhmudov, Solar System Research 24 (1990) 12.

Dust trails along asteroid 3200 Phaethon's orbit

S. Urakawa[a], S. Takahashi[a], Y. Fujii[a], M. Ishiguro[a], T. Mukai[a] and R. Nakamura[b]

[a]Graduate School of Science and Technology, Kobe University, Rokko-dai-cho 1-1,Nada, Kobe 657-8501, Japan

[b]National Space Development Agency of Japan, Earth Observation Research Center, Roppongi 1-9-9, Minato, Tokyo,Japan

We performed observations of the zodiacal light on November 16–18, 1999 at Mauna Kea (4200 m, Hawaii) by using a cooled CCD camera to search for the dust trails along the orbit of asteroid 3200 Phaethon, which is thought to be the parent object of the Geminid meteoroid stream. No significant enhancement of the brightness larger than $1S_{10\odot}$ appears along a line of sight passing through the orbit of Phaethon.

1. INTRODUCTION

The asteroid 3200 Phaethon, discovered as 1983TB using IRAS [1] has the orbital elements: semi-major axis a=1.271 AU, eccentricity e=0.890, inclination i=22.1°, longitude of the ascending node Ω=265.6° and argument of perihelion ω=321.8°. Since the Earth approaches this orbit in the middle of December, Phaethon is thought to be the parent object of the Geminid meteoroid stream. Furthermore, Phaethon is expected to be a comet in a dormant phase as the parent objects of meteoroid streams are generally thought to be comets.

A faint glow at an ecliptic latitude β of 0° and a helioecliptic longitude $\lambda - \lambda_{\odot}$ of 78° has been found in our zodiacal light observations at Hale Pohaku (2800 m, Hawaii) on November 2nd, 1997 (see Figure 1b in [2]). The analysis of this faint structure suggested that this feature related to the orbit of asteroid 3200 Phaethon. However, since its brightness of about $1S_{10\odot}$ is near our detection limit, we could not conclude the existence of dust trails associated with asteroid 3200 Phaethon. The confirmation of this structure was the motivation of the observations presented below.

2. OBSERVATION AND DATA ANALYSIS

We made photometric observations of a region in Cancer, where the orbit of asteroid 3200 Phaethon could be seen on November 17th, 1999. We used a wide-angle lens (Sigma 24 mm lens, F=2.8) attached to a cooled CCD camera (Mutoh CV-16) at Mauna Kea (4200 m, Hawaii). The angular resolution and the field of view are 2.50' pixel^{-1} and 32°×21° (768×512 pixels) respectively. A special filter is designed to fit the broadest window of visible airglow and artificial sky lines, between Hg at 435 nm and NI at 524 nm (see Figure 2 in [3]). The exposure time was set to 3 mins, and the temperature of the CCD chip was kept at $-29°C$. In order to check the dark and readout noise during the observations, 53 dark frames were taken throughout the night. The frames for the flat fielding were taken using the same instrument inside the integrating sphere at the National Institute of Polar Research (NIPR), Tokyo, Japan.

The observed frames allow us to estimate the optical depth $\tau(z)$ of atmospheric extinction via the photometry of standard stars, by changing the zenith angle with time i.e. $\tau(z)=0.113/\cos z$. From the photometry of solar analog stars, we found that 1 ADU in our system is equal to $2.84 S_{10\odot}$, where $1 S_{10\odot}=1.28\times10^{-8}$ W m^{-2} sr^{-1} μm^{-1} at 500 nm.

2.1. The brightness of the predicted dust trail

The expected brightness of the dust trail is listed in Table 1. It is assumed that the dust trail can be considered as a cylindrical dust tube with a diameter 0.03 AU, existing along the orbit of Phaethon. Furthermore, the number density of meteoroids (dust grains) is constant inside the tube and the dust grains are spheres with uniform diameters. The brightness of possible dust trails at the observation time was calculated based on the scattering properties of dust grains shown given in [4], where the phase angle (Sun–dust grain–Earth angle) is $120° \sim 140°$. It was found [5] that an enhancement of brightness caused by the dust trails associated with comet 55P/Tempel-Tuttle in 1998. It was detected when the Earth entered into the dust trail of the Leonid meteoroid stream, and it was estimated that the number density of dust grains with a typical diameter of 10 μm was 1.5×10^{-10} m^{-3}.

Table 1
The expected brightness of dust trails associated with the orbit of asteroid 3200 Phaethon

Number density (m^{-3})	Dust diameter (μm)	Brightness ($S_{10\odot}$)
1.5×10^{-10}	10	0.1
1.5×10^{-12}	100	0.1
1.5×10^{-8}	10	10
1.5×10^{-10}	100	10

It is seen from Table 1 that the detection of dust trails associated with asteroid Phaethon would be difficult, assuming the meteoroids in the dust trail had a similar size and number density to those in the dust trail of 55P/Tempel-Tuttle. A pessimistic prediction also comes from the fact that we will see the dust trails of Phaethon nearly perpendicular to its orbit from outside of the dust tube, while the Leonid meteoroid stream was observed by [5] along its orbit from the inside the dust tube. However, we expect that the hypothetical dust trail of Phaethon may be denser in number density than that of normal meteoroid streams because of the very active appearance of Geminid meteor shower. In addition, it is assumed that the parent body (Phaethon) is in a dormant cometary phase, and consequently the meteoroids existing along its orbit have already spent a longer time in space after leaving the parent body in its active cometary phase. In that case, since larger grains have longer lifetimes in the interplanetary space against the Poynting-Robertson effect, the size of meteoroids surviving near the orbit of Phaethon is expected to be larger. These conjectures lead us to an optimistic view of observations.

3. RESULTS

The sky region of the observations is shown in Figure 1, where the horizontal axis denotes the $\lambda - \lambda_\odot$ between 90° and 120°, and the vertical axis means the β between 0° and 20°. It was expected that the dust trails appear in $6° < \beta < 11°$ (150< pixels <275), as shown by the solid line in Figure 1. After removing other contributions to the sky brightness from the observed raw data, the zodiacal light was extracted by using the same method as described in [2]. Then we used Fourier filtering to enhance the structure with an angular scale between 0.7° and 5°. This filtering process removes the large scale structure like the smooth component of the zodiacal light.

We have performed the scan of signals along the line perpendicular to the orbit of Phaethon, and in order to enhance the significance of the signals, we summed up the scan results of brightness along the line parallel to the orbit (see Figure 2). In Figure 2, the horizontal axis denotes the separation distance from the orbit of Phaethon, and the vertical axis means the averaged brightness over the Fourier-filtered image from Figure 1. No significant enhancement of brightness larger that 1 $S_{10\odot}$ appears near the orbit of Phaethon. The scattering of the resulting data may be caused by noise, with a magnitude of about $0.6 S_{10\odot}$.

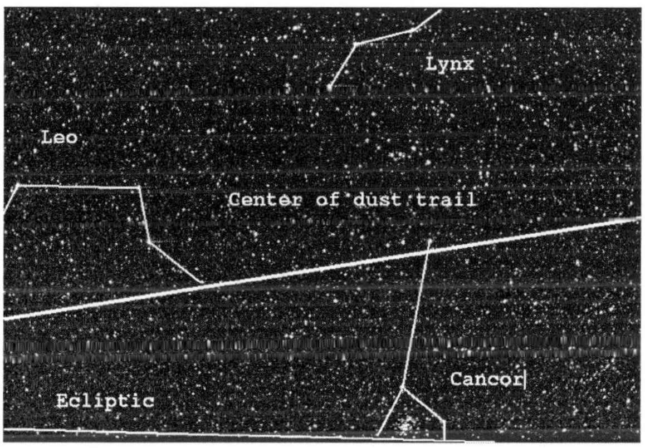

Figure 1. A sky region of observation in November 17th, 1999. A horizontal axis denotes the helioecliptic longitude of 90°–120°, and a vertical axis means the ecliptic latitude of 0°–20°. The orbit of asteroid 3200 Phaethon is illustrated by a thick solid line.

4. SUMMARY

No significant brightness enhancement associated with the orbit of asteroid 3200 Phaethon (potential parent body of the Geminid meteoroid stream) appears in our CCD photometry of the zodiacal light. It should be noted that our search was done for the trail structure with an angular scale of between 0.7° and 5.0° by applying a Fourier filtering

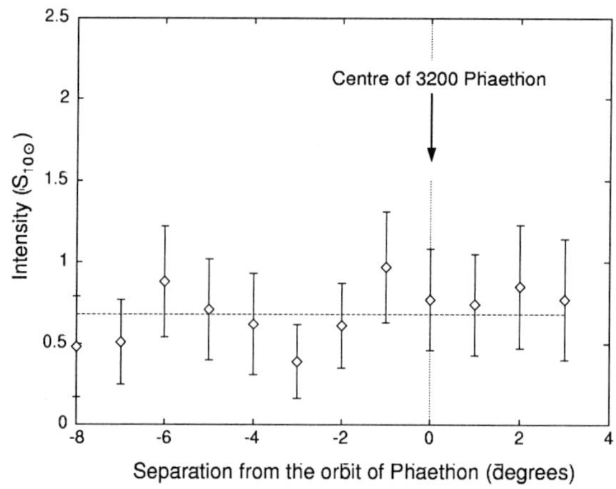

Figure 2. Brightness profile of Fourier-filtered image along the line perpendicular to the orbit of Phaethon, where the brightness is in unit of $S_{10\odot}$, and it is averaged over the direction parallel to the orbit of Phaethon.

process. If the dust trails had a narrow structure, less than our angular resolution of observation (0.62° or ~15 pixels) we would have been able to distinguish the structure from the background zodiacal light. Up to now, we have not yet determined the origin of faint structure detected at $\beta = 6°$ and $\lambda - \lambda_{\odot} = 78°$ in [2]. One of the reasons for our negative result may come from the difference in the solar elongation angle between 90°–120° in this observation and 78° in [2]. The brightness of dust trails may depend on the solar elongation angle. Further searches for dust trails associated with active meteor showers will be done using a new CCD instrument developed in [6].

REFERENCES

1. J.K. Davies, S.F. Green, B.C. Stewart, A.J. Meadows and H.H. Aumann, Nature 309 (1984) 315.
2. M. Ishiguro, R. Nakamura, Y. Fujii, K. Morishige, H. Yano, H. Yasuda, S. Yokogawa and T. Mukai, Astrophys. J. 511 (1999), 432.
3. T. Mukai and M. Ishiguro, this volume.
4. S.S. Hong, Astron. Astrophys. 146 (1985) 67.
5. R. Nakamura, Y. Fujii, M. Ishiguro, K. Morishige, S. Yokogawa, P. Jenniskens and T. Mukai, Astrophys. J. 540 (2000) 1172.
6. M. Ishiguro, T. Mukai, R. Nakamura and M. Ueno, this volume.
7. W.T. Reach, B.A. Franz and J.L. Weiland, Icarus, 127 (1997) 461.
8. I.P. Williams and Z. Wu, Mon. Not. R. astr. Soc. 262 (1993) 231.
9. B.Å.S. Gustafson, Astron. Astrophys. 225 (1989) 533.
10. H.U. Keller, M.L. Marconi and N. Thomas, Astron. Astrophys. 227 (1990) L1.

II Observations of the Zodiacal Light

CCD imaging of the zodiacal light

T. Mukai[a] and M. Ishiguro[a]

[a]Graduate School of Science and Technology, Kobe University,
Nada, Kobe 657-8501, Japan

We review recent developments in ground-based observations of the zodiacal light at visible wavelengths. These developments are largely due to the introduction of refrigerated charge coupled device (CCD) detectors. CCD images show not only the global structure of the zodiacal cloud, e.g. its symmetry plane, but also faint structures, such as dust bands and dust trails with a brightness of a few $S_{10\odot}$, which were previously only seen in satellite data. We also briefly mention CCD spectroscopy of the zodiacal light, and emphasize the importance of laboratory calibration of the detector system, and application of state-of-the-art data reduction methods in the analysis of such faint and diffuse objects.

1. INTRODUCTION

The IRAS, and later COBE, satellites initiated a new era in the zodiacal cloud sciences, starting in the 1980s (e.g. [1], [2], [3]). We have learned that the zodiacal cloud has a complex structure, including localized enhancement of interplanetary dust particles (IDPs) associated with asteroid families (dust bands), along cometary orbits (dust trails), and near planets (resonance rings or circumsolar rings). All of these new findings have come from satellite observations at infrared wavelengths. Since the late 1990s, new detector systems, such as cooled charge coupled device (CCD) cameras, have made it possible to detect faint structure in the zodiacal cloud at visible wavelengths from ground-based observation (e.g. [4], [5]). The resulting CCD images taken at high altitude sites have enabled us to study the structure of the zodiacal cloud in detail.

In this review, we will first mention the application of CCD detectors to the spectroscopy of the zodiacal light (ZL) in order to examine the dynamical behaviour of IDPs; we will then summarize recent results from CCD photometry of the ZL, including the Gegenschein.

2. CCD SPECTROSCOPY OF THE ZODIACAL LIGHT

Doppler shifts observed in sunlight scattered by IDPs (ZL) have been used to examine the dynamical behaviour of IDPs. Due to the low surface brightness of the ZL, it is difficult to measure the Doppler shifts reliably. Two different experimental strategies have been adopted in previous work, as noted in [6], namely using: (i) a scanning Fabry-Pérot interferometer to measure the profiles of Fraunhofer lines in the ZL (e.g. [7]), and (ii) a correlation mask radial velocity spectrometer to measure the line median or centre of gravity (e.g. [8]). It is demonstrated[6] that the line profiles in ZL spectra are important diagnostics of the dynamical behaviour of the IDPs. Consequently, they recommended observations "exploiting the optical advantages of the Fabry-Pérot instrument in association with the 2D/high quantum efficiency of the charge coupled device (CCD) detector".

CCD spectroscopy for Doppler shift measurements was performed in 1995 at Mt. Haleakala (altitude 3000 m, Hawaii) using the Fabry-Pérot étalon spectrometer developed by [9], with a CCD camera cooled to the temperature of liquid nitrogen (see [10]). Doppler shifts of the MgII line at 5183.6 Å in the morning ZL were detected. Although the resulting data, as shown in Figure 1, cited from [10], have a rather large scatter, those at a solar elongation angle, ϵ, smaller than 40° confirm the previous conclusion that IDPs have prograde orbits, i.e. the Doppler shifts in the morning ZL occur on the red side. The error bar in the Doppler shifts is estimated as about ± 0.1Å in Figure 1, while that in ϵ is $\pm 2°$. The Doppler shift values detected however, seem to be different from those reported by [8], and the data for $\epsilon \geq 40°$ appear on the violet side. No reasonable interpretation of these data has yet been presented.

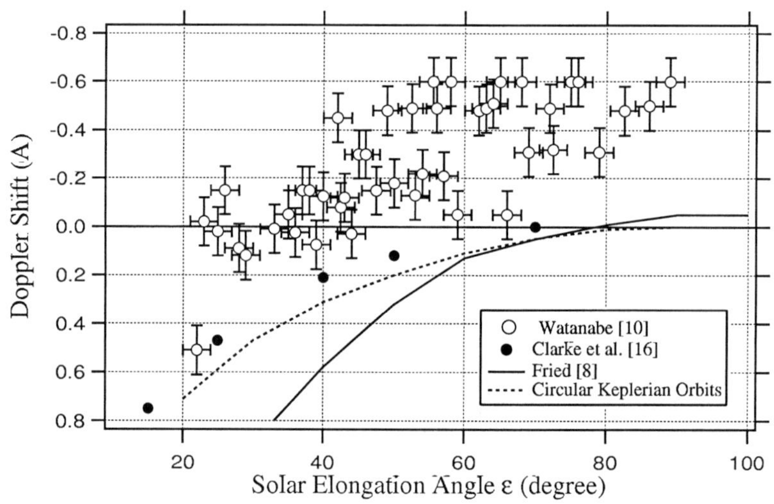

Figure 1. Doppler shifts of the zodiacal light as a function of solar elongation angle. The observed results are cited from [10] (morning ZL) and [8] (average of the morning and evening ZL). The computed results are cited from [6] for the morning ZL, and for a single grain in a circular Keplerian orbit.

It is generally realized that it is not easy to estimate the dynamical behaviour of IDPs from observed Doppler shifts. Each observed spectrum results from the combined scattering of sunlight by myriads of dust particles along a line of sight. These dust particles have different radial velocities relative to the observer, and also relative to the Sun. Consequently, we detect a mixture of the Doppler shifts caused by different particles located at different heliocentric distances. Thus, the observed shifts are complex functions of the number density and spatial density of the IDPs, as well as of their scattering properties.

It was anticipated that CCD spectroscopy of the ZL might yield significant data on line profiles in the ZL spectra, allowing this complex situation to be disentangled. However, the spectra are extremely faint, especially at larger ϵ, as shown by [10], and consequently very noisy, which prevents the line profiles from being studied in any detail. To measure Doppler shifts in the ZL, substantial improvement will be required in the next generation of CCD spectroscopic instruments.

3. CCD Photometry of the Zodiacal Light

To obtain a reliable 'snapshot' of faint and diffuse objects with a cooled CCD camera, we have to allow for many problems with the stability of the detection system, i.e. the linearity of the sensitivity of the CCD chips in the weak intensity region, dark current, read-out noise, and flat fielding (see e.g. [5], [11]).

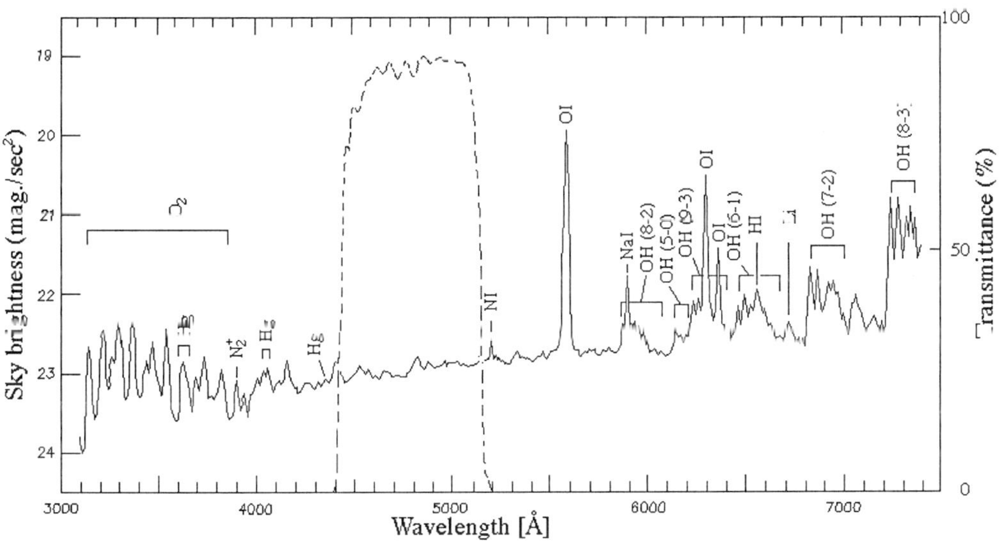

Figure 2. The transmission spectrum of the filter used from the 1998 observations (dotted curve), where the emission spectrum of the airglow came from the CFHT Observers' Manual for the average zenith sky brightness.

Previously, we used an illuminated sheet of white acrylic plastic and/or a milk tank to take flat field images (reference frames). Although this method provided reasonable reference frames, the introduction of reference frames taken inside an integrating sphere has dramatically improved the quality of reference frames for flat fielding. The integrating sphere can supply uniform radiation with a controlled intensity to the CCD camera of interest, where the error in the absolute flux of the calibration radiation is less than 5%. This technique, coupled with applying a proper filter to reduce the contamination of airglow emission (see Figure 2), was the key improvement that led to success in finding the localized faint structures in the zodiacal cloud that are described below.

3.1. SMOOTH COMPONENT

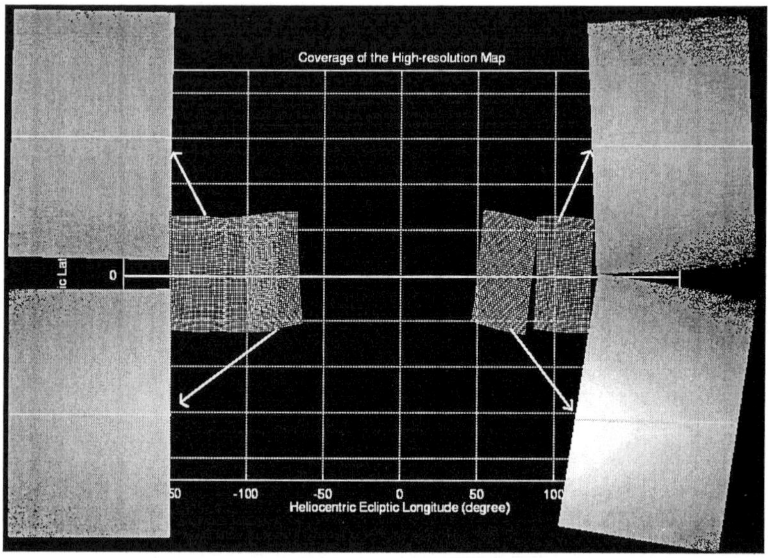

Figure 3. The mosaic of the zodiacal light observed in December 15 (evening; right images) and 16 (morning; left images), 1998 at Mauna Kea (4200 m, Hawaii).

Figure 3 shows a mosaic of four pieces of our CCD images of the ZL, obtained at Mauna Kea (4200 m, Hawaii) on December 15 (evening) and 16 (morning; local time), 1998 using a cooled CCD camera. The reference report on the diffuse night sky brightness [12] was a useful guide to excluding other contributions to the sky brightness from our observed data. A recent analysis [13] of the smooth component of the evening ZL shows that the inclination of the plane of symmetry of the ZL is close to i=2.2°±0.9° and the longitude of its ascending node is Ω=53°±7°. There is a large discrepancy in Ω between this value and the 96°±15° deduced from photometry of the ZL [14] (DLR), while i is close to DLR's value of 1.5°±0.4°. Since DLR's result was based on data averaged over a fairly long time period, whereas a 'snapshot' of the zodiacal cloud taken in the evening direction in December was used in [13], the difference in Ω may suggest that the symmetry plane varies with the season.

To obtain a reliable 3-D model for the zodiacal cloud, we have to compare the observed 2-D brightness distribution with that calculated along a line of sight based on a cloud model. A sophisticated method for model calculation was presented in [15], and Yoshishita et al. [13] similarly addressed model fitting. The absolute intensity of the airglow in the zenith direction $I_{AG}(zenith)$ is a key parameter in estimating the airglow brightness at a certain zenith distance based on the van Rhijn function (see [12]). The optimized model[13] yields $I_{AG}(zenith) = 41S_{10\odot}$, while $I_{AG}(zenith) = 100S_{10\odot}$ was estimated in [5]. This reduction of $I_{AG}(zenith)$ in [13], compared with [5], arose from the effective removal of airglow lines in new filter shown in Figure 2, whereas the old filter used in [5] has an effective wavelength of 440±60 nm. One of the scientific motivations at Kobe University for building a new detection system is to establish the absolute values of the ZL and the airglow, as reported in [11].

An advantage of CCD photometry of the ZL is that we can take a 'snapshot' of the zodiacal cloud at any season. From comparison of the previous observations noted above, we suspect that the zodiacal cloud cannot be represented by a simple structure with a single plane of symmetry. We will continue to take 'snapshots' of the zodiacal cloud in different seasons to examine its spatial structure in detail.

3.2. GEGENSCHEIN

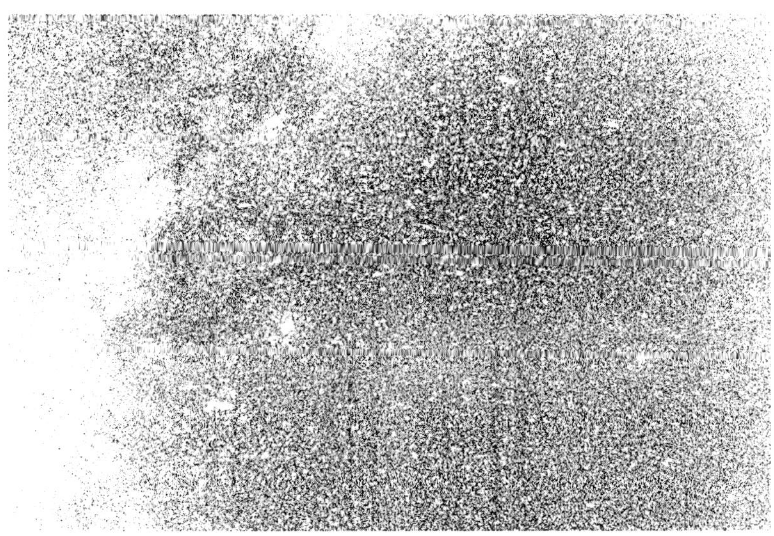

Figure 4. A CCD image of the Gegenschein, taken on 1st November 1997 at Hale Pohaku (2800 m, Hawaii) with a 10 min exposure time and a field of view of 101°×68°. The image was taken at the highest altitude of the Gegenschein (near the center of the frame). Our Galaxy is located on the left side, and airglow and atmospheric light at lower altitudes can be seen near the edge of the frame.

The Gegenschein is a faint patch of the ZL seen near the anti-solar direction. A comprehensive review of the observational history of the Gegenschein in the pre-CCD photometry era was given in [16]. The advantages of the CCD instrument are apparent when taking images of a faint and diffuse object, such as the Gegenschein. That is, CCD photometry of the Gegenschein can be obtained using shorter exposures than are required for conventional photography: compare the 50 s exposure of [4] or the 20 min one of [17] using CCD instruments, to the 1 hour exposure of [18] which required a photographic plate. In photoelectric observations of the Gegenschein, as reported by [19], the 2D-brightness distribution of the Gegenschein was deduced from a combination of data measured at different points, e.g. 5° intervals in ecliptic longitude and 3° intervals in ecliptic latitude, whereas CCD photometry can image a larger region of the sky in one exposure, i.e. 46°×31° [4], 98°×66° [17] (see 101°×68° in Figure 4).

The position of maximum brightness of the Gegenschein provides useful information on the spatial distribution of the zodiacal cloud beyond 1 AU. It is generally believed that the latitude of the Gegenschein shows a seasonal variation. Figure 5, cited from [17], shows considerable scatter, suggesting that the symmetry plane of the zodiacal cloud cannot be defined by one simple plane, such as the invariant plane of the Solar System. Furthermore, the existence of dust bands found in [5] in the region of the Gegenschein, described later, may cause such scattering of the observation results. Searches for seasonal variation in the position, size, shape and brightness distribution of the Gegenschein will be important scientific objectives for the new CCD instrument[11].

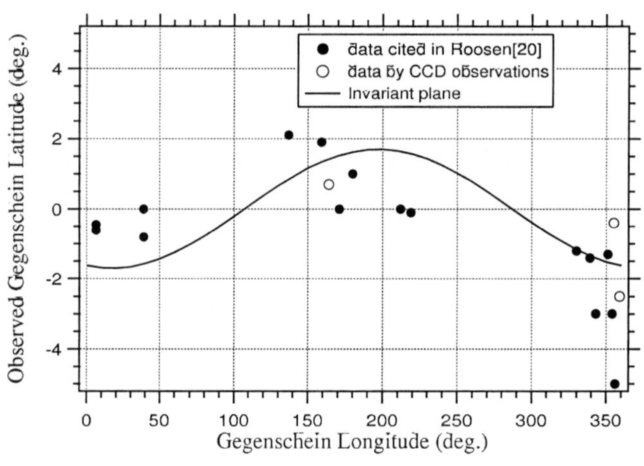

Figure 5. The position of the maximum brightness of the Gegenschein cited from [17].

3.3. LOCALIZED STRUCTURE

The inhomogeneous structure of the zodiacal cloud was initially revealed by applying the Fourier filtering method (see e.g. [3]) to data acquired by infrared satellites. A question to be addressed is how such structure can be found using ground-based instruments operating at visible wavelengths?

3.3.1. Dust bands

As shown in Figure 6, the surface brightness profile of the Fourier-filtered morning zodiacal light demonstrates the existence of dust bands at visible wavelengths. The brightness of the dust bands detected at a solar elongation angle of 78° and an ecliptic latitude of 0° (the Themis/Koronis dust bands) is about $5S_{10\odot}$ which corresponds to 2% of the surface brightness of the ZL. The ability to detect such a faint feature was attained when fluctuation of the sky brightness across the CCD was reduced to less than $0.99S_{10\odot}$.

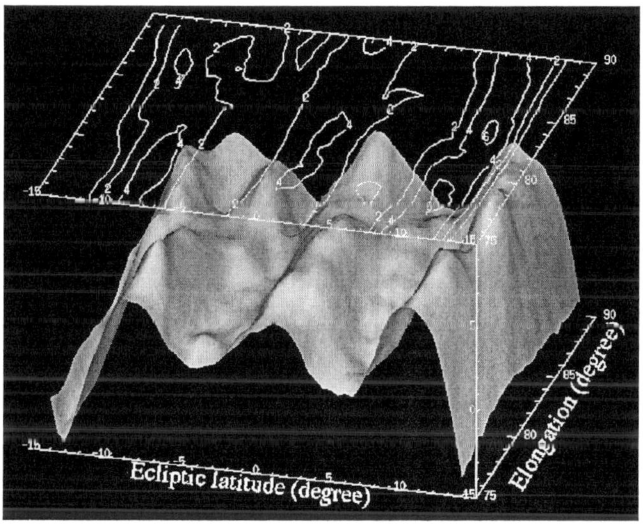

Figure 6. Three-dimensional profile of the Fourier-filtered profile, where the smooth component of the ZL has been removed. The isophotomap in the upper plane denotes the excess brightness in units of $S_{10\odot}$ [5].

Several undefined dust bands were reported by [3], based on analysis of infrared satellite data. To find the corresponding features at visible wavelengths, 'snapshots' of the ZL taken in different seasons covering a wide range of solar elongation angles will become a powerful tool. The limitations of the IRAS data were noted in [21], e.g. limited coverage in elongation angle ϵ of $60°\leq\epsilon\leq120°$. DIRBE shifted the coverage to $64°<\epsilon<124°$. In principle, however, CCD photometry can take images of the ZL in the range $30°\leq\epsilon\leq180°$ from the ground. Specifically, an extension of the dust bands in the region of the Gegenschein was discovered in [5]. We will continue to elucidate the ϵ-dependence of the brightness of the dust bands in order to study their origin and dynamical evolution over time.

3.3.2. Dust Trails

The results of a survey of cometary dust trails based on IRAS data [22] have found eight trails associated with known short-period comets. It is hard to recognize any slight enhancement of the IDPs along a cometary orbit at visible wavelengths, because less enhanced dust particles in the foreground/background contribute to the scattered light and mask the dust trails. This effect arises because the brightness of scattered light depends on the Sun–target–observer angle (the phase angle α), as well as on the number density of the IDPs. The negative result of a search for a dust trail along the orbit of asteroid 3200 Phaethon [23] may be caused by the observed condition of less efficiency in the scattering function observed in α=120°$\sim$140°.

A meteor shower is believed to be when the Earth is passing through a dust tube associated with a dust trail. If we can see the sky near the location of the radiant of the meteor shower at this time, it should be possible to see enhancement of dust grains inside the tube. On 17th November 1998, CCD photometry of the Leonid meteoroid stream region was performed [24]. Thanks to a CCD instrument similar to that used by [5], they found an excess of brightness about (2$\sim$3)% above the background ZL in the direction of the dust trail of comet 55P/Tempel-Tuttle, the parent of the Leonid meteoroid stream. This result was the first detection of dust trails at visible wavelengths, and from inside a dust trail.

4. SUMMARY

We have shown that the CCD instrument is a powerful tool for detecting faint and diffuse objects in the night sky. Some of the CCD images in our gallery are presented here with their scientific interpretation (for more detail see zodi.planet.sci.kobe-u.ac.jp/). Finally, we would like to emphasize that these images were taken using a portable and inexpensive detector system. The total weight of the CCD detector system is less than 10 kg, and the cost of the primary CCD camera was about 10,000 US dollars. We hope that some readers are interested in these CCD images and will join us in finding new features in zodiacal clouds by using their own CCD instruments.

We thank J.F. James, R. Nakamura, T. Watanabe, S. Takahashi, Y. Fujii, C. Yoshishita, and S. Urakawa for their collabolations in observations and data analyses.

REFERENCES

1. F.J. Low et al. Astrophys. J. 278 (1984 L19.
2. M.V. Sykes, L.A. Lebofsky, D.M. Hunten and F.J. Low, Science 232 (1986) 1115.
3. W.T. Reach, B.A. Franz and J.L. Weiland, Icarus 127 (1997) 461.
4. J.F. James, T. Mukai, T. Watanabe, M. Ishiguro and R. Nakamura, Mon. Not. R. astr. Soc. 288 (1997) 1022.

5. M. Ishiguro, R. Nakamura, Y. Fujii, K. Morishige, H. Yano, H. Yasuda, S. Yokogawa and T. Mukai, Astrophys. J. 511 (1999) 432.
6. D. Clarke, S.A. Matthews, C.G. Mundell and A.S. Weir, A&A 308 (1996) 273.
7. J.F. James and M.J. Smeethe Nature 227 (1970) 588.
8. J.W. Fried, A&A 68 (1978) 259.
9. J.F. James, Mon. Not. R. astr. Soc. 280 (1996) 1055.
10. T. Watanabe, Analysis of Doppler Shifts observed in the Zodiacal Light, Ph.D. thesis, Kobe University, 1997.
11. M. Ishiguro, T. Mukai, R. Nakamura and M. Ueno (2000) this volume.
12. Ch. Leinert, S. Bowyer, L. Haikala, M. Hanner, M.G. Hauser, A.C. Levasseur-Regourd, I. Mann, K. Mattila, W.T. Reach, W. Schlosser, J. Staude, G.N. Toller, J.L. Weiland, J.L. Weinberg and A.N. Witt, A&A Suppl. 127 (1998) 1.
13. C. Yoshishita, M. Ishiguro, T. Mukai and R. Nakamura (2000) this volume.
14. R. Dumont and A.C. Levasseur-Regourd A&A 64 (1978), 9.
15. T. Kelsall, J.L. Weiland, B.A. Franz, W.T. Reach, R.G. Arendt, E. Dwek, H.T. Freudenreich, M.G. Hauser, S.H. Moseley, N.P. Odegard, R.F. Silverberg and E.L. Wright, Astrophys. J. 508 (1998) 44.
16. R.G. Roosen, Rev. Geophys. & Space Phys. 9 (1971) 275.
17. M. Ishiguro, H. Fukushima, D. Kinoshita, T. Mukai, R. Nakamura, J. Watanabe, T. Watanabe and J.F. James, Earth, Planets & Space 50 (1998) 477.
18. S. Suyama, A&A 52 (1976) 145.
19. H. Tanabe, Publ. Astron. Soc. Japan 17 (1965) 339.
20. R.G. Roosen, Icarus 13 (1970) 523.
21. S.F. Dermott, S. Jayaraman, Y.L. Xu, K. Grogan and B.Å.S. Gustafson, AIP 348 (1996) 25.
22. M.V. Sykes and R.G. Walker Icarus 95 (1992) 180.
23. S. Urakawa, S. Takahashi, Y. Fujii, M. Ishiguro, T. Mukai and R. Nakamura (2000) this volume.
24. R. Nakamura, Y. Fujii, M. Ishiguro, K. Morishige, S. Yokogawa, P. Jenniskens and T. Mukai, Astrophys. J. 540 (2000) 1172.

WIZARD: New observation system of zodiacal light in Kobe University

M.Ishiguro[a] T.Mukai[b], R.Nakamura[c], F.Usui, and M.Ueno[d]

[a]The Institute of Space and Astronautical Science(ISAS), Yoshinodai 3-1-1, Sagamihara, Kanagawa 229-8510, Japan

[b]Graduate School of Science and Technology, Kobe University, Rokko-dai-cho 1-1,Nada, Kobe 657-8501, Japan

[c]National Space Development Agency of Japan (NASDA), Harumi 1-8-10, Chuou-ku, Tokyo 104-6023, Japan

[d]Graduate School of Arts and Sciences, University of Tokyo, Komaba 3-8-1, Meguro, Tokyo 153-8902, Japan

We describe a new system(WIZARD[1]) for zodiacal light observations, developed by a group in Kobe University, Japan. Since the zodiacal light is faint and widespread all over the sky, the system consists of a very sensitive CCD camera with a quantum efficiency of 90% at 500 nm and a wide angle lens with a FOV of $92° \times 46°$. WIZARD will be able to measure the absolute brightness of diffuse sky in visible wavelengths. The zodiacal component will be separated from the integrated star light, airglow continuum and the scattered light in the atmosphere, in the data reduction procedure. We report the design of WIZARD and the expected performance[2].

1. INTRODUCTION

The zodiacal light is the diffuse sunlight scattered by interplanetary dust particles. It is a difficult task to do high quality photometry of zodiacal light from a ground-based observatory because of its faintness as well as the contamination of diffuse light sources. Active ground-based studies were done in the 1960's and 1970's by using a photo-multiplier attached to a telescope on a high altitude mountain (see e.g. [1]). Their efforts have provided us with an overview of the zodiacal light, but they have also revealed limits of the ground-based observation with photomultiplier mounted on a telescope i.e. low spatial resolution and enormous observation time, compared with the spaceborne observations (see e.g. [2]).

About 20 years later, the high sensitivity and imaging capability of the Charge Coupled Device (CCD) has enabled us to obtain a 'snapshot' of diffuse faint objects with a portable

[1]Wide-field Imager of Zodiacal light with ARray Detector

[2]We obtained the first image from WIZARD in 2001 at Mauna Kea(4200 m, Hawaii) in collaboration with the SUBARU Telescope. In this paper, we describe the design and expected performance of WIZARD in this developing phase.

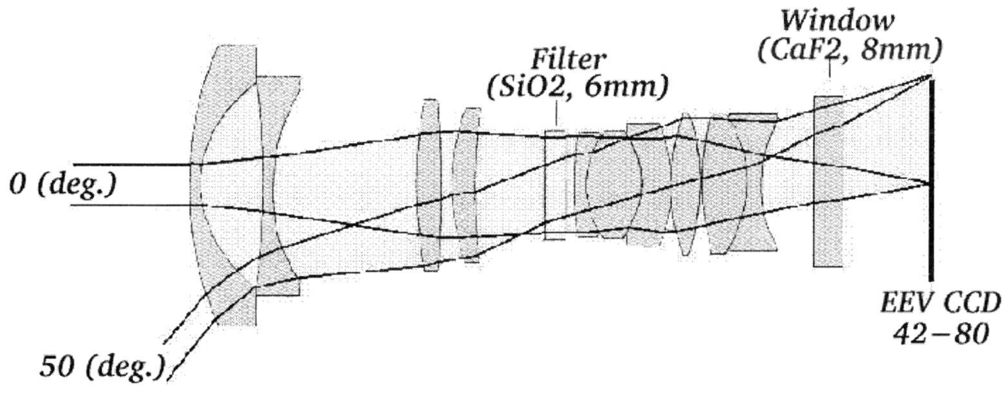

Figure 1. Plan for the lens unit. Rays at the center and edge of frame are drawn.

Table 1
Comparison of the observation systems

	New Instruments	Previous Instruments*
f (mm)(F)	32.5(2.8)	24.0(2.8)
QE(%) at 500 nm	90	20
FOV(°)	92 × 46	32 × 21
Spatial resolution (arcmin pixel^{-1})	1.35	2.50
Peak wavelength (nm)	480	440
Uncertainty of the zero point (e^-)	< 5	10–100

*cited from [4]

and inexpensive system [3]. It appeared however, that there were many problems in a CCD's properties i.e. dark current, read-out noise, flat fielding, and so on. By taking flat field frames inside a integrating sphere, we have succeeded to find the dust bands from the ground [4]. It is known that there still remains a problem of how to reduce the noise originating from the CCD itself. In this paper, we will report the development of new system with much smaller instrumental noise.

2. DESCRIPTION OF THE SYSTEM

2.1. CCD camera head

As described above, the stability of the CCD camera is a critical factor in measuring the absolute brightness. A fluctuation in the temperature of the CCD chip will cause uncertainty in the zero point of the photometric system, and its sensitivity.

We have employed a liquid-N_2 cooled CCD chip (EEV CCD 42-80) with a compact dewar ($\sim$ 1 kg) for WIZARD. The chip has a huge imaging area (27.6×55.3 mm), 2048×4096 pixels, as well as high sensitivity (90% at 500 nm). Since our previous CCD chip was cooled electronically, that system was consequently subject to the unclear temperature

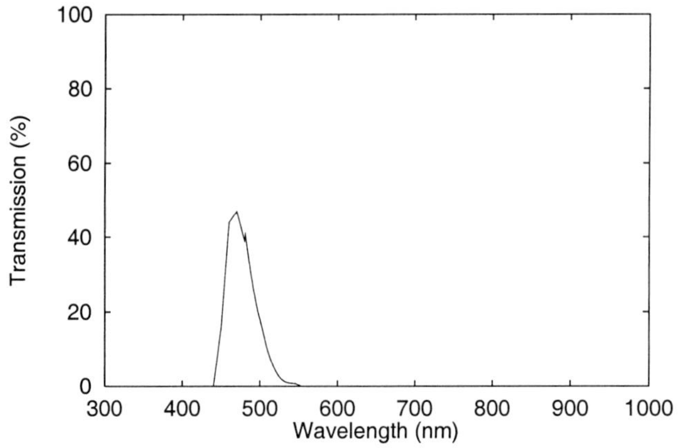

Figure 2. Transmission of the filter. The peak wavelength and FWHM are 480 nm and ~45 nm respectively.

variation. Dark noise of the new chip will be negligible. Also, the readout noise will be significantly reduced by adopting an excellent device control system; COGITO3 [5]. At 190 K, measured readout noise and dark current is 20 e^- and $\ll$1 e^- respectively.

2.2. OPTICS

We used a small F number in a design of lens unit (where F is the ratio of focal length to aperture) because an observed flux of diffuse light source is proportional to F^{-2}. Moreover, it is preferable to cover the sky from the zenith to the horizon in the same frame in order to estimate the brightness of atmospheric diffuse light at the same time. The lens is developed by the Genesia Corporation (Figure 1). Its F value and a focal length f are 2.8 and 32.5 mm respectively. With the EEV CCD 42-80, it gives a FOV of 92°×46°. The optical filter is designed to avoid the prominent airglow emissions (see Figure 2). Both of the optical filter and shutter are set at pupil position. Due to an oblique path from off-optical axis, the wavelength of light transmitted through the interference filter is shifted toward a shorter wavelength (e.g. roughly 10 nm for the 50 degree field position). To avoid this effect, the filter is made of colored glass. The whole lens system is mounted on the thick honeycomb plate and supported by the shutter and filter box. The dewar is also mounted on the plate (see Figure 3). This honeycomb plate is fixed on an equatorial mount, which is useful to track the stars in the region between 48° N and 48° S terrestrial latitude.

3. ANTICIPATED PERFORMANCE AND FUTURE PLANS

In this section, we estimate an expected count of zodiacal light observed by WIZARD based on the table provided in [6]. The table represents an annually averaged brightness

Figure 3. Side view of the CCD system.

with low spatial resolution. Kelsall et al. [2] reported a detail model of the zodiacal cloud based on COBE infrared observations. We assume the dust distribution of the visible zodiacal light is the same as that of the infrared emission, and employ this model as a visible zodiacal cloud model. We use a volume scattering phase function given by Hong [7], and adjust the absolute brightness referring to Levasseur-Regourd and Dumont [6]. As a result, we can simulate not only the absolute brightness in the visible band, but also the minute spatial structures, which are absent in the original table.

Taking into account the photon and read-out noise, we estimate an observed intensity, which includes the zodiacal light, by the new CCD system at Mauna Kea, Hawaii (4200 m). An optical depth for the diffuse light is assumed to be 0.10, and the brightness of the airglow and scattered light by the Earth's atmosphere are assumed to be 30 and 12 $S_{10\odot}$ respectively, at the zenith. These are the typical values deduced from our previous observations. The brightness of airglow and scattered light are extrapolated by the van Rhijn function and Dumont's formula (see e.g. [4]). Integrated star brightness is assumed to be $20S_{10\odot}$, which is independent of the galactic coordinate. The observed time is set to the end of astronomical twilight (19:08 HST on December 22).

The dots in Figure 4 show the anticipated brightness of the night sky observed by the new CCD system with an exposure time of 5 min. It should be noted that these are the results of a simulation with 1.35'×1.35' spatial resolution. In the case where the pixels are combined to give same resolution as IRAS ZOHF (20'×20'), the noise was reduced to $\sim 1/\sqrt{220}$ (lines in Figure 4). It is apparent that even a five minute exposure will allow us to detect asteroidal dust bands near the ecliptic plane (bump structures around $\beta \sim -1°$) and the north-south asymmetry of zodiacal light brightness. This new observation system will also yield information on the brightness of airglow

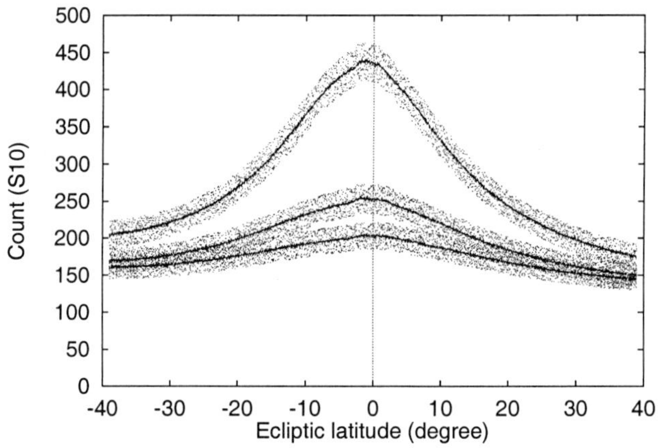

Figure 4. Scattered dots denote the expected intensity observed by the new CCD system. Each line corresponds to the combined data with 20'×20' resolution.

continuum resulted from chemiluminescence of NO_2 [8], and atmospheric diffuse light produced mainly by the scattering from byproduct aerosol particles [9].

Acknowledgements. First light observations by WIZARD were supported by a colleague from the SUBARU telescope. We thank Dr. M. Nakagiri, K. Sekiguchi and A. Miyashita (NAOJ) for help with our observations at Mauna Kea. We also thank Dr. N. Takeyama (Genesia Corporation) and S.M. Kwon for useful comments and discussion.

REFERENCES

1. R. Dumont, Astron. Astrophys. 38 (1975) 405.
2. T. Kelsall, J.L. Weiland, B.A. Franz, W.T. Reach, R.G. Arendt, E. Dwek, H.T. Freudenreich, M.G. Hauser, S.H. Moseley, N.P. Odegard, R.F. Silverberg and E.L. Wright, Astrophys. J. 508 (1998) 44.
3. J.F. James, T. Mukai, T. Watanabe, M. Ishiguro and R. Nakamura, Mon. Not. R. astr. Soc. 288 (1997) 1022.
4. M. Ishiguro, R. Nakamura, Y. Fujii, K. Morishige, H. Yano, H. Yasuda, S. Yokogawa and T. Mukai, Astrophys. J. 511 (1999) 432.
5. M. Ueno and T. Wada, COGITO-3; A flexible imaging device control system, in prep.
6. A.C. Levasseur-Regourd and R. Dumont, Astron. Astrophys. 84 (1980) 277.
7. S.S. Hong, Astron. Astrophys. 146 (1985) 67.
8. S.B. Mende, G.R. Swenson, S.P. Geller, R.A. Viereck, E. Murad and C.P. Pike, J. Geophys. Res. 98 (1993) 19117.
9. S.S. Hong, S.M. Kwon, Y.-S. Park and C. Park, Earth Planets Space 50 (1998) 487.

Brightness distribution of zodiacal light observed by a cooled CCD camera at Mauna Kea

C. Yoshishita[a], M. Ishiguro[a], T. Mukai[a] and R. Nakamura[b]

[a]Graduate School of Science and Technology, Kobe University, Kobe 657-8501, Japan

[b]Earth Observation Research Center, NASDA, Tokyo 106-0032, Japan

We performed the photometric observations of the zodiacal light at Mauna Kea (4200 m, Hawaii) during $5^h14^m - 7^h35^m(UT)$ in December 16, 1998 by using a cooled CCD camera at an effective wavelength of about 500 nm (439-524 nm) with a fish-eye lens. Focusing on the smooth component of the zodiacal light, we have derived the inclination $i = 2°.2 \pm 0°.9$ and the longitude of ascending node $\Omega = 53° \pm 7°$ of the symmetry plane. A significant discrepancy in Ω between our new result and the previous reported values may suggest that the symmetry plane of zodiacal cloud cannot be fitted by a simple single plane. This suggestion comes from a fact that we used one snapshot taken in December, while the previous values came from the data averaged over the observations in a different season.

1. INTRODUCTION

The zodiacal light is sunlight scattered by the cloud of interplanetary dust particles. The plane of symmetry of zodiacal light contains the information for the spatial distribution of the zodiacal dust cloud, and hence the dynamical evolution of the interplanetary dust particles. The previous observations of zodiacal cloud in the inner solar system [1] show that the observed plane of symmetry of zodiacal light is close to the orbital plane of Venus ($i = 3°.4, \Omega = 76°.7$), while those outside a solar distance of 1 AU indicate it is close to the invariable plane of the solar system ($i = 1°.6, \Omega = 107°$), where i and Ω are inclination and longitude of ascending node of the plane of symmetry, respectively. In Table 1, the values of i and Ω are summarized based on the previous observations. Figure 1 shows the geometry of the plane of symmetry, relative to the ecliptic plane, and the sky region of our observation.

Table 1
The i and Ω for the symmetry plane deduced from the visible observations.

method	$i(°)$	$\Omega(°)$	references
Helios 1 and 2	3.0 ± 0.3	87 ± 4	[1]
Gengenschein model	2.0 ± 0.5	100 ± 20	[2]
D2A spacecraft and Tenerife data	1.5 ± 0.4	96 ± 15	[3]
Modified fan model	2.2 ± 0.9	53 ± 7	This work

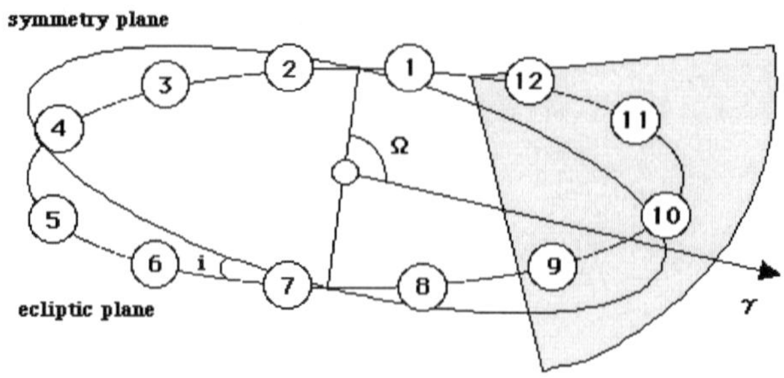

Figure 1. The circle mark in the center is the Sun. The numbers indicate the position of the Earth at the beginning of the respective month. The γ is vernal equinoctial direction, and the shaded fan-shape denotes the sky area of observation in the evening of December 16th.

2. OBSERVATIONS

We observed the evening zodiacal light at Mauna Kea ($4200m$, Hawaii) during $5^h14^m - 7^h35^m(UT)$ in December 16, 1998. The resulting images are shown in Figure 3 of [4]. We used a cooled CCD camera with a fish-eye lens(f=16mm,F=2.8). To reduce the contamination of the airglow, we used a special filter, which has no prominent airglow emission lines in its available wavelength region of 39–524 nm [4]. Our observations cover the range of helioecliptical longitude of $50° < \lambda - \lambda_\odot < 120°$, and ecliptic latitude of $-25° < \beta < 25°$. The field of view is $51° \times 34°$ and the angular resolution is 2.0' pixel^{-1} without a binning. The exposure time was set to 10 mins, and the temperature of the CCD chip was kept at $-30°$C.

3. DATA REDUCTION

First, we applied the basic data reduction procedures to the raw data, such as dark current subtraction and flat fielding. It is well-known that a frame taken by a fish-eye lens shows strong 'vignetting', that is, the detected intensity in the central part of the frame is brighter than that in the edge of frame. The vignetting, as well as the pixel-to-pixel variation in sensitivity, can be corrected by flat-fielding. The reference frames for flat-fielding were taken inside the integrating sphere, which can provide a uniform illumination on the wide FOV of our fish-eye lens. Furthermore, we remove the stars by using the computer program and eliminate the inhomogeneous structure of the zodiacal light, such as the dust bands existing around the ecliptic plane, by applying the Fourier-filtering procedure. Thanks to this technique, we got rid of the structure with angular scales between 5° and 20°, i.e. dust bands components.

The residual components inside the image frame, i.e. the sky brightness (I_{obs}) observed from the ground-based site, consists of the light from several different sources. Namely, they are the zodiacal light(I_{ZL}), the airglow from upper atmosphere(I_{AG}), the integrated starlight of unresolved stars(I_{ISL}), and the light scattered by Earth's atmosphere(I_{sca}). It is shown that

$$I_{obs} = (I_{ZL} + I_{AG} + I_{ISL})\exp(-\tau_{eff}(z)) + I_{sca} \quad (1)$$

where $\tau_{eff}(z)$ denotes the effective optical depth for the extinction of diffuse light source at zenith distance z. We have derived the optical depth $\tau(z)$=0.16/cos z for point source by the photometry of standard stars. Then, we assume $\tau_{eff}(z) \simeq 0.75\tau(z)$ [5]. The I_{ISL} is deduced from an interpolation of the Pioneer's data cited in [6], and the I_{sca} comes from [7]. The zenith angle dependence of I_{AG} is assumed from the van Rhijn function [8].

4. MODEL FITTING

To obtain the proper zodiacal cloud model, we compare to the brightness distribution $I(\lambda - \lambda_{\odot})$ calculated along the line of sight. That is,

$$I(\lambda - \lambda_{\odot}) = C \int_0^{\infty} N(r,\zeta)\Phi(\theta)\frac{1}{r^2}dl \tag{2}$$

$$\begin{aligned} N(r,\zeta) &= N_o r^{-\alpha} \exp(\beta' g^{\gamma}) \\ g &= \frac{\zeta^2}{2\mu} \; (for \; \zeta < \mu), \qquad \zeta - \frac{\mu}{2} \; (for \; \zeta \geq \mu) \end{aligned} \tag{3}$$

where the modified fan model is applied for the number density of zodiacal cloud $N(r,\zeta)$ [9] and for the scattering phase function $\Phi(\theta)$ defined in [10]. $\alpha = 1.2$ is assumed for the density power-law exponent. To avoid the edge effect derived from Fourier-filtering, the region of $55° < \lambda - \lambda_{\odot} < 115°$ and $-20° < \beta < 20°$ are gridironed by $5° \times 5°$ and the data at each crosspoint are used in the model fitting. The total of 111 data points are available. The eight variables are required for fitting the parameters in the models, i.e. i, Ω, a level of bias, scaling factor(C), three shape parameters(β',γ,μ) for the modified fan model, and the intensity of airglow emission at the zenith $I_{AG}(z = 0)$. We have determined such eight parameters by the simplex method so as to minimize the difference between our photometric data and the calculated brightness. The followings are the values of parameters of interest obtained here. That is, $i = 2°.2$, $\Omega = 53°$, $\beta' = 4.62$, $\gamma = 0.937$, $\mu = 0.216$ and $I_{AG}(z = 0) = 41\ S_{10\odot}$.

5. RESULTS AND SUMMARY

We have derived the brightness distribution of the smooth zodiacal light based on our observation by a cooled CCD camera at Mauna Kea in December 1998 in the sky region of $55° < \lambda - \lambda_{\odot} < 115°$ and $-20° < \beta < 20°$. As a result of the model fitting procedure, we have found that the brightness distribution of our observation can be explained reasonably by the zodiacal cloud model with a symmetry plane of $i = 2°.2$ and $\Omega = 53°$. Figure 2 shows the results of model fitting, where the observed data (the dots) fit well to the curve of model calculation at almost all cases of $\lambda - \lambda_{\odot}$, but some disagreement appears around the ecliptic latitude of $\beta = 0°$ in $\lambda - \lambda_{\odot} = 55°$. This may arise from a fact that the dust band components around the ecliptic plane are removed too much from the original photometric data. This disagreement causes, however, no serious effect in a determination of i and Ω.

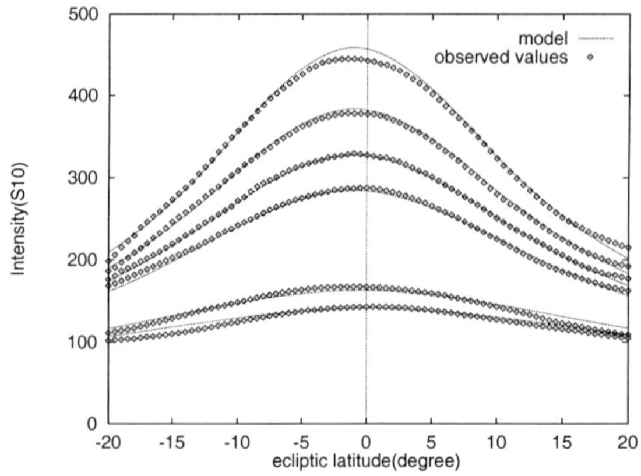

Figure 2. The result of model fitting in case of $i = 2^\circ.2$ and $\Omega = 53^\circ$. The six pairs from top to bottom are the profiles in the heliocentric longitude of $\lambda - \lambda_\odot = 55^\circ, 60^\circ, 65^\circ, 70^\circ, 100^\circ$ and 115°, respectively.

The resulting values of i and Ω are not completely compatible with the invariable plane($i = 1^\circ.6, \Omega = 107^\circ$) of the solar system and the orbital plane of Venus($i = 3^\circ.4, \Omega = 76^\circ.7$). They agree in i with the previous results deduced from the Helios data [1] and Tenerife data [3], but significantly differ in Ω. The previous results, e.g. [1] and [3], came from the data averaged over the observations in a different season, while we used a snapshot obtained in December. Therefore, this discrepancy in Ω may suggest a complex structure of the zodiacal clouds. Further observations from the ground-based site in a different season will reveal the spatial distribution of zodiacal cloud more detail.

REFERENCES

1. C. Leinert, M. Hanner, and I. Richter, Astron. Astrophys. 80 (1980) 328.
2. S.S. Hong and S.M. Kwon in Origin and Evlution of Interplanetary Dust, (eds. A.C. Levasseur-Regourd and H. Hasegawa) Kluwer, Dordrecht (1991) 147.
3. R. Dumont and A.C. Levasseur-Regourd, Astron. Astrophys. 64 (1978) 9.
4. T. Mukai and M. Ishiguro, this volume.
5. S.S. Hong, S.M. Kwon, Y.-S.Park and C.Park, Earth, Planets and Space 50 (1998) 487.
6. C. Leinert, S. Bowyer, L.K. Haikala, M.S. Hanner, M.G. Hauser, A.C. Levasseur-Regourd, I. Mann, K. Mattila, W.T. Reach, W. Schlosser, H.J. Staude, G.N. Toller, J.L. Weiland, J.L. Weinberg and A.N. Witt, Astron. Astrophys. Suppl. 127 (1998) 1.
7. H.J. Staude, Astron. Astrophys. 39 (1975) 325.
8. P.J. Van Rhijn, Astrophys. J. 50 (1919) 356.
9. T. Kelsall, J.L. Weiland, B.A. Franz, W.T. Reach, R.G. Arendt, E. Dwek, H.T. Freudenreich, M.G. Hauser, S.H. Moseley, N.P. Odegard, R.F. Silverberg and E.L. Wright, Astrophys. J. 508 (1998) 44.
10. S.S. Hong, Astron. Astrophys. 146 (1985) 67.

High spatial resolution distribution of the zodiacal light brightness

S.M. Kwon,[a] S.S. Hong[b] and J.L. Weinberg[c] *

[a]Dept. of Science Education, Kangwon National University, Chunchon 200-701, KOREA

[b]Astronomy Program, SEES, Seoul National University, Seoul 151-742, KOREA

[c]Space Astronomy Laboratory, Snellville, GA 30039, U. S. A.

By adopting a semi-empirical method for correcting the atmospheric diffuse light and using an improved technique for subtracting the discrete starlight, we have isolated the zodiacal light (ZL) surface brightness in ground-based, photo-polarimetric observations of the night sky. Advantages of this new reduction methodology compared to previous schemes are pointed out. A two-dimensional distribution of the ZL is presented with a spatial resolution better than 2 degrees over most of the sky that can be observed from the ground. The symmetry plane is found to have an inclination $i \simeq 2°$ and longitude of ascending node $\Omega \sim 80°$ by comparing the observed brightness distribution with that expected for a 3-dimensional cosine model.

1. INTRODUCTION

The zodiacal light (ZL) is well known to be sunlight scattered by interplanetary dust particles. Therefore, by observing the ZL one is able to probe the overall large scale properties of the interplanetary dust complex, including its 3-dimensional density structure.

Major contributions to the night sky brightness derive from four sources: BS – bright stars that are individually resolved by a given telescope, IS – integrated starlight and the diffuse Galactic light, ZL – zodiacal light, and AG – airglow emitted by the Earth atmosphere. Each of these is affected by both extinction and scattering by atmospheric constituents. The telescope's field of view intercepts not only the attenuated, directly transmitted brightnesses of the four sources but also that part of IS, ZL and AG that is scattered into the field of view from other parts of the sky (ADL - atmospheric diffuse light). Lack of accurate information about the BS distribution, difficulties in correcting for the diffusely-scattered light of extended astronomical sources by the Earth's atmosphere, and the changing and uncertain nature of the AG have made it very difficult to obtain an accurate, high spatial resolution map of the ZL brightness from ground observations.

A machine-readable star catalogue has been prepared that includes all stars resolvable by the telescope, thereby permitting accurate determination of their brightnesses in the telescope's reference system and removal of their brightness contributions, BS. Photo-polarimetric measurements of the IS by the Pioneer 10 and 11 space probes from beyond

*This work was supported by Korea Research Foundation Grant KRF-2000-DP-0441.

the asteroid belt have provided high resolution maps of diffuse starlight over most of the sky [1,2]. A semi-empirical reduction methodology has recently been developed [3] to make time-dependent corrections for the ADL. Combining these elements with atmospheric extinction measurements into a single reduction methodology, we have been able to isolate the ZL over extended areas of the sky.

2. OBSERVATIONAL DATA AND REDUCTION

From archives of night sky observations by Weinberg and Mann [4] at Mt. Haleakala, Hawaii, we selected observations from one night, 21/22 August 1968, to demonstrate this new reduction methodology and to derive a high spatial resolution map of ZL brightness. The telescope was used in an almucantar scan mode to repeatedly observe at 5080Å and 5300Å over a full 360° of azimuth at 8 zenith distances, at 5° interval from 45° to 80°. In this way, 11 sets of almucantar scans (88 scans in total) were obtained for each of the wavelengths. In our scheme of data reduction, parameters such as the zenith extinction optical depth τ_o, the telescope's effective field of view (FOV), and the calibration factor C are all simultaneously determined from the same set of data from which the ZL brightnesses are derived. This ensures internal consistency in the reduction procedure and also minimizes errors in the subtraction process.

By using 93 bright stars which are identified with distinct peaks in the scan profiles, values were determined for τ_o and C·FOV simultaneously. To take account of the changing atmospheric properties, we divided the 11 sets of data into 5 subgroups and determined τ_o for each group. The average values of the time-dependent τ_os are 0.183 and 0.173 for 5080Å and 5300Å, respectively, and the relative amplitude of the time variation amounts to about 10%. Details of the procedure can be found in Kwon et al. [5,6].

3. REMOVAL OF THE BS, IS, AND ADL COMPONENTS

Positional and photometric information have been assembled on 8372 stars resolved by the instrument that are brighter than 6.5 mag in the visual, and this information has been stored in a single database called STARSUB. The STARSUB Catalogue is searched for those bright stars which come into the FOV along each scan path, and the brightness profiles of these stars as seen in the telescope reference system are synthesized. We also synthesized the IS profiles using the aforementioned Pioneer 10 and 11 data set. The sum of the two profiles is compared graphically with the total observed profile. Synthesized profiles of IS+BS were constructed for each of the 176 scans by adjusting trial values for the zenith extinction optical depth τ_o and effective extinction optical depth τ_{eff} [3] until the comparison was satisfactory. Subtraction of the synthetic profile from the observed profile removes the directly-transmitted contributions of BS and IS from the observed brightness. But, in addition to ZL, the residual still contains the directly-transmitted AG and diffusely-scattered components of IS, ZL, and AG.

ADL is one of the most difficult components to remove in night sky observations, with numerous attempts but limited success [7,8,9 and others]. In this study we utilized an effective optical depth τ_{eff} and calculated the ADL brightness as a function of zenith distance using the quasi-diffusion method (QDM), which solves the problem of radiative transfer in an anisotropically scattering spherical atmosphere [3].

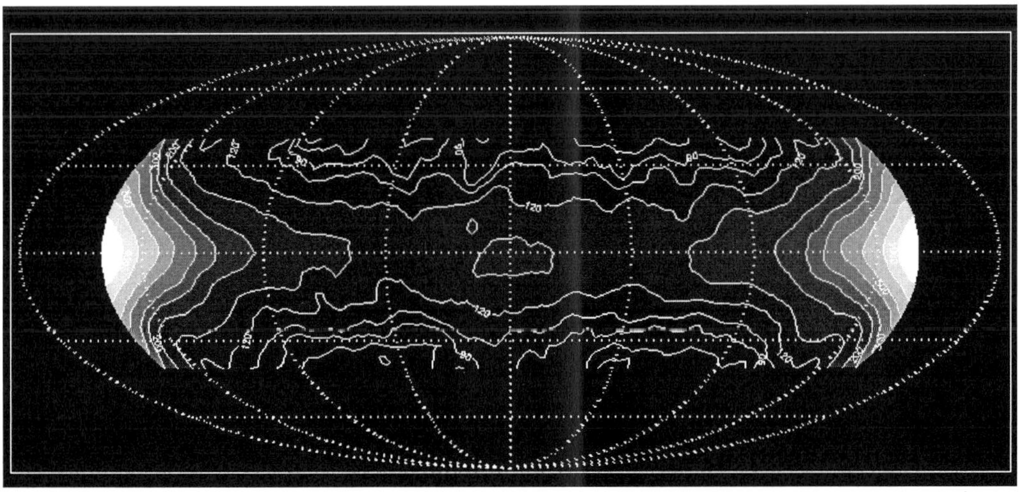

Figure 1. Isophotal contours of observed ZL brightness plotted in an Aitoff projection in the ecliptic coordinate system. Contour levels are in S_{10} brightness units 70, 80, 90, 100, 120, 150, 200, 250, 300, 400, 500, 700, and 1000. The ZL distribution clearly shows asymmetries between the northern and southern hemispheres (ecliptic latitudes) and between the morning and evening (east and west of the Sun) sides.

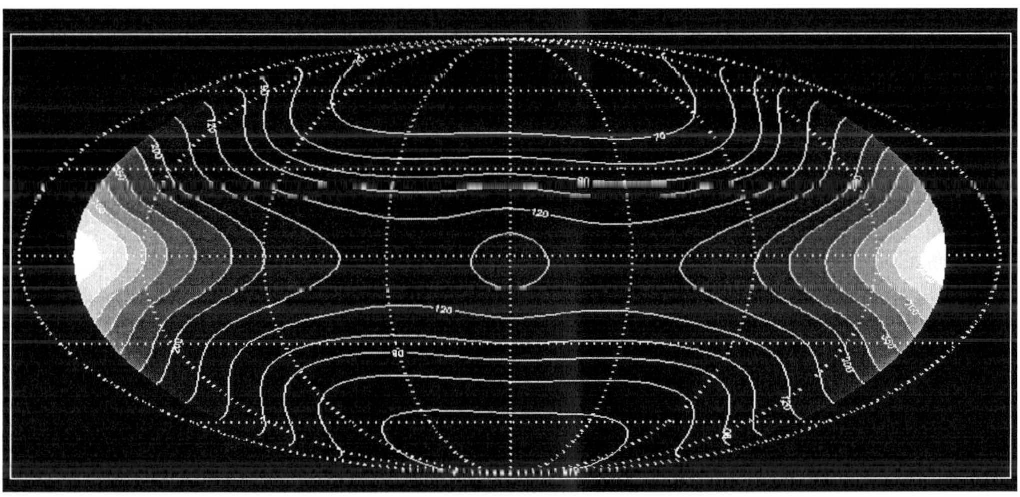

Figure 2. ZL brightness calculated using the 3-dimensional cosine model with the symmetry plane being placed at inclination angle i=2° and ascending node Ω=80°. Brightness contours are on the same level as in Figure 1. The model calculations agree with major observational details, especially with the southward shifts of the Gegenschein peak and morning ZL cone.

4. RESULT AND CONCLUSION

To further check for internal consistency in our reduction methodology, we compared ZL brightness maps at 5080Å and 5300Å and found excellent agreement between the two independently-reduced data sets. Therefore, we decided to combine the two data sets and produced in Figure 1 the merged isophotal contour map of ZL over the sky. This distribution clearly shows asymmetries between the morning ($\lambda - \lambda_\odot > 180°$) and evening ($\lambda - \lambda_\odot < 180°$) regions and between the north and south regions of the ecliptic. The distribution also shows small scale structure not seen in some earlier studies [e.g. 10].

We have also calculated ZL brightness using 3-dimensional optical models of zodiacal cloud [11]. By comparing the overall morphology of the isophotal contours of the observed and calculated ZL distributions, we found the symmetry plane inclination i to be $\sim 2°$ and the longitude of the ascending node Ω to be $\sim 80°$. As seen in Figure 2, use of these symmetry plane parameters with a cosine model shows excellent agreement with the observed distribution, reproducing the observed southward shifts of the peak Gegenschein and of the morning side ZL cone.

A new reduction methodology for photometric observations of the night sky radiation is summarized and illustrated, a methodology with the following advantages: 1) Since all calibration parameters are determined from the same set of observations, an internal consistency is assured. This also reduces errors in the resulting brightnesses to about 10% or less. 2) Calculations using the QDM code enable us to determine the zenith distance dependence of the ADL brightness, even near the horizon. This permits the use of near-horizon observations without the heretofore very large uncertainties associated with the reduction of such observations.

REFERENCES

1. G.N. Toller, Ph.D. Thesis, State University of New York at Stony Brook, 1981.
2. G.N. Toller, H. Tanabe and J.L. Weinberg, Astron. Astrophys. 188 (1987) 24.
3. S.S. Hong, S.M. Kwon, Y.-S. Park and C. Park, Earth, Planets and Space 50 (1998) 487.
4. J.L. Weinberg and H.M. Mann, in Proceedings of the Symposium on the Zodiacal Light and Interplanetary Medium (ed. J.L. Weinberg) (1967) 3.
5. S.M. Kwon, Ph.D. Thesis, Seoul National University, 1990.
6. S.M. Kwon, S.S. Hong, J.L. Weinberg and N.Y. Misconi, in Origin and Evolution of Interplanetary Dust (eds. A.C. Levasseur-Regourd and H. Hasegawa) Proc. of IAU Colloq. 126, Dordrecht, Reidel (1991) 179.
7. R. Dumont, Ann d'Astrophys. 28 (1965) 265.
8. A.C. Levasseur-Regourd and R. Dumont, Astron. Astrophys. 84 (1980), 277.
9. J.L. Weinberg, Ph.D. Thesis, University of Colorado, 1963.
10. S.S. Hong, N.Y. Misconi, M.H.H. van Dijk, J.L. Weinberg and G.N. Toller, in Properties and Interactions of Interplanetary Dust, (eds. R.H. Giese and P.L. Lamy) Proc. of IAU Colloq. 85, Dordrecht, Reidel (1985) 33.
11. S.M. Kwon and S.S. Hong, Earth, Planets and Space 50 (1998) 501.

Zodiacal light observations with the Infrared Space Observatory*

P. Ábrahám[ab], Ch. Leinert[a], and D. Lemke[a]

[a]Max–Planck–Institut für Astronomie, Königstuhl 17, D-69117 Heidelberg, Germany

[b]Konkoly Observatory of the Hungarian Academy of Sciences, P.O.Box 67, H-1525 Budapest, Hungary

ISO performed an extensive observing programme on the infrared zodiacal light, including multi-filter photometry of the global brightness distribution, observations of the asteroidal bands and cometary dust trails, and investigation of the small scale brightness fluctuation. The detailed determination of the spectral energy distribution of the zodiacal light, including mid-infrared spectrophotometry, opens the possibility for the separation of the main components of the infrared sky. The mid-infrared spectrum may provide information on the nature of the constituents, and on the size distribution of the interplanetary grains. Two years after the ISO mission we review the observations and summarize the results of ISOPHOT and ISOCAM.

1. INTRODUCTION

For optical astronomers 'zodiacal light' is a dim glow visible with the naked eye after sunset or before sunrise. This conspicuous phenomenon, first studied by Cassini in 1683, is caused by the scattering of sunlight by dust particles orbiting in the interplanetary space. For infrared astronomers, however, zodiacal light is the thermal re-radiation of sunlight absorbed by the same particles. The cloud formed by these particles, called the Interplanetary Dust Cloud (IDC), occupies the inner solar system extending out to at least the asteroidal belt. It contains 10^{16} – 10^{17} kg of dust, equivalent to the mass of a large comet. In spite of the low density of the IDC ($\approx 10^{-19}$ kg m^{-3}), in most directions the zodiacal light dominates the brightness of the infrared sky in the 3–70 μm wavelength range. The absorption and scattering of solar radiation decreases the orbital velocities of the particles due to the Poynting-Robertson effect, limiting the lifetime of individual grains to $10^4 - 10^5$ years. The most likely source to replenish the evaporated dust is the destruction of comets and asteroids. The basic observational facts on the zodiacal light and interplanetary dust are reviewed in [1,2].

Since the Earth is orbiting within the IDC, there are direct ways to study individual dust particles, such as optical and radio observations of meteors, collection of particles in the upper atmosphere or in the Antarctic, in-situ capture of dust by space probes,

*Based on observations with ISO, an ESA project with instruments funded by ESA member states (especially the P/I countries France, Germany, the Netherlands and the United Kingdom) with participation of ISAS and NASA.

and the study of lunar micro-craters [1]. To determine the large scale structure of the IDC, however, the analysis of the global brightness distribution of the zodiacal light is the only available tool. Although the observed zodiacal light is an integral along the line-of-sight, via repeated mapping of the zodiacal light at different orbital positions of the Earth, i.e. at different dates during the year, the 3-dimensional structure of the IDC can be determined. The brightness distribution of the zodiacal light contains also substructures (asteroidal bands, cometary trails, rings of resonantly trapped dust) which carry important information on the origin of the interplanetary dust particles. Finally, the shape of the spectral energy distribution of the zodiacal light gives information on the properties of the dust particles averaged over the whole IDC. In this respect the mid-infrared spectrum has a special importance because in the 5–15 μm range the spectral shape is very sensitive to dust properties, and in case individual spectral features were identified even the dust composition can be determined.

The zodiacal light, however, is only one component of the infrared sky brightness, and it has to be separated from the galactic cirrus emission and from the extragalactic background light (not to mention the possible instrumental straylight). The all-sky multi-filter photometric surveys performed by the IRAS satellite and by the DIRBE instrument on-board the COBE satellite have produced extensive databases of the brightness distribution of the infrared sky [3,4]. Based on their data, both the IRAS and the COBE/DIRBE teams developed methods to extract the zodiacal light from the total sky brightness [3–5]. The DIRBE data have special importance due to their high quality and due to the dedicated observing strategy of repeated mapping of a significant fraction of the sky every week during the 9 months cooled operational phase of the instrument. To extract the zodiacal light the DIRBE method used the fact, that the zodiacal light is the only component which exhibits annual variation due to the Earth's orbital motion. The results from the two satellites have changed the traditional view of a smooth and symmetric zodiacal light distribution to the picture of a structured cloud containing substructures and asymmetries (e.g [6]).

ESA's Infrared Space Observatory (ISO) [7], operated between Nov. 1995 and April 1998, was not designed for all-sky surveys. It was an observatory performing pointed observations of mainly point sources. However, the cold focal plane allowed to perform absolute sky brightness measurements, and the two and half year mission resulted in a large set of individual observations of the extended sky brightness at mid- and far-infrared wavelengths, in both photometric and spectrophotometric modes. The beam of $\leq 3'$ was considerably smaller than the beam of DIRBE ($42'$), and could efficiently avoid point sources and cirrus structure. ISO's good filter coverage in the 3–200 μm range as well as the possibility of performing mid-infrared spectrophotometry provide better spectral coverage than the previous satellites had done (recently the IRTS satellite could also observe the near- and mid-infrared low-resolution spectrum of the infrared zodiacal light [8]). The rejection of straylight from the Sun, Earth, and Moon has been proved to be excellent [9]. Via its absolute photometric flux calibration, ISO can give an independent determination of the absolute brightness level of the zodiacal light (between the IRAS and DIRBE absolute calibrations in the far-infrared there is an unresolved zero point difference [3]).

ISO had two focal plane instruments sensitive enough to measure the extended emission of the sky: the spectro-photopolarimeter ISOPHOT in the 2.5–240μm wavelength range [10], and the mid-infrared camera ISOCAM at 2.5–18μm [11]. The sky background turned out to be too faint for the on-board spectrometers SWS and LWS. The highest quality data were obtained by dedicated measurements of selected dark sky regions, but also off-source observations in the vicinity of point sources, the serendipitious survey of ISOPHOT [12], and the parallel mode observations of ISOCAM [13] provided valuable data. Most ISOCAM measurements of compact sources could also be used by taking pixels off the source. So far mainly the dedicated observations have been analysed, but we will set up a database of all suitable ISOPHOT off-source measurements. This database will also be useful for studies of the galactic cirrus and the extragalactic background light. Data from the ISOPHOT Serendipity Survey at 170μm and the ISOCAM parallel mode in the mid-infrared are reduced, but have not yet been used for zodiacal light studies.

In this contribution we discuss the ISO results at the present status of calibration as well as the expected future contribution from the new data. Since the authors are related to the ISOPHOT team, the main emphasis will be on the ISOPHOT results.

2. GLOBAL BRIGHTNESS DISTRIBUTION OF THE ZODIACAL LIGHT

The low sky coverage of ISO does not allow us to create an independent global model for the zodiacal light. Nevertheless, ISO can contribute to the study of the large scale structure of the zodiacal light in two ways. First, combining the new ISO measurements with the existing DIRBE all-sky maps, the ISO photometry can be used to cross-check and supplement the DIRBE data at many individual positions. This could refine our knowledge on the distribution of the extended sky brightness, possibly leading to a refinement of the DIRBE zodiacal light model, too. The advantages of the ISO data are their higher spatial resolution and the existence of observations within the September-December period when DIRBE was out of operation. The second way is to take advantage of ISO's good filter coverage in the 2.5-240μm wavelength range, and separate the main components of the infrared sky by their spectral characteristics. This spectral decomposition offers an independent way to determine the zodiacal light contribution in selected line-of-sights, and the results can be confronted with the zodiacal light extraction made by the DIRBE team. In the following subsections we investigate how realistic is it to combine the ISOPHOT and DIRBE databases; the spectral energy distribution of the zodiacal light will be discussed in Sect. 5.

2.1. Transformation between the ISOPHOT and DIRBE photometric systems

Surface brightness measurements with ISOPHOT and DIRBE may give different results on the same position due to differences in the photometric calibrations of the two instruments, and due to the fact that the ISOPHOT calibration has not been finalized yet. Combining DIRBE and ISOPHOT measurements requires therefore precise transformation relationships between the two photometric systems. In this subsection we create such relationships for 5 selected ISOPHOT filters, and check the accuracy of the transformation. We will use the ISOPHOT calibration files available in the PHT Interactive Analysis (PIA) package V7.3.2 [14]. The transformation relations will have to be updated any time when a new version of the ISOPHOT calibration is released.

For the purpose of a systematic comparison of the two photometric systems we observed

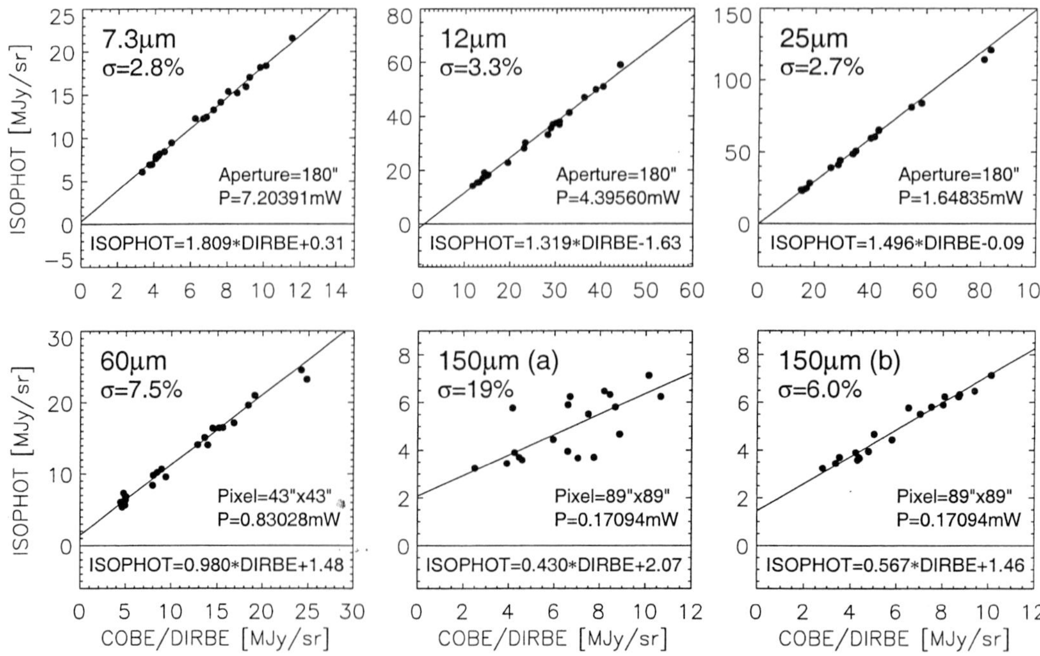

Figure 1. ISOPHOT versus DIRBE surface brightness values. For each filter the beam, the heating power of the internal fine calibration source, and the standard deviation of the data points from a linear fit are given. At 150μm the DIRBE values are computed via interpolation; in panel (a) the DIRBE 140μm data are included in the interpolation, while in panel (b) they are not.

with ISOPHOT a network of 30 dark fields covering the sky accessible by ISO ($60° < \varepsilon < 120°$, where ε is the distance from the sun). In these observations at least 7 and at most 13 filters were used in the 3.6–200 μm wavelength range. The fields were carefully examined for low cirrus levels in the IRAS 100 μm maps and do not contain infrared point sources. As far as possible, we also avoided stars visible on the Palomar Observatory Sky Survey in the beam. Absolute photometry mode was selected, in many cases including dark current measurements and checks for the zero level of the fine calibration source (FCS) measurements. All observations in a given filter were performed with the same aperture and FCS power.

The data was analysed with PIA V7.3.2 in the standard way, with no signal linearization. Surface brightness values were extracted from the DIRBE Weekly Sky Maps (nssdcftp.gsfc.nasa.gov/spacecraft_data/cobe/) for the dates/positions of the ISOPHOT observations, by taking the median of typically 66 pixels in a 1.5° radius field around the intended position. The ISOPHOT and DIRBE data sets were appropriately colour-corrected. To predict DIRBE values for the 7.3 and 150 mm ISOPHOT filters, interpola-

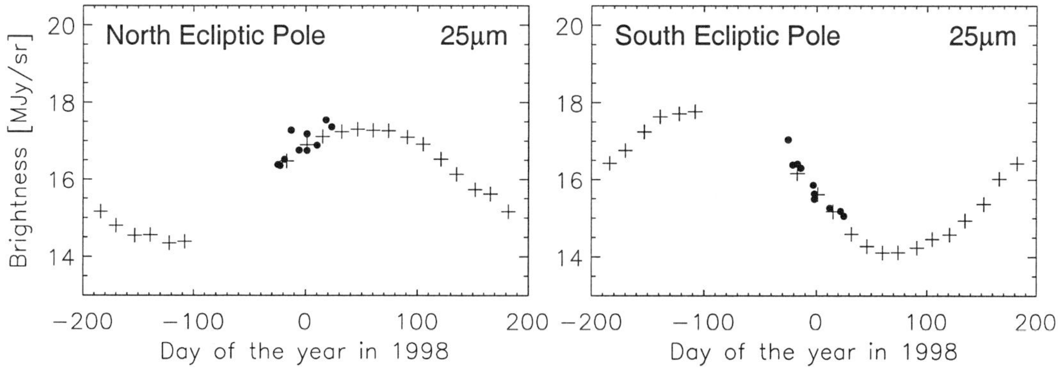

Figure 2. Annual brightness variations towards the North and South Ecliptic Poles at 25μm [15]. *Black dots* are ISOPHOT observations transferred to the DIRBE photometric system (see text), *plus signs* are values extracted from the DIRBE weekly maps. Both the ISOPHOT and DIRBE values are colour corrected.

tion was used: at 7.3μm the DIRBE 4.8 and 12μm points were connected with a Planck function; at 150μm a spline interpolation in the logarithmic space among the 60, 90, and 240μm DIRBE bands was performed. The DIRBE 140μm band was not used in this interpolation, because it introduced higher noise (see Fig. 1).

The resulting ISOPHOT vs. DIRBE relationships for the selected 5 ISOPHOT filters are shown in Fig. 1. The 60 and 150μm values are averages over the 3x3 and 2x2 pixels of the C100 and C200 camera, respectively. The figure shows that over the whole brightness range of the zodiacal light the calibrations of the two instruments are related by simple linear transformations. Note that for apertures and FCS powers different from those given in Fig. 1 the coefficients of the linear fits may be different. At 7.3, 12, and 25μm ISOPHOT and DIRBE have almost identical zero points. At 60 and 150μm a zero point difference seems to be present. In most filters the slope of the linear relationships differ significantly from unity. The random scatter is a few percent of the absolute brightness, and this is the level of precision which can be expected when ISOPHOT data are transformed into the DIRBE photometric system or vice versa. When the DIRBE 140μm data point was included in the interpolation (Fig. 1 lower middle panel) the scatter was significantly higher than in the case when the 140μm point was ignored (Fig. 1 lower right panel), suggesting that the calibration of this DIRBE band is of somewhat lower accuracy.

These results demonstrate that the ISOPHOT and DIRBE databases of extended sky brightness can be transformed into each other's photometric system with an accuracy of a few percent. This result offers the possibility to supplement the DIRBE data with ISOPHOT observations from the Sept-Dec period when DIRBE was not cooled, and to check the DIRBE 140μm photometry by using the less noisy ISOPHOT data.

2.2. Example: 25μm photometry of the North and South Ecliptic Poles

As a test of the photometric transformations described above, we re-analysed a set of ISOPHOT 25μm observations of the North and South Ecliptic Poles, analysed already by Holmes & Dermott [15]. In Fig. 2 we plotted the DIRBE observations for the 9 months when DIRBE was operational, and overplotted the ISOPHOT measurements transformed into the DIRBE photometric system using the relationship in Fig. 1. The two data sets are in good agreement, with a remaining scatter around a smooth fit to the DIRBE data of 0.31 MJy/sr (2.0%) and 0.18 MJy/sr (1.1%) for the North and South Ecliptic Poles, respectively.

2.3. Concept of a zodiacal light model for ISOPHOT

The derivation of a zodiacal light model for all ISOPHOT filters is a calibration task, since observations of faint extended emission (diffuse nebulae, cirrus, extragalactic background) has to be corrected for the interplanetary contribution. Based on the ISOPHOT data alone we cannot create an independent global model for the zodiacal light; instead we use the following strategy: we (1) adopt the DIRBE zodiacal light model [4] consisting of a fan shape cloud, the asteroidal bands, and the Earth's resonant dust ring; then (2) apply the relationships of Fig. 1 (or actually a set of similar relationships for different apertures and FCS powers) to transform the DIRBE model predictions to the ISOPHOT photometric system. Values of the DIRBE zodiacal light model for a given position and date can be extracted from the DIRBE Sky and Zodi Atlas (DSZA) (available at http://cobe.gsfc.nasa.gov/cobe1/DIRBE/DSZA/). Derivation of a complete set of transformation relationships between ISOPHOT and DIRBE is in progress.

The adapted DIRBE model can be considered as the first version of the ISOPHOT zodiacal light model. The predictions of this model can be checked by spectral decomposition (zodiacal light + galactic cirrus + extragalactic background) of spectral energy distributions observed by ISOPHOT towards dark fields (Sect. 5).

3. SUBSTRUCTURES IN THE ZODIACAL DUST: ASTEROIDAL BANDS AND COMETARY TRAILS

The main contribution to the zodiacal light comes from dust grains in the 20–200 μm size range [16]. The lifetime of individual particles of this size at 1 AU is about 10^4 yrs, and to maintain the zodiacal cloud in a steady state an average dust input of $\approx 9 \times 10^6$ g s^{-1} is required [17]. Possible dust sources are collisions within the asteroidal belt, active comets, satellites and planetary rings, as well as a stream of interstellar dust particles. The relative contributions of these sources are not well determined, and may depend on the position of the observer within the IDC. ISO observed the asteroidal bands and cometary trails, the most important places where fresh dust is released. The new measurements contribute to our knowledge on the dust ejection mechanism, and can help to quantify the fraction of dust particles of asteroidal and cometary origin.

3.1. Asteroidal Bands

The asteroidal bands were discovered as two local maxima in the IRAS scans across the Ecliptic Plane at $\beta \approx 0°$ and as shoulders at $\beta \approx \pm 10°$ [18]. They were further investigated using the COBE/DIRBE database [19]. The amplitude of the bands is 1–3%

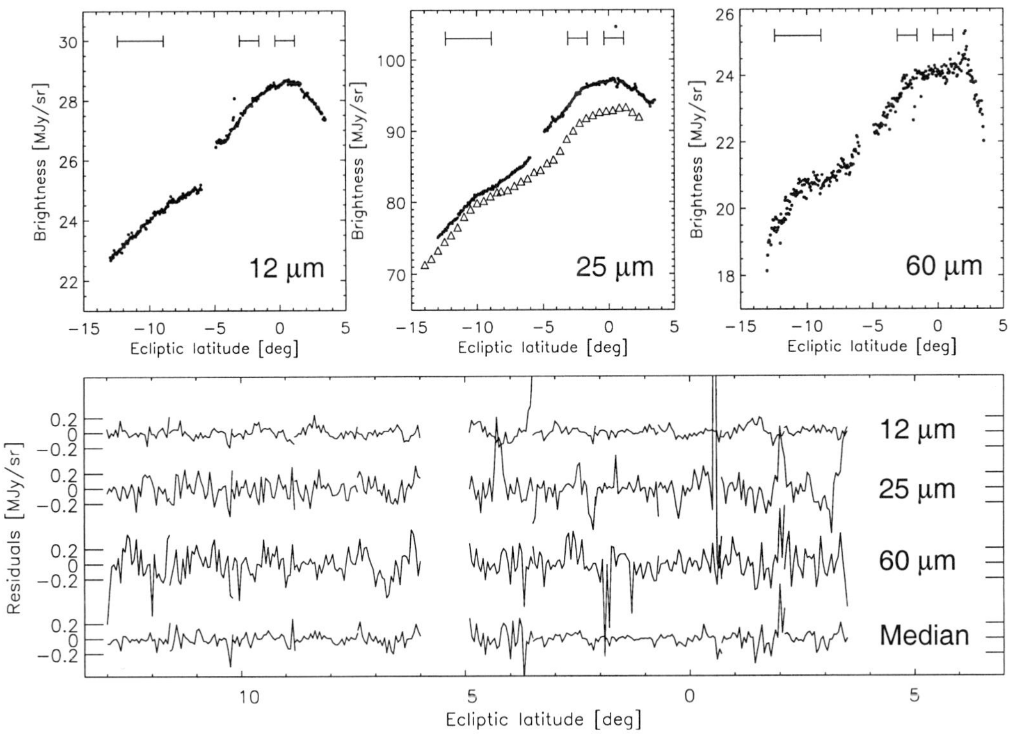

Figure 3. *Upper panels:* high spatial resolution ($\Delta\beta = 3'$) scans across the Ecliptic taken by ISOPHOT at $\lambda - \lambda_{\odot} = +90°$ on Jan. 8, 1997 (*dots*). At $\lambda \leq 25$ μm the 180″ circular aperture was selected; at 60μm the C100 camera (3×3 pixels, 43″ × 43″/pix) was used. The β=[-5,+5] degree scans were performed two days later than the [-13,5] scans, causing an offset in the absolute brightness. The 25μm scan was repeated on Jan. 8, 1998 with lower spatial resolution ($\Delta\beta = 30'$, *triangles*). The positions of the $-10°$ and $\pm1.4°$ asteroidal bands, as predicted by [19], are marked. *Lower panel:* Residuals of the high resolution scans after subtracting the smooth component.

of the zodiacal light brightness, their width at half maximum is 1.5°–3.5° [19,20]. The bands are interpreted as the result of collisions in the asteroidal belt, and are associated with the asteroidal families Themis, Koronis, and Eos [19,21,22]. The discovery of the dust bands showed unambiguously that there is an important asteroidal contribution to the IDC. A detailed modelling of the bands indicates that their dust content constitutes 5–10% of the zodiacal cloud [22]. Another estimate of the fraction of asteroidal dust comes from modelling of the Earth's resonant circumsolar ring and leads to a value of about 30% [23].

With ISOPHOT we observed the asteroidal bands by performing a multi filter high spatial resolution scan ($\lambda = 12, 25, 60\mu$m, $\Delta\beta = 3'$) across the Ecliptic at $\lambda - \lambda_{\odot} = 88-90°$.

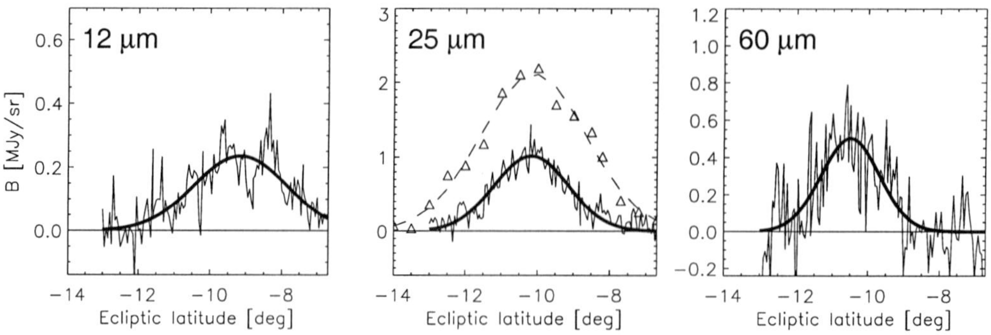

Figure 4. ISOPHOT brightness profiles of the $-10°$ asteroidal band. At $25\mu m$ the profile extracted from the lower spatial resolution scan, observed on Jan. 8, 1998, is plotted with triangles. The band profiles are fitted with Gaussians, whose parameters are given in Tab. 1.

The purpose of the high resolution was to take advantage of ISO's small beam and search for substructures within the bands, similar to those indicated by the IRAS data within the $\pm1.4°$ bands (see Fig.9 in [19]). The $25\mu m$ scan was repeated one year later with a lower spatial resolution of $30'$. Figure 3 shows the observed brightness profiles (upper panels). The β=[-5,+5] and β=[-13,5] scans were performed with a difference of 2 days, introducing an offset in the absolute brightness. Since the observed ecliptic latitudes of the bands depend on the annual motion of the observer as well as on the distance from the sun, we predict and mark in Fig. 3 the expected positions of the $\pm1.4°$ and $-10°$ bands, using Figs. 4 and 5 of Reach [19].

The plots demonstrate that at 25 and $60\mu m$ ISOPHOT detected both the $\pm1.4°$ and the $-10°$ bands in the form of shoulders in the smooth profile. At $12\mu m$ the $-10°$ band is marginally detected, but the presence of the $\pm1.4°$ pair of bands is not obvious. In the lower panel of Figure 3 we plotted the residuals of the high resolution scans after subtracting a smoothed baseline. Though peaks and dips are present in the individual 12,25, and $60\mu m$ scans, there are no significant features which are present at all three wavelengths (see the median of the 3 individual scans at the bottom). These results seem to exclude the presence of any arcminute-scale structure of the asteroidal bands at the 1σ level of 0.067, 0.157, and 0.180 MJy/sr at 12, 25, and $60\mu m$, corresponding to 0.25%, 0.18%, and 0.82% of the total sky brightness, respectively.

In order to study the bands in details, we extracted the brightness profile of the $-10°$ band by fitting a 4th order polynomial to the β=[-13.5,-13.0] and β=[-7.0,-4.0] sections of the scans which are not contaminated by asteroidal bands. The profiles are shown in Fig. 4, details of the scans as well as parameters of Gaussian fits to the brightness profiles are given in Tab. 1.

At $12\mu m$ both the location and the width of the band differ from the corresponding values at longer wavelengths, and it is not clear if we see the same physical structure or

Table 1
ISOPHOT scans across the $-10°$ asteroidal band, and parameters of the band derived from Gaussian fits to the observed profiles. The peak intensity of the band is colour corrected assuming a blackbody of 175 K. At 60μm the 3x3 pixels of the C100 camera were averaged, thus as aperture the full size of the array is given.

Wavelength [μm]	Aperture	λ	$\lambda - \lambda_\odot$	Step	Date	B [MJy/sr]	β_{centre}	FWHM
12	180″	18°	+90°	3′	8-Jan-1997	0.27	-9.2°	3.0°
25	180″	18°	+90°	3′	8-Jan-1997	1.00	-10.2°	2.4°
60	135″x135″	18°	+90°	3′	8-Jan-1997	0.49	-10.5°	1.9°
25	180″	18°	+90°	30′	8-Jan-1998	2.08	-10.1°	3.5°

not. The measured central position of the band at 25 and 60μm is close to the expected $\beta = -11.6$ deg derived from DIRBE data, but the width of the band in the high resolution scans is significantly less than the 3.5 deg measured by DIRBE. The amplitudes of the Gaussians are close to the numbers derived from DIRBE [19]. The repeated 25μm observation, observed one year later, outlines a rather different profile with higher amplitude and larger width. It is not clear if we see real physical changes on a timescale of one year, or if the differences are introduced by the slightly different observing strategy and data reduction of the two scans. The 25/60μm colour temperature of the band is 208 K, placing the $-10°$ asteroidal band at a heliocentric distance of about 1.8 AU.

3.2. Cometary trails

Crossing the orbits of a few periodic comets (Tempel 2, Encke, Kopff, Tempel 1, Gunn, Schwassmann-Wachmann 1, Churyumov-Gerasimenko, and Pons-Winnecke) IRAS detected long narrow streams of dust particles [24]. These dust 'trails' typically extend 10° behind and 1° ahead of the comet, and they should not be confused with the visually more prominent 'tails' formed by sub-micrometer-sized particles blown away by the solar radiation pressure. Rather they correspond to the anti-tails observed in the visible for particular geometric arrangements of the Sun, Earth and comet, and apparently extending towards the Sun (a spectracular unpublished picture of comet Encke and its trail was taken at mid-infrared wavelengths with the ISOCAM camera by Reach). The trails consist of mm-sized dust particles ejected from the comet during active times extending over many years [25]. Theoretical investigations of dust ejection from comets predict that the common dust production of 85 short-period and 101 long-period comets is of the order of only 3×10^4 g s^{-1} [26]. This production rate is by far too low to maintain the zodiacal cloud. Recent infrared observations indicate, however, that due to the very low albedo of the cometary dust, the mass input rate may be higher than suspected [27]. In addition, the dust injection rate is not necessarily constant in time, and the contribution from active comets in the past is not known. A higher dust supply rate, in the form of comet showers at 37 and 50 million years ago, is indicated by measurements of the ^{3}He content in interplanetary dust particles from oceanic core samples [28].

A key parameter in the calculations of the dust supply rate from comets is the timescale on which the ejected grains are mixed with the general interplanetary medium. Davies

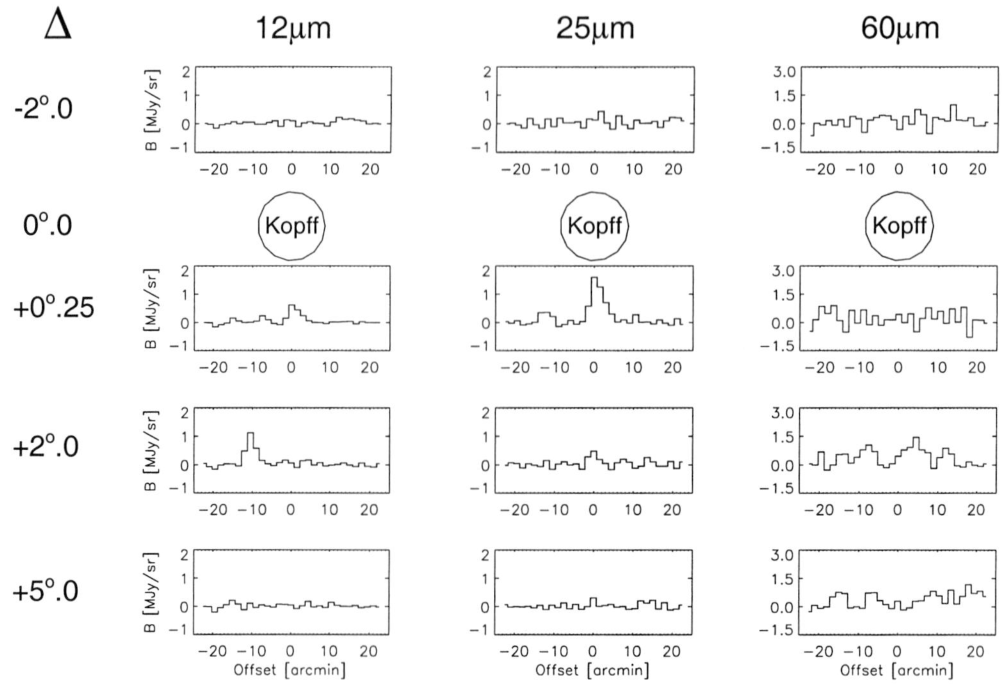

Figure 5. ISOPHOT observations of the dust trail of comet P/Kopff in October 1996. The 1-dimensional multi-filter (12, 25, and 60μm) scans were performed perpendicular to the orbit at separations of $-2°$, $+0.25°$, $+2°$, and $+5°$ ahead and behind of the nucleus.

et al. [27] observed the dust trail of comet P/Kopff at mean anomalies of 0.5° and 1.0° behind the nucleus, and compared the results with measurements from 13 years earlier obtained with IRAS [27]. The new observations were performed at 12 μm with ISOCAM. The trail was clearly detected towards both positions with a surface brightness of $\sim 0.33\,\mathrm{MJy\,sr^{-1}}$. This value is about a factor of 2 lower than that predicted from the IRAS data, indicating different levels of activity between perihelion passage before and after the IRAS observations. Comparing the sizes of the P/Kopff trail observed by the two satellites, a significant broadening was observed during this 13 yr period.

With ISOPHOT we also observed the trail of comet P/Kopff at 12, 25, and 60 μm in October 1996. Figure 5 shows both the observed positions and the derived brightness profiles taken perpendicular to the orbit at different distances behind and ahead of the comet. The figure shows that the trail was clearly detected at 0.25° behind the comet at 12 and 25μm, and it was clearly resolved by the $3'$ aperture exhibiting an asymmetric profile. The trail, however, was only marginally detected at larger distances behind the nucleus, and no anti-tail was seen, showing that the intensity of the trail drops relatively quickly behind the comet. The colour temperature determined from the 12 and 25μm

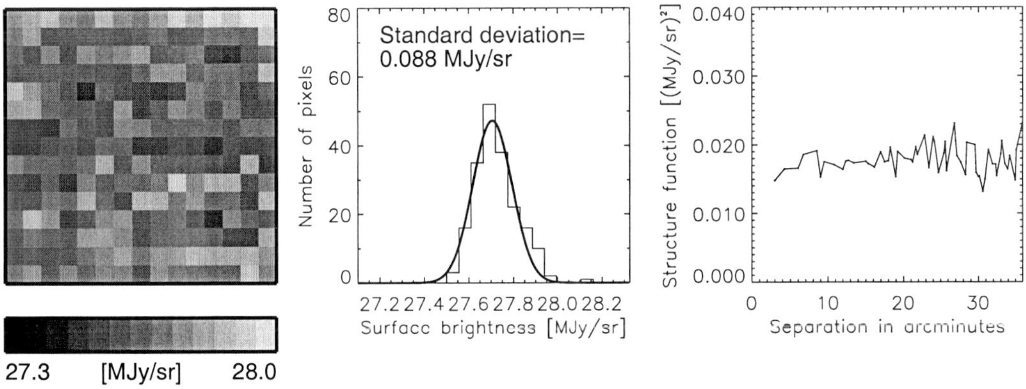

Figure 6. Map of 45′×45′ obtained at the North Galactic Pole at 25 μm (left), the histogram of the brightnesses measured in the central 13×13 pixels (middle), and the structure function (right).

values measured towards the orbit of the comet gives 215 K (or 225 K if the surface brightness values were transformed into the DIRBE photometric system, see Sect. 2). Since during these observations P/Kopff was located at about 2 AU from the sun, the derived temperature is somewhat higher than the expected ≈ 200 K at this heliocentric distance. A similar temperature increase was also found from IRAS measurements.

4. ARCMINUTE STRUCTURE OF THE ZODIACAL LIGHT

One of the main characteristics of the zodiacal light is the large-scale smoothness of its brightness distribution over the sky. Little is known of the inhomogeneities in the zodiacal light brightness at small spatial scales. Structure must exist at some level because the zodiacal light is produced by particles which are replenished from localised sources, comets and asteroids.

We mapped a few fields of ≈0.5°×0.5° at 25 μm with ISOPHOT in order to look for fluctuations in the zodiacal light at low, intermediate and high ecliptic latitudes. These fields were selected for low cirrus emission, and bright infrared point sources were avoided. Figure 6 shows one of the maps with pixel size of 3′. The image smoothness is quantified in the central panel of Fig. 6 which shows the histogram of the map. The width of the almost perfectly Gaussian distribution - after removing the instrumental noise contribution - is about 0.2% of the total brightness level. The right panel of Fig. 6 displays the structure function $<(I_i - I_j)^2>$ calculated from the measured single-pixel brightnesses in the map. The flat distribution with no dependence on the spatial scale indicates the complete lack of large scale coherent structures in the brightness distribution. The very low fluctuation and the lack of correlated structures support the concept of a generally smooth zodiacal light distribution. More details of this study can be found in [29].

5. SPECTRUM OF THE ZODIACAL LIGHT

5.1. 3–200μm spectral energy distribution

The spectral energy distribution of the zodiacal light is dominated by scattered sunlight at $\lambda \leq 3.5$ μm and by thermal emission from dust grains at longer wavelengths. The thermal spectrum, which is of primary interest for ISO, cannot be observed from the ground. Starting in 1971 [30], broad-band photometry from rockets and balloons were performed in order to define the general shape of the zodiacal light in the thermal infrared [31–33]. The most extensive data sets are derived from the satellite missions IRAS and COBE [4,5].

ISO observed the spectrum of the infrared sky brightness by performing multi-filter photometry with ISOCAM at $\lambda \leq 16$ μm and by ISOPHOT at wavelengths up to 200μm. In Fig. 7 we plot the broad band spectrum measured towards a dark position at $\lambda - \lambda_{\odot} = 90°$ and $\beta = 0°$ using the absolute photometric mode of ISOPHOT and the calibration in PIA V7. The mid-infrared ISOPHOT-S spectrum measured at the same position and date is overplotted with a solid line. The discrepancy between the DIRBE data, the ISOPHOT-S spectrum and the broad-band photometry reflects the unot yet finalised status of ISOPHOT (PIA V7) extended source calibration. Nevertheless, the figure reveals the main advantage of the ISOPHOT zodiacal light observations: the good filter coverage at the most interesting mid-infrared and $\lambda > 100\mu$m wavelengths, including mid-infrared spectrophotometry. With this spectral information one can try to decompose the main components of the infrared sky: zodiacal light, galactic cirrus, and extragalactic background light using their different spectral characteristics.

5.2. Mid-infrared spectrophotometry

One of the most important contributions of ISO to zodiacal light studies is the mid-infrared (6–16 μm) spectrophotometry, where the brightness of the zodiacal light increases by two orders of magnitude. Before the ISO mission only one rocket measurement of the mid-infrared spectrum had been attempted [34]. Reach et al. [35] observed the spectrum at one sky position with the ISOCAM circular variable filters from 5–16 μm. They found that the spectrum is remarkably well-fit by a Planck curve of T=262 K, and that there are no indications of sharp lines. They also found a good agreement with the COBE/DIRBE broad-band photometry at 4.9 and 12μm.

With ISOPHOT we also observed the mid-infrared spectrum, although the spectral coverage of the spectrophotometer ISOPHOT-S is limited to the slightly smaller 5.9–11.7 μm range. We collected 27 high quality ISOPHOT-S measurements distributed over the sky region accessible by ISO (Fig. 8*a*). Some positions are identical with those where ISOPHOT multi-filter photometry was also performed. We defined six sky segments by considering the symmetry of the zodiacal light (outlined by long-dashed lines in Fig. 8*a*), and created template spectra by averaging the individual spectra of each segment. The six templates are plotted on the left hand side of Fig. 9. The templates can be well approximated by Planck curves whose temperatures show systematic variations along the sky: increasing temperature with the ecliptic latitude and decreasing temperature with the distance from the sun (see Figure 8*b*). The residuals of the blackbody fits, plotted on the right hand side of Fig. 9, do not show any definite spectral features like PAHs or silicate bands. The lack of a strong 10μm feature in the observed spectrum, similar to

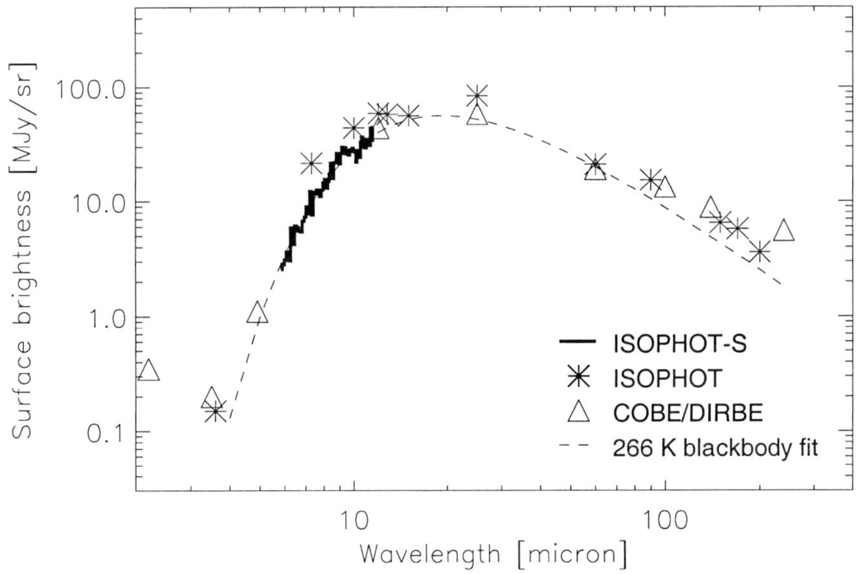

Figure 7. ISOPHOT broad-band photometry and mid-infrared spectrophotometry of the zodiacal light at $\lambda - \lambda_{\odot} = 90°$ and $\beta = 0°$, using the PIA V7 calibration. For comparison the corresponding DIRBE values are overplotted. All values are colour corrected.

that observed in comets, rules out the presence of an important population of very small silicate grains.

The observed spectrum of the zodiacal light is an integral of dust emission along the line of sight. The volume emissivity depends on the optical properties of the grains, on their size distribution, and on the temperature. Though the infrared observations cannot be used to derive directly these parameters, the spectrum of the zodiacal light provides a very strong constraint on models which use different optical constants and size distributions. The mid-infrared spectral range is very sensitive to dust temperature, because it corresponds to the Wien part of a Planck curve for large particles at 1 AU. The temperature, on the other hand, depends strongly on the wavelength dependent absorption properties of the dust grains.

Reach et al. [35] compared the mid-infrared spectrum of the zodiacal light observed by ISOCAM with synthetic spectra calculated for typical materials like obsidian, olivine, andesite, graphite, magnetite, water ice, and different size distributions. They found that none of the assumed constituents or size distributions could reproduce the observed spectrum. The best results were achieved for andesite with a "lunar" size distribution (derived from lunar micro-crater studies by [16]), and for "astronomical silicates" [37] with an "interplanetary" size distribution [16].

Our motivation to study the variation of the mid-infrared spectrum of the zodiacal light over the sky was to learn if interplanetary dust particles of different origin were distinguishable by their observed properties. It was expected that the spectrum of dust at high

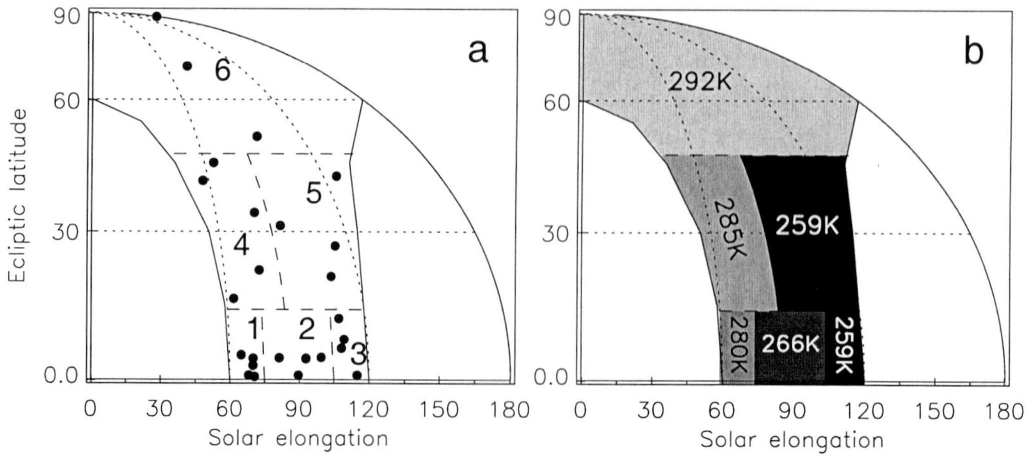

Figure 8. (*a*) distribution of the 27 ISOPHOT-S spectra over the sky accessible by ISO, assuming a symmetry of the zodiacal light wrt. the sun. The sky was divided into six segments (*long-dashed lines*), and the individual spectra of each segment were averaged. (*b*) temperature values derived by Planck-fitting to the average spectra of the segments.

ecliptic latitudes should show similarities to the spectrum of comets of high inclination, while properties of dust close to the Ecliptic Plane should resemble those of the asteroids. Our ISOPHOT-S data taken at 27 different positions over the sky, however, do not show any significant variation in the spectra other than temperature effects. It appears that the origin of the particles cannot be determined from their mid-infrared spectrum, because of the lack of spectral features, and because the present IDC appears to be well mixed.

ACKNOWLEDGEMENT

The ISOPHOT Data Centre is supported by Deutsches Zentrum für Luft- und Raumfahrt e.V. (DLR) with funds of Bundesministerium für Bildung und Forschung, grant no. 50 QI 9801 3. The responsibility for the content of this publication is on the authors. The ISOPHOT data presented in this paper were reduced using the ISOPHOT Interactive Analisys package PIA, which is a joint development by the ESA Astrophysics Division and the ISOPHOT Consortium, lead by the Max-Planck-Institut für Astronomie (MPIA).

REFERENCES

1. Ch. Leinert and E. Grün, in *Physics and Chemistry in Space* (Eds.: R. Schwenn & E. Marsch) Springer-Verlag, Berlin (1990) 207.
2. Ch. Leinert, S. Bowyer, L.K. Haikala et al., A&A Supp. 127 (1998) 1.
3. S.L. Wheelock, T.N. Gautier, J. Chillemi et al., *IRAS ISSA Explanatory Supplement*, JPL, Pasadena, (1994).
4. T. Kelsall, J.L. Weiland, B.A. Franz, W.T. Reach, R.G. Arendt, E. Dwek, H.T.

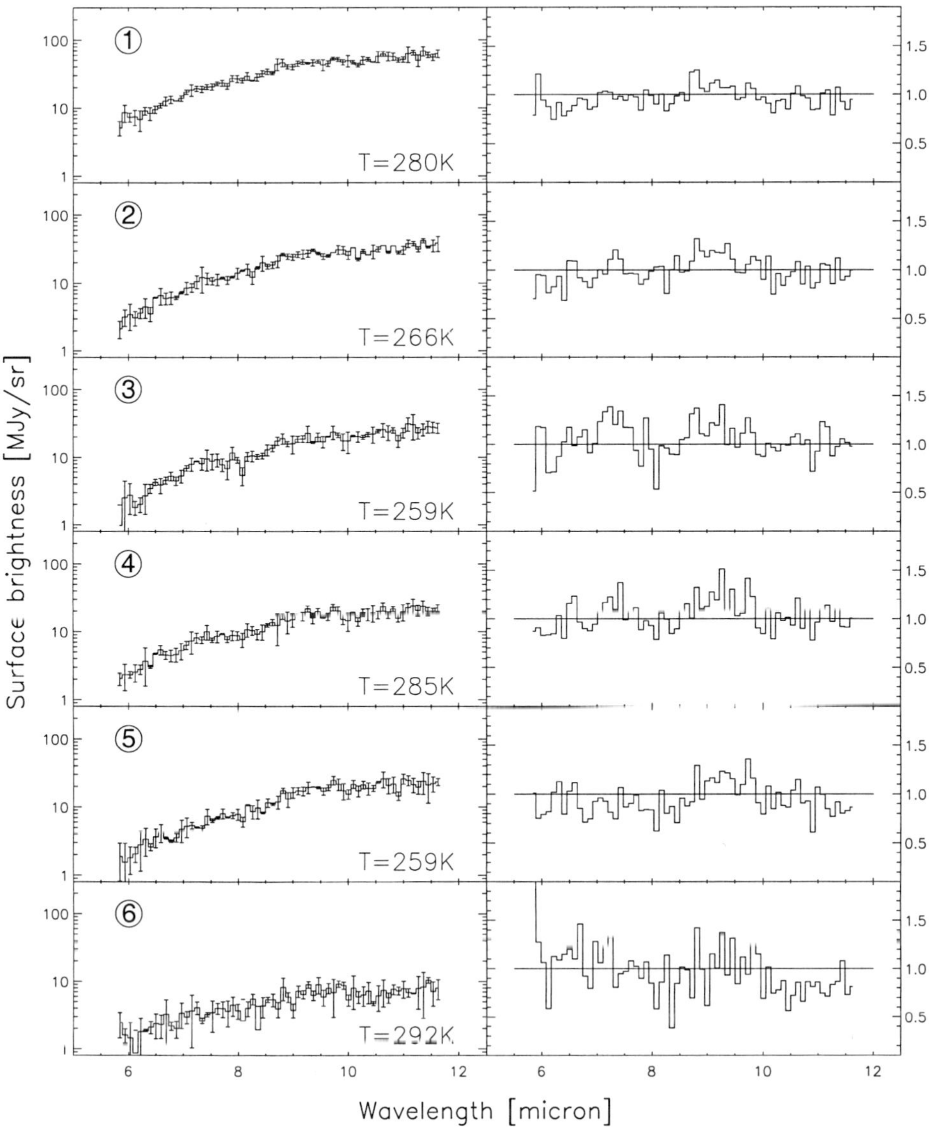

Figure 9. ISOPHOT-S spectra of the zodiacal light. *Left panel*: template spectra derived by averaging the individual spectra in each of the sky segments defined in Fig. 8a. The temperature values are derived from blackbody fits. *Right panel*: residuals of the blackbody fits.

Freudenreich, M.G. Hauser, S.H. Moseley, N.P. Odegard, R.F. Siverberg, and E.L. Wright, ApJ 508 (1998) 44

5. J.M. Vrtiled, M.G. Hauser, ApJ 455 (1995) 677
6. W.T. Reach, in *Proc. of Diffuse Infrared Radiation and the IRTS*, ASP Conference Series 124 (1997) 33
7. M.F. Kessler, J.A. Steinz, M.E. Anderegg et al., A&A 315 (1996) L27.
8. T. Ootsubo, T. Onaka, I. Yamamura et al., Advances in Space Research 25 (2000) 2163
9. U. Klaas, D. Lemke, T. Kranz et al., in *Proc. of the SPIE Infrared Astronomical Instrumentation Conference* (Ed.: A.M. Fowler) SPIE 3354 (1998) 996.
10. Lemke, D. et al. A&A 315 (1996) L64.
11. C.J. Cesarsky et al. A&A 315 (1996) L32.
12. S. Bogun, D. Lemke, U.Klaas et al. A&A 315 (1996) L71.
13. R. Siebenmorgen, A. Abergel, B. Altieri et al. A&A 315 (1996) L169.
14. C. Gabriel, J. Acosta-Pulido, I. Heinrichsen, D. Skaley, H. Morris, W.-M. Tai, in *Proc. of the ADASS VI conference* (Eds.: G. Hunt, H.E. Payne), ASP Conf.Ser. 125 (1997) 108
15. E.K. Holmes, S.F. Dermott, in *Proc. of ISO Beyond Point Sources* (Eds. R.J. Laureijs, K.Leech, M.F. Kessler), ESA SP-445 (2000) 67
16. E. Grün, H.A. Zook, H. Fechig, R.H. Giese, Icarus 62 (1985) 244
17. G.J. Flynn, Nature 376 (1995) 114
18. F. Low, D. Beintema, F.N. Gautier et al., ApJ 278 (1984) L19.
19. W.T. Reach, B.A. Franz, J.L. Weiland, Icarus 127 (1997) 461
20. W.T. Reach, ApJ 392 (1992) 289
21. S.F. Dermott, P.D. Nicholson, J.A. Burns, J.R. Houck, Nature 312 (1984) 505.
22. S.F. Dermott, D.D. Durda, B.A.S. Gustafson et al., in *Asteroids, Comets, Meteors* (Eds.: A. Milani, M. Martini, A. Cellino) Kluwer, Dordrecht (1993) 127
23. S.F. Dermott, S. Jayaraman, Y.L. Xu, B.A.S. Gustafson, L.C. Liou, Nature 369 (1994) 719
24. M.V. Sykes, L.A. Lebofsky, D.M. Hunten, F. Low, Science 232 (1986) 1115.
25. M.V. Sykes, D.J. Lien, R.G. Walker, Icarus 86 (1990) 236
26. T. Mukai, in *Highlights of Astronomy* 8 (1989) 305
27. J.K. Davies, M.V. Sykes, W.T. Reach et al., Icarus 127 (1997) 251
28. K.A. Farley, Nature 376 (1995) 153
29. P. Ábrahám, Ch. Leinert, D. Lemke, A&A 328 (1997) 702
30. T.B. Soifer, J.R. Houck, M. Harwit, ApJ 168 (1971) 73
31. S.D. Price, T.L. Murdock, L.P. Marcotte, AJ 85 (1980) 765
32. T.L. Murdock, S.D. Price, AJ 90 (1985) 375
33. A. Salama, P. Andreani, G. Dall'Oglio et al., AJ 93 (1987) 467
34. D.A. Briotta Jr., Ph D. Thesis, Cornell Univ., Cornell, (1976) 1
35. W.T. Reach, A. Abergel, F. Boulanger et al., A&A 315 (1996) L381
36. W.T. Reach, ApJ 335 (1988) 468
37. B.T. Draine, H. Lee, ApJ 285 (1984) 89

III Interplanetary Dust

Light scattering and the nature of interplanetary dust

A.Chantal Levasseur-Regourd

Université Paris 6 / Aéronomie CNRS, BP 3, Verrières, 91371 France

Observations of the solar light scattered by interplanetary dust particles provide information about their physical properties. It is clear, from polarisation observations, that the nature of the dust particles is not the same everywhere, and that it depends upon their distance to the Sun. It is quite likely, from the shapes of the polarisation phase curves, that the dust cloud is built up of irregularly shaped compact particles or aggregates, the average size of which is greater than a few micrometers.

Laboratory measurements represent a most promising approach to derive physical properties of the dust from some key parameters, such as the minimum, inversion and maximum regions in the phase curves, as well as the polarisation-wavelength dependence or the polarisation-albedo dependence. From comparisons with other dust populations, with light scattering computations, and with laboratory measurements, it may already be assumed that the interplanetary dust is built up of both compact particles (possibly of asteroidal origin) and of absorbing fluffy aggregates (possibly of cometary origin). Changes in the properties of the dust most likely result from the relative location of these sources, and from the evolution with time of the fragile dust particles.

1. INTRODUCTION

Information about the physical properties (i.e. shape, size distribution, porosity, albedo) of the interplanetary dust particles is essential, together with the knowledge of their chemical composition and of their dynamics, to infer their origin, and to allow some comparisons between the different dust populations, which replenish the interplanetary dust cloud. The determination of the physical properties from dust samples analysis is restricted to interplanetary dust particles collected in the near-Earth environment, while the estimation of the physical properties by dust impact studies is mainly performed on-board satellites and space probes. The physical (and chemical) properties of dust particles belonging to still inaccessible regions of the interplanetary dust cloud can thus only be approached through observations of the light scattered (or emitted) by these particles.

This review first summarises the light scattering properties of the interplanetary dust, with emphasis on local information retrieved by inversion techniques. These scattering properties are then compared with those of other (real or virtual) dust particles. Finally, the various laboratory tools that allow light scattering measurements by dust clouds to be interpreted are presented, and the constraints that may be derived about the physical properties of interplanetary and cometary dust are discussed.

2. LIGHT SCATTERING PROPERTIES OF INTERPLANETARY DUST

The faint cone of zodiacal light, seldom visible with the naked eye in the night sky, results from the scattering of solar light on interplanetary dust. Its shape indicates that the dust cloud spatial density increases towards the Sun and towards the ecliptic plane. Numerous quantitative observations have been performed from Earth or space observatories during the second half of the XXth century (e.g [1,2,3]). Although ground-based observations suffer from the atmospheric airglow contamination, CCD imaging techniques (e.g. [4,5]) have recently been successfully applied to studies of the interplanetary dust cloud.

Zodiacal light has a solar type spectrum, at least from the near ultraviolet to the near infrared, and is linearly polarised, as expected for solar light scattered by an optically thin medium. The polarisation values can immediately be used to compare, without any normalization to a constant distance to the Sun and to the observer, data obtained for different locations and directions.

2.1. Observations

From the Earth and the Earth's orbit, the scattering properties mainly depend upon the direction of observation, defined by the ecliptic latitude (β) and the helio-ecliptic ($\lambda - \lambda_o$) longitude. For a given direction, small temporal fluctuations are pointed out, which originate in the slight inclination of the symmetry surface upon the ecliptic, in the eccentricity of the Earth's orbit, and possibly in local heterogeneities of the cloud (e.g. dust trails, meteor streams).

Tables providing at 1 AU the line-of-sight integrated brightness and polarisation are available (see e.g. [6,7]). They are given as a function of the latitude above the symmetry plane and of the angle between the direction of the Sun and that of the projection of the line-of-sight on the symmetry plane. The brightness is either indicated in magnitude related units, or in SI units at a given wavelength. The polarisation is a ratio, smaller than one in absolute value. Although the direction of polarisation (electric field vector) is usually perpendicular the scattering plane (Sun, observer, line of sight), it may be parallel to it, leading then to a negative value of the polarisation in the gegenschein (backscattering) region.

Deep space measurements have been performed from Helios 1/2 [3] and from Pioneer 10/11 [8,9] space probes, providing observations from 0.3 AU to the outer edge of the asteroid belt. For measurements performed towards a constant direction, the brightness decreases with increasing solar distance (r) of the observer, approximately as $r^{-2.45 \pm 0.15}$, and the polarisation increases approximately as $r^{+0.3}$. This latter result demonstrates that the properties of the dust change with the solar distance. Although it is difficult to compare data retrieved line of sight data along different directions and from different locations, a slight decrease of the polarisation with increasing wavelength is suspected, at least in the near infrared domain [7]. An extensive review on the observations and on their interpretation can be found in [10].

2.2. Local values

The scattering takes place along a significant part of the line of sight, on which each point corresponds to a specific region of the dust cloud, and to a different phase (α) or scattering ($\theta = \pi - \alpha$) angle. Inversion methods are thus required, to provide local bulk information about the dust properties. Since rigorous inversions are only feasible tangentially to the direction of motion of the Earth or of a moving probe, and for the section of the line of sight where the observer is located, some assumptions need to be done.

Taking into account the above-mentioned change in polarisation, the dust properties cannot be assumed to be the same everywhere. Mathematical methods have thus been developed to retrieve local information in remote regions. Results have mainly been obtained in the symmetry plane, for instance from the nodes of lesser uncertainty method (e.g. [11,12]), or from the kernel of Voltera integral method [13]. The results provided by these two methods, applied to two different data sets, are remarkably consistent.

The solar distance (*R*) dependence of the local properties is determined between about 0.3 and 1.5 AU. From the nodes of lesser uncertainty method, the local polarisation at 90° phase angle, *P*(90°), is found to slowly increase with increasing solar distance, while the mean local albedo at 90°, *A*(90°), decreases with solar distance, and the local dust density varies approximately as 1/*R*. For a wavelength of about 0.5 μm:

$$P_{0.5\,\mu m}(90°) = (0.30 \pm 0.03)\, R^{+0.5\pm0.1}, \tag{1}$$

$$A(90°) = (0.07 \pm 0.03)\, R^{-0.34\pm0.05}. \tag{2}$$

Similarly, from the kernel of Voltera integral method, the local polarisation is found to be higher at 1 AU than at 0.3 AU, at least for phase angles above 30°. In other words, when the particles get closer to the Sun under Poynting-Robertson drag, that is to say when they get warmer and older, their polarisation gets smaller and their albedo gets higher.

The dependence of the local polarisation upon the phase angle is also derived. The phase curves, obtained at 0.3, 1.0, and 1.5 AU, are found to be smooth, with a slight negative branch in the backscattering region, and a wide positive branch with a near 90° maximum. Near 1.5 AU, the transition from the negative to the positive branch, so-called the inversion region, takes places at (15 ± 5)° phase angle, and the slope at inversion is of about 0.2 percent per degree.

These smooth polarisation phase curves, which can be described by functions such as:

$$P(\alpha) = p_0\,(\sin\alpha)^a\,(\cos\alpha/2)^b\,\sin(\alpha - \alpha_0), \tag{3}$$

have shapes similar to those obtained for cometary dust and asteroidal regoliths [14,15], as well as for circumplanetary dust and some atmospheric aerosols. The smoothness of the polarisation phase curves strongly suggests that the scattering dust is built up of irregularly shaped compact particles or aggregates, the size of which is greater than the wavelength.

3. COMPARISON WITH OTHER DUST PARTICLES

3.1. Collected samples

Direct laboratory analysis of dust particles collected in the near-Earth environment (IDPs) indicates that most of the particles collected in the Earth environment are roughly ellipsoidal aggregates, in a 1 μm to 1 mm size range, of complex smaller grains (see e.g. [16]). Although the particles collected at 1 AU in the ecliptic represent a biased sample, and have suffered some atmospheric heating, their shapes agree well with the shapes suggested by the polarisation phase curves.

Under close inspection, the IDPs are found to be aggregates of mostly black material (including carbon) with occasional clear (silicate) grains and reflective (sulphide) grains.

Their low albedos [17] are attributed both to their porous structure and to the presence of strongly absorbing material. The density of the 10 micrometer-sized particles is a 0.3 to 6.2 g cm^{-3} range, with an average at 2.0 g cm^{-3} [18], but larger particles might have densities smaller than 1 g cm^{-3}.

3.2. Cometary and asteroidal dust

The interplanetary dust complex is continually replenished by dust particles from comets and asteroids. Besides, some dust particles originating from the Kuiper belt and from the local interstellar cloud progressively reach the asteroid belt and later the inner solar system. Comets are a major source of dust, with heated nuclei releasing dust particles, including large ones that build up meteor streams. An other significant source comes from the asteroid belt, where impacts and collisions generate numerous fragments with a very wide size distribution. The local interplanetary dust bulk albedo at 0° and 1 AU in the ecliptic actually has a value, of the order of (0.15 ± 0.08), which seems intermediate between that of comets and of asteroids.

As already mentioned, the shape of the phase curve of interplanetary dust is quite comparable to that of cometary dust or asteroidal regolith. Different classes of asteroids and of comets can be pointed out through the differences noticed in the values of the inversion angle, of the slope at inversion, and of the maximum in polarisation (e.g. [15,19]). Three cometary classes, mainly corresponding to different sizes distributions, have typically been defined, corresponding to (i) comets with a near 90° maximum of 0.10 to 0.15, (ii) comets with a higher maximum of 0.25 to 0.30, and (iii) comet Hale-Bopp, whose polarisation at a given phase angle is always the highest (see e.g. Hanner, this volume).

The anti-correlation noticed, for interplanetary dust, between polarisation and albedo, is also noticed for asteroids, but does not seem to be observed for comets. From a comparison between comets Halley and Hale-Bopp, and possibly between various cometary regions, a higher polarisation actually seems to be correlated to a higher albedo [20].

It may be of interest to notice that the polarisation at a given phase angle (greater than about 30°) decreases with increasing wavelength for asteroidal regoliths. On the opposite, it increases with the wavelength for cometary dust, which exhibit a clear trend towards higher polarisations at longer wavelengths. For the well-documented comet Hale-Bopp, the polarisation at a given phase angle clearly increases with the wavelength, with possibly a maximum near 1600 nm [21]. However, from observations of comets Hale-Bopp and Halley, the trend seems to be reversed for distances to the nucleus smaller than about 2000 km [22,23], suggesting that the physical properties of fresh dust particles are not the same than those of older particles, which have already suffered evaporation or fragmentation.

3.3. Virtual particles

The constraints imposed by the various sets of results, including the discrepancies between the polarisation-albedo dependence and the polarisation-wavelength dependence, should give clues to the physical properties of the different sets of scattering particles. It is natural to compare the values observed to those derived from light scattering computations, and it makes sense to perform computations with realistic irregular particles, which are neither spheroidal nor cylindrical.

Irregular compact particles have actually been used for light scattering computations, as well as aggregates, which could be fractal particles build up of smaller elementary spheres or gaussian solids (see e.g. [24,25]). Significant results have already been obtained, which suggest the existence of both fluffy aggregates of absorbing material and compact particles in the interplanetary dust cloud. The negative branch in the polarisation phase curves could be a clue to the existence of absorbing material or of dust aggregates, in which some multiple scattering takes place. However, working with realistic "virtual" particles represents a difficult task, since the computational times are usually long, and since questions about the validity range and the unicity of the solutions may be raised.

4. LABORATORY TOOLS

Direct light scattering measurements on irregularly shaped particles and aggregates are likely to provide the missing link to interpret the light scattering observations. Elaborate techniques are required, in order of obtaining results representative of low-density dust clouds, and of avoiding multiple scattering on gravity-packed layers. Microwaves facilities, as well as polar nephelometers, are now fully operational. These latter instruments can be used under microgravity conditions to avoid any sedimentation of the dust particles [26].

Microwave analogue measurements scale the light scattering up to centimetre-sized particles (see e.g. [27,28,29]). The University of Florida facility allows measurements to be performed at two wavelengths, on artificially constructed complicated particles of various indices. The polarisation at a wavelength equivalent to the red is usually lower than at a wavelength equivalent to the blue, except for some absorbing polydisperse aggregates, that could be representative of cometary dust [30].

Systematic nephelometer type measurements have been carried out at LAS in Marseilles, at two wavelengths with irregular compact particles levitating in a gas flow [31]. Smooth polarisation curves, with a near 90° maximum, are obtained for some dielectric particles.

Two nephelometer type instruments, operating at various wavelengths in the visible and near infrared, have recently been designed in France to allow measurements on low-density dust clouds of natural and industrial particles, of Moon and Mars dust analogues, as well as of powdered dust of meteoritic origin. The PROGRA^2 experiment, developed at Aéronomie and at LPCE, has been used in the laboratory and in reduced gravity conditions, on board airplanes devoted to parabolic flight campaigns [32,33]. The CODAG-LSU experiment, developed at Aéronomie, has been used in reduced gravity conditions, mainly for calibration purposes, and in microgravity conditions, during an ESA rocket flight [34].

Polarisation phase curves have already been retrieved for numerous samples with PROGRA^2. The polarisation at a given wavelength (greater than about 30°) usually decreases with the wavelength for compact particles, while, on the opposite, it seems to increase with the wavelength for highly porous aggregates of submicron-sized particles [35,36]. Such aggregates have an important scattering efficiency, and only a small fraction of them drastically changes the scattering properties of the sample.

Calibration brightness and polarisation phase curves obtained with CODAG-LSU have emphasized the slight discrepancies that may occur between computational models and laboratory measurements, due to the fact that experimental micron-sized spheres are neither perfectly spherical nor smooth. Monitoring of the phase curves during the ESA MASER-8 rocket flight has demonstrated the possibility of obtaining naturally formed aggregates through ballistic aggregation processes, under microgravity conditions [25,37].

It should indeed be possible to accurately measure the scattering properties of well-documented naturally formed aggregates and regoliths under long duration microgravity conditions. It is expected that the ICAPS (Interactions in Cosmic and atmospheric particles Systems) European project will, on board the International Space Station, provide the information still needed to interpret the scattering properties of the interplanetary dust cloud in terms of physical properties.

5. CONCLUSIONS

The scattering properties of the interplanetary dust are quite well known, at least in the ecliptic plane. The polarisation phase curves are smooth, with a slight negative branch in the backscattering region, a near 15° inversion and a wide positive branch with a maximum at about 90°. For a fixed phase angle (greater than 30°), the local polarisation increases with increasing solar distance, while the local albedo decreases with solar distance. The polarisation seems to slightly decrease with increasing wavelength, at least in the near infrared domain, but the wavelength dependence of the local polarisation is not yet be known.
Comparisons with other observations (IDPs, cometary dust, asteroidal regolith), together with the results of numerical simulations (irregular compact particles, aggregates), and those of laboratory measurements (microwave analogue technique, polar nephelometers), lead to the same conclusions: the interplanetary dust cloud is built of a mixture of irregular compact particles and of fluffy aggregates, mainly of asteroidal and cometary origin, which are likely to suffer some weathering and to evolve with time.
More constraints about the significance of some key parameters (characteristics of the polarisation phase curves, polarisation-wavelength dependence, polarisation-albedo dependence), and thus about the nature of the interplanetary dust, should be provided in a near future through new observations (including out-of-ecliptic observations), together with the development of sophisticated numerical and laboratory measurements under microgravity conditions.

REFERENCES

1. R. Dumont and F. Sanchez, Astron. Astrophys. 38 (1975) 397.
2. A.C. Levasseur-Regourd and R. Dumont, Astron. Astrophys. 84 (1980) 277.
3. C. Leinert, I. Richter, E. Pitz and M. Hanner, Astron. Astrophys. 110 (1982) 355.
4. M. Ishiguro, R. Nakamura, Y. Fujii, K. Morishige, H. Yano, H. Yasuda, S. Yokogawa and T. Mukai, Astrophys. J. 511 (1999) 432.
5. R. Nakamura, Y. Fujii, M. Ishiguro, K. Morishige, S. Yokogawa, P. Jenniskens and T. Mukai, Astron. Astrophys. 540 (2000) 1172.
6. A.C. Levasseur-Regourd, in Physics, chemistry, and dynamics of interplanetary dust, (eds. B.A.S. Gustafson and M.S. Hanner) ASP 104, San Francisco (1996) 301.
7. C. Leinert, S. Bowyer, L. Haikala, M. Hanner, M.G. Hauser, A.C. Levasseur-Regourd, I. Mann, K. Mattila, W.T. Reach, W. Schlosser, J. Staude, G.N. Toller, J.L. Weiland, J.L. Weinberg and A. Witt.,Astron. Astrophys. Suppl. 127 (1988) 1.
8. M.S. Hanner, J.G. Sparrow, J.I. Weinberg and D.E. Beeson, in Interplanetary dust and zodiacal Light. Lecture Notes in Physics 48, (eds. H. Elsässer and H. Fechtig) Springer-Verlag, Berlin (1976) 29.
9. G.N. Toller and J.L. Weinberg, in Properties and interactions of interplanetary dust, (eds. R.H. Giese and P. Lamy) Reidel, Dordrecht (1985) 21.

10. A.C. Levasseur-Regourd, I. Mann, R. Dumont and M.S. Hanner, in Interplanetary dust, (eds. E. Grün, B. Gustafson, S. Dermott, and H. Fechtig) Springer-Verlag, Berlin (2001) 57.
11. R. Dumont and A.C. Levasseur-Regourd. Planet. Space Sci. 33 (1985) 1.
12. A.C. Levasseur-Regourd, J.B. Renard and R. Dumont, in Origin and evolution of interplanetary dust, (eds. A.C. Levasseur-Regourd and H. Hasegawa), Kluwer, Dordrecht (1991) 131.
13. K. Lumme, in Light scattering by non spherical particles, (eds. M.I. Mishchenko, J.W. Hovenier and L.D. Travis) Academic Press, San Diego (2000) 555.
14. A.C. Levasseur-Regourd, R. Dumont and J.B. Renard. Icarus, 86 (1990) 264.
15. B. Goidet-Devel, J.B. Renard and A.C. Levasseur-Regourd. Planet. Space Sci. 43 (1995) 779.
16. E.K. Jessberger, T. Stephan, D. Rost, P. Arndt, M. Maetz, F.J. Stadermann, D.E. Brownlee, J.P. Bradley and G. Kurat, in Interplanetary dust, (eds. E. Grün, B. Gustafson, S. Dermott, and H. Fechtig) Springer-Verlag, Berlin (2001) 253.
17. J.P. Bradley, L.P. Keller, D.E. Brownlee and K.L. Thomas, Meteoritics Planet. Sci. 31 (1996) 394.
18. S.G. Love, D.J. Joswiak and D.E. Brownlee, Icarus, 111 (1994) 227.
19. A.C. Levasseur-Regourd, E. Hadamcik and J.B. Renard, Astron. Astrophys. (1996) 327.
20. C.G. Mason, R.D. Gehrz, T.J. Jones, M.S. Hanner, D.M. William and C.E. Woodward, Astrophys. J. 549 (2001) 635.
21. E. Hadamcik and A.C. Levasseur-Regourd, Dust evolution of comet C/1995 O1 (Hale-Bopp) by imaging polarimetric observations, Astron. Astrophys., submitted.
22. K. Jockers, V.K. Rosenbush, T. Bonev and T. Credner, Earth Moon Plan. 78 (1997) 373.
23. E. Hadamcik. Thesis, Université Paris VI (1999).
24. K. Lumme and J. Rahola, J. Quant. Spectrosc. Radiat. Transfer 60 (1998) 355.
25. V. Haudebourg, Thesis, Université Paris VI (2000).
26. A.C. Levasseur-Regourd, M. Cabane, V. Haudebourg and J.C. Worms, Earth, Moon, Planets, 80 (1998) 343.
27. J.M. Greenberg, N.E. Pedersen and J.C. Pedersen, J. Appl. Phys. 32 (1961) 233.
28. R.H. Giese, K. Weiss, R.H. Zerull and T. Ono, Astron. Astrophys. 65 (1978) 265.
29. B.A.S. Gustafson, J.M. Greenberg, L. Kolokolova, Y. Xu and R. Stognienko, in Interplanetary dust (eds. E. Grün, B. Gustafson, S. Dermott, and H. Fechtig) Springer-Verlag, Berlin (2001) 510.
30. B.A.S. Gustafson and L. Kolokolova, J. Geophys. Res. 104 (1999) 31711.
31. P. Combet, P. Lamy and J. Loesel, Astron. Astrophys., submitted.
32. J.C. Worms, J.B. Renard, A.C. Levasseur-Regourd and E. Hadamcik, Adv. Space Res., 23 (1999) 1257.
33. J.C. Worms, J.B. Renard, E. Hadamcik, N. Brun-Huret and A.C. Levasseur-Regourd, Planet. Space Sci. 48 (2000) 493.
34. A.C. Levasseur-Regourd, M. Cabane, E. Chassefière, V. Haudebourg and J.C. Worms, Adv. Space Res. 23 (1999) 1271.
35. J.C. Worms, J.B. Renard, E. Hadamcik, A.C. Levasseur-Regourd and J.F. Gayet, Icarus, 142 (1999) 281.
36. E. Hadamcik, J.B. Renard, J.C. Worms, A.C. Levasseur-Regourd and M. Masson, Icarus, 155, (2002) 497.
37. A.C. Levasseur-Regourd, V. Haudebourg, M. Cabane and J.C. Worms, In First int. symposium on microgravity research and applications, ESA SP-454 (2001) 797.

The size-frequency distribution of zodiacal dust band material

K. Grogan[a] and S.F. Dermott[b]

[a]NASA Goddard Space Flight Center, Code 681, Greenbelt, MD 20771, USA

[b]Department of Astronomy, University of Florida, Gainesville, FL 32611, USA

Recent observations of the size-frequency distribution of zodiacal cloud particles obtained from the Long Duration Exposure Facility (LDEF) cratering record reveal a significant large particle population (100 micron diameter or greater) near 1 AU. While the Solar System Dynamics group at the University of Florida has had much success in modeling the zodiacal cloud from a dynamical standpoint, our work has been limited by the fact that only small particles (25 micron diameter and smaller) have been considered. This was due to the prohibitively large amount of computing power required to numerically analyze the dynamical history of larger particles. The recent availability of cheap, fast processors has finally made this work possible. For the first time, we are able to produce models for the components of the zodiacal cloud that include a size-frequency distribution able to match IR observations in multiple wavebands. An analysis of the Solar System dust bands yields a size-frequency distribution in agreement with LDEF, in that large particles are shown to dominate.

1. INTRODUCTION

The Solar System dust bands are features of the zodiacal cloud emission which have been unambiguously related to the comminution of main-belt asteroids [1]. As such, they offer an important insight into the replenishment of the zodiacal cloud, and an opportunity to cast light upon a fundamental question which has yet to be answered satisfactorily: is the zodiacal cloud predominantly asteroidal or cometary?

The approach of the Solar System Dynamics group at the University of Florida towards zodiacal cloud modeling [2] differs in a fundamental way to methods found elsewhere in the literature. This is perhaps due in most part to motivation: many authors are interested in zodiacal emission only as a source of noise, which needs to be modeled only to be subtracted from a given dataset to facilitate other investiagtions. To achieve this, empirical formulations of the various components of the foreground zodiacal emission are produced from observations [3]. This is in contrast to our physically motivated approach, which follows the orbital evolution of a set of dust particles numerically from source to sink, allowing us to place constraints on the origin of the various emission signatures.

Our previous attempts to model the dust bands have been based on the dynamical history of 10 micron astronomical silicate particles [4]. This is because the numerical approach is extremely numerically intensive, and until recently the dynamics of larger parti-

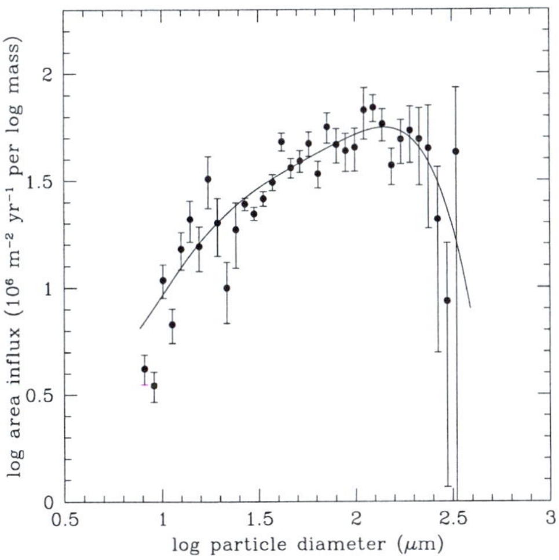

Figure 1. The terrestrial influx of zodiacal dust particles, as measured from the cratering record on the Long Duration Exposure Facility, LDEF.

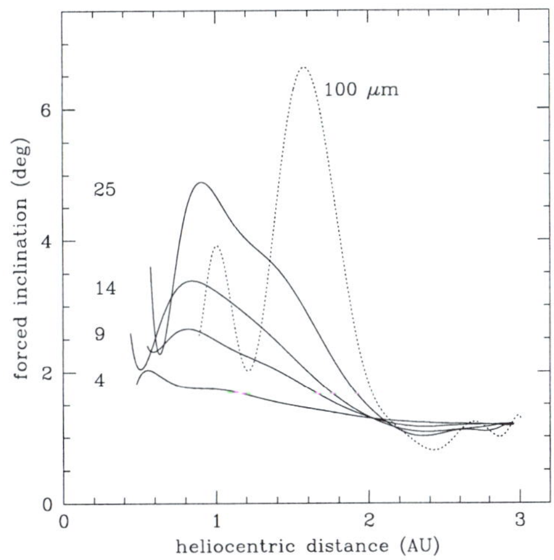

Figure 2. The variation with heliocentric distance of the forced inclination of dust particles in the zodiacal cloud, shown for five different particle diameters.

cles have been out of reach (the dynamical lifetime of a particle in a Poynting-Robertson (P-R) drag dominated system scales with its size [5]). However various evidence, including that of the LDEF cratering record [6], points to the fact that large particles contribute significantly to the zodiacal emission (Figure 1), and these particles need to be taken into account.

2. METHODS

We use the RADAU fifteenth order integrator program with variable time steps taken at Gauss-Radau spacing [7] with which we integrate the full equations of motion of interplanetary dust particles (IDPs) of various sizes. Our simulations include seven planets (Mercury and Pluto excluded) and account for both P-R drag and solar wind drag. In this way we are able to build a description of both the proper elements (intrinsic, reflecting the formation process) and forced elements (imposed by planetary perturbations) of the particles and their variation with heliocentric distance to create a snapshot of the present day forced and proper element distribution.

The distribution of orbital elements originating from a given source, once produced from the numerical integrations, is visualized by the FORTRAN algorithm SIMUL [8]. SIMUL sets up a large three-dimensional array, and calculates the contribution of surface area to each cell from all orbits in the distribution. Line of sight integrations are then performed to simulate the observing strategy of any given satellite.

3. RESULTS

Figure 2 shows the variation of the forced inclination of dust particles originating in the Eos family over a range of diameter spanning from approximately 1-100 microns. It is clear that the dust band material is dispersed into the background cloud as it passes through secular resonances located near the inner edge of the asteroid belt. A similar dispersal is found for eccentricity, making asteroidal particles appear more cometary in nature. The dust bands therefore have a natural inner edge, and explains why previous attempts to model the dust bands have been most successful when confining the dust band material to the asteroid belt.

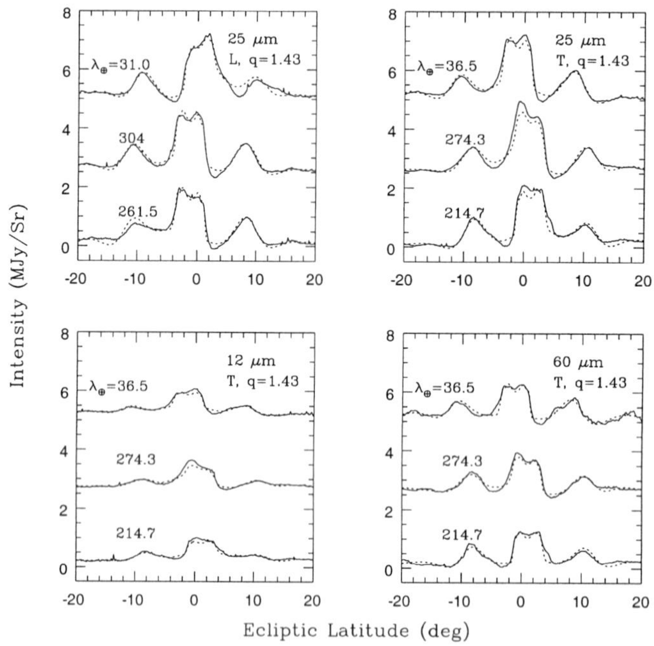

Figure 3. IRAS dust band profiles (solid lines) in three wavebands are compared with models with a size-frequency index q=1.43. All profiles were made at 90° solar elongation angle in a direction either leading (L) or trailing (T) the Earth in its orbit. The low value of q implies that the dominant particles are large (100 micron diameter or greater).

Once the distributions of the various particle sizes have been calculated, they can be combined to investigate the nature of the dust band size-frequency distribution. A system in collisional equilibrium has a size-frequency distribution index $q = 11/6$, in which particles at the small end of the distribution dominate. As q is reduced, more large particles are introduced until $q \leq 5/3$ at which point large particles dominate. We find that a q of approximately 1.4 produces the best simultaneous fit to the 12, 25 and 60 micron IRAS observations (Figure 3), providing further evidence of the preponderance of large particles in the zodiacal cloud. However, whereas most existing information on the size-frequency distribution of the cloud is based on evidence near 1 AU, our results are valid in the main-belt region. The results shown correspond to astronomical silicate material, but are largely insensitive to reasonable choices for the particle composition. Themis and Koronis dust comprises the central band, whereas the ten degree band is best modeled from material originating from the inner edge of the inclination distribution of the Eos family (9.35°) with relatively high dispersion (1.5°). This is perhaps suggestive of a catastrophic origin for the ten degree band, rather than being an equilibrium feature associated with the grinding down of the Eos family as a whole. However, the mean inclination of the model dust band is dependent on the nature of the dispersion of the orbital inclinations, and further work is required to characterize the dispersion of particles larger than 100 microns. This is an ongoing effort, and early results we have obtained using a modified symplectic algorithm are given elsewhere in these proceedings [9].

A direct result of the dust band modeling is the fraction of dust band material in the cloud. These models suggest a fraction of around 25%. Further work needs to be done to relate this number to the total asteroidal contribution to the cloud, but assuming that collision rates in family and non-family asteroids are similar, and that the ratio of non-family to family members is about 3:1, our estimate for the asteroidal contribution is approximately 75%. However this will change as we obtain more information on the large particle dynamics and refine our models.

4. CONCLUSIONS

- Large particles in the main-belt region seem to dominate the dust band emission.
- Dust band features do not persist into the inner Solar System but have a natural inner edge produced by the effect of the near 2 AU secular resonances.
- It may be difficult to distinguish between asteroidal and cometary particles in the inner Solar System on dynamical grounds alone.
- The 'ten degree band' emission does not correlate with Eos family members, but is produced at the inner edge of the inclination distribution.

REFERENCES

1. S.F. Dermott, P.D. Nicholson, J.A. Burns and J.R. Houck Jr, Nature 312 (1984) 505.
2. S.F. Dermott et al., in Formation and Evolution of Solids in Space (eds. J.M. Greenberg and A. Li) Kluwer, Dordrecht (1999) 565.
3. T. Kelsall et al., Astrophys. J. 508 (1998) 44.
4. K. Grogan, S.F. Dermott, S. Jayaraman and Y.L.Xu, Planet. Space Sci. 45 (1997) 1657.
5. J.A. Burns et al., Icarus 40 (1979) 1.
6. S.G. Love and D.E. Brownlee, Science 262 (1993) 550.
7. E. Everhart, in Dynamics of Comets (eds. A. Carusi and G.B. Valsecchi) Reidel, Dordrecht (1985) 185.
8. S.F. Dermott et al., in Comets to Cosmology, (ed. A. Lawrence) Springer-Verlag, Berlin (1988) 3.
9. T.J.J. Kehoe, S.F. Dermott and K. Grogan, this volume.

A Dissipative Mapping Technique for Integrating Interplanetary Dust Particle Orbits

Thomas J. J. Kehoe, Stanley F. Dermott [a] and Keith Grogan [b]

[a]Department of Astronomy, University of Florida, Gainesville, FL 32611, USA

[b]Jet Propulsion Laboratory, Mail Stop 169-327, Pasadena, CA 91109, USA

The LDEF (Long Duration Exposure Facility) cratering record suggests a significant population of large interplanetary dust particles (100 μm diameter and greater) near 1 AU, implying that particles with diameters as large as 500 μm may be significant sources of the infrared flux that we receive from the asteroid belt. However, integration of the full equations of motion of these very large particles, including radiation pressure, Poynting-Robertson drag and solar-wind drag, is extremely numerically intensive. As a result, our previous efforts to determine the dynamical history of main-belt, asteroidal dust particles were limited to particles with diameters less than 100 μm. We have recently developed an integration code based on a modified symplectic algorithm which, when combined with the availability of cheap, fast processors, provides us with the opportunity to extend our models of the zodiacal cloud to include this important, and possibly dominant, large particle population. Here, we present initial results from our numerical simulations.

1. INTRODUCTION

The aim of this research is to better understand the dynamical behaviour of asteroidal dust particles over a wide range of particle sizes. Our previous attempts to model the orbital evolution of these particles using RADAU [1], a fifteenth order numerical integrator with variable time steps, were limited to particles with diameters less than 100 μm [2]. This is because the orbits of interplanetary dust particles with diameters greater than this decay into the Sun, under the influence of Poynting-Robertson and solar-wind drag, on time scales of the order of millions of years or longer [3]. This puts any numerical investigation of the dynamical behaviour of a reasonably numerous (hundreds to thousands) distribution of large dust particles beyond the reach of currently available computational resources, when using traditional integration techniques. However, empirical evidence, such as the LDEF (Long Duration Exposure Facility) cratering record [4], strongly suggests the existence of a significant population of large interplanetary dust particles (100 μm diameter and greater) near 1 AU, implying that particles with diameters as large as 500 μm may be significant sources of the infrared flux that we receive from the asteroid belt [5]. It is therefore important to extend our current models of the zodiacal cloud to include this large particle population.

To overcome this problem, we have developed a unique integration code specifically designed for evolving the orbits of large populations of dust particles under the effects of radiation pressure, Poynting-Robertson drag and solar-wind drag, as well as point-mass gravitational forces. To achieve this, we have applied the dissipative mapping technique introduced by Malhotra [6] to the specific problem of deriving a MVS (Mixed Variable Symplectic) type integration code [7] that also incorporates the effects of these non-gravitational forces [8]. The development and testing of this dissipative code is described in detail elsewhere [9]. This new integration code is significantly faster than more conventional integration techniques and will allow us to investigate, for the first time, the dynamics of asteroidal dust particles with diameters up to 500 μm or even larger.

2. NUMERICAL SIMULATIONS

Forced elements are the component of a dust particle's osculating (or instantaneous) orbital elements imposed by the presence of the planets. These forced elements vary with heliocentric distance, time and particle diameter, and are responsible for the large-scale asymmetries observed in the distribution of dust particles in the zodiacal cloud. To calculate the current distributions of these forced elements, we employed the new dissipative code discussed above to evolve representative samples of asteroidal dust particles forward in time to the present epoch, along with the planets Jupiter, Saturn, Uranus, and Neptune, from a number of different epochs in the past. As the time scale for a dust particle orbit to decay under the effect of Poynting-Robertson and solar-wind drag is dependent on the particle size, each set of past epochs chosen was dependent on the size of the dust particles considered, and a separate set of integrations had to be carried out for each different particle size. Up to 80 past epochs were selected for each particle size, in order to provide a comprehensive picture of the forced element distribution of asteroidal dust particles across a wide range of semimajor axis values in the inner solar system at the present time. In this paper, we consider asteroidal dust particles (originating in this case from the Eos family, although this is not critical) composed of astronomical silicate of density 2,500 kg m^{-3} with diameters 10, 100, and 200 μm, for which we calculated β values (the ratio of radiation pressure to solar gravity) of 0.04871, 0.00446, and 0.00221 respectively, using Mie theory. The longest integrations performed for the 10-, 100-, and 200-μm diameter dust particles were for time scales of 0.06, 0.6, and 1.2 Myr, respectively.

To obtain initial orbital element distributions for our forward integrations we first employed a standard MVS integration code (incorporating point-mass gravitational forces only) to evolve Eos family asteroids, along with the gas giant planets, backwards in time from the present. Initial osculating orbital elements for 444 Eos family asteroids were obtained from The Asteroid Orbital Element Database [10] for the epoch of Julian Date 2450700.5, using the family classification of Zappalà *et al.* [11]. Osculating orbital elements for the planets were obtained for the same epoch using the data from Standish *et al.* [12]. Using a low-order secular perturbation theory 'particle on a circle' approximation [13], we then generated initial osculating orbital elements for 124 dust particles, representative of the whole Eos asteroid family, at each of the past epochs required.

The results of the integrations presented in the next section represent a total of over 4 months CPU time running on a variety of Pentium processors.

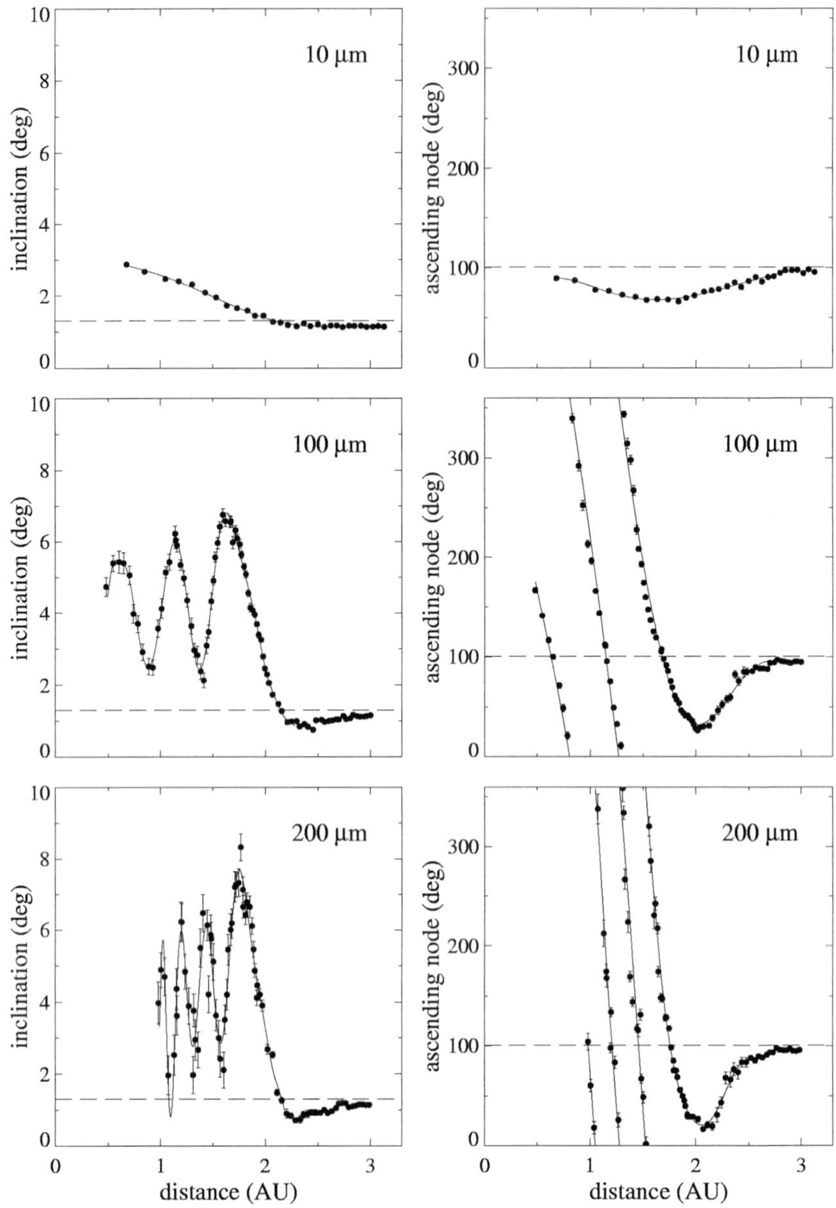

Figure 1. Variation of the forced inclination (left) and the forced longitude of ascending node (right) with heliocentric distance at the present epoch (Julian Date 2450700.5) for Eos family dust particles with diameters 10, 100, and 200 μm. The dashed lines show the present osculating inclination (left) and osculating longitude of ascending node (right) for Jupiter. Reprinted with permission from Dermott *et al.* [14]. Copyright 2001, Springer-Verlag.

3. RESULTS and CONCLUSIONS

The forced inclinations and longitudes of ascending node of some small (10 μm diameter) and large (100 and 200 μm diameter) Eos family dust particles, obtained using the new dissipative code, are shown in Fig. 1. All orbital elements are heliocentric and given with respect to the mean ecliptic and equinox of the standard J2000 reference frame. In the region of the main asteroid belt (between 2.5 and 3 AU), the forced elements of the large particles display similar behaviour to that of the small particles. That is, their forced inclinations and nodes are locked onto Jupiter's inclination and node, respectively. The low dispersion of the inclinations and nodes in this region of the main belt, regardless of particle size, is the fundamental reason why dust bands are observed at those heliocentric distances. However, as the large dust particles encounter the ν_{16} secular resonance at the inner edge of the asteroid belt (at about 2 AU), the effect of the resonance disperses their forced inclinations and nodes, diffusing the dust band particles into the broad-scale zodiacal background. The ν_6 secular resonance (also at about 2 AU) produces analogous behaviour in the forced eccentricities and longitudes of pericentre of the dust particles. The effects of these secular resonances are more pronounced for the large dust particles because they are acted on by the resonances for longer periods of time. The orbital element distributions of large asteroidal dust particles produced by intra-family collisional attrition therefore lose their characteristic family signatures in the inner region of the main belt and become indistinguishable from the general background cloud of zodiacal dust. Consequently, the IRAS (Infrared Astronomical Satellite) dust bands have a natural inner edge at about 2.5 AU. The action of these secular resonances also means that large asteroidal dust particles in the inner solar system have orbits with significant eccentricities and inclinations and may be comparable to some cometary orbits.

REFERENCES

1. Everhart, E. 1985. In *Dynamics of Comets: Their Origin and Evolution* (A. Carusi and C. B. Valsecchi, Eds.), pp. 185–202. Reidel, Boston
2. Grogan, K., *et al.* 2001. *Icarus* **152**, 251–267.
3. Wyatt, S. P., Jr., and F. L. Whipple 1950. *Astrophys. J.* **111**, 134–141.
4. Love, S. G., and D. E. Brownlee 1993. *Science* **262**, 550–553.
5. Wyatt, M. C., *et al.* 1999. *Astrophys. J.* **527**, 918–944.
6. Malhotra, R. 1994. *Celest. Mech. Dynam. Astron.* **60**, 373–385.
7. Wisdom, J., and M. Holman 1991. *Astron. J.* **102**, 1528–1538.
8. Burns, J. A., *et al.* 1979. *Icarus* **40**, 1–48.
9. Kehoe, T. J. J. 1999. Ph.D. thesis, University of London.
10. Bowell, E. L. G. 1997. The Asteroid Orbital Elements Database, Lowell Observatory.
11. Zappalà, V., *et al.* 1995. *Icarus* **116**, 291–314.
12. Standish, E. M., *et al.* 1992. In *Explanatory Supplement to the Astronomical Almanac* (P. K. Seidelmann, Ed.), pp. 279–323. University Science Books, Mill Valley.
13. Dermott, S. F., *et al.* 1992. In *Chaos, Resonance and Collective Dynamical Phenomena in the Solar System* (S. Ferraz-Mello, Ed.), pp. 333–347. Kluwer, Dordrecht.
14. Dermott, S. F., *et al.* 2001. In *Interplanetary Dust* (E. Grün, B. Å. S. Gustafson, S. F. Dermott, and H. Fechtig, Eds.), pp. 569–639. Springer-Verlag, Berlin.

Dust en-route to Jupiter and the Galilean satellites

Harald Krüger, Eberhard Grün

Max-Planck-Institut für Kernphysik, Postfach 10 39 80, 69029 Heidelberg, Germany

Spacecraft investigations during the last ten years have vastly improved our knowledge about dust in the Jovian system. All Galilean satellites, and probably all smaller satellites as well, are sources of dust in the Jovian system. In-situ measurements with the dust detectors on board the Ulysses and Galileo spacecraft have for the first time demonstrated the electromagnetic interaction of charged dust grains with the interplanetary magnetic field and with a planetary magnetosphere. Jupiter's magnetosphere acts as a giant mass-velocity spectrometer for charged 10-nanometer dust grains. These dust grains are released from Jupiter's moon Io with typical rate of $\sim 1\,\mathrm{kg\,s^{-1}}$. The dust streams probe the plasma conditions in the Io plasma torus and can be used as a potential monitor of Io's volcanic plume activity. The other Galilean satellites are surrounded by tenuous impact-generated clouds of mostly sub-micrometer ejecta grains. Galileo measurements have demonstrated that impact-ejecta derived from hypervelocity impacts onto satellites are the major – if not the only – constituent of dusty planetary rings. We review the in-situ dust measurements at Jupiter and give an update of most recent results.

1. INTRODUCTION

Until the 1970s, when the Pioneer 10 and Pioneer 11 spacecraft passed by Jupiter, the exploration of the giant planet and its satellites was restricted to remote astronomical observations from the Earth. It was pure speculation whether dust would exist in the environment of Jupiter. Pioneer 10/11 were equipped with in-situ dust detectors which recorded several impacts when the spacecraft flew by Jupiter [1]. Due to the relatively high detection threshold of the penetration detectors, however, only particles larger than several micrometers could be recognized.

The next spacecraft to visit Jupiter were Voyager 1 and Voyager 2. Although they did not carry dedicated dust detectors on board, they drastically changed our knowledge of dust in the Jovian system. Jupiter's rings were discovered by remote sensing with Voyager 1, although earlier hints that this faint dusty ring might exist came from a dip in the density of charged particles measured near Pioneer 11's closest approach to Jupiter [2,3], as well as from the impact events recorded by the Pioneer dust detectors. Typical grain sizes derived from the Voyager images were a few micrometers for the faint gossamer ring, whereas the main ring turned out to be composed of macroscopic rocky material. Another discovery by Voyager was tidally driven active volcanism on Io, Jupiter's innermost Galilean moon. At the time it was speculated that small dust grains entrained in Io's plumes may get accelerated away from Io by electromagnetic forces [4,5].

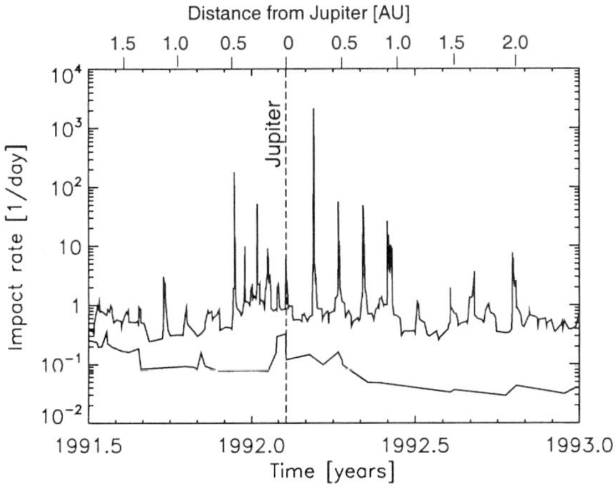

Figure 1. Impact rate of dust particles observed by Ulysses around Jupiter fly-by. The curves show all impacts recorded (upper curve) and impacts of dust particles with masses greater than 10^{-15} kg (lower curve). The impact rates are means always including six impacts. The distance from Jupiter is indicated at the top. Note that after Jupiter fly-by the craft receded from the planet at about 35° jovigraphic latitude.

The next major step forward in the investigation of Jovian dust came from the Ulysses spacecraft which flew by the planet in 1992. Ulysses is equipped with a highly sensitive impact-ionization dust detector capable of measuring dust grains down to sizes of 0.1μm [6]. With Ulysses, periodic collimated streams of dust particles with up to 2000 impacts per day were discovered while the spacecraft was within 2 AU from the giant planet [7,8] (Figure 1). The streams occurred at approximately monthly intervals (28 ± 3 days) and their impact directions implied that the grains originated from the Jovian system. No periodic phenomenon for small dust grains in interplanetary space was known before.

Confirmation of the Jupiter dust streams came from the Galileo spacecraft which carries a twin of the Ulysses dust detector on board [9]: dust 'storms' with up to 10,000 impacts per day were recorded while Galileo was within 0.5 AU from the planet. [10,11]. Since December 1995, Galileo has been the first man-made spacecraft in orbit about a giant planet of our Solar System. It explores Jupiter, its satellites and its huge magnetosphere. With the Galileo dust detector the dust streams seen in interplanetary space were also detected within the planet's magnetosphere. The grains showed a strong electromagnetic interaction with the Jovian magnetic field (see Sect. 2).

In December 2000 the Cassini spacecraft flew by Jupiter on its way to Saturn and provided a unique opportunity for simultaneous two-spacecraft measurements of the Jovian dust streams. The Cassini cosmic dust analyser [12] measured the chemical composition of dust stream particles in-situ for the first time.

Apart from the Jovian dust streams, Galileo allowed for studies of impact-generated dust clouds surrounding the Galilean satellites [13] (Sect. 3), a tenuous dust ring in the region between the Galilean satellites [14,15] and further out from the satellites [16] as well as interplanetary and interstellar particles captured by the Jovian magnetosphere [17,18] (Sect. 4). The detection of most of the observed features was unexpected and their discovery has greatly expanded our knowledge about dust in the Jovian magnetosphere.

Comprehensive reviews of more than 10 years of dust measurements with Ulysses and Galileo focussing on Jovian dust as well as interplanetary and interstellar dust have also been given by Grün et al. [19] and Krüger et al. [20].

2. JUPITER DUST STREAMS

2.1. Electromagnetically interacting dust

The impact directions of the dust stream particles measured with Galileo and Ulysses in interplanetary space were close to the line-of-sight direction to Jupiter. The approach direction of most streams, however, deviated too much from the direction to Jupiter to be explained by gravitational forces alone. This deviation was correlated with the magnitude and the direction of the interplanetary magnetic field [10] which implied that strong non-gravitational forces must have been acting on the grains. The observed 28 day period in the impact rate (Figure 1) was most likely caused by changes in the tangential component of the solar wind magnetic field which periodically accelerated the particles towards and away from the ecliptic plane [21,22]. Numerical simulations showed that only particles with velocities in excess of 200 $\mathrm{km\,s^{-1}}$ and radii in the range $5\,\mathrm{nm} \leq s \leq 15\,\mathrm{nm}$ were compatible with the observations [23]. Larger (smaller) grains did not interact enough (interacted too strongly) with the interplanetary magnetic field to explain the observed impact directions. This demonstrated that the solar wind magnetic field acts as a giant mass-velocity spectrometer for charged dust grains.

Strong electromagnetic interaction of dust grains was also found with the Galileo detector within the Jovian magnetosphere. Figure 2 shows an example of the impact rate measured with Galileo in the inner part of the magnetosphere. During this and most other times when Galileo collected data in this spatial region, the impact rate fluctuated with 5 and 10 h periodicities and the fluctuations typically exceeded 2 orders of magnitude. Furthermore, the impact directions of the grains and the measured charge rise times and charge amplitudes which were used to derive particle speeds and masses showed similar fluctuations [24]. These fluctuations were correlated with the position of Galileo in the Jovian magnetic field (cf. bottom panel of FIgure 2). Due to a 9.6° tilt of Jupiter's magnetic axis with respect to the planet's rotation axis the magnetic equator sweeps over the spacecraft in either up- or downward direction every 5 h.

In addition to the 5 and 10 h periods which are compatible with Jupiter's rotation period, a modulation of the impact rate with Io's orbital period (42 h) could also be recognized during some time intervals (e.g. Galileo orbits E4 [24], G7 [25] and C9 [26]) while at other times an Io modulation was missing (e.g. Galileo orbit G2, Figure 2). A detailed frequency analysis of a two year dataset showed Io's orbital frequency as a "carrier frequency" and primary source of the Jovian dust streams [27]. Jupiter's magnetic field frequency modulates Io's frequency signal, giving rise to modulation sidelobe products seen around first order (10 h) and harmonic (5 h) Jupiter magnetic field frequencies. These modulation products confirm Io's role as a primary source of the Jovian dust streams. Io as a source can best explain the time series analysis results showing Io's orbit periodicity.

An Io source is also compatible with the deduced particle sizes of $\sim 10\,\mathrm{nm}$: photometric observations of the Io plumes obtained with Voyager imply a size range of 5 to 15 nm [28], in agreement with numerical simulations [23]. Recent Hubble Space Telescope (HST) observations constrained the grains to be smaller than 80 nm [29]. Hence, given the ejection speeds of more than 200 $\mathrm{km\,s^{-1}}$, Io turned out to be a source for interplanetary and interstellar dust!

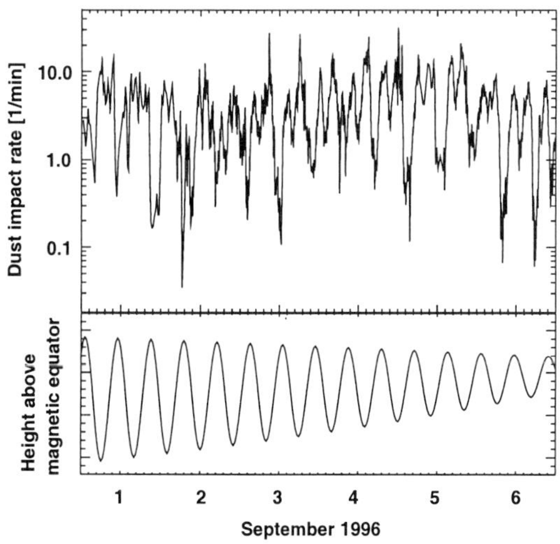

Figure 2. Top: Impact rate of Jovian dust stream particles measured in September 1996 (Galileo's G2 orbit). In the time period shown the spacecraft approached the planet from 60 to 10 R_J jovicentric distance (Jupiter radius, R_J = 71,492 km). Note the strong fluctuations with 5 and 10 h periods. Bottom: Height of Galileo spacecraft above Jovian magnetic equator. A dipole tilted by 9.6° with respect to Jupiter's rotation axis has been assumed.

The suggested mechanism to eject dust grains from within the Jovian magnetosphere matched the size and velocity range of the observed stream particles by recognizing that these grains become positively charged in the Io plasma torus and can get accelerated by Jupiter's corotational electric field [22,30,31]. As grains traverse the various plasma regions in the torus, however, their charge will not remain constant. Dust grains escaping Io's plumes first enter the cold plasma torus where they become negatively charged (~ -3 V). Grains that reach the outer hot regions of the torus change their sign of charge to positive ($\sim +3$ V) because of secondary electron emission. Once positively charged, grains will be accelerated by the outward pointing corotational electric field. They will leave the Jovian system if their radii are between about 9 and 180 nm [24]. Smaller grains remain tied to the magnetic field lines and gyrate around them like ions do, whereas bigger grains move on gravitationally bound orbits which are – depending on the particle size – more or less affected by the Lorentz force. Recent investigations showed that a higher secondary electron yield which leads to potentials of -5 V in the cold torus and $+5$ V elsewhere gives better agreement with the observations [32].

Since Io is located very close to Jupiter's equatorial plane, the particles are to a first order approximation accelerated outward along this plane. Because of the 9.6° tilt of Jupiter's magnetic field with respect to the planet's rotation axis, however, the particles also experience a significant out-of-plane component of the Lorentz acceleration: particles continuously released from Io move away from Jupiter in a warped dust sheet which has been nick named 'Jupiter's dusty ballerina skirt' [30]. A detector attached to a spacecraft moving in Jupiter's equatorial plane detects an increased number of particles when this dust sheet sweeps over its position. The 5 and 10 h fluctuations in the dust impact rate as well as the impact directions of grains observed by Galileo [24] can be explained with this scenario of electromagnetically coupled dust grains. However, only grains within a narrow size range around 10 nm are in agreement with the observed features. Smaller and larger stream particles were not detected with the Galileo dust instrument.

The charge of a particle escaping from the Io torus strongly depends on variations in the plasma density and temperature in space and time and thus is a function of Io's position at the time of particle release. In fact, the position where a particle is released from the torus is correlated with Io's position (Graps, personal comm.). In addition, the torus shows a strong dawn-to-dusk asymmetry in the plasma conditions that influence the escape of the dust particles. Grain charges are more negative on the dawn side of the torus where a lower electron temperature leads to a reduced secondary electron production. Particles on the dawn side remain captured in the torus for longer times because of their lower positive charge. Six years of Galileo dust stream measurements clearly show a variation of the flux with Jovian local time: significantly higher dust fluxes were measured on the dawn and on the dusk sides than on the noon side of Jupiter (Krüger et al., in prep.) as predicted by numerical modelling [31]. Thus, the Jovian dust streams serve as tracers of the plasma conditions in the Io torus.

The fly-by of the Cassini spacecraft at Jupiter in December 2000 provided a unique opportunity for a two-spacecraft time-of-flight measurement (Cassini-Galileo) of particles from one collimated stream from the Jovian dust streams. Particles in a stream were detected with Galileo as the spacecraft was inside the Jovian magnetosphere close to the orbit of Europa (about $12\,R_J$), and then particles in the same stream were detected by Cassini outside the magnetosphere (at $140\,R_J$). The Cassini data imply that particles of different sizes have different phases with respect to Jupiter's rotation (Kempf et al., in prep.), a result which was also seen in earlier Galileo data [24]. The comparison of the measurements from both dust instruments, however, is hampered by the higher detection sensitivity of the Cassini detector with respect to the Galileo detector. Both instruments have detected stream particles with different sizes and, hence, different phases. The analysis is ongoing and more detailed modelling to describe the phase relation of different-sized particles is in progress. The present analysis indicates particle speeds of about $400\,\mathrm{km\,s^{-1}}$. This value is in agreement with speeds for 5 nm particles as derived from dynamical modelling and earlier studies of the Jovian dust stream dynamics [23].

The Cassini dust instrument is equipped with a time-of-flight mass spectrometer which measures the elemental composition of dust grains with a mass resolution $M/\Delta M \approx 100$. During Cassini's approach to Jupiter impact spectra of a few hundred dust stream particles have been measured and their chemical composition reflects the chemistry found on Io. With the Cassini instrument the surface composition of a satellite other than our Moon has been measured directly.

2.2. Io as a source of dust in the Jovian system

How significant is Io as a source of cosmic dust? How does the amount of dust ejected compare with other dust sources in the Solar System? With a simple calculation we can derive the total dust production rate of Io. Given the spread of Io dust along and away from Jupiter's equatorial plane, we assume a cone-shaped emission pattern of dust originating at Jupiter. We assume a cone opening angle of 35° and isotropic dust emission towards all jovigraphic longitudes. Although Galileo measurements were obtained only along the Jovian equatorial plane, this opening angle is justified by the Ulysses measurements. Ulysses measured the dust streams at 35° jovigraphic latitude after Jupiter fly-by (cf. Figure 1). For a given impact rate R, particle density $\rho = 2\,\mathrm{g\,cm^{-3}}$, particle radius $a = 10\,\mathrm{nm}$, a detector sensitive area of $A = 0.02\,\mathrm{m^2}$ and a cone radius $r = 30\,R_J$ the

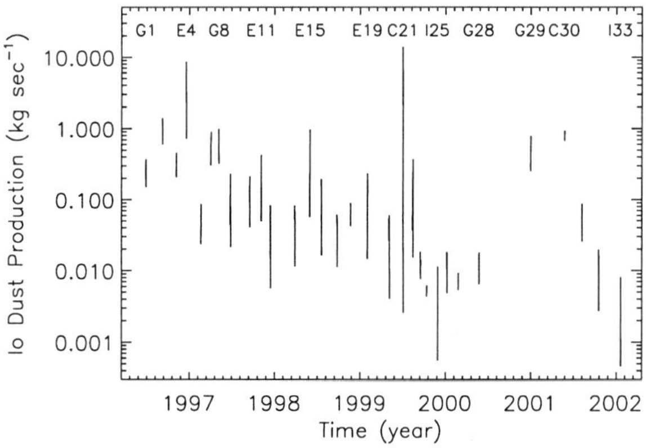

Figure 3. Total dust production rate of Io assuming that the grains are ejected into a cone with 35° opening angle centered at Jupiter. Each vertical bar represents data from one Galileo orbit. The height of the bar shows the dust production rate derived from measurements between 10 to 30 R_J from Jupiter. The data have been corrected for a Jovian local time variation of the dust emission from the Io torus and for a long-term change of the dust instrument sensitivity (Krüger et al. in prep.). The labels of individual Galileo orbits are indicated at the top. No dust stream measurements were collected during Galileo orbits 5 and 13.

total amount of dust emitted from Io per second can be calculated. With $R = 0.1\ldots100$ impacts per minute detected from 1996 to 2001 the average dust ejection rate is $10\,\mathrm{g\,s^{-1}}$ to $10\ \mathrm{kg\,s^{-1}}$(Figure 3). If we take a typical value of $1\mathrm{kg\,s^{-1}}$ of dust and compare it with $1\,\mathrm{ton\,sec^{-1}}$ of plasma ejected from Io into the torus, the dust amounts to only 0.1% of the total mass released. These numbers indicate that Io's volcanic plumes are also a minor source for interplanetary dust compared with comets or main belt asteroids [33]. Io, however, turns out to be a major dust source for the Jovian system itself. The total mass of dust produced by Io as 10 nm-sized particles is comparable to the mass of dust ejected as micrometer-sized particles by the other Galilean satellites, which have no volcanic activity (Sect. 3).

The Jovian dust stream measurements can serve as a monitor of Io's volcanic plume activity. With Galileo imaging ten active plumes have been observed which is comparable with nine plumes seen by Voyager [34]. At least tow types of plumes can be distinguished: large, faint ones, with short-lived or intermittent activity (Pele-type) or small, bright, long-lived ones (Prometheus-type). The most powerful plume ever detected on Io, Pele, is the archetype of the first category and was observed to an altitude of more that 400 km [29]. Pele is also the location of the most stable high temperature hot spot on Io and is probably related to an active lava lake. Plumes are normally related to hot spots but not vice versa. The Pele plume is known to be rich in S_2 gas as well as SO_2 [35]. Although it has been suggested that the Pele plume may be a pure gas plume, plume observations can also be interpreted as due to very fine ($\leq 80\,\mathrm{nm}$) particulates [29].

It is of special interest to see whether variations in the dust production rate deduced from the dust stream measurements can be related to the activity of the Pele or other plumes on Io, or to the total thermal output of the satellite. A correlation with the activity of the Pele plume seems most promising because only the most powerful plumes are expected to accelerate the grains to sufficient altitudes so that they can finally escape from the satellite [4,36].

The dust production of Io for individual orbits of Galileo is shown in Figure 3. Here, the vertical bars indicate the variation in the derived dust production rate if we vary the jovicentric distance at which the dust flux is taken between 10 and 30 R_J during one orbit. This reveals a strong variation in the dust production rate from orbit to orbit which is up to two orders of magnitude. If the plasma conditions in the Io torus and the Jovian magnetic field did not change too drastically from orbit to orbit, it reflects the variation of the activity of the Io plumes.

We have compared the dust production rate shown in Figure 3 with the total thermal output of Io deduced from Galileo near-infrared measurements (Spencer, personal comm.). Unfortunately, this did not give a clear picture. This negative result, however, is not too surprising because Io's overall thermal output is not very well correlated with plume activity. The Pele plume was observed in July 1995, July 1996, December 1996 and possibly July 1997 [34]. It was absent in June 1996, February 1997, June 1997 and July 1999. Although, the strong drop in the dust impact rate from December 1996 to February 1997 (E4 to E6 orbit) is consistent with these detections/non-detections, for other measurements it is not. Especially, the non-detection of the plume on 2 July 1999 is in contradiction with the large measured dust emission.

A correlation of the in-situ dust measurements with either Galileo or Earth-based imaging observations turns out to be very difficult because the imaging observations represent only sporadic glimpses. Many more observations would be needed to establish a firm link between the Galileo dust measurements and the activity of (an) individual plume(s) on Io. The picture is further complicated by the fact that the plume activity sometimes changes on timescales of days to weeks. Ideally, one would need imaging observations at exactly the same time as the dust measurements.

We have also estimated the Io dust production from the measurements of Galileo and Ulysses in interplanetary space out to 1 AU from Jupiter assuming again that the dust is uniformly distributed into a cone of 35°. This leads to unrealistically high dust production rates of more than 10^7 $\mathrm{kg\,s^{-1}}$. It indicates that this simple picture cannot be extrapolated to interplanetary space and that the dust is not distributed uniformly to such large distances. Rather, the dust particle trajectories must undergo some focussing effect due to electromagnetic interaction with the interplanetary magnetic field.

Additional evidence for such a focussing effect came from Galileo measurements in 2000 when the spacecraft has left the Jovian magnetosphere for the first time since 1995. Measurements outside the magnetosphere at a distance of $\sim$ 280 R_J (0.13 AU) from Jupiter gave a surprisingly high impact rate of up to 10 impacts per minute (Figure 4). This value was comparable with the rates detected both in interplanetary space (Figure 1) and close to Jupiter during Galileo's early orbital mission (Figure 2).

In May and June 2000 (days 145 to 170), while Galileo was receding from Jupiter (from 10 to 170 R_J), the impact rate dropped by more than two orders of magnitude (from 0.05 to 0.0005 impacts per minute). This drop was close to the inverse square of the source

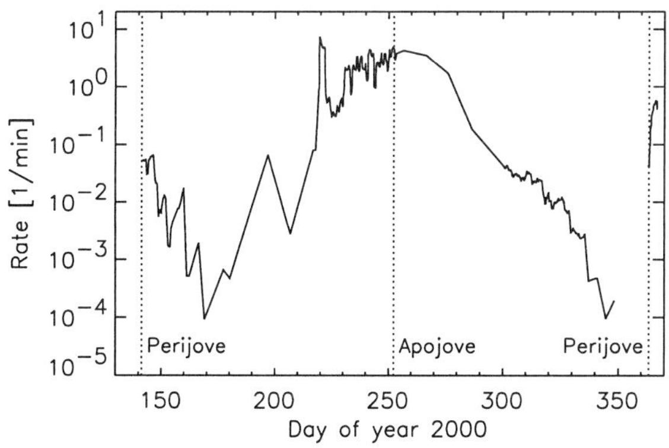

Figure 4. Impact rate of dust stream particles measured with Galileo between May and December 2000 (Galileo G28 orbit). Galileo perijove (at $\sim 8\,R_J$ distance from Jupiter) and apojove ($\sim 280\,R_J$) are indicated. The impact rate has been averaged over a 40 h time interval. Due to different modes of dust instrument read-out the data have a time resolution varying from less than an hour to more than 20 days (28). The dust streams were outside the field of view of the dust detector before day 141 and again between day 350 and 363.

distance. When Galileo was outside the magnetosphere, beyond $\sim 200\,R_J$ from Jupiter (after day 180), the impact rate increased by about four orders of magnitude. Between August and October 2000 (days 230 to 280), Galileo remained more or less stationary with respect to Jupiter and Io, and the impact rate remained remarkably constant for about two months with roughly 1 impact per minute. Assuming – as before – that dust particles get ejected into a 35° cone, this leads to a dust production of Io of $\sim 100\,\mathrm{kg\,s^{-1}}$. It seems unlikely that such a high dust production is maintained over such a long time period. More likely is a focussing effect of the grains due to the boundary between the Jovian magnetosphere and the interaction with the interplanetary magnetic field. Interestingly, the impact directions measured with Galileo indicate that the grains approached the sensor from a direction very close to the ecliptic plane. Similarly high impact rates were was also detected with the Cassini dust instrument [12] in September 2000 at ~ 0.3AU from Jupiter when the spacecraft was approaching the planet (Kempf et al., in prep.).

Frequency analysis of the Galileo dust impact rates measured beyond $\sim 250\ R_J$ did not reveal 5 and 10 h periodicities as was seen within the magnetosphere. Instead, a strong peak at Io's orbital period showed up in the frequency spectrum (A. Graps, personal comm.), much stronger than seen close to Jupiter.

3. DUST-ENSHROUDED SATELLITES

Between December 1995 and January 2002 the Galileo spacecraft had a total of 31 targeted encounters with all four Galilean satellites. During many of these fly-bys the impact rate of dust grains showed a sharp peak within about half an hour centered on closest approach to the satellite [38,24,25]. This indicated the existence of dust concentrations in

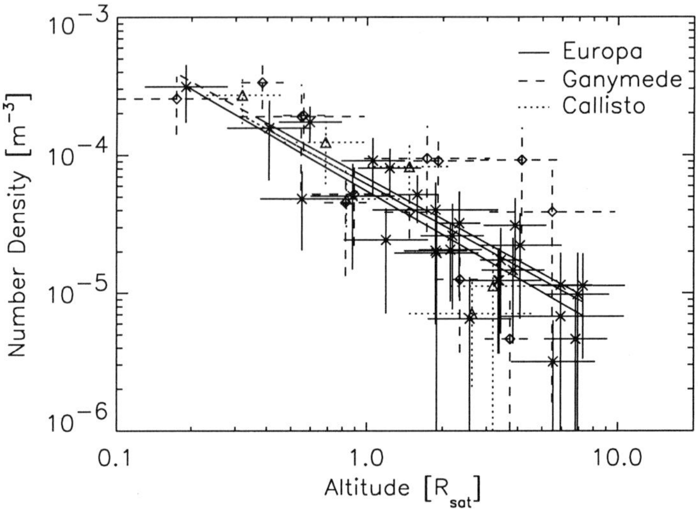

Figure 5. Number density of dust as a function of altitude above the surface of Ganymede (data from 4 fly-bys), Europa (8 flybs) and Callisto (3 fly-bys). The altitude is shown in units of the satellite radius $R_{sat} = 1560, 2634, 2409\,km$ in the case of Europa, Ganymede and Callisto, respectively. Vertical error bars reflect statistical uncertainty due to the small number of impacts. The solid lines are least squares fits to the measured number densities.

the close vicinities of Europa, Ganymede and Callisto. No dust cloud could be measured close to Io because the spacecraft orientation prevented the detection of dust particles during all fly-bys at this satellite.

Analysis of the impact directions and impact speeds showed that the grains belonged to steady-state dust clouds surrounding the satellites [13,39]. The measured radial density profiles of the dust clouds (Figure 5) together with detailed modelling of the impact-ejection process implied that the particles had been kicked up by hypervelocity impacts of micrometeoroids onto the satellite's surface [16]. The projectiles were most likely interplanetary dust particles.

The measured mass distribution of the grains was consistent with such an ejection mechanism with grain sizes being mostly in the range $0.5\ \mu m \leq s \leq 1.0\ \mu m$. It implied that the particle dynamics was dominated by gravitational forces, whereas non-gravitational, especially electromagnetic forces were negligible. Most ejected grains follow ballistic trajectories and fall back to the surface within minutes after they have been released. Only a small fraction of the ejecta has sufficient energy to remain at high altitudes for several hours to a few days. Although they eventually strike the satellite's surface, these short-lived but continuously replenished particles form a tenuous steady-state dust cloud which entirely envelopes the satellite. The total amount of debris contained in such a steady-state cloud is roughly 10 tons.

The optical thickness of the cloud is by far too low to be detectable with imaging

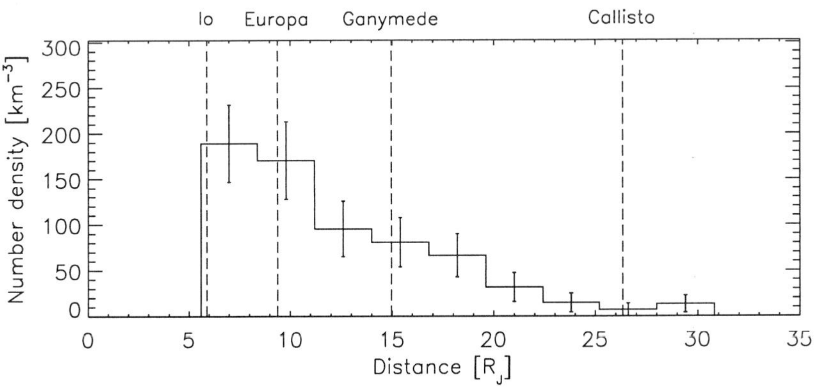

Figure 6. Number density of micrometer-sized dust particles between 6 and 30 R_J constructed from the Galileo measurements. The orbits of the Galilean satellites are indicated by vertical dashed lines. Only data from Galileo orbits I24 to I33 are shown because earlier orbits only partially traversed the region between Io and Europa. Dust cloud particles identified in the close vicinity of the moons are not shown.

techniques. Only a highly sensitive detector of the Galileo/Ulysses type could recognize a sufficient number of grains to detect these clouds. The low dust density is illustrated by the fact that only 35 cloud particles impacted the detector during 4 fly-bys at Ganymede [13].

A detailed analysis of the entire Galileo dataset for the three Galilean satellites is ongoing. One goal is to check for signatures of a leading-trailing asymmetry of the ejecta clouds, which can be expected from the orbital motion of the satellite with respect to the field of impactors [40].

The Galileo measurements are the first successful in situ detection of satellite ejecta in the vicinity of a source moon. All celestial bodies without gaseous atmospheres (asteroids, planetary satellites of all sizes) should be surrounded by an ejecta dust cloud. Before Galileo, there were few attempts of direct in-situ detections of ejecta close to satellites — most notably, near the Moon [41]. These experiments, however, did not lead to definite results.

4. DUSTY JOVIAN RINGS

Apart from Io dust streams (Sect. 2) and circum-satellite ejecta-clouds (Sect. 3) the in-situ Galileo measurements have revealed additional populations of Jovian dust (Table 1): since the beginning of Galileo's orbital tour about Jupiter the dust detector has measured more than 400 impacts of mostly micrometer-sized grains widely distributed in circum-jovian space. Although the highest fluxes of grains occurred in the region between Io's and Callisto's orbit ($\sim$ 6 to 26 R_J from Jupiter, [24,25], Figure 6) impacts were also detected out to 200 R_J and beyond. These grains form a tenuous dust ring around Jupiter with a number density of $\approx 2 \cdot 10^2\,\text{km}^{-3}$ at Europa's orbit. The spatial locations where these grains were detected, the impact directions and the charge signals imply that these are actually two populations: besides a population of particles on prograde orbits about Jupiter, another population on retrograde orbits must exist as well [14]. The grains on retrograde orbits are most likely interplanetary or interstellar grains captured by the Jovian magnetosphere [17,18]. Numerical models show that a tiny fraction of the impact debris released from the surface of the satellites by hypervelocity impacts (Sect. 3) is ejected at speeds sufficient to escape from the satellites entirely [15] (an amount of $10\,\text{g}\,\text{sec}^{-1}$ has been estimated to leave Ganymede). The ejected material goes into orbit about Jupiter and forms a tenuous ring of dust particles mostly on prograde orbits. This ring extends at least from Io's orbit ($5.9\,R_J$ from Jupiter) out to Callisto's orbit ($26\,R_J$) but the dust detections indicate that it continues further out and further in (see below).

In the outer region of the Jovian system, between 50 and 300 R_J, about 100 dust impacts were detected. Their orbits are compatible with prograde and retrograde jovicentric orbits with a wide range of inclinations [16]. The number densities of $\sim 10\,\text{km}^{-3}$ are more than an order of magnitude lower than those found in the region between the Galilean satellites but, on the other hand, by about an order of magnitude larger than the interplanetary background. Sources for these grains are Jupiter's outer regular and irregular moons.

Indications for the existence of the ring can already be found in earlier measurements by the Pioneer 10/11 and Ulysses spacecraft: 12 meteoroid penetrations have been recorded with Pioneer within $45\,R_J$ (Jupiter radius, $R_J = 71{,}492\,\text{km}$) from Jupiter [1] and Ulysses has recorded 9 impacts of micrometer-sized dust grains in this spatial region. Two-third of the Ulysses impacts were detected at $\sim 35°$ jovigraphic latitude after Jupiter fly-by.

Between Io's orbit at $5.9\,R_J$ and the outer extension of the gossamer ring at about $3.1\,R_J$, extremely little is presently known about the dust environment. Although Galileo has traversed part of this region during orbit insertion in December 1995, dust measurements were very patchy because the instrument had to be saved from the hazards of Jupiter's radiation environment. However, a few probably micrometer-sized dust impacts were detected within Io's orbit [42,11].

Still closer to Jupiter lies the region of Jupiter's prominent ring system which consists of three components: the main ring, the halo and the tenuous gossamer rings. Here, the dust densities are so large that dust investigations habe been performed with remote sensing techniques. The vertical extension and density profiles of the rings imply that a significant fraction if not all of the dust forming the rings is impact-ejecta derived from the inner moons Adrastea and Metis (in the case of the main ring), and Amalthea and Thebe (in the case of the gossamer rings [43]). These satellites orbit Jupiter inside the ring system. The motion of the dust grains in a certain size range contained in the gossamer ring is most probably dominated by the Poynting-Robertson drag force, indicating that the plasma density in this region is much lower than previously thought [44].

5. CONCLUSIONS AND OUTLOOK

The Galileo dust measurements have drastically expanded our knowledge about dust in the Jupiter system. In fact, Jovian dust has been studied to at least a similar extent as cosmic (i.e. non-artificial) dust in the Earth environment. The properties of the various

Table 1
Physical parameters of dust populations detected in-situ at Jupiter. Column 2 gives typical particle sizes (radii) assuming spherical particles, col. 3 the mean measured impact speeds, col. 4 lists the radial distance range where the particles have been detected, and col. 5 gives derived particle number densities in space.

Population (1)	Particle size (μm) (2)	Impact speed ($\mathrm{km\,s^{-1}}$) (3)	Jovicentric distance (4)	Number density ($\mathrm{m^{-3}}$) (5)
Stream particles	$\sim 0.01^*$	$\leq 400^*$	$6\,R_J$– 2 AU	$10^{-3} - 10^{-8}$
Ejecta clouds	0.3 – 1	6 – 8	$\leq 10\,R_{sat}{}^{\dagger}$	$10^{-4} - 10^{-5}$
Ejecta ring	0.6 – 2	~ 7	6 – 30 R_J	$10^{-6} - 10^{-7}$
Captured particles	0.5 – 1.5	~ 20	6 – 20 R_J	$\sim 10^{-7}$
Outskirts ring	1 – 2	~ 5	$\geq 50\ R_J$	$\sim 10^{-8}$

$*$: derived from dynamical modelling.
$\dagger$: altitude above satellite surface.

Jovian dust populations studied in-situ with Galileo are summarised in Table 1.

All Galilean satellites are sources of dust in the Jovian system. The Galileo measurements have for the first time demonstrated the electromagnetic interaction of charged dust grains with a planetary magnetosphere. Jupiter's magnetosphere acts as a giant mass-velocity spectrometer for charged dust grains in space. The Io dust streams can be used as a potential monitor of the activity of Io's plume activity.

The Io dust stream particles probe the conditions in the Io plasma torus. Since in a completely radially symmetric plasma and magnetic field configuration no 10 h period should show up in the impact rate, only the 5 h period should be there. The prominent modulation of the rate with the 10 h period points to variations in the acceleration mechanism of the grains correlated with Jovian local time which are presently not completely understood.

In February 2004, Ulysses will approach Jupiter to 0.8 AU again. Additional dust stream measurements with Ulysses in interplanetary space at high jovigraphic latitudes and for varying Jovian local times will be beneficial to test our understanding of this new phenomenon.

The Galileo measurements of impact-generated dust clouds surrounding the Galilean satellites can be considered as unique natural impact experiments to study the dust ejection mechanism due to hypervelocity impacts onto celestial bodies without atmospheres. They complement laboratory experiments in an astrophysically relevant environment. Although far from being perfect impact experiments, the Galileo results offer two extremely important improvements over laboratory experiments: 1) the projectile and target materials and projectile speeds are astrophysically relevant, and 2) the masses and speeds of the ejecta particles can be determined in an important region of parameter space (micrometre sizes and $\mathrm{km\,s^{-1}}$ impact speeds). This is especially important in view of the Cassini mission. Cassini will start its exploration of the Saturnian system in 2004 and

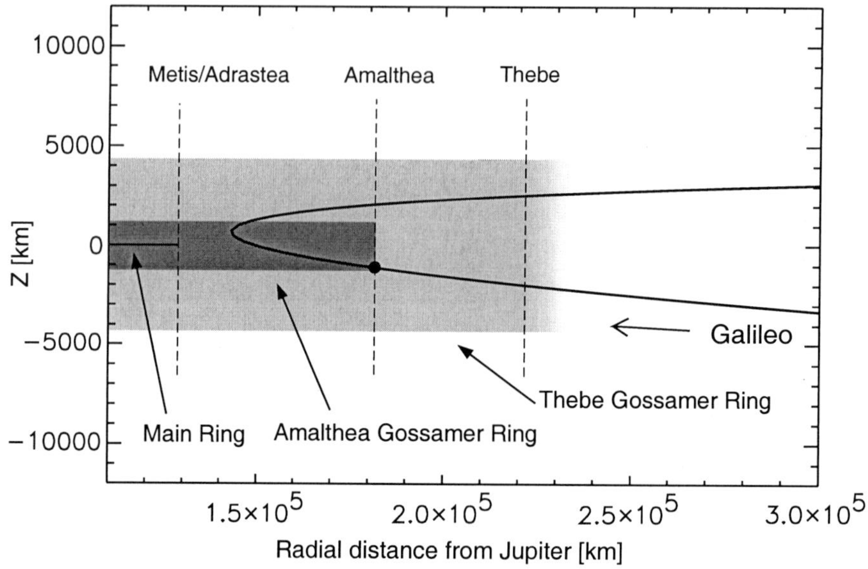

Figure 7. Galileo trajectory during the passage of the Jovian gossamer rings in November 2002. The position of Amalthea during closest approach is shown as a filled circle.

will fly by several of Saturn's satellites during its orbital tour about the giant planet. It will provide a unique opportunity to study the dust environments of many of the small Saturnian satellites.

Although considered to be the archetype of an ethereal dusty planetary ring, the Jovian gossamer and main ring system has been relatively incompletely studied to date. The in-situ measurements of ejecta grains escaping from the circum-satellite dust clouds and images of Jupiter's main and gossamer rings have demonstrated that impact-ejecta derived from hypervelocity impacts onto satellites is the major – if not the only – constituent of these dusty planetary rings. The details of the complex dynamics of grains over a large size range and under the various forces acting on the grains are as yet only poorly understood.

In November 2002 – during its final orbit about Jupiter – Galileo will traverse the gossamer ring system and fly by Amalthea (Figure 7). Detailed in-situ studies of the dust grains in the gossamer rings will provide a better understanding of the forces dominating the grain dynamics in the rings (gravity, Lorentz force, plasma drag, Poynting-Robertson drag, radiation pressure). The relative importance of each force varies strongly with grain size and distance from the planet and leads to drastically different size distributions at different locations along the gossamer rings and in the main ring. Investigation of why the Poynting-Robertson drag dominates over the other forces will lead to a comprehensive picture of the grain dynamics in the gossamer ring, a necessary step in deriving a full picture of the dust dynamics throughout the Jovian magnetosphere. In-situ studies of the ring material can provide valuable information about the surface properties of the source moons. Comparative studies of ejecta from the large Galilean moons and the smaller ones embedded in the gossamer rings will provide information about the ejection process over a large range in speed not accessible in the laboratory. Especially the close fly-by at Amalthea (< 300 km) will allow to test on a small moon the models for the impact-ejection process which have been developed for the much larger Galilean satellites.

In-situ dust measurements provide information about the physical properties of the dust environment not accessible with imaging techniques. Since all dusty planetary rings in our Solar System are most likely dominated by impact-ejecta, studies of Jupiter's gossamer ring provide valuable information not only about the mechanism feeding this ring system but also about the processes that govern planetary rings in general. Studies of the Jovian ring with Galileo and of the Saturnian ring with Cassini will lead to a vastly improved understanding of the formation and evolution of dusty planetary rings.

Acknowledgements. We wish to thank the Ulysses and Galileo projects at ESA and NASA/JPL for effective and successful mission operations. We also wish to express our gratefulness to our co-investigators. Without their continuous efforts over the years the results presented here would not have been achievable. We are also grateful to John R. Spencer and John Stansberry who provided thermal near-infrared data of Io. This research has been supported by the German Bundesministerium für Bildung und Forschung through Deutsches Zentrum für Luft- und Raumfahrt e. V. (DLR, grant 50 QJ 9503 3). Support by Deutsche Forschungsgemeinschaft (DFG) and Max-Planck-Institut für Kernphysik (MPIK) are also gratefully acknowledged.

REFERENCES

1. D.H. Humes, J.M. Alvarez, R.L. O'Neal and W.H. Kinard, J. Geophys. Res. 79 (1974) 3677.
2. R.W. Fillius, C.E. McIlwain and A. Mogro-Campero, Science 188 (1975) 465.
3. M.H. Acuña and N.F. Ness, J. Geophys. Res. 81 (1976) 2917.
4. T.V. Johnson, G. Morfill, and E. Grün, Geophys. Res. Lett. 7 (1980) 305.
5. G.E. Morfill, E. Grün and T.V. Johnson, Planet. Space Sci. 28 (1980) 1087.
6. E. Grün, H. Fechtig, J. Kissel, D. Linkert, D. Maas, J.A.M. McDonnell, G.E. Morfill, G. Schwehm, H.A. Zook and R.H. Giese, Astron. Astrophys. Supp. 92 (1992) 411.
7. E. Grün, H.A. Zook, M. Baguhl, A. Balogh, S.J. Bame, H. Fechtig, R. Forsyth, M.S. Hanner, M. Horányi, J. Kissel, B.A. Lindblad, D. Linkert, G. Linkert, I. Mann, J.A.M. McDonnell, G.E. Morfill, J.L. Phillips, C. Polanskey, G. Schwehm, N. Siddique, P. Staubach, J. Svestka and A. Taylor, Nature 362 (1993) 428.
8. M. Baguhl, E. Grün, G. Linkert, D. Linkert and N. Siddique, Planet. Space Sci. 41 (1993) 1085.
9. E. Grün, H. Fechtig, M.S. Hanner, J. Kissel, B.A. Lindblad, D. Linkert, D. Maas, G.E. Morfill and H. A. Zook, Space Science Reviews 60 (1992) 317.
10. E. Grün, M. Baguhl, D.P. Hamilton, R. Riemann, H.A. Zook, S. Dermott, H. Fechtig, B.A. Gustafson, M.S. Hanner, M. Horányi, K.K. Khurana, J. Kissel, M. Kivelson,

B.A. Lindblad, D. Linkert, G. Linkert, I. Mann, J.A.M. McDonnell, G.E. Morfill, C. Polanskey, G. Schwehm and R. Srama, Nature 381 (1996) 395.
11. H. Krüger, E. Grün, D.P. Hamilton, M. Baguhl, S. Dermott, H. Fechtig, B.A. Gustafson, M.S. Hanner, M. Horányi, J. Kissel, B.A. Lindblad, D. Linkert, G. Linkert, I. Mann, J.A.M. McDonnell, G.E. Morfill, C. Polanskey, R. Riemann, G. Schwehm, R. Srama and H. Zook, Planet. Space Sci. 47 (1999) 85.
12. R. Srama, J.G. Bradley, E. Grün, T.J. Ahrens, S. Auer, M. Cruise, H. Fechtig, A. Graps, O. Havnes, A. Heck, S. Helfert, E. Igenbergs, E.K. Jeßberger, T.V. Johnson, S. Kempf, H. Krüger, P. Lamy, M. Landgraf, D. Linkert, F. Lura, J.A.M. McDonnell, D. Möhlmann, G.E. Morfill, G.H. Schwehm, M. Stübig, J. Svestka, A.J. Tuzzolino, R. Wäsch and H. A. Zook, Space Science Reviews, submitted.
13. H. Krüger, A.V. Krivov, D.P. Hamilton and E. Grün, Nature 399 (1999) 558.
14. K.-U. Thiessenhusen, H. Krüger, F. Spahn and E. Grün, Icarus 144 (2000) 89.
15. A.V. Krivov, H. Krüger, E. Grün, K.-U. Thiessenhusen and D. P. Hamilton, J. Geophys. Res. 107 (2002) in press.
16. A.V. Krivov, I. Wardinski, F. Spahn, H. Krüger and E. Grün, Dust on the outskirts of the Jovian system. Icarus (2002) in press.
17. J. E. Colwell, M. Horányi and E. Grün, Science 280 (1998) 88.
18. J.E. Colwell, M. Horányi and E. Grün, J. Geophys. Res. 103 (1998) 20023.
19. E. Grün, H. Krüger and M. Landgraf, in Cosmic Dust (eds. A. Balogh, R. Marsden and E. Smith) Springer Praxis (2001) 373.
20. H. Krüger, M. Horányi, A.V. Krivov and A. Graps, In Jupiter: Planet, Satellites & Magnetosphere (eds. Bill McKinnon, Fran Bagenal and Tim Dowling) Cambridge University Press (2002) submitted.
21. D.P. Hamilton and J.A. Burns, Nature 364 (1993) 695.
22. M. Horányi, G. Morfill and E. Grün, Nature 363 (1993) 144.
23. H. Zook, E. Grün, M. Baguhl, D.P. Hamilton, G. Linkert, D. Linkert, J.-C. Liou, R. Forsyth and J.L. Phillips, Science 274 (1996) 1501.
24. E. Grün, H. Krüger, A.L. Graps, D.P. Hamilton, A. Heck, G. Linkert, H.A. Zook, S. Dermott, H. Fechtig, B.A. Gustafson, M.S. Hanner, M. Horányi, J. Kissel, B.A. Lindblad, G. Linkert, I. Mann, J.A.M. McDonnell, G.E. Morfill, C. Polanskey, G. Schwehm and R. Srama, J. Geophys. Res. 103 (1998) 20011.
25. H. Krüger, E. Grün, A. Graps and S. Lammers, In Proceedings of the VII. International Conference on Plasma Astrophysics and Space Physics (eds. E. Marsch, J.Büchner, I. Axford and V. Vasyliunas) Kluwer Academic Publishers 264 (1999) 247.
26. H. Krüger, E. Grün, A. Heck and S. Lammers, Planet. Space Sci 47 (1999) 1015.
27. A.L. Graps, E. Grün, H. Svedhem, H. Krüger, M. Horányi, A. Heck and S. Lammers, Nature 405 (2000) 48.
28. S.A. Collins, J. Geophys. Res. 86 (1981) 8621.
29. J.R. Spencer, P. Sartoretti, G.E. Bellester, A.S. McEwen, J.T. Clarke and M.A. McGrath, Geophys. Res. Lett. 24 (1997) 2471.
30. M. Horányi, G. Morfill and E. Grün, J. Geophys. Res. 98 (1993) 221.
31. M. Horányi, E. Grün and A. Heck, Geophys. Res. Lett. 24 (1997) 2175.
32. M. Horányi, Physics of Plasmas 7(10) (2000) 3847.
33. F. L. Whipple, Astron. Astrophys. 187 (1987) 852.

34. A.S. McEwan, L. Keszthelyi, P. Geissler, D.P. Simonelli, M.H. Carr, T.V. Johnson, K.P. Klaasen, H.H. Breneman, T.J. Jones, J.M. Kaufman, K.P. Magee, D.A. Senske, M.J.S. Belton and G. Schubert, Icarus 135 (1998) 181.
35. J.R. Spencer, K.L. Jessup, M.A. McGrath, G.E. Ballester and R. Yelle, Science 288 (2000) 1208.
36. W.H. Ip, Geophys. Res. Lett. 24 (1996) 3671.
37. H. Krüger, E. Grün, A. Graps, D. Bindschadler, S. Dermott, H. Fechtig, B.A. Gustafson, D.P. Hamilton, M.S. Hanner, M. Horányi, J. Kissel, B.A. Lindblad, D. Linkert, G. Linkert, I. Mann, J.A.M. McDonnell, G.E. Morfill, C. Polanskey, G. Schwehm, R. Srama and H. Zook, Planet. Space Sci. 49 (2001) 1285.
38. E. Grün, P. Krüger, S. Dermott, H. Fechtig, A. Graps, B.A. Gustafson, D.P. Hamilton, A. Heck, M. Horányi, J. Kissel, B.A. Lindblad, D. Linkert, G. Linkert, I. Mann, J.A.M. McDonnell, G.E. Morfill, C. Polanskey, G. Schwehm, R. Srama and H.A. Zook, Geophys. Res. Lett. 24 (1997) 2171.
39. H. Krüger, A.V. Krivov and E. Grün, Planet. Space Sci. 48 (2000) 1457.
40. M. Sremčević, A.V. Krivov and F. Spahn, Planet. Space Sci. in prep.
41. H. Iglseder, K. Uesugi and H. Svedhem, Adv. Space Res. 17 (1996) 177.
42. E. Grün, D.P. Hamilton, R. Riemann, S. Dermott, H. Fechtig, B.A. Gustafson, M.S. Hanner, A. Heck, M. Horányi, J. Kissel, M. Kivelson, H. Krüger, B.A. Lindblad, D. Linkert, G. Linkert, I. Mann, J.A.M. McDonnell, G.E. Morfill, C. Polanskey, G. Schwehm, R. Srama and H. A. Zook, Science 274 (1996) 399.
43. M.E. Ockert-Bell, J.A. Burns, I.J. Daubar, P.C. Thomas, J. Veverka, M.J.S. Belton and K.P. Klaasen, Icarus (1999) 138.
44. J.A. Burns, M.R. Showalter, D.P. Hamilton, P.D. Nicholson, I. de Pater, M.E. Ockert-Bell and P.C. Thomas, Science 284 (1999) 1146.

CDA cruise science: Comparison of measured dust flux at 1AU with models

M. Müller[a,b], B.J. Goldsworthy[a], N. McBride[a,b], S.F. Green[a,b], J.A.M. McDonnell[a,b], R. Srama[c], S. Kempf[c] and E. Grün[c]

[a]Unit for Space Sciences and Astrophysics, University of Kent at Canterbury, CT2 7NR, UK.

[b]Planetary and Space Science Research Institute, The Open University, Walton Hall, Milton Keynes MK7 6AA, UK.

[c]Max-Planck-Institut für Kernphysik, Saupfercheckweg 1, 69117 Heidelberg, Germany

The data gathered by the Cosmic Dust Analyser (CDA) on board Cassini close to 1 AU are investigated. To compare the measured flux with models, the sensitivity of the targets is derived from calibration data. The interplanetary model by Staubach [1] underestimates the measured flux by more than one order of magnitude. Our attempt to classify the measured events as impacts on the large and the small target, indicates that there are impacts on the side walls which lead to measurable signals on the targets. This brings the measured flux closer to the model, but cannot explain the large discrepancy. It is shown that interstellar particles could fill the gap between the interplanetary flux model and the measured impact rate.

1. INTRODUCTION

In the time from April 12th and June 22nd 1999 Cassini approached the Sun from 1.2 AU to 0.74 AU. CDA was triggered 57 times. 27 of these events show clear enough signals to identify them to be due to dust impacts. CDA has two different targets [2]. The charge of positive ions produced by an impact on the targets is measured by an ion collector grid in the middle of the detector. The negative ions are detected at the targets. We compare the measured impact rate on both targets with models for the interplanetary and interstellar dust flux. To do this the sensitivity of both targets needs to be determined first. Finally, we report on the attempt to classify the 27 impacts in impacts on the large and the small target.

2. CDA SENSITIVITY

To compare the dust flux measured by CDA with models it is necessary to estimate the minimum dust particle mass which can be detected by CDA. CDA is triggered if the charge measured at one of the targets, one of the grids or the multiplier, exceeds a certain threshold. The charge produced at a target due to a particle impact is usually described by a power law dependence on mass and velocity, $Q_t = Cm^{\alpha}v^{\beta}$ (see e.g. [3]). The coefficients α, β and C are determined from calibration data taken with the CDA flight spare unit. If the mass and velocity in the power law are specified in units of 10^{-16} kg and 10 km s^{-1}, respectively, a fit to the power law gives $\alpha = 1.25$, $\beta = 4.52$ and $C = 8 \times 10^{-14}$ C and $\alpha = 0.85$, $\beta = 3.3$ and $C = 1.7 \times 10^{-13}$ C for the large and the small target, respectively.

In Figure 1 it can be seen that the data only cover a narrow strip in mass-velocity space. The cut-off towards higher masses and high velocities is due to the limitations of the Van de Graaff accelerator (VdG). The cut-off towards low masses and velocities has a different slope for low and high particle masses. The cut-off at low masses can be approximated by a line of constant kinetic particle energy. In a VdG with given acceleration voltage the kinetic energy is proportional to the particle charge. Since the charge of the particle needs to be detected in the accelerator to determine its mass and velocity [4], this cut-off is given by the minimum detectable charge and is therefore not due to CDA. The slope of the cut-off at higher masses can be explained with a curve of constant charge produced at a target and shows the sensitivity of CDA. The cut-off corresponds to the target charges of 2×10^{-14} C and 5×10^{-14} C for the large and the small target, respectively.

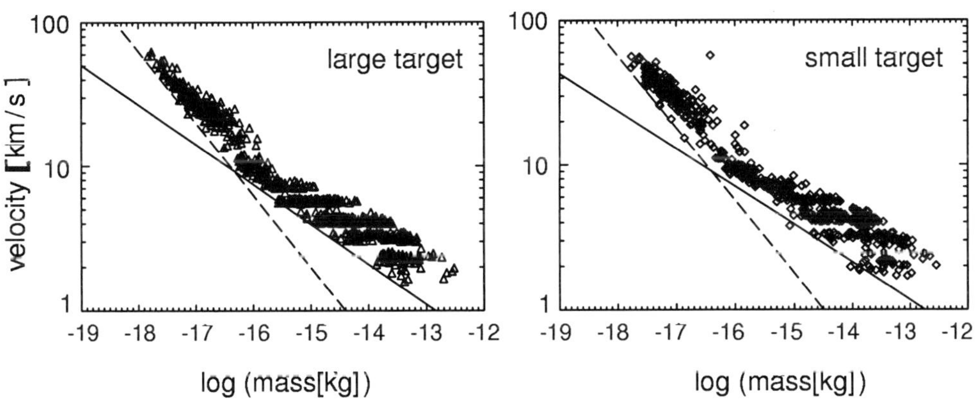

Figure 1. The particle velocity versus mass is shown for the calibration data. The dashed line corresponds to line of constant particle kinetic energy. The solid line is a line of constant target charge produced on the respective target

3. COMPARISON OF MEASURED FLUX WITH MODELS

The 27 dust impacts measured in a period of 72 days leads to a dust flux of 0.4±0.1 impacts per day. To compare this dust flux with models, the sensitive area of the large and the small target for different angles from the symmetry axis was computed. We assume 70% transmission for the entrance grid, independent of the impact direction.

In Figure 2 the measured flux is compared with different models. Divine's interplanetary dust flux model [5] is in good agreement with the measured flux. However the model by Staubach [1] underestimates the flux. Comparing the two models at 1 AU with the isotropic model by Grün et al. [6], which gives a flux of 0.8/day, we find that the Divine model is quite close to the isotropic flux (0.3/day) whereas the Staubach model is more than an order of magnitude (0.03/day) less than the isotropic model. It is important to note that the Divine model was constrained by absolute count rates of the Galileo and Ulysses dust detector only, whereas Staubach took directional information into account. As Cassini is not a spinning spacecraft, the Staubach model is therefore the most appropriate model. We must conclude that the dust flux measured by CDA cannot be explained with the interplanetary dust flux alone.

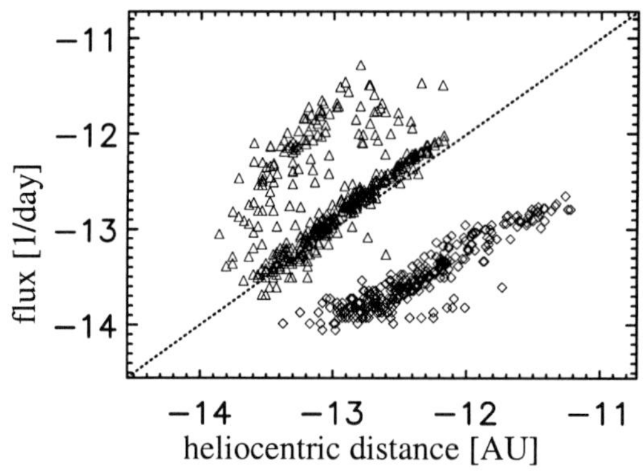

Figure 2. The measured flux rate (solid) is compared with models: Divine (short dash), Staubach (long dash), interstellar model (dotted).

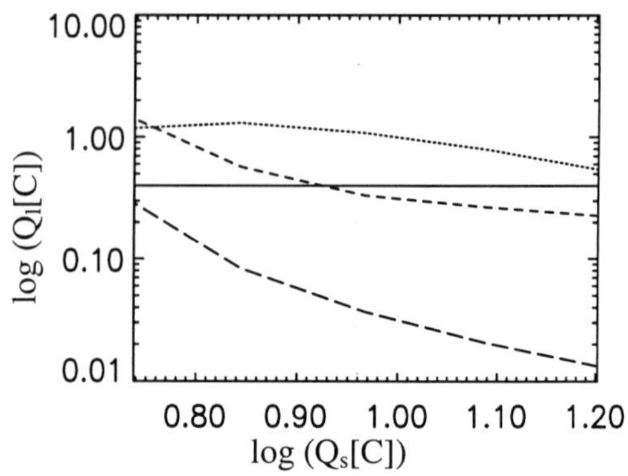

Figure 3. The charge measured at the small target (trianagles) versus the charge measured at the large target (diamonds) is shown.

Figure 2 also shows the interstellar flux of dust particles assuming that the interstellar particles move on straight trajectories through the inner solar system. According to this model, the interstellar particle flux varies between 0.5 and 2 particles per day and overestimates the measured flux slightly. However, the interstellar flux model was constructed from data measured outside 2.5 AU [7]. Since solar radiation pressure and the magnetic field in the heliosphere, tend to decrease the flux in the inner solar system [7], the interstellar flux shown in Figure 2 must be understood as an upper bound of the real flux. Therefore the gap between measured data and the Staubach model can be explained with interstellar particles.

4. IMPACT CLASSIFICATION

Ejecta produced during a primary dust impact can produce further impact plasma elsewhere in the detector. Therefore it is common that both targets show signals after an impact. Figure 3 shows for the calibration data the charges liberated on the large and the small target. Figure 3 suggests that from the ratio of the target charges it can uniquely be decided whether the primary impact took place at the large or the small target. However, it can also be seen in Figure 3 that an impact on the large target can cause as much charge on the small target as it produced on the large one. Therefore it must be suspected that impacts anywhere inside CDA can produce signals at the targets. Therefore there are three instead of two possibilities for the place of the primary impact (large target, small target and none of the targets) and the target ratio cannot uniquely decide the place of the primary impact. The target ratio can at most rule out one of the targets as the place of the primary impact:

$$\frac{Q_l}{Q_s} > 0.32 \Rightarrow \text{not small target} \qquad \frac{Q_l}{Q_s} < 0.32 \Rightarrow \text{not large target} \tag{1}$$

To decide whether the primary impact occurred at the remaining target or the wall another criterion is required. From the calibration data, it was found that the charge measured at the grid is at most 40% of the charge measured at the target, where the primary impact occurred. Therefore, if the ratio of charges measured at the grid and the target exceeds 0.4 the impact did not occur at this target. With this criterion, for 6 out of the 27 measured impacts, the targets are ruled out as the place of the primary impact. For one of these 6 events the ions arrive at the ion collector grid before a signal is observed at the target. This confirms independently that the primary impact did not hit one if the targets.

6. CONCLUSIONS

The sensitivity of both CDA targets was determined from calibration data. On the basis of this the measured impact rate was compared with existing models for the interplanetary and interstellar dust flux. Since Cassini is not a spinning spacecraft, the measured event rate cannot be compared with an isotropic model. The only interplanetary dust flux model which was constrained by directional information of impacts is the model by Staubach [1]. However, at 1AU the Staubach model underestimates the measured flux by more than one order of magnitude and the interplanetary flux cannot explain the impact rate.

Our attempt to classify the measured events into impacts on the large and small target, indicated that impacts elsewhere in the detector (e.g. the side walls) can lead to measureable signals on the targets. A preliminary criterion classified 6 out of 27 impacts not to be due to a primary impact on one of the targets. The classification as it is described here, is not able to uniquely identify whether the primary impact hit one of the targets. To do this calibration data with shots on the side-walls are required.

The impacts which did not hit one of the targets bring the measured flux closer to the flux predicted by the Staubach model. However, they cannot explain the large discrepancy between model and measurements. Therefore the comparison of the model with the measurements suggests that interstellar particle impacts were recorded. It is difficult to judge how the presented model calculations are influenced by dust impacts measured onboard Galileo and Ulysses. Therefore more direct comparison of measured dust fluxes at 1 AU, but at other ecliptic longitudes, could decide more quantitatively how many of the measured impacts are likely to be of interstellar origin. Ultimately, velocity estimations for individual events could prove that the orbit of the impacting particle was hyperbolic.

REFERENCES

1. P. Staubach, Ph.D. thesis, Univ. of Heidelberg, Germany (1996).
2. R. Srama, E. Grün, in Physics, Chemistry of Interplanetary Dust, ASP Conference Series, 104 (1996) 227.
3. F.R. Krüger, ESA SP-224, (1984) 49.
4. M.J. Burchell, M.J. Cole, J.A.M. McDonnell and J.C. Zarnecki, Meas. Sci. Technol., 10 (1999) 41.
5. N. Divine, J. Geophys. Res. 98 (1993) 17029.
6. E. Grün, H.A. Zook, H. Fechtig, R.H. Giese, Icarus 62 (1985) 244.
7. P. Landgraf, Ph. D. thesis, Univ. of Heidelberg, Germany (1998).

Halo orbits around Saturn

J. E. Howard[a] and M. Horányi[a]

[a]Laboratory for Atmospheric and Space Physics and Department of Physics, Campus Box 0392, University of Colorado, Boulder, CO 80309

We investigate the effects of planetary oblateness, quadrupole magnetic field, and solar radiation pressure on nonequatorial dust grain orbits about Saturn. Radiation pressure is found to be a strong perturber for small ($\approx$ 100 nm) positively charged conducting grains.

1. INTRODUCTION

Recently analytic equilibrium and stability conditions were obtained for charged dust grains orbiting Saturn [1,2], including both positive and negatively charged grains in prograde or retrograde, equatorial or nonequatorial ('halo') orbits. The single particle Hamiltonian model included Keplerian gravity, co-rotating magnetic field (taken to be an aligned centered dipole), and corotational electric field. Planetary oblateness (J_2), quadrupole magnetic field terms (g_2) as well as non-axisymmetric effects such as plasma drag, radiation pressure, and time-dependent charging [3,4] were all neglected. The results gave simple existence conditions and stability bounds for arbitrary circular orbits. Equatorial orbits were parametrized by cylindrical radius (ρ_0) and charge-to-mass ratio (q/m) conveniently measured by the quantity $\hat{\Phi} = \Phi_s/a_\mu^2$, where Φ_s is the surface potential of the grain in Volts and a_μ is its radius in microns. For nonequatorial orbits the spherical radius (r_0) was employed. Here we extend this model to include planetary oblateness (J_2), quadrupole magnetic field (g_2), and radiation pressure. A well depth is defined, showing that halo orbits are as deeply trapped as their equatorial cousins. Consequently these grains are not greatly perturbed by J_2 and g_2; their primary effect is to make the motion more ergodic and occasionally chaotic. Radiation pressure breaks the axisymmetry so that an effective potential no longer exists except in a average sense. Its effects are found to be much more pronounced for conducting than for dielectric grains. In general the relative strength of radiation pressure increases quadratically with distance from the planet, so that outer E-ring particles are most highly perturbed. As in our previous studies the surface potential is fixed at $\Phi_s = \pm 10V$; magnetospheric effects [5] are described elsewhere [6].

2. HAMILTONIAN MODEL

The motion of a charged dust grain orbiting Saturn may be described in an inertial frame by the scaled Hamiltonian, in cylindrical coordinates $(\rho,\ \phi,\ z)$ [2]

$$H = \frac{1}{2}(p_\rho^2 + p_z^2) + U^e(\rho, z) - k\omega_k^2 \rho \cos\phi \tag{1}$$

where

$$U^e = \frac{1}{2\rho^2}\left(p_\phi - \omega_c \Psi\right)^2 + U + \omega_c \Omega \Psi \tag{2}$$

is the two dimensional effective potential. Here distances are measured in units of the planetary radius (R_p), Ω is the planetary spin rate, $p_\phi = \rho^2\omega + \omega_c\Psi$ is the scaled canonical momentum, $\omega_c = qB_0/mc$ is the cyclotron frequency, and $\omega_k = \sqrt{GM_p/R_p^3}$ is the Kepler frequency, both evaluated on the planetary equator. The x-axis is directed from Saturn *toward* the Sun. The gravitational potential is, in scaled coordinates, including planetary oblateness

$$U = -\frac{\omega_k^2}{r}\left[1 + \frac{1}{2}J_2\frac{(\rho^2 - 2z^2)}{r^3}\right] \tag{3}$$

where $r = \sqrt{\rho^2 + z^2}$ and $J_2 = 0.01667$. The magnetic stream function, including both dipole and quadrupole terms, is

$$\Psi = \frac{\rho^2}{r^3}\left[1 + \frac{3g_2 z}{2r^2}\right] \tag{4}$$

with $g_2 = 0.01642$. Note that the quadrupole part vanishes on the equatorial plane. Radiation pressure is measured by the dimensionless coefficient

$$k = \beta\left(\frac{d}{R_p}\right)^2\left(\frac{M_{sun}}{M_p}\right) \tag{5}$$

where d is the Sun-planet distance and $\beta = J_0 Q_{pr}/\rho_g r_g$, with solar constant $J_0 = 5.7\times10^{-5}$ and Q_{pr} ranges from ≈ 0.3 for a 100 nm dielectric grain to more than 2.0 for a 1 μm conducting grain [7]. We shall take $\rho_g = 1$ g cm^{-3} throughout.

The equilibrium (circular) orbits are given by $\nabla U^e = 0$; stability is determined by the Hessian det D^2U^e, which gives the type of each critical point. Each stable critical point is surrounded by a potential well, bounded by one or more saddle points. Thus, one speaks of *saddle point confinement* [8], in which all orbits with total energy $E < E_s$ are confined. Figure 1a depicts a positively charged conducting grain in a prograde equatorial orbit at $\rho_0 = 2$ with $\hat{\Phi} = 100$, corresponding to grain radius $a = 158$ nm. As shown in [1] these orbits destabilize at a critical value of q/m which is easily calculated analytically for $J_2 = g_2 = k = 0$. In each case either a pitchfork or a tangent bifurcation of the effective potential occurs. In [2] we showed that non-equatorial 'halo' orbits exist for both positively and negatively charged grains, but that retrograde orbits were impossible for negatively charged grains. Analytic stability boundaries were derived and corresponding bifurcations identified, again all with $J_2 = g_2 = k = 0$. Figure 1b shows a typical halo orbit for a prograde positively charged conducting grain, with $r_0 = 5$ and $\hat{\Phi} = 1100$ ($a = 95$ nm).

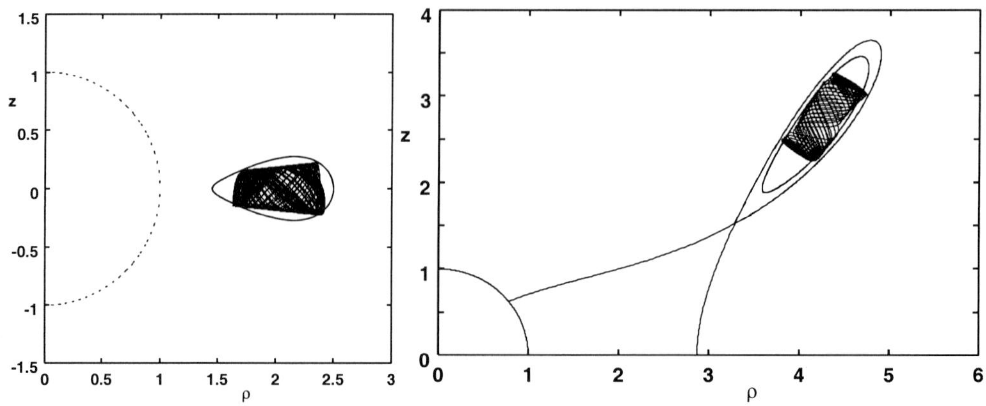

Figure 1. Trapped orbits for conducting grains near Saturn: (a; left) equatorial $\rho_0 = 2$, $\hat{\Phi} = 400$, (b; right) halo, $r_0 = 5$, $\hat{\Phi} = 1100$

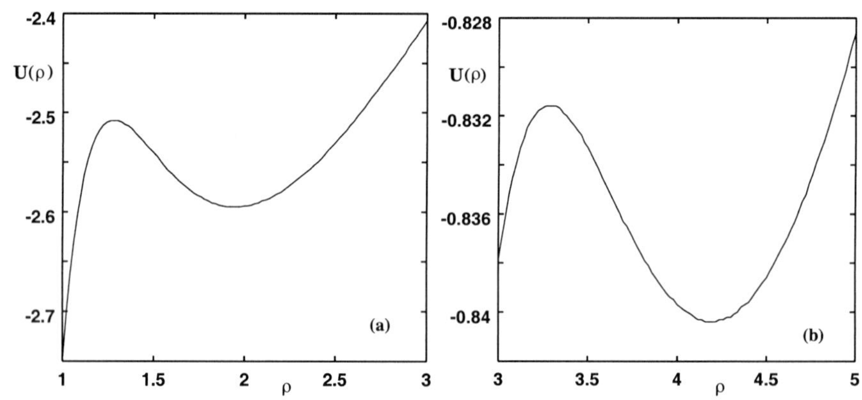

Figure 2. Potential profiles (arbitrary units) for the two orbits of Figure 1: (a; left) equatorial, (b; right) halo.

3. WELL DEPTH

It is well known that equatorial dust grains are deeply trapped in their potential wells. However, it is not obvious that the much smaller halo grains are sufficiently deeply trapped to survive the perturbations discussed in the following sections. To this end we define the *well depth*

$$\Delta = E_{max}/E_{min} - 1 \tag{6}$$

which varies with q/m, becoming vanishingly small at a bifurcation point. While care must be taken in defining E_{max}, this detail need not concern us here; usually it suffices to take $E_{max} = min(E_s, E_\infty)$, where E_s is the saddle point energy. Figure 2 compares potential profiles for the equatorial and halo orbits shown in Figure 1, with that for the halo well taken along a line joining the elliptic and hyperbolic critical points.

4. OBLATENESS AND QUADRUPOLE FIELD

First consider nonzero J_2 and g_2, for which the Hamiltonian remains axisymmetric. For equatorial orbits the dynamics is primarily Keplerian for $a_\mu \geq 0.2$ μm. In such cases it is well known that the dipole field and oblateness both cause Keplerian ellipses to precess and can in fact cancel, with important consequences for E-Ring particles [7]. However, this precession is of minor importance for particle trapping, since the locations of the critical points of U^e are only slightly perturbed. Furthermore, U^e is *structurally stable*, i.e. its critical points do not change type under small perturbations [9]. In fact the orbits in Figures 1 and 2 all include J_2 and g_2 and are seen to be firmly confined to their ideal two dimensional potential wells. Extensive orbital calculations show that the effects of g_2 are miniscule compared to J_2 and we therefore omit it from further discussion.

5. RADIATION PRESSURE

The effects of radiation pressure are more subtle and can have large long-term effects. non-magnetic Mars and Venus. Dynamically the presence of radiation pressure breaks the axisymmetry of (1), so that the motion becomes truly three dimensional, and the canonical momentum p_ϕ is no longer conserved. Nevertheless, for magnetic planets with $k << 1$, p_ϕ is still conserved on the average, as guaranteed by the KAM theorem, with the result that orbits can be trapped in 3D potential wells. For the orbit of Figure 1a p_ϕ varies by only 0.24%. In general only positively charged conducting grains at large distance from the planet are significantly perturbed from their two dimensional potential wells by radiation pressure.

6. DISCUSSION

We have seen that stable nonequatorial ('halo') orbits may exist about Saturn. These orbits are composed of very small ($\approx$ 100 nm) grains and are insensitive to the influence of J_2 and g_2. Radiation pressure is relatively small for dielectric grains but can be large for distant conducting grains. Work is currently underway to determine the effects of time-dependent charging on these intriguing orbits.

REFERENCES

1. J.E. Howard, M. Horányi and G. A. Stewart, Phys. Rev. Lett. 83 (1999) 3993.
2. J.E. Howard, H.R. Dullin and M. Horányi, Phys. Rev. Lett. 84 (2000), 3244.
3. E. Grün, G.E. Morfill and D. A. Mendis, in Planetary Rings (eds. R. Greenberg and A. Brahic) University of Arizona Press, Tucson, 1984.
4. M. Horányi, Ann. Rev. Astr. Astrophys. 34 (1996) 383.
5. J.D. Richardson and E.C. Sittler Jr., JGR 95 (1990) 12019.
6. J.E. Howard and M. Horányi, Geophys. Res. Lett. 28 (2001) 1907.
7. M. Horányi, J. Burns and D. Hamilton, Icarus 97 (1992) 248.
8. J.E. Howard, Phys. Rev. A51 (1995) 3934.
9. T. Poston and I. Stewart, Catastrophe Theory and its Applications, Pitman, Boston, 1978.

Charging processes for dust particles in Saturn's magnetosphere

A.L. Graps[a] and E. Grün [a]

[a]Max-Planck-Institut für Kernphysik, Saupfercheckweg 1, 69117 Heidelberg, Germany.

We investigated the electrostatic charging behavior of submillimeter-sized dust particles located in Saturn's magnetosphere. The charging effects we considered included electron/ion capture from the magnetospheric plasma, electron/ion capture from the solar-wind plasma, the photoelectric effect from solar radiation, and secondary electron emission from energetic electrons. In our results we show charging times and equilibrium charges for particles located in different regions of Saturn's magnetosphere. We find that charging in Saturn's magnetosphere is not particularly sensitive to the dust particle's material properties. The equipotential ranges from ~ -2 V at 3.5 R_S, decreasing to ~ -5 V at 6 R_S, and then increasing to ~ -1.5 V at 10 R_S. The charging time for one micron-sized particles is a few minutes, and for 0.01 micron-sized particles the charging time is 6 hours (or more). The latter is a significant fraction of Saturn's rotation period.

1. INTRODUCTION

The dynamics of dust, submillimeter-sized dust grains, are a fascinating area of study of our Solar System's dynamical evolution. Small particles, especially charged particles, respond to forces other than gravitational, in particular, electromagnetic forces. The dust particle's properties and dynamics fall into a complex regime between nuclear physics and electromagnetic physics and gravitational physics. In order to calculate charges on a dust particle around a planet, we must characterize:

- The planet's magnetospheric features: its magnetic field and plasma,
- The physical processes onto the dust particle that generate currents, and
- The material properties of the dust particle.

In this paper, we make reference to results of charging of dust particles in Earth's magnetosphere [1][2]. Earth is an interesting charging environment for dust particles, in part, because the dynamic magnetospheric plasma shows steep changes in the electron and ion energies and densities, therefore, the electron energy can be quite high (e.g. a few thousand eV). Table 1 lists some basic parameters comparing Saturn's and Earth's magnetosphere.

Table 1
Basic parameters of Saturn's and Earth's magnetosphere

Parameters	Saturn	Earth
Rotation Period (day)	0.44	1.0
Dipole Moment (Gauss cm^3)	2.4×10^{28}	7.9×10^{28}
Field at Equator (Gauss)	0.22	0.305
Dipole Axis (deg)	0.0	+10.8
Magnetopause Distance	20 R_S	10 R_E
Plasma Source	Solar Wind, Atmosphere, Rings, Moons	Solar Wind, Atmosphere

Data from [3].

Table 2
Saturn plasma representative numbers

Component	Location (R_S)	Energy (eV)	Density (cm^{-3})	Debye Length (m)
Cold electrons	10	8.6	1	22
	3	0.005	52	0.073
Hot electrons	10	862	0.6	280
	3	27	0.6	50
Hydrogen ions	10	17	0.3	56
	3	6	6	7.4
Oxygen ions	10	250	0.9	120
	3	31	46	6.1

2. SATURN'S PLASMA ENVIRONMENT

The electrostatic potential of a dust particle not only depends on the physical properties of the particle, but also on the plasma environment, such as the plasma number density, temperature (energy), velocity distribution of the plasma particles, intergrain distance, and the relative motion between the dust particles and the plasma [4].

Inside of Saturn's plasmasphere, the plasma density increases towards the planet from $\sim$ 1 electron cm^{-3} at Saturn radius 10 R_S to $\sim$ 100 electrons cm^{-3} at 3 R_S, and the electron energy kT_e decreases from $\sim$ 100 eV to $\sim$ 10 eV.

To characterize Saturn's plasma, we utilized plasma data from M. Horányi. This plasma data is a four component plasma (hydrogen, oxygen, hot electrons, and cold electrons) fit to the Voyager data described in [5]. The Debye screening length is the distance that the Coulomb field of an arbitrary charge of the plasma is shielded. We can calculate the charge for an isolated grain if we have only one grain within a sphere of radius Debye length. Figures 1 and 2 show the energy, density our plasma data, and we state, in Table 2, some representative plasma values for our plasma data.

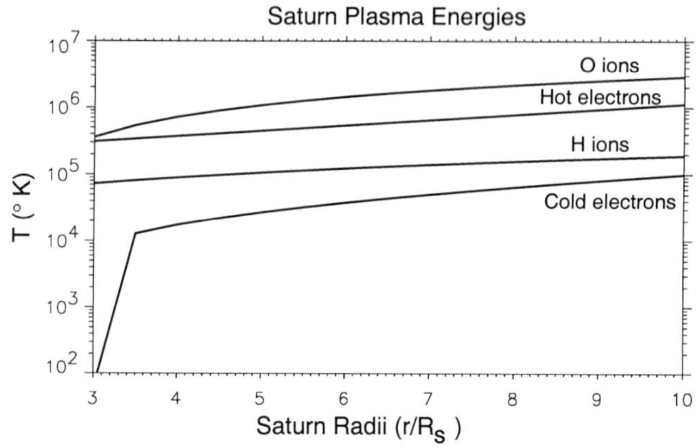

Figure 1. Saturn plasma energies of our four component plasma: hot electrons, cold electrons, oxygen ions, and hydrogen ions.

3. CHARGING PROCESS

We calculate the time-varying charge due to currents acting on a dust particle in a planetary magnetosphere using the following expression:

$$\sum_k I_k = I_{\text{i,e,moving}} + I_{\text{sec}} + I_\nu \tag{1}$$

where I_k is the current of the k-th charging processes [4]. We consider 3 charging currents. The first charging current is: $I_{\text{i,e,moving}}$, which is the collection of ions and electrons onto the dust particle from the ambient plasma. The second current: I_{sec}, the secondary electron current, occurs when a high energy electron impacts the dust particle, some of the dust material is ionized, and electrons are ejected from the particle. The third current, I_ν, photoelectron emission current, occurs when a UV photon impacts the dust particle and photoelectrons are released. For a more complete treatment, one should add reflected electrons from the secondary electron emission [6] and the small particle effect [7].

The secondary electron current is dependent on the dust particle material. If one wants to characterize different dust material properties, then one applies the secondary electron emission maximum yield δ_{m} and the primary energy E_{m} at which the maximum yield occurs, acquired from laboratory measurements. The yield is the ratio of the secondary current to the primary current, given the energy of the impacting electron or ion. Example yield and energy values for relevant solar system material is shown in Table 3.

4. CHARGING RESULTS

We choose, as our canonical example, a hybrid dust particle with material properties similar to a conducting graphite particle δ_{m}=1.5, E_{m} = 250 eV, but with photoelectron

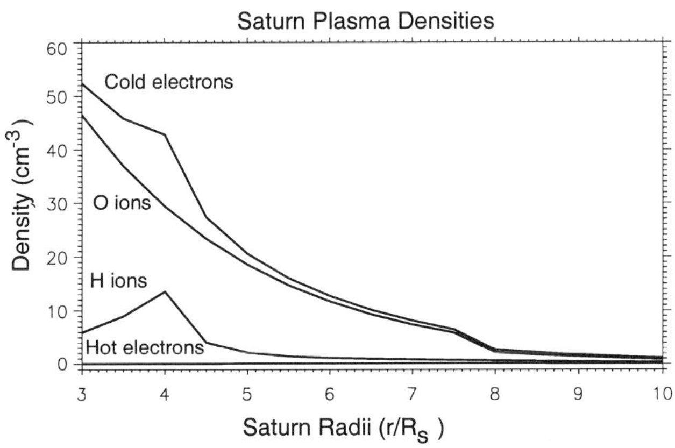

Figure 2. Saturn plasma densities of our four component plasma: hot electrons, cold electrons, oxygen ions, and hydrogen ions. The highest energy and density components are the oxygen ions and hot electrons.

yield properties similar to a dielectric particle (in Horányi et al.'s, modeling work, the photoelectron yield is denoted χ and ranges from χ=1.0 for conducting magnetite dust particles to χ=0.1 for dielectric olivine particles). These properties were chosen in order to compare with charging results we have obtained for a dust particle in Earth's magnetosphere.

For our canonical dust particle, we calculated the equilibrium potential ('equipotential'), the charging time, and examined the dominant currents for a particle at Saturn radii locations of 3 R_S to 10 R_S. Equilibrium potential for the dust particle is reached when the sum of the charging currents is zero. The charging time is the time for a particle's potential to reach an equilibrium. The currents that we examined are the electron and ion collection currents, the photoelectron current and the secondary electron current. Figures

Table 3
Examples of dust particle material properties.

Material	density (g cm^3)	δ_m	E_m (eV)
Graphite	2.26	1	250
SiO_2	2.65	2.9	420
Mica	2.8	2.4	340
Fe	7.86	1.3	400
Al	2.70	0.95	300
MgO	3.58	23	1200
Lunar dust	3.2	~1.5	500

Data from [7].

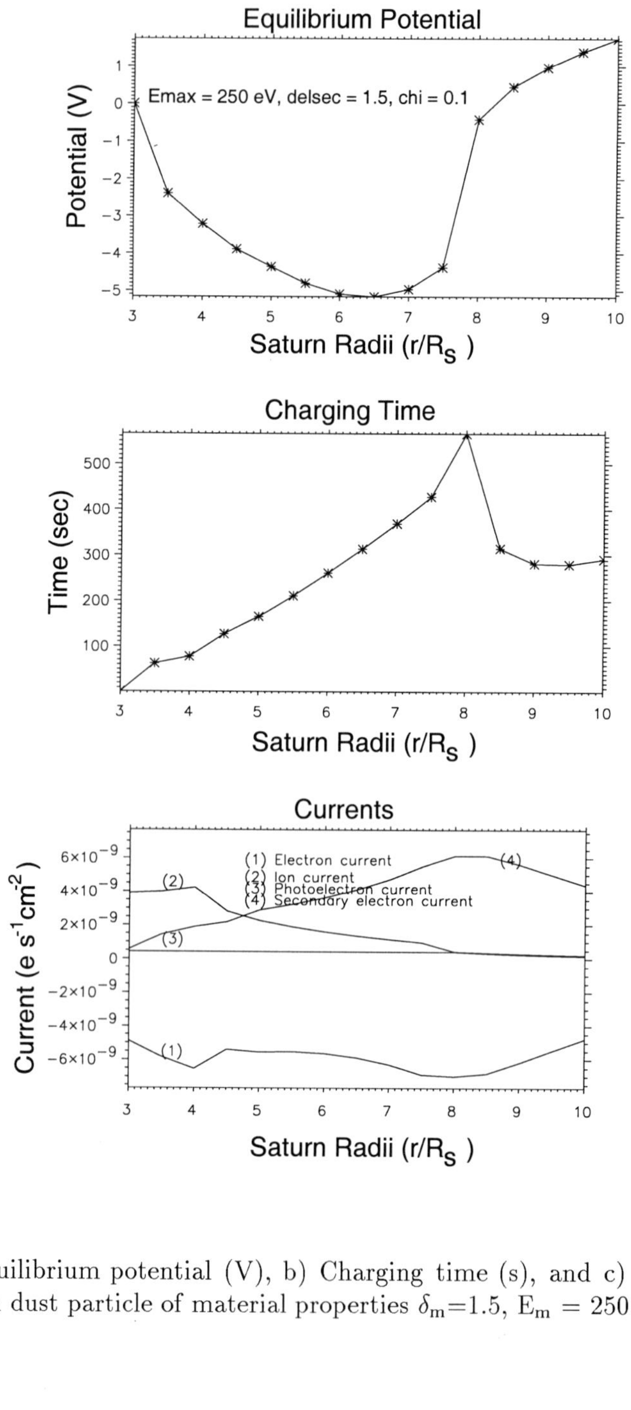

Figure 3. a) Equilibrium potential (V), b) Charging time (s), and c) Currents (e s^{-1} cm^{-2}) for a 1 μm dust particle of material properties δ_m=1.5, E_m = 250 eV, χ=0.1.

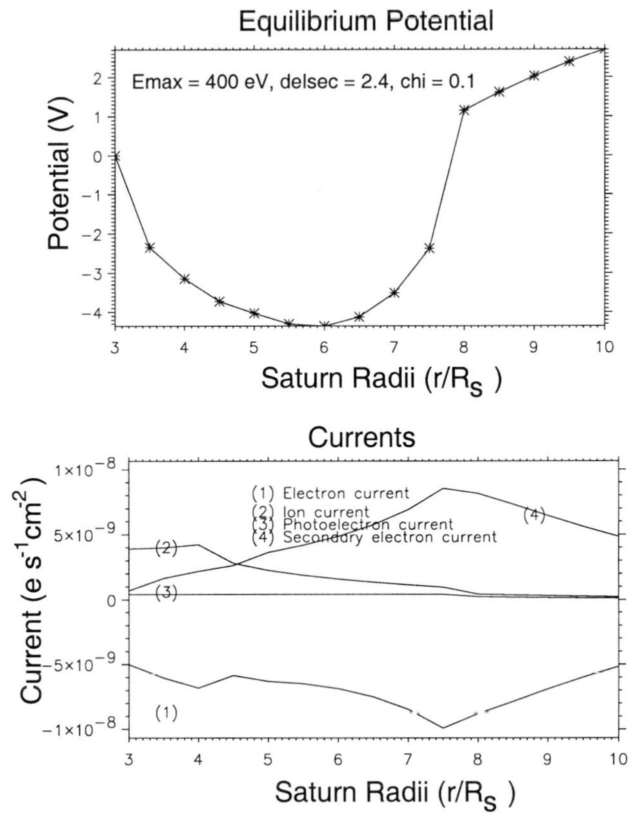

Figure 4. a) Equilibrium potential (V), and b) Currents (e s^{-1} cm^{-2}) for a 1 μm dust particle of material properties δ_m=2.4, E_m = 400 eV, χ=0.1.

3a–c display the results for the equilibrium potential, the charging time, and the dominant currents for our canonical case, 1 μm dust particle. Here, the equipotential ranges from ~ -2 V at 3.5 R_S, decreasing to ~ -5 V at 6 R_S, and then increasing to ~ -1.5 V at 10 R_S. The charging time for the starred positions is $\sim$ 1 minute. If we perform the same calculations for a 100 times smaller particle with the same material properties, then we find charging times on order of a few hours, which is a signficant fraction of Saturn's rotation period. Also, for a 100 times smaller particle, the secondary electron emission current will be more efficient, causing the smaller particle to charge more positively than for the larger (1 μm-sized) dust particle.

For an identical dust particle in a geostationary location in Earth's magnetosphere, we calculated dramatic differences in the equipotential values and the charging times when we applied plasma conditions appropriate to 'disturbed' and 'quiet' Earth magnetosphere conditions, and when we slightly varied the material properties from δ_m=1.5, E_m = 250

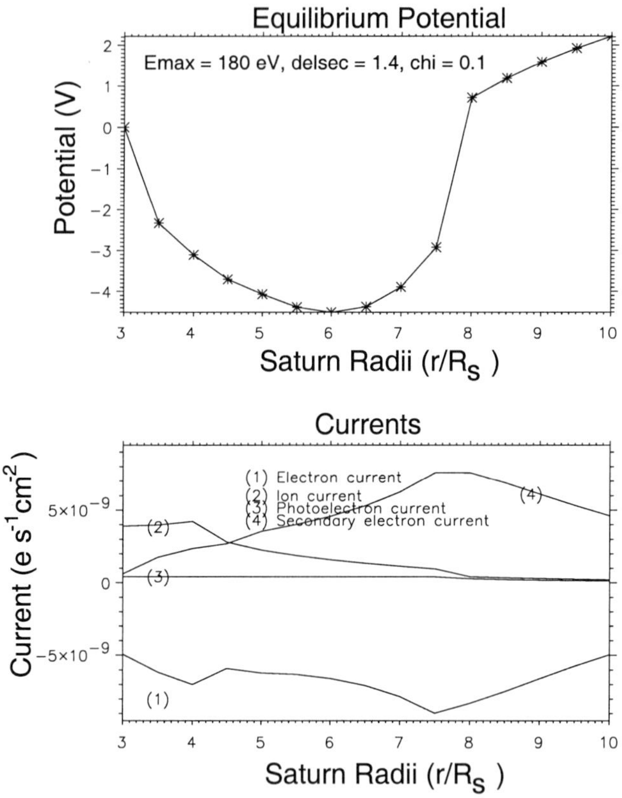

Figure 5. a) Equilibrium potential (V), and b) Currents (e $s^{-1}cm^{-2}$) for a 1 μm dust particle of material properties δ_m=1.4, E_m = 180 eV, χ=0.1.

eV, χ=0.1 to δ_m=1.4, E_m = 180 eV, χ=0.1. For the first set of material properties in quiet Earth plasma conditions, the equilibrium potential was $\sim$ +5 V. However, for the second set of material properties in disturbed Earth plasma conditions, the equilibriuim potential was ~ -3000 V. The charging time was about 10 seconds for a 1 μm dust particle in quiet Earth plasma conditions, and one-third that time for a 1 μm dust particle in active Earth plasma conditions. The charging time generally *increases* with *decreasing* particle radii.

What happens when we vary the material properties δ_m, E_m, and χ=0.1 for a 1 μm dust particle in the same way in Saturn's magnetosphere, using the same charging processes as we applied for a particle in Earth orbit? Surprisingly, we find very little change in the resulting potentials, charging times, and currents, as seen in Figures 4 and 5. Removing each of the currents, one by one however, *does* have an effect, in particular the secondary electron emission. If we calculate equipotentials without the secondary electron emission current, then the dust particle potential stays negative throughout the magnetosphere, and doesn't reach positive potentials beyond 8 R_S. On the other hand, removing the photoelectron emission current doesn't alter the equilibrium potentials in a significant way.

The charging time for a 1 μm dust particle is on the order of a few minutes, while for a 0.01 μm dust particle, the charging time is on the order of $\gtrsim$ few hours.

5. SUMMARY

- Charging in Saturn's magnetosphere is not particularly sensitive to the dust particle's material properties. This is a large contrast to dust particles in Earth's magnetosphere, where small material property changes have a big effect on the equilibrium potential.
- The charging time for one micron-sized particles is a few minutes, and for 0.01 micron-sized particles the charging time is 6 hours (or more). The latter is a signficant fraction of Saturn's rotation period.

Acknowledgements. We are very grateful to M. Horányi, Laboratory for Atmospheric and Space Physics, University of Colorado, Boulder, for his Saturn plasma data and helpful guidance on dust particle charging mechanisms.

REFERENCES

1. A., Juhász and M. Horányi, JGR 102 (1997) 7237.
2. A.L. Graps and E. Grün, Properties, Charging, and Dynamics of Interplanetary Dust Particles in Earth's Magnetosphere, Final Report for ESA contract 13145/98/NL/WK, Unispace Kent 2000.
3. J. Van Allen and F. Bagenal, in The New Solar System (eds. J.K. Beatty, C. Petersen and A. Chaikin), and also Appendix, Sky Publishing, Cambridge (1999).
4. H. Kimura and I. Mann, Astrophys. J. 499 (1998) 454.
5. J.D. Richardson and E.C. Sittler Jr., JGR 95 (1990) 12019.
6. S. Jurac, R.A. Baragiola, R.E. Johnson and E.C. Sittler Jr., JGR 100 (1995) 14821.
7. B.T. Draine and E.E. Salpeter, Astrophys. 231 (1979) 77.

Mars Dust Counter (MDC) on board NOZOMI: Initial results

S. Sasaki [a], E. Igenbergs[b], R. Münzenmayer[b$], H. Ohashi [c], G. Hofschuster [b], W. Naumann[b#], G. Färber[d], F. Fischer[d], A. Fujiwara[e], A. Glasmachers[f], E. Grün[g], Y. Hamabe[a,e], H. Iglseder[h], H. Miyamoto[a&], T. Mukai[i], K. Nogami[j], G. Schwehm[k], H. Svedhem[k], M. Born[b], T. Kawamura[j], D. Klinge[k], K. Morishige[e], T. Naoi[a], R. Peeks[b], H. Yano[e] and K. Yamakoshi[l]

[a]Department of Earth and Planetary Science, University of Tokyo, Tokyo 113-0033, Japan

[b]Fachgebiet Raumfahrttechnik, Technische Universität München, 85748 Garching, Germany

[c]Department of Ocean Sciences, Tokyo University of Fishery, Tokyo 108-0075, Japan

[d]Lehrstuhl für Prozeßrechner, Technische Universität München, 80333 München, Germany

[e]Institute of Space and Astronautical Sciences, Kanagawa 229-8510, Japan

[f]Universität / GH Wuppertal, FB13 Lehrstuhl für Meßtechnik, 42097 Wuppertal, Germany

[g]Max Planck Institut für Kernphysik, Saupfercheckweg 1, 69117 Heidelberg, Germany

[h]Wilkhahn, Eimbeckhausen Im Landerfeld 8, 31848 Bad Münder, Germany

[i]Department of Earth and Planetary Sciences, Kobe University, Kobe 657-0013, Japan

[j]Dokkyo University School of Medicine, Mibu, Tochigi 312-0207, Japan

[k]ESA-ESTEC, PO Box 299, 2200 AG Noordwijk, The Netherlands

[l]Deceaced. Formerly at Institute for Cosmic Ray Research, University of Tokyo

[$]Now at Astrium Space, Postfach 1420 , 88039 Friedrichshafen, Germany

[#]Now at Kayser-Threde GmbH, Wolfratshauser Str. 48, 81379 München, Germany

[&]Now at Department of Geosystem Engineering, University of Tokyo, 113-0033, Japan

Mars Dust Counter (MDC) is a light-weight (730g) impact-ionization dust detector on board NOZOMI, a Japanese Mars mission, which was launched on July 4th 1998. The main aim of MDC is to detect the predicted Martian dust rings / tori. It can also cover velocity-mass ranges of interplanetary and interstellar dust particles. By August 2000, MDC had detected more than 60 dust particles. In 1999, it detected five fast particles probably of interstellar origin. For five years from 1999 to 2003, NOZOMI will orbit the sun and MDC can measure interplanetary and interstellar dust between the Earth's and Mars' orbits.

1. INTRODUCTION

To measure the basic characteristics – mass, velocity, and direction - of dust particles in space, an impact ionization dust detector is one of the most powerful tools. Mars Dust Counter (MDC) is an impact-ionization dust detector, whose mass is only 730g. MDC is on board the Japanese Mars mission NOZOMI (PLANET-B), which was launched on July 4th,

1998. The main objective of NOZOMI is to study Martian aeronomy and outer environment, especially the interaction between the Martian upper atmosphere and the solar wind. NOZOMI's apoapis distance is larger than Deimos' orbital radius. The primary objective of MDC is to discover Martian dust rings or tori whose particles are ejected from Phobos and Deimos [2-11]. To sustain the dust number density of dust rings, there is a proposed self-sustaining mechanism, where satellite-dust collisons should supply additional dust. If the self-sustaining mechanism is effective, the dust abundance would be high enough to be detectable by MDC [8]. MDC NOZOMI started its measurement just after the launch and MDC has been measuring continuously dust particles in space.

2. MDC-NOZOMI

MDC is a revised type of the impact-ionization dust detectors on board HITEN and BREMSAT which successfully measured dust particles around the Earth [12]. The dimensions of MDC are 136 x 127 x 181 mm^3, its sensor aperture is 124 x 115 mm^2, and its power consumption is 3.78 W [1]. The sensor is directed 45 degrees from the opposite direction of the spin axis of NOZOMI. Since the spin axis points to the Earth during cruise and circum-martian orbits, the sensor box can avoid direct solar light which would increase noise signals from photoelectrons.

The sensor has two charge collector plates which are biased by positive (+220V) and negative (-220V) voltages. All surfaces of the sensor interior are gold plated. When a dust particle impacts on the sensor interior, generated impact plasma is separated into positive ions on the negative collector (ion channel) and electrons on the positive collector (electron channel). In MDC-NOZOMI, charge signal on the grounded neutral target is also recorded to distinguish an impact signal from a noise signal. Charges are recorded by three channels for 200 μs through logarithmic amplifiers (Figure 1). From the total charge and risetime of the

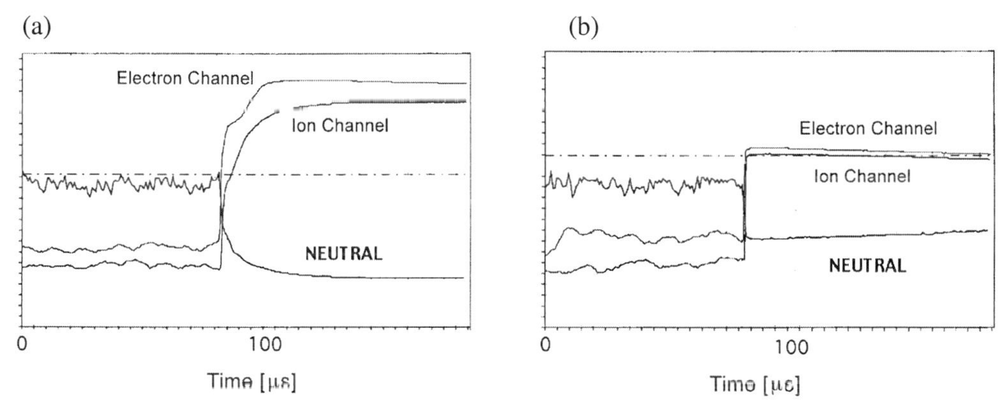

Figure 1. The impact signals of MDC NOZOMI. There are three signals corresponding to ion, electron, and neutral channels. (a) Impact on 98-07-28. According to preliminary estimates, mass and velocity are $1.1\text{x}10^{-13}$ kg and 7.5 km s^{-1}, respectively. There is a two-stage rise of the impact charge in the electron channel signal, which would indicate the secondary impact of impact ejecta in the sensor box. (b) Impact on 99-03-01 (B in Figure 3(b)). According to preliminary estimates, mass and velocity are $1.1\text{x}10^{-19}$ kg and 60 km s^{-1}, respectively. This is an interstellar particle.

signal, mass and velocity of the particle can be obtained using laboratory calibration data. For example at velocity 10 km s^{-1}, we can determine the mass of particles approximately between 5 x 10^{-18} and 10^{-13} kg, corresponding to a particle diameter range between 0.1 and 10μm.

3. MDC OPERATION IN SPACE

3.1. Initial Operations Around the Earth

NOZOMI ("hope" in Japanese) was successfully launched on July 4th (JST) 1998 from Kagoshima Space Center of ISAS (Institute of Space and Astronautical Science). For five months, NOZOMI took elliptic orbits around the Earth with two gravitational swing-bys with the moon. The largest distance between NOZOMI and the Earth was 1,200,000 km in November. On July 10th, MDC was first turned on and subsequently it detected the first impact signal. Figure 1 shows examples of impact signals. Impact number for each observation epoch is shown in Table 1. So far MDC has detected more than 60 sure signals from dust impacts. Figure 2 shows mass and velocity of detected particles.

Table 1
Detected number of dust particles by MDC

Observation Duration	Impacts
98-7 to 98-12 (circumterrestrial)	26
99-1 to 99-6	15
99-7 to 99-12	8
00-1 to 00-8	15

On 18th November 1998, NOZOMI passed through the Leonid meteor stream, which was produced by Comet 55P/Tempel-Tuttle. Timing of the encounter between NOZOMI and the Leonid meteor tube was predicted to be about one day later than that of the Earth-Leonids encounter. Leonid dust could be easily distinguished by the estimated velocity and sensor direction. Just before the NOZOMI-Leonids encounter, MDC detected two clear impact signals. Preliminary estimates of impact velocity of both particles are 13 and 20 km s^{-1}. NOZOMI is a spin-stabilized satellite and MDC is placed on one of its side panels. From pointing considerations the particle at 154 degree may belong to the Leonids, although the velocity is not as high as the encounter velocity with the Leonids of 70 km s^{-1}.

3.2. Cruise Phase

After the fly-by with the Earth on December 20th 1998, NOZOMI entered a transfer cruise orbit toward Mars. A problem during the powered fly-by on the day, resulted in a change to the initial orbital plan to enter the circum-martian orbit in October 1999. Now NOZOMI will enter the circum-martian orbit on 1st January 2004 after a five-year cruise phase between the Earth's and Mars' orbits involving two fly-bys with the Earth. MDC detected nearly 40 dust particles in the cruise phase up to August 2000 (Table 1). As seen in Figure 2, MDC detected relatively high-velocity and low-mass particles compared with the circum-terrestiral phase.

Impact data in 1999 are projected into the ecliptic plane (Figure 3). Figure 3(a) shows the measured velocity of detected particles with sensor aperture direction. Figure 3(b) shows estimated velocity and direction of dust particles in space. MDC has detected at least 8 dust particles whose velocity is higher than 40 km s^{-1}. This velocity, faster than the Keplerian

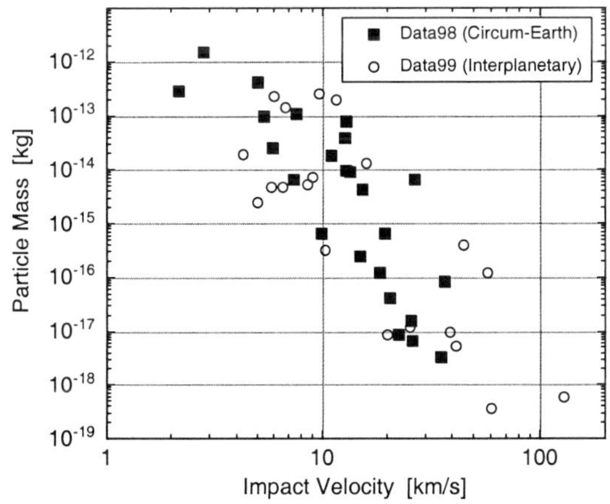

Figure 2. Velocity and mass of detected particles. Filled marks are impacts in 1998 (circum-terrestrial phase) and open marks denote impacts in 1999 (cruise phase). The detected velocity range is 2 to 100 km s^{-1} and the mass range is $3x10^{-19}$ to 10^{-12} kg, which correspond well to MDC-HITEN [12].

velocity, would be explained by interstellar particles since the MDC aperture avoids particles from the solar direction such as beta meteoroids. The direction of interstellar particles (Figure 3) should correspond to the relative direction of the solar system against the ambient gas cloud with relative velocity as high as 26 km s^{-1} [13]. Among the high-velocity particles, two should correspond to the typical interstellar particles. One example of dust impact is shown in Figure 1(b). Here, high velocity is characterized by a sharp rise of charge signals. Moreover, a couple of high-velocity particles have neither Keplerian or interstellar gas directions. They might be ascribed to unknown interstellar sources. As seen in Figure 3, when NOZOMI is around –x axis, i.e., around the vernal equinox, the relative velocity between NOZOMI and the interstellar flow is higher than 50 km s^{-1}. This would be the reason why MDC detected more high-velocity particles around this season. A change of detection rate in 1999 (15 in the first half and 8 in the latter half) is probably due to the change of MDC aperture direction, since NOZOMI's spin axis is directed always towards the Earth for the telemetry.

4. CONCLUSION

After launch in July 1998, more than 60 dust particles were detected by MDC on the Japanese Mars Mission NOZOMI (formerly PLANET-B) by August 2000. In 1999 during the cruise phase, MDC detected at least five high velocity dust particles of interstellar origin. Interstellar dust particles entered the region inside the Martian orbit. MDC-NOZOMI will continue observing interplanetary and interstellar dust particles. From the beginning of 2004 for at least two years, MDC will enter Martian orbit and investigate proposed existence of dust rings or tori around Mars.

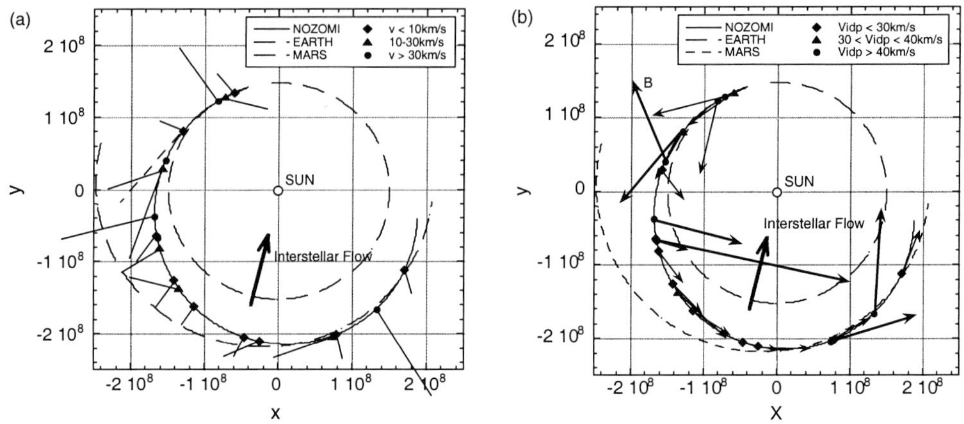

Figure .3. Dust impacts detected by MDC in 1999. Orbits of NOZOMI, the Earth, and Mars are shown on the ecliptic plane. +x direction is the direction of the sun at vernal equinox. NOZOMI takes an elliptic orbit whose perihelion is the Earth's orbit and aphelion is Mars' orbit. NOZOMI reached Mars' orbit in October. For each impact, the direction of the MDC sensor is expressed by a line from the impact mark. The length of each line corresponds to velocity. The big arrow denotes the direction of the interstellar medium towards the solar system. (a) Direction of MDC sensor aperture and impact velocity (i.e. relative velocity between NOZOMI and a dust particle). (b) Velocity of dust in the interplanetary space calculated from impact velocity and orbital velocity. Dust particles with high velocity (>40 km s^{-1}) are expressed by bold lines.

REFERENCES

1. E. Igenbergs, S. Sasaki, R. Muenzenmayer, H. Ohashi, G. Faerber, F. Fischer, A, Fujiwara, A. Glasmachers, E. Grün and Y. Hamabe, Earth Planets Space 50 (1998) 241.
2. S. Soter, Rep. Center Radiophys. Space Res. 462 (1971).
3. M. Horányi, J.A. Burns and M. Tatrallyay, Geophys. Res. Lett. 17 (1990) 853.
4. A. Juhász, M. Tatrallyay, G. Gevai and M.J. Horanyi, Geophys. Res. 98 (1993) 1205.
5. S. Sasaki, in Proc. 26th ISAS Lunar Planet. Symp. (1993) 40.
6. S. Sasaki, in Proc. 27th ISAS Lunar Planet. Symp. (1994) 47.
7. H. Ishimoto and T. Mukai, Planet. Space Sci. 42 (1994) 691.
8. S. Sasaki, in Physics, Chemistry, and Dynamics of Interplanetary Dust. ASP Conf. 104, (1996) 187.
9. D. P. Hamilton, Icarus 119 (1996) 153.
10. H. Ishimoto, Icarus 122 (1996) 153.
11. A. V. Krivov and D. P. Hamilton, Icarus 128 (1997) 335.
12. E. Igenbergs, A. Hüdepohl, K. Uesugi, T. Hayashi, H. Svedhem, H. Iglseder, G. Koller, A. Glasmachers, E. Grün, G. Schwehm, H. Mizutani, T. Yamamoto, A. Fujimura, N. Ishii, H. Araki, K. Yamakoshi and K. Nogami, in Origin and Evolution of Interplanetary Dust (eds. A. C. Levasseur-Regourd and H. Hasegawa) Kluwer, Dordrecht (1991) 15.
13. E. Grün, B. Gustafson, I. Mann and M. Baguhl, Astron. Astrophys. 286 (1994) 915.

Dust Telescope: A new tool for dust research.

Eberhard Grün[a], Harald Krüger[a], Ralf Srama[a], Sascha Kempf[a], Siegfried Auer[b], Luigi Colangeli[c], Mihaly Horanyi[d], Peter Withnell[d], Jochen Kissel[e], Markus Landgraf[f], and Hakan Svedhem[g].

[a]Max Planck Institut für Kernphysik, Postfach 10 39 80, 69029 Heidelberg, Germany

[b]A&M Associates, Basye, VA 22810-0421, USA

[c]Osservatorio Astronomico di Capodimonte, Via Moiariello 16, 80131 Napoli, Italy

[d]Laboratory for Atmospheric and Space Physics, University of Colorado, Boulder, CO 80309-09392, USA

[e]Max Planck Institut für Extraterrestrische Physik, Giessenbachstrasse, D-87540 Garching, Germany

[f]ESA/ESOC, Robert-Bosch-Strasse 5, 64293 Darmstadt, Germany

[g]ESA/ESTEC, PB 299, NL-2200AG Noordwijk, The Netherlands

Dust particles in space carry information about their birth at a remote site in space and time not accessible to direct investigation. When we know where dust particles come from, we can derive from their state and composition important knowledge about the processes by which they were formed. This information can be gained by a combination of trajectory analysis together with the physical and chemical analysis of dust particles. Potential targets of a dust telescope can be interstellar dust phenomena (e.g. local interstellar medium or dusty stellar systems like beta-Pictoris), interplanetary phenomena (e.g. meteor stream dust, cometary, or asteroidal dust, or dust from the moon), or even space debris (e.g. fine grains from solid rocket burns). It is proposed to use a 1 m² dust telescope with 50° aperture. Such an instrument would detect 5 and 0.5 interplanetary dust grains of 10^{-15} g and 10^{-12} g per day, respectively. A state-of-the-art dust telescope consists of an array of parallel mounted dust analyzers. Potential components are a high resolution impact mass spectrometer, a dust analyzer for the determination of physical and chemical dust properties, a dust momentum sensor, and a large-area impact detector with trajectory analysis. A first example of such a dust telescope is carried by the proposed ***Galactic DUNE*** mission. The goal of DUNE is the analysis of interstellar grains near Earth.

1. INTRODUCTION

A dust telescope is a dust analyzer that provides chemical and physical information on dust particles together with directional information that allows the reconstruction of the trajectory of the dust particle, all the way back to its source region. Already some of the first dust detectors in space on the OGO (Orbiting Geophysical Observatory) satellites and on the Lunar Explorer 35 (launched in 1966 and 1967, [1]) were time-of-flight (TOF) systems aiming at the trajectory analysis of the recorded particles. They consisted of a thin film front sensor and a rear sensor 10 cm apart. However, misled by the early false reports of a very high micro-meteoroid flux in the Earth's environment only 5 cm^2 as sensitive area of a single detector were chosen. This had as consequence that no or only very few impacts were recorded during these missions. Berg and Richardson [2] extended this idea by combining 16 TOF tubes into a single 100 cm^2 dust detector flown on Pioneer 8 and 9. This detector made the first important dust observation in interplanetary space and discovered new dusty phenomena [3], however, the total number of TOF events for which trajectories could be derived was low [4].

Because of the very low dust flux in space, later dust detectors had a wide field-of-view (FOV > 1 sterad which corresponds to an acceptance cone of > 60° half-angle) and a detection area of 100 cm^2. The number of dust impacts recorded by these detectors at 1 AU from the Sun were less than 100 per year. Modern dust detectors, like the ones on the Galileo, Ulysses and Cassini missions employ a detection area of 1000 cm^2. Despite their wide FOVs these detectors provided useful directional information on dust streams, like interstellar grains [5] and Jupiter dust streams [6]. Because of the large numbers of particles in the stream (some 100 interstellar grains and some 10,000 Jupiter stream particles) statistical methods could be employed and the stream directions were derived with 10° and better than 2° accuracy, respectively.

In the early 1990s a Cosmic Dust Collection Facility (CDCF, [7]) was considered for implementation on the International Space Station. This CDCF combined the idea of trajectory analysis together with the subsequent intact collection of cosmic dust particles in low-density aerogel. For trajectory analysis several approaches were studied: segmented thin-film penetrations and pick-up charge measurements of the electrically charged dust particles [8] for position determination in at least two planes. The design goal was the determination of the impact direction with an accuracy of about 1° for tracing back the trajectories from low-Earth orbit to their cometary or asteroidal sources. However, because of NASA's budget cuts this project was cancelled before the technology was fully developed.

It was Cassini's Cosmic Dust Analyzer, CDA, that proved for the first time an important technique of accurate non-contacting trajectory measurement in space: the measurement of the pick-up charge of dust particles that entered CDA. Another important element was initially developed for the Halley missions: an impact ionization mass spectrometer [9] that provided compositional information on the impacting grain. A follow-up version of it is currently flying on the Stardust mission to comet Tempel 2. These instruments provided mass spectra of the impact generated ions from the dust particles at mass resolution $M/\Delta M$ of better than 100 atomic mass units. Thereby, these instruments are able to resolve most elements relevant for the characterization of extra-terrestrial materials, like meteorites.

It is the purpose of this paper to show how all these elements are brought together forming a true dust telescope that employs only state-of-the-art technology (section 3). However, before we do that, we will discuss the science questions that can be addressed with such an instrument (section 2). In section 4 we describe a specific mission proposal to study interstellar dust near Earth. This "***Galactic DUNE***" (Galactic DUst measurements Near Earth) mission is being proposed to space agencies for future implementation. Section 5 concludes by summarizing the major new features of the exciting opportunity for novel dust research.

2. SCIENCE

Two main types of dust populate interplanetary space: interplanetary dust that originates from asteroids and comets and interstellar dust. Between these different types there exists a genetic relation: interstellar dust is the basic component from which stars and planets form today. Cometary dust is the most primitive material that still exists in the planetary system - it is suspected that it partially consists of unaltered interstellar dust. Dust from the asteroid belt was generated by collisions between bigger objects, the material of which has been modified during their formation.

There exist several links between meteorites (the best studied extraterrestrial material) and asteroids: (1) Photographically determined orbits of four ordinary chondrites resemble near-Earth-asteroid orbits, (2) mineralogical evidence indicates an origin of most meteorites from differentiated asteroids sized objects, and (3) optical properties of meteorite classes resemble the spectral reflectance data of asteroid classes. But a tiny (10^{-5}) fraction of some primitive meteorites has a strange isotopic signature that deviates significantly from the 'average solar system abundance' of the bulk meteoritic material: these presolar grains have retained their interstellar identity. However, it must be noted that the collection of meteorites is a biased sample and does not represent the meteoroid population in near-Earth space. Some of the very friable material may not survive the passage through the Earth's atmosphere as is evidenced in the meteor records.

Besides the 'meteorite-window' there is another 'window' through which extraterrestrial material reaches the surface in more or less undisturbed state. Small interplanetary dust particles of a few to 10 μm diameter are decelerated in the tenuous atmosphere above 100 km height. At this height the deceleration is so gentle that the grains will not reach the temperature of substantial evaporation. These dust particles subsequently sediment through the atmosphere and become accessible to collection and scientific examination. According to their elemental composition IDPs come in three major types: 1. chondrite 60% 2. iron-sulfur-nickel 30%, and 3. mafic silicates (iron-magnesium rich silicates, i.e. olivine and pyroxene) 10%. Most chondritic IDPs are porous aggregates but some smooth chondritic particles are found as well. Aggregates of 0.1-0.5 μm-sized grains may contain varying amounts of carbonaceous material of unspecified composition. A significant enrichment in volatile (low condensation temperature) elements is found if compared to primitive chondrites. This observation is being used for the argument that these particles consist of some very primitive solar system material that had never seen temperatures above about 500°C.

A remarkable feature of IDPs is their large variability in isotopic composition. Extreme isotopic anomalies have been found in some IDPs (e.g. factors of 1000 off the solar hydrogen isotope ratio). Under typical solar system condition only fractions of a percent isotopic variations can occur. These huge isotopic variations indicate that some grains are not homogenized with other solar system material but have preserved much of their presolar character. These aggregate IDPs are believed to originate from comets, based on their porous structure, high carbon content, and relatively high atmospheric entry velocities. The match between the mineral identifications in cometary spectra and the silicates seen in the IDPs strengthens the link between comets and anhydrous chondritic aggregate IDPs.

In-situ measurements of the Halley missions and modern infra-red observations brought a wealth of information on cometary dust. It was found that the cometary particulates are intimate mixtures of two end-member components, one refractory organic component rich in the elements H, C, N, and O, and the other one rich in rock-forming elements as Si, Mg, Fe, respectively. The bulk abundance of the rock-forming elements in Halley's dust are indistinguishable from the solar and chondritic abundance within a factor of two. However, the presence of the wide range of isotopic compositions of carbon [17] excludes any equilibration processes affecting the carbon carrier during comet formation or later in its history. Infrared-spectrometry has identified the mineral assemblage in Comet Hale-Bopp [33] dust that resembles closely anhydrous chondritic aggregate IDPs: it consists of crystalline Mg-rich, Fe-poor enstatite and forsterite minerals that are formed at high temperatures plus lower-temperature glassy or amorphous grains. Mg-rich, Fe-poor particles in turn dominate Halley's dust.

The solid material that makes up the Earth and other planets resided in galactic interstellar dust (ISD) grains 5 10^9 years ago, before it was mixed and altered during the planetary formation process. However, information on galactic dust in the solar vicinity is extremely limited.

ISD research began around 1930, when the observed reddening of starlight was attributed to the presence of dust in interstellar space. With the advent of infrared astronomy it became possible to study the thermal emission of ISD and to detect it at lower column densities. Infrared spectroscopy allowed the analysis of molecular features and thus gave insight into the chemical composition of ISD. Prominent silicate features near 10 and 20 μm were detected. The infrared spectra also contained features at 3.3, 6.2, 7.7, and 11.3 μm which were tentatively attributed to polycyclic aromatic hydrocarbons (PAHs).

The depletion of some chemical elements in the gas phase of the interstellar medium compared to their cosmic abundance, was interpreted as a consequence of the condensation of these elements onto dust grains. A grain model by Mathis [24] suggests three grain types: small graphite grains; silicate grains; composite grains containing carbon (amorphous, hydrogenated, or graphitic), silicates, and oxides. On the basis of the same elementary composition, Li Greenberg [25] propose three different grain types: large grains consisting of a silicate core and an organic refractory mantle; small carbonaceous grains; very small grains (or large molecules) of PAHs.

What is known about the origin and evolution of ISD grains? Evolved stars continuously lose mass. About 90% of the stellar mass loss is provided by cool, high-luminosity stars, in particular by asymptotic giant branch (AGB) and post-AGB stars. As the ejected gas cools in expanding stellar winds, and also in supernova remnants, solid particles condense out of the gas phase. This so-called stardust provides the seeds for ISD grains that grow in cool interstellar clouds by accretion of atoms and molecules and by agglomeration. Interstellar shock waves provide an effective destruction mechanism for ISD grains as small grains are caught in the shock and collide with bigger grains that don't follow the shock's motion due to their larger inertia. In diffuse interstellar clouds, the grains lose their volatile constituents due to ultraviolet (UV) irradiation and thermal sputtering in interstellar shock fronts. Ultimately, an ISD grain can be incorporated (and destroyed) in a newly forming star, or else, it can become part of a planetary system. In this way, ISD grains are repeatedly recycled through the galactic evolution process.

The first goal of a dust telescope is to distinguish by their kinem atic properties dust particles from different sources: interstellar grains from the different types of interplanetary dust grains. Differences in the chemical and isotopic composition can then be used to determine their formation and history.

2.1 Distinction of interstellar from interplanetary dust

The kinematics of interstellar dust (ISD) is determined by the solar motion through the local interstellar cloud, which thus defines the encounter velocity of ISD in the heliocentric frame and its dynamics in the immediate solar vicinity. The Sun moves through the local interstellar cloud (LIC) which consists of warm partially ionized material. The diffuse interstellar gas in the Sun's vicinity is ablated from the parent molecular cloud complex when impacted by the combined stellar winds and supernova shock fronts from star formation and destruction in the Scorpius-Centaurus association [10]. The LIC has an extent of a few pc and is located close to the edge of the local bubble, a region of about 100 pc in extent which is believed to be excavated by supernova shocks.

From the in situ measurements of the Ulysses and Galileo dust instruments, the upstream direction of ISD particles was found to be close to ecliptic coordinates $\lambda_{ecl} = 259°$, $\beta_{ecl} = 8°$ [5]. This direction is approximately the upstream direction of interstellar gas as measured by the Ulysses GAS experiment [11]. The initial velocity of ISD grains with respect to the Sun is compatible with 26 km/s, which is also the initial velocity of the gas. Close to the Sun, the motion of ISD grains is modified by their interaction with solar gravity, radiation pressure, and the solar electromagnetic field.

The zodiacal cloud of interplanetary dust has a flattened, lenticular shape which extends along the ecliptic plane about 7 times further from the Sun than perpendicular to the ecliptic plane. Zodiacal light observations found this cloud to extent out to the asteroid belt. Most interplanetary dust particles orbit the Sun in prograde elliptical orbits just like the planets do. Meteor and satellite observations provided the present dynamical state of the interplanetary dust cloud. Models developed by Divine [12] and Staubach [13] give orbital element distributions for various dust populations that describe the measurements. Cometary and asteroidal dust orbits can partially be distinguished by their orbital characteristics.

A dust telescope can use two different, complementary strategies to establish the interstellar origin of small ($m < 10^{-11}$ g), and big ($m > 10^{-11}$ g) dust grains. For big ISD grains, the primary electrostatic charge is sufficiently high in order to determine the impact velocity and direction, by use of a large area charge and trajectory sensing instrument [14]. ISD grains can then be easily distinguished from interplanetary grains, because their velocity exceeds the local solar system escape velocity.

An ensemble of small ISD grains can be distinguished from the interplanetary dust population by statistical arguments, even if the exact trajectories of the individual particles are not known. This method has been demonstrated by the identification of an interstellar subset of the Ulysses and Galileo dust impacts [15] [16]. The Galileo and Ulysses instruments can only determine the impact direction of the dust particles with an accuracy of 70°. However, the statistical analysis of the distribution of impact directions of a large set of impacts, gives the interstellar upstream direction within 10° accuracy in ecliptic latitude and 20° accuracy in ecliptic longitude [5]. With a similar argument, the mean velocity of dust impacts on the Ulysses sensor after Jupiter flyby has been found to exceed the solar system escape velocity at Jupiter [15]. Compared to the Ulysses and Galileo measurements, a dust telescope can provide an improved identification capability of small ISD grains. This is achieved by instruments with a narrow field of view of only 25° half-width.

2.2 Chemical and isotopic analysis

The dust telescope will use time-of-flight mass spectrometers to analyze the chemical and isotopic dust composition. The method of in situ mass spectrometry was demonstrated by the PUMA and PIA instruments that took data in the coma of comet Halley on board the spacecraft Vega 1, Vega 2, and Giotto, respectively. The data collected by PIA/PUMA demonstrate that each individual event detected by the mass spectrometer contains a wealth of scientific information.

The statistical analysis of spectra obtained from many individual dust particles gives information about the composition of their source. In the case of Halley, it was found that the abundances of elements more refractory than O, resemble solar composition and the composition of C1 chondrites alike. H, C and N on the other hand, are less depleted than in C1 chondrites, indicating the even more primitive (unequilibrated) character of Halley's dust compared to C1. For C and N, the composition of dust from comet Halley is nearly solar [17]. The organic component of Halley dust has been inferred from coincidence analysis of molecular lines [18]. The organic component consists mainly of highly unsaturated hydrocarbon polymers containing C-H and C-N-H compounds.

The dust telescope will use techniques for the analysis of dust which have previously successfully been applied to cometary dust. For the isotopic analysis of the detected grains, the mass spectrometer must have a mass resolution on the order of $M/\Delta M \geq 100$. A lower mass resolution of $M/\Delta M$ 20 - 50 is sufficient to obtain an inventory of heavy elements in the detected grains.

2.3 Measurement of the interstellar dust flux near Earth

At large heliocentric distances, the Ulysses and Galileo dust instruments measured an average ISD flux for 10^{-15} g particles of $1.5\ 10^{-4}\ m^{-2}s^{-1}$ [15]. The local ISD flux measured on a satellite moving with the Earth around the Sun exhibits an annual variation. Assuming a constant ISD flux of $1.5\ 10^{-4}\ m^{-2}s^{-1}$ and a velocity of 26 km/s from a direction of $\lambda_{ecl} = 259°$, $\beta_{ecl} = 8°$, the maximum interstellar flux of $3\ 10^{-4}\ m^{-2}s^{-1}$ occurs in February and remains more than 4 times larger than the interplanetary flux during 6 months around this time [14]. This flux gives a rough estimate for the expected impact rates onto dust instruments with sensitive areas of 0.01 m^2, 0.1 m^2 and 1 m^2. The corresponding rates of 10^{-15} g particles, are $10^{-6}\ s^{-1}$ (1 per 12 days) for 0.01 m^2 sensitive area, and $10^{-5}\ s^{-1}$ (1 per day) for 0.1 m^2 sensitive area, or $10^{-5}\ s^{-1}$ (1 per day) of 10^{-12} g particles for 1 m^2 sensitive area, respectively.

The dynamics of ISD grains in the solar system is governed by solar gravity, radiation pressure, and the Lorentz force induced by the radially expanding solar wind magnetic field. These forces shape the trajectory of an individual grain and determine the ISD flux distribution at any given location in the solar system.

If electromagnetic forces are neglected, the distribution of ISD in the solar system can be calculated analytically [19]. In this case, and if gravity dominates radiation pressure ($\beta < 1$), ISD is concentrated in a wake downstream of the Sun. In the case $\beta > 1$, an interstellar dust-free cavity forms around the Sun [20]. The influence of the solar wind magnetic field on the dynamics of ISD grains is important for grains smaller than 0.4 μm in diameter. It was shown numerically [21] and confirmed by the in situ measurements of Ulysses [19] that the flux of small ISD grains varies with the 22-year polarity cycle of the solar magnetic field. As a result of the electromagnetic interaction, the ISD flux in the outer solar system dropped in mid-1996 from $1.5\ 10^{-4}\ m^{-2}s^{-1}$ to $5\ 10^{-5}\ m^{-2}s^{-1}$. According to this model, the best time to measure a large number of ISD grains will be between 2004 and 2012, when the flux, and thus the impact rate, of 10^{-14} g grains at 1 AU will be enhanced by up to a factor of 10. The flux of large ($m > 10^{-11}$ g) grains is permanently increased at 1 AU by gravitational focusing.

2.4. Measurement of the interstellar dust size distributions

While the size distribution of interplanetary dust is well understood [22] [23] the size distribution of ISD grains is less well established. In order to match the observed wavelength dependence of interstellar extinction, ISD models introduce a grain size distribution [24] [25]. This provides a basic test for astrophysical ISD models. The measurements of local ISD by Ulysses and Galileo have already shown that the local ISD population contains more large grains than expected by these models [5]. However, because of their limited target area of 0.1 m^2, Ulysses and Galileo detected fewer than 10 ISD grains with masses above 10^{-11} g, which corresponds to a particle diameter of 2 μm.

3. INSTRUMENTATION

The dust telescope described here consists of four state-of-the-art dust instruments and a plasma monitor, PLASMON. The dust instruments have a field of view of 25° half-width implemented by extra baffles. The relatively narrow field of view of the dust telescope allows pointed observations of the dust flow. The four instruments are, a high resolution impact mass spectrometer (Cometary and Interstellar Dust Analyzer, CIDA), a dust analyzer for the determination of physical and chemical dust properties (Cosmic Dust Analyzer, CDA), a large area impact detector with trajectory analysis (Dust Detector System, D2S), and the momentum sensor ISIDE (Impact Sensor for Interstellar Dust Exploration). This selection of instruments is not mandatory. When new instruments with improved capabilities become available they should be used.

Three dust instruments share a common impact plane of 1 m^2 that is divided into 9 segments. Seven segmented polyvinylidene fluoride (PVDF) targets of D2S surround the central opening for CDA. ISIDE is mounted in one corner. The positions of CDA and ISIDE can be exchanged with any D2S element, depending on the interface to the spacecraft. Figure 1 shows the dust telescope with CIDA and PLASMON attached to the baffle of the D2S dust collimators and CDA and ISIDE below. Seven of the nine collimators belong to D2S, while CDA and ISIDE have their own collimators. PLASMON monitors the plasma environment and thus supports the interpretation of the dust charge measurements. The total mass of the dust telescope is 52.2 kg. The instrument properties are given in Table 1. The instruments employ various detection techniques and have different sensitivities. CIDA and CDA have the highest sensitivity for fast particles.

3.1. The Cosmic Dust Analyzer, CDA

The Cosmic Dust Analyzer (CDA) is a modified version of the instrument flying on board the Cassini spacecraft [26]. The Cassini CDA has been operated successfully since January 1999.

The CDA instrument is an impact ionization detector: A high speed particle impacts a target and becomes partially ionized. The generated impact plasma is separated in an electric field such that the electrons are detected at the target and the ions are detected at a central charge collector and a multiplier. The signals are converted to analogue voltages, which are digitized, processed, and stored in the CDA instrument computer.

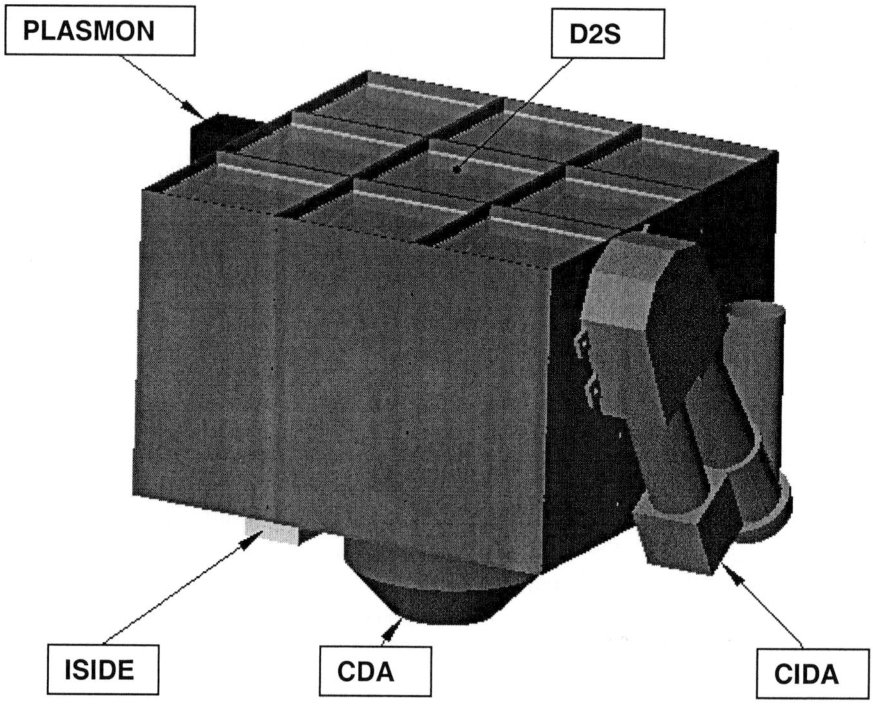

Figure 1. Dust telescope. On top are 9 dust collimators containing the charge sensing grids, seven of which belong to D2S. CDA is mounted below the central collimator, whereas ISIDE is mounted underneath a corner collimator. The overall height of the dust telescope is about 0.7 m. CIDA and PLASMON are attached to the baffles of D2S.

Table 1
Instrument characteristics (sensitive area and mass threshold at 20 km/s impact speed of the dust instruments) and consumables (mass and power) of the model payload. The mass for each instrument includes the mass for the electronics box.

Instrument	Sensitive Area [m^2]	Mass Sensitivity [g]	Power [W]	Mass [kg]
D2S	7x0.11	10^{-12}	11	23.7
CIDA	0.009	10^{-14}	15	11
CDA	0.1	10^{-14}	11	8.6
ISIDE	4x0.015	$5\ 10^{-13}$	15	6
PLASMON	./.	./.	1	0.7

CDA carries an integrated time-of-flight mass spectrometer (22 cm drift length) in order to analyze the elemental composition of an impinging dust particle. The fast sampling rate of 100 MHz at the multiplier enables this channel to record a time of flight mass spectrum of mass resolution $M/\Delta M \approx 50$. The resolution is high enough to identify the major elemental composition.

3.2. Cometary and Interstellar Dust Analyser, CIDA

The CIDA instrument on board Stardust [27] is a high resolution time-of-flight mass spectrometer. The drift tube of the time-of-flight spectrometer CIDA is folded and has a total length of 56 cm. The reflector compensates for different initial starting energies of the ions and thus increases the mass resolution. Amplifiers connected to the multiplier allow the measurement of the TOF spectrum. The useful target size is 120 mm in diameter. For an impact angle of 40° from the target normal, this corresponds to a sensitive area of 0.009 m^2. CIDA is a direct derivative of the highly successful PIA and PUMA instruments flown on the Giotto and Vega spacecraft [9]. However, it has about 10 times larger sensitive area (0.009 m^2) and twice its mass resolution ($M/\Delta M = 250$).

3.3. Dust Detector System, D2S

The D2S instrument consists of an impact detector of the type flown on Vega [28] and a trajectory analyzer of charged dust grains, similar to the charge measurement sub-system of Cassini CDA [26]. The proposed large surface area dust detector consists of an array of 7 individual D2S detectors. Each D2S detector will apply two sets of parallel grids for position, charge and velocity sensing, a shield grid in front of the charge sensing stage, and a PVDF film for mass determination at the bottom of the instrument.

The detection principle of the charge-sensing trajectory detectors is based on charge induction by the dust particle's electrical charge [8] [29] [30]. Two parallel grids are connected to a single charge sensitive amplifier. When a charged dust grain passes through these grids, it induces a charge on the grids, and this can be measured as a function of time. The amplitude of the signal is proportional to the charge on the dust particle, and the duration of the signal contains information on the velocity of the grain. Two sets of parallel grids are inclined in perpendicular directions. Thus, a particle's instantaneous position is measured in two perpendicular planes, and the trajectory of the charged particle can be reconstructed. Several dust charge signals (of > 10^{-15} C) obtained by Cassini CDA proved the principle.

The detection principle of the PVDF film detectors is based on the depolarisation signal a dust particle generates, when penetrating a permanently polarised PVDF thin film [28] [31]. Dust grains penetrating the thin PVDF film remove dipoles along their trajectory producing a fast electric charge pulse without requiring bias voltages. The produced signal is a function of the particle mass and velocity. PVDF film dust detectors have been extensively tested and calibrated in laboratory experiments.

3.4. Momentum sensor, ISIDE

The Impact Sensor for Interstellar Dust Exploration (ISIDE) is aimed at monitoring the dust mass flux, measuring the momentum of the impacting grains and their velocity. ISIDE is formed by an array of four sensor plates. ISIDE sensing elements are similar to the sub-systems used in the GIADA experiment, for ESA's Rosetta mission.

The instrument uses piezoelectric transducers (PZT) as impulse sensors. PZTs have been used successfully in high velocity impact sensors since the Explorer I mission and have given excellent results in the Giotto cometary mission (DIDSY experiment) and on the Mir space station. As an example, thanks to DIDSY, it was possible to monitor for the first time the dust flux at comet 1P/Halley during flyby with a relative velocity of 68 km/s [32].

Each ISIDE sensing element is formed by a 12 x 12 cm aluminum diaphragm equipped with five piezoelectric sensors placed at the four corners and at its center, and thus able to detect the grain impact. The geometry of the plate has been chosen to have a propagation frequency close to the resonating frequency of the sensing crystals and to damp reflected waves from the sensor edges, which would increase the noise.

3.5. Plasma Monitor, PLASMON

PLASMON is a plasma monitor consisting of a magnetometer, a simple electrostatic plasma analyser and a Faraday cup. All three sensors are integrated in the spherical PLASMON sensor which has a diameter of 15 cm. The PLASMON instrument will be mounted at the dust baffle. The sensors are controlled by a single DPU which serves as the interface to the spacecraft. All three sensor types are well space-proven. A similar instrument, the ROMAP, was developed and tested for the Rosetta Lander. Assuming a radial solar wind bulk velocity, all major plasma parameters such as electron and proton densities, and temperatures and proton bulk velocity can be derived.

4. MISSION

A first example of a dust telescope has been described in the proposal for the ***Galactic DUNE*** mission which is being proposed by an international team of 66 scientists (http://mpi-hd.mpg.de/galileo/DUNE/). The goal of the mission is the in-situ characterization of galactic interstellar dust, in order to provide crucial information not achievable with astronomical methods.

In order to meet this scientific goal, the ***Galactic DUNE*** mission was designed to allow the distinction between ISD and interplanetary dust, to measure the chemical and isotopic composition of ISD, and to monitor the ISD flux for various directions and masses over the duration of the mission. To minimize the contamination by man-made solid particles near Earth, ***Galactic DUNE***'s trajectory lies outside the debris belts at the distance of geosynchronous orbit. The selected trajectory is a high Earth orbit close to one libration point of the Earth-Sun system (Figure 2).

The ***Galactic DUNE*** satellite exploits the effect of the Earth's motion ($v_E \approx 30$ km/s) on the local ISD velocity ($v_{ISD} \approx 26$ km/s, assuming $\beta = 1$) and flux. This can be used to confirm the interstellar origin of the measured dust population. During the Earth's upwind motion, the velocities add up to 56 km/s. When the dust telescope points into this direction, the flux of ISD grains is almost an order of magnitude larger than the flux of interplanetary grains (Figure 3). During the downwind motion the relative velocity and consequently the ISD flux are low. At least two years are needed in order to characterize the ISD flux along the Earth's orbit about the Sun in a statistically significant way and to determine the various components of interplanetary dust particles. During a two-year mission more than 1,000 galactic and interplanetary grains will be recorded, 60 of which with high resolution mass spectra.

In Figure 3 (bottom) we suggest an observation strategy for the dust telescope. The average impact velocity measured by the charge sensors of D2S or CDA can then be analyzed and compared to the local escape velocity. An additional identification criterion for small ISD grains is provided by the annual variation of the ISD flux, due to the Earth's upstream and downstream motion in spring and autumn respectively.

During the first two years of the mission the interstellar dust flux along the Earth's orbit around the Sun, as well as the interplanetary dust flux, will be determined. In order to provide statistically meaningful data, the dust telescope will perform blocks of observations, each of duration 4 weeks, with a fixed instrument pointing in inertial space. The observations will be linked by maintenance periods of about two days.

The observation strategy for the second year can be modified depending on the findings from the first year. In particular, deviations of the flux direction of big interstellar grains from the nominal direction are important to characterize the coupling of big galactic grains to the LIC, or larger structure s, and to identify potential sources of these particles.

Observations during the later phases of the mission could address other related scientific questions. Potential targets are specific interstellar phenomena. For example, sampling could be undertaken, of the directions from which an interstellar dust flux is suspected (e.g. β–Pictoris). In addition, interplanetary phenomena (e.g. dust in meteoroid streams, cometary dust or dust from the Moon), or even debris studies (fine grains from solid rocket burns) could be studied.

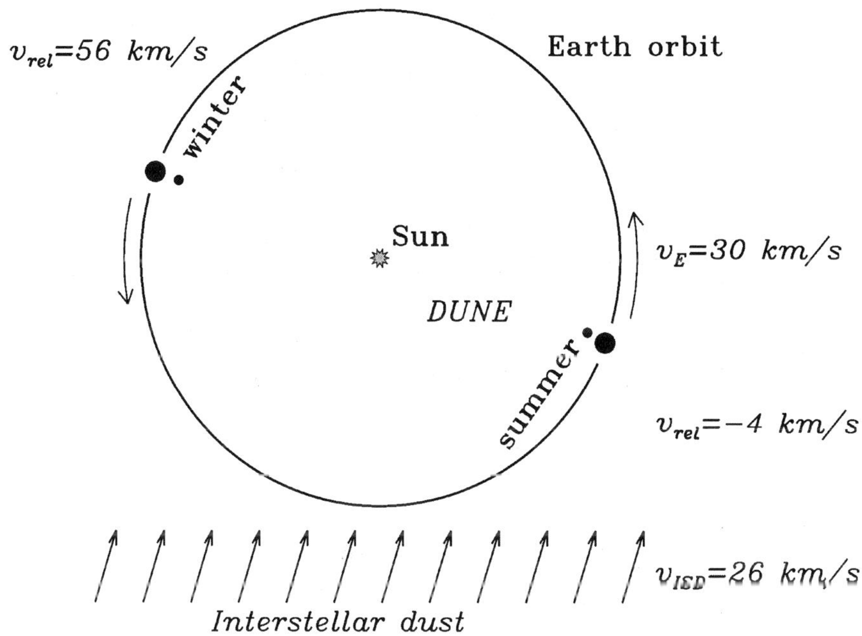

Figure 2. ***Galactic DUNE*** mission scenario. Earth's orbit around the Sun (v_E = 30 km/s), and the direction of the interstellar gas and dust flow at speed v_{ISD} = 26 km/s are shown. Two positions of the Earth and the satellite are shown (right: late summer, and left: late winter, not to scale). The corresponding fluxes are $F_W = 4.5\ 10^{-4}\ m^{-2}\ s^{-1}$ in winter and $F_S = 6.6\ 10^{-5}\ m^{-2}\ s^{-1}$ in summer. We have assumed that radiation pressure and solar gravity for ISD grains cancel each other, therefore ISD traverses the solar system on straight trajectories.

Orbits about the collinear L1 or L2 points are of particular interest for Galactic DUNE, as they meet the requirement of being at large distance from Earth's orbital debris belts, and they provide a very stable thermal environment. An additional advantage of orbits about L1 or L2 is that they allow easy orbit manipulations by means of lunar flybys as demonstrated by the Japanese Hiten and Nozomi missions. This feature of L1 and L2 orbits is particularly interesting for possible follow-up missions to return samples of ISD to Earth.

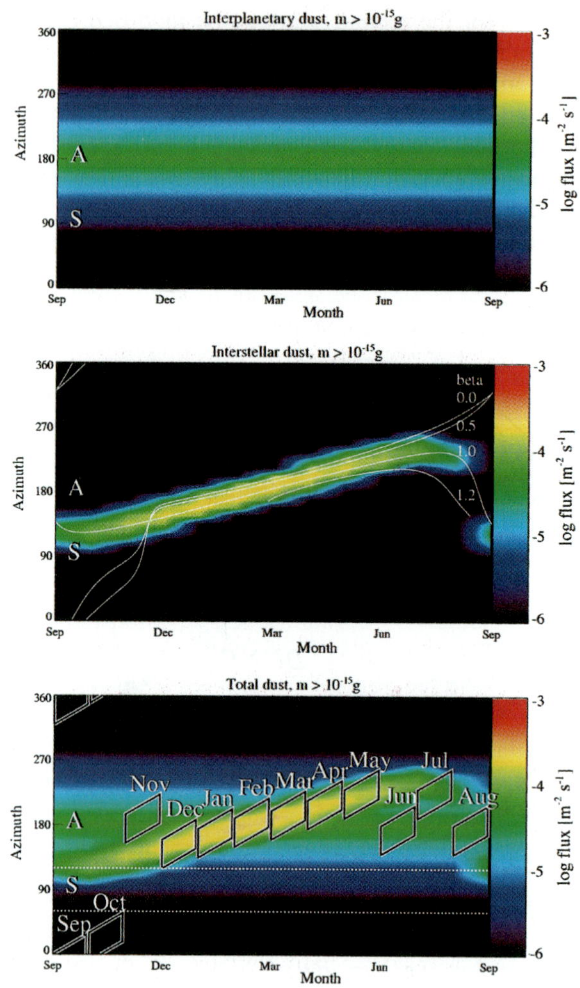

Fig. 3. Model dust fluxes and observation strategy. We assume an ISD upstream direction of $\lambda_{ecl} = 259°$, $\beta_{ecl} = 8°$. The top figure shows the interplanetary dust flux as derived from the interplanetary dust flux model (divine1993, Grün1997a). The assumed spacecraft orbit is a circular orbit about the Sun at 1 AU with the spacecraft position characterized by its ecliptic longitude. The assumed detector sensitivity is 10^{-15} g and the detector has a 50° wide aperture. Azimuth 90° corresponds to the solar direction (S) and 180° is the apex direction of the Earth's motion (A). The middle figure shows the flux of ISD grains with an assumed β value of 1. Center flux directions of ISD grains of different β values (0, 0.5, 1, and 1.2) are shown by white lines. The bottom figure displays the combined fluxes. Overlain is a sample of possible observations represented by the diamonds. Individual observations last about a month (the months indicated represent the start of an observation at about the 20th day of the month). The azimuth range corresponds to ± 20°. It is assumed that during each observation the spacecraft with the dust telescope, points to a fixed direction in inertial space. Within 30° from the solar direction is the Sun avoidance zone.

5. CONCLUSIONS

The goal of the proposed dust telescope mission is the in-situ characterization of galactic interstellar dust, in order to provide crucial information not achievable with astronomical methods. Galactic dust constitutes the solid phase of interstellar matter, from which stars and planetary systems form. Information on this fundamental material is extremely sparse. However, following the discovery by Ulysses (and recent confirmation by Cassini) that micron-sized galactic dust grains pass through the solar system, the analysis of galactic dust with an Earth orbiting spacecraft is now within our reach. ***Galactic DUNE*** establishes the next logical step beyond NASA's Stardust mission, to a full-fledged galactic dust analysis program, with an orbit being easily accessible for follow-up sample return missions. ***Galactic DUNE*** provides the following major advancements in interstellar dust research beyond Stardust:

- Reliable distinction between interstellar and interplanetary dust and other dust sources,
- Analysis of the elemental, molecular and isotopic composition of interstellar grains,
- Characterization of the interstellar dust flow through the planetary system,
- Determination of mass, speed and electrical charge of individual particles in order to derive the size distribution of interstellar dust from 0.1 to 10 μm.

The science payload is a dust telescope consisting of a complement of instruments that use various dust detection techniques (having heritage from instruments on Giotto, Vega, Cassini, Stardust, and Rosetta). They are selected for 1) large area impact detection and trajectory analysis of micron sized and larger dust grains, 2) determination of physical properties of small grains, like flux, mass, speed, electrical charge, and coarse chemical composition, and 3) high resolution chemical analysis of galactic dust. A plasma monitor supports the dust charge measurements. The viewing directions of all dust instruments shall be co-aligned with narrow fields of view. About 600 interstellar grains are expected to be recorded by this payload every year, with 30 or more providing high resolution mass spectra.

The total payload mass is 53 kg and the power consumption is 53 W. The mission requires a 3 axis stabilized spacecraft of 1° pointing accuracy. The orbit shall be outside the debris belts, orbits around L1 or L2 being especially suitable. A mission duration of 2 years is required, with a possible 1 year extension. 50 Mbits of scientific data are transmitted to ground once every week. Monthly command uploads and spacecraft attitude changes are envisioned.

REFERENCES

1. W.M. Alexander, C.W. Arthur and J.L. Bohn, Space Research XI (eds. K. YA. Kondratyev, M. J. Rycroft and C. Sagan) Akademie Verlag, Berlin (1971) 279.
2. O.E. Berg and F.F. Richardson, Rev. Sci. Instrum. 40 (1969) 1333.
3. O.E. Berg and E. Grün, Space Research XIII (eds. M. J. Rycroft and S. K. Runcorn), Akademie-Verlag, Berlin (1973) 1047.
4. J.W. Rhee, O.E. Berg, F.F. Richardson and S. Auer, Nature 252 (1974) 555.
5. P.C. Frisch, J. Dorschner, J. Geiß, J.M. Greenberg, E. Grün, M. Landgraf, P. Hoppe, A.P. Jones, W. Krätschmer, T.J. Linde, G.E. Morfill, W.T. Reach, J. Slavin, J. Svestka, A. Witt and G.P. Zank, Astrophys. J. 525 (1999) 492.

6. H. Krüger, E. Grün, A. Heck, and S. Lammers, Planet. Space. Sci. 47 (1999) 1015.
7. M.E. Zolensky, R.A. Barrett, L. Hrubesh, F. Hörz and D Lindstrom, Abstracts of the Lunar and Planetary Science Conf. 20 (1989) 1251.
8. S. Auer, in Physics, Chemistry and Dynamics of Interplanetary Dust, (eds.B.A.S. Gustafson and M.S. Hanner) ASP Conference Series 104 (1996) 251.
9. J. Kissel, ESA SP-1077 (1986) 67.
10. P.C. Frisch, in Lecture Notes in Physics, vol 506, Springer, Berlin (1988) 269.
11. M. Witte, M. Banaszkiewicz and H. Rosenbauer, in The Heliosphere in the Local Interstellar Medium (eds. R. von Steiger, R. Lallement and M.A. Lee) Space Science Reviews, Vol. 78, Kluwer Academic Publishers (1996) 289.
12. N. Divine, J. Geophys. Res. 98 (E9) (1993) 17029.
13. P. Staubach, E. Grün and R. Jehn. Adv. Space Res. 19 (1997) 301.
14. E. Grün, M. Landgraf, M. Horanyi, J. Kissel, H. Krüger, R. Srama, H. Svedhem and P. Withnell, J. Geophys. Res. 105 (2000) 10403.
15. E. Grün, B.A.S. Gustafson, I. Mann, M. Baguhl, G.E. Morfill, P. Staubach, A. Taylor and H.A. Zook, Astron. Astrophys. 286 (1994) 915.
16. M. Baguhl, E. Grün, D.P. Hamilton, G. Linkert, R. Riemann and P. Staubach, Space Sci. Rev. 72 (1995) 471.
17. E.K. Jeßberger and J. Kissel, In Comets in the post-Halley era, Proc. IAU Coll. 116, Kluwer (1991) 1075.
18. J. Kissel and F.R. Krüger, Nature 326 (1987) 755.
19. M. Landgraf, M. Müller and E. Grün, Planetary and Space Science 47 (1999) 1029.
20. M. Landgraf, K. Augustsson, E. Grün and B.A.S. Gustafson, Science 286 (1999) 2319.
21. M. Landgraf. J. Geophys. Res. 105 (2000) 10303.
22. E. Grün, H.A. Zook, H. Fechtig and R.H. Giese, Icarus 62 (1985) 244.
23. J.A.M. McDonnell and D.J. Gardner, Icarus 133 (1998) 25.
24. J.S. Mathis, Astrophys. J. 472 (1996) 643.
25. A. Li and J.M. Greenberg, Astron. Astrophys. 323 (1997) 566.
26. R. Srama, J.G. Bradley, E. Grün, T.J. Ahrens, S. Auer, M. Cruise, H. Fechtig, A. Graps, O. Havnes, A. Heck, S. Helfert, E. Igenbergs, E.K. Jeß berger, T.V. Johnson, S. Kempf, H. Krüger, P. Lamy, M. Landgraf, D. Linkert, F. Lura, J.A.M. McDonnell, D. Möhlmann, G.E. Morfill, G. H. Schwehm, M. Stübig, J. Svestka, A.J. Tuzzolino, R. Wäsch and H.A. Zook, Space Sci. Rev. in press.
27. D.E. Brownlee, P. Tsou, D. Burnett, B. Clark, M.S. Hanner, F. Horz, J. Kissel, J.A.M. McDonnell, R.L. Newburn, S. Sandford, Z. Sekanina, A.J. Tuzzolino and M. Zolensky, Meteoritics and Planetary Science 32 (1997) A22.
28. J.A. Simpson and A.J. Tuzzolino, Nucl. Instrum. and Methods A 236, (1985) 187.
29. S. Auer, Rev. Sci. Instrum. 46 (1975) 127.
30. S. Auer and F.O. Bun, Highly transparent and rugged sensor for velocity determination of cosmic dust particles, LPI Tech. Rep. 94-05, (1994) 25.
31. A.J. Tuzzolino, Nucl. Instrum. and Methods, A 316 (1992) 223.
32. J.A.M. McDonnell, P.L. Lamy and G. Pankiewicz, in Comets in the Post-Halley Era (eds. R.L. Newburn, M. Neugebauer and J. Rahe) Kluwer. (1991) 1043.
33. M.S. Hanner, Space Sci. Rev. 90 (1999) 99.

Planetary Aerosol Monitor/ Interplanetary Dust Analyzer

B. Å. S. Gustafson[a], F. Giovane[b], T. Waldemarsson[a], L. Kolokolova[a], Yu-l. Xu[a] and J. McKisson[a]

[a] Dept. of Astronomy, University of Florida, Gainesville, FL, USA

[b] Space Sc. Div., Naval Research Lab., Washington, DC, USA

The Laboratory for Astrophysics at the University of Florida has developed a dust analyzer capable of measuring key particle physical parameters without destruction or alteration of the dust grains. A laser illuminates dust grains passing through the device and a set of detectors obtains intensity and polarization of the scattered light at specific scattering angles. The choice of scattering angles, a sophisticated light collecting assembly and analysis algorithms derived from our data base of microwave analogue scattering measurements allow classification of the dust particles according to not only their light-scattering properties, but also according to size, shape (equidimensional, elongated, smooth or angular), composition (dielectric, absorbing, or metallic) and internal structure (compact or porous). The results of tests of the laboratory version of the PAM/IDA are presented.

1. INTRODUCTION

The Planetary Aerosol Monitor/Interplanetary Dust Analyzer is a light-scattering device of a new generation developed for remote sensing of single dust particles of micron size. The choice of scattering angles, a sophisticated light collecting assembly, and analysis algorithms (derived from microwave analog scattering measurements), allow ***classification*** of the dust particles according not only to their light-scattering properties and speed, but also to:

- size
- shape (equidimensional, elongated, smooth or angular)
- composition (dielectric, absorbing, or metallic)
- internal structure (compact or porous)

The processing software is able to classify the particles autonomously and to generate a particle parameter matrix (including correlation parameters). A compact data set can thus be stored and the already reduced data can be transmitted at any time.

The schematics of the instrument is shown on Figure 1. The instrument wakes up from its power saving mode within a few microseconds after a dust particle intercepts a light curtain in the instrument's ***alert stage***. This triggers a second light curtain in the ***scattering stage*** where the detector optics are designed to obtain intensity and polarization averaged over specific scattering angle ranges. Time of flight between the alert and scattering stages gives the particle speed. Particles exit unaltered so that a momentum, mass, or any other type of analysis or collection device may operate in tandem with PAM/IDA

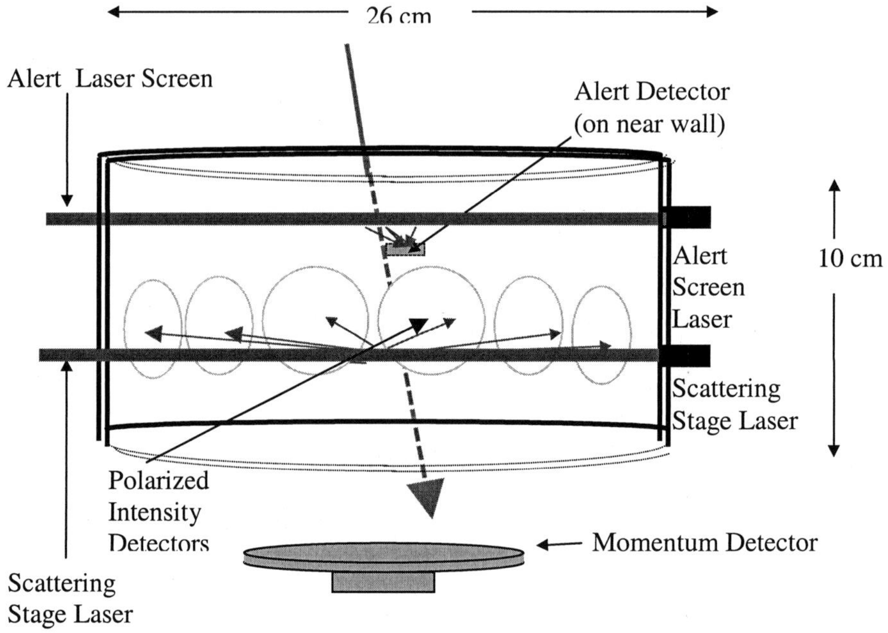

Figure 1: Schematics of the 2.5 kg, 1.5 W operating power, low bit rate, PAM/IDA

2. IDEOLOGY OF PAM/IDA MEASUREMENTS

2.1 Particle size

The basis of the particle size determination is the idea that the sum of intensities of the scattered light in all scattering angles PAM/IDA scattering stage detectors is nearly independent of particle composition, shape, and internal structure but sensitive to the particle size. This idea was checked with theoretical calculations for spherical particles and with microwave analog experiments for aggregated particles. Figure 2 shows how the idea works.

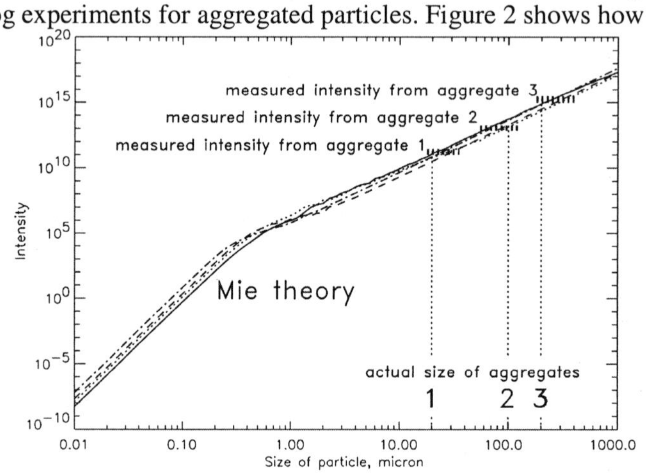

Figure 2: Sum of detected intensities for spheres (— ice, - - - silicate, ···· organics, - · - · - carbon) and from microwave analog measurements for the aggregated particles (see Fig. 3).

From Mie calculations one can see that the sum intensity for the given scattering angles has almost no dependency on the composition of spherical particles. Results of the microwave measurements show that even for so complicated particles as the aggregates shown on Figure 3 the size of aggregate, determined independently by direct measurements of aggregate dimensions, can be well determined by measurements of the sum intensity of light, scattered by the aggregates. The range of measured intensities corresponds to the measurements for different orientations of aggregates. The vertical lines correspond to the average size of aggregates.

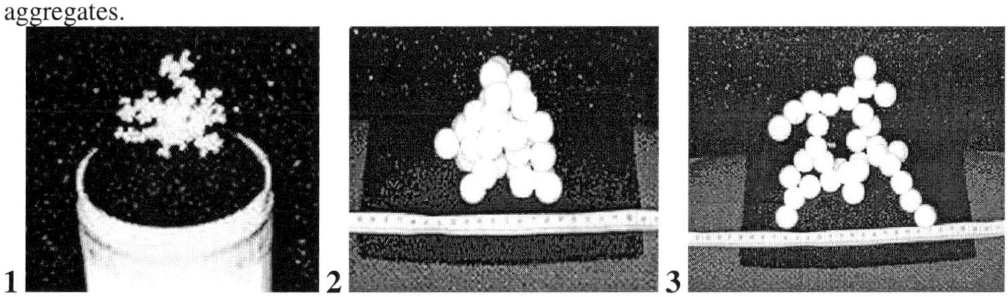

Figure 3: Aggregate 1: 500 spheres of size parameter (circumference to wavelength ratio) x= 1.5 each. Scaled to λ=0.8 μm, the size of each sphere is 0.38 μm and the dimension of the aggregate is 20 μm. Aggregate 2: 37 spheres of size parameter x= 20 each. Scaled to λ=0.8 μm, the size of each sphere is 5 μm and the dimension of the aggregate is 100 μm. Aggregate 3: 43 spheres of size parameter x= 20 each. Scaled to λ=0.8 μm, the size of each sphere is 5 μm and the dimension of the aggregate is 200 μm.

2.2. Classification according to composition

Figure 4 illustrates the idea how PAM/IDA measurements of polarization of scattered light can be used to classify particles according to their composition.

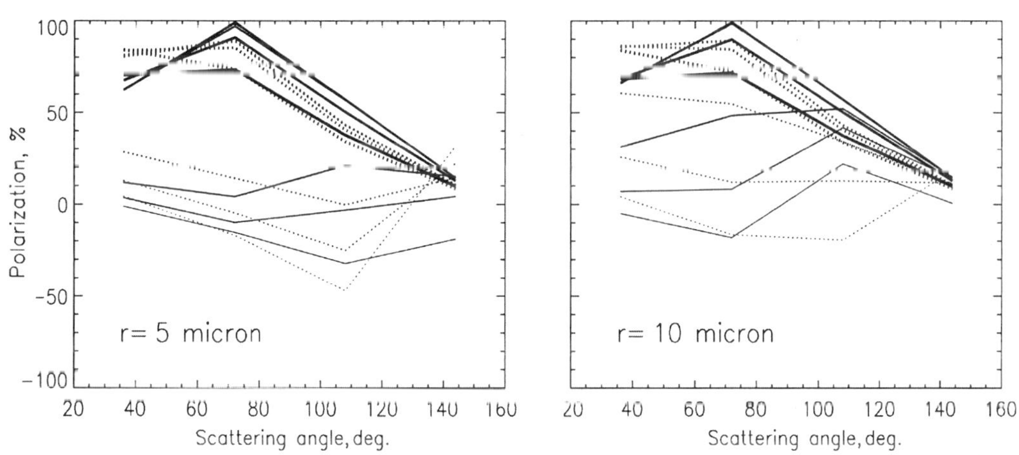

Figure 4: Transparent and absorbing particles can be distinguished using as few as four scattering angles. The optical constants are: n = 1.5(—), 1.9 (⋯⋯); κ = 0.001, 0.005, 0.01, 0.05, 0.1, 0.5, 1 (from thin to thick lines).

Shown on Figure 4, results of theoretical (Mie theory) simulation of PAM/IDA measurements demonstrate that transparent and absorbing particles show qualitatively different trends in polarization as a function of scattering angle. For transparent particles polarization is usually low, even negative and has a minimum at medium scattering angles, whereas for absorbing particles polarization is high, positive and gets the maximum at medium scattering angles. The position of the maximum or minimum of polarization depends on the real part of the refractive index. This allows the refractive index $m=n+i\kappa$ of the particle material to be estimated.

3. Laboratory tests of the PAM/IDA

PAM/IDA was tested using transiting water drops. The main purpose of the tests was to show the sensitivity and accuracy of the measurements. Figure 5 shows the results of the measurements in comparison with Mie theory calculations for water spheres.

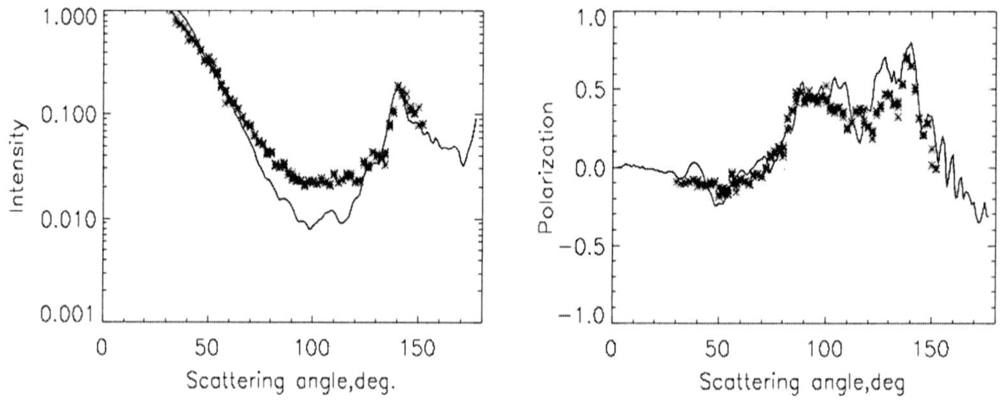

Figure 5: Particle sizes are in the 36-40 micron diameter range. The solid line is Mie calculations for a 38-micron diameter spherical water drop.

4. CONCLUSIONS

PAM/IDA is a new generation light-scattering device to measure physical parameters of single particles PAM/IDA provides non-disturbed analysis of particles, i.e. particles exit intact, so that PAM/IDA can be mounted in front of other instrumentation. At 2.5 kg, 1.5 W operating power, and low bit-rate due to onboard processing, PAM/IDA is designed for low impact on spacecraft resources. A laboratory version of the PAM/IDA was successfully tested and PAM/IDA is ready for development of a flight version. Even though it as designed to study interplanetary and cometary dust with small alterations PAM/IDA is suitable for studying planetary aerosols.

ACKNOWLEDGEMENTS

The work was supported by NASA grant NAG5-4547.

IV Dust in the Outer Solar System and Other Planetary Systems

Dust in young solar systems

N.A. Krivova [a] [b]

[a]Max-Planck-Institut für Aeronomie, D-37191 Katlenburg-Lindau, Germany

[b]Astronomical Institute, St. Petersburg University, 198504 St. Petersburg, Russia

Recent spectacular discoveries of extrasolar planets and circumstellar discs around main-sequence stars have offered a fertile ground for comparative studies of these objects and of the Solar System. There is no question that the planets and discs are manifestations of the same phenomenon called a 'solar system', even though the detection of planets around stars with discs is as yet hampered by observational limitations. In this review we concentrate on the main- and pre-main-sequence stars with circumstellar discs representing younger analogues of the Solar System. The emphasis is placed on the dust component of the discs. Properties of the dust, its spatial distribution and some physical processes in which it is involved are covered. Some general evolutionary tendencies are also touched upon.

1. DISCS AND/OR PLANETS?

The origin of the Solar System is one of the oldest and still one of the most fascinating problems of astronomy. Real impetus to the research was given in the 1980s when IRAS observations evidenced the presence of dust around a few main-sequence (MS) stars [1,2] and the disc of β Pic was imaged by Smith and Terrile [3]. The next wave of interest to the problem was driven by recent discoveries of planets around other solar-type stars ([4–8]; see [9,10] for recent reviews). These intriguing discoveries have given us an opportunity to look at our Solar System from the outside and, in addition, at its past. Whereas the first offers a clearer view into the global properties of a dust disc, the latter provides insight into the process of the system's formation.

The presence of dusty discs around stars during the planet formation stage is required by all plausible theories (e.g. [11]). Being a by-product of the star formation, discs make an excellent reservoir of the material for the formation of planets and smaller bodies. "Discs, then, are the thread connecting protostars and planets" [12]. The bulk of the material left over after star formation is completed disappears by the time when the planetary system is formed. The remainder is present mostly in a solid form. The Edgeworth-Kuiper Belt (EKB; e.g. [13–15]) provides a vivid illustration of that. Further abundant evidence is found that at some evolutionary stages discs and planets may coexist:
1. The giant gas planets are thought to be formed rapidly, with a solid core being developed in $\lesssim 10^6$ yr and accretion of a massive gaseous envelope being accomplished in $\sim 10^7$ yr (see Zuckerman [16] and references therein). This is less than the ages of typical

disc systems.
2. Decreasing with time, the masses of discs even around 'old' pre-main-sequence (PMS) stars are typically still higher than $0.01 M_{\odot}$ [12,17–19] — a minimum mass required for planet formation in a system (e.g. [11]). At the same time the total masses of circumstellar material around main-sequence stars are much less (e.g. [20–24] among others) evidencing that planets, if any, must already have been formed there.
3. A low amount of gas around MS stars is another argument in favour of the already completed planet formation (e.g. [20,22,25–28]). When compared to the standard gas to dust ratio (≈ 100), the gas is typically depleted by a factor of up to $100 - 1000$.
4. Inner depletion zones in the discs of MS stars (see Sect. 3.1) are also thought to be a sequel of the presence of a planet.

Because of many observational restrictions the Solar System was for a long time the only example showing a co-existence both of planets and a disc. Improved observational techniques, however, made it possible to image even tenuous debris discs around nearby stars. As a result, three other solar-type stars known to be orbited by a planetary-mass companion (55 Cnc, ρ CrB and HD 210277) were recently found to also possess discs [29,30].

Nevertheless, the discs as tenuous as our EKB dust disc are still beyond the observational limitations [30], so that the discs accessible to observations and study are typically more massive and younger than the EKB. That is why we direct our attention now to these younger analogues of the Solar System (see Figure 1). By this the group of so-called Vega-type systems is typically meant. First recognized as a separate class of objects by IRAS observations, they have been intensively studied during the last 15 years. The four famous IRAS prototypes — β Pic , Vega, Fomalhaut and ε Eri — belong to this class. Generally speaking, the group includes all MS stars surrounded by dust discs. In this sense, our Solar System may also be considered as a Vega-type system.

The discs are not, however, something that suddenly develops when the star reaches the main sequence. Just the opposite: young stars deeply embedded in residues of the protostellar clouds gradually get free of the remnant material, the clearing being particularly effective in the inner parts. This primordial material is lost from the interior of the shells well before the stars reach the main sequence (e.g. [31,32]). The discs around older stars, inevitably evidenced by different observational data, are thus second-generation, or debris, discs. They are maintained by a collisional cascade of dust grains released by comets and/or planetesimals (Sect. 3.3). It is clear now, that circumstellar shells of the 'oldest' PMS stars, though somewhat more extended and dense than the discs in Vega-type systems, are actually of the same nature. So by "young solar systems" we mean here systems, in which a MS or PMS star older than a few Myr is surrounded by *debris* dust material.

2. CIRCUMSTELLAR DUST: OBSERVATIONAL MANIFESTATIONS

The presence of dust around MS stars was first indicated by the mid- and far-*IR excesses* detected by IRAS for four sources [1,2]. Later on the excesses were found for many other MS stars (e.g. [33–38]). The excesses are typically seen in far-IR, although in a few cases mid-IR emission is also detectable. The younger the system, the larger is the excess in

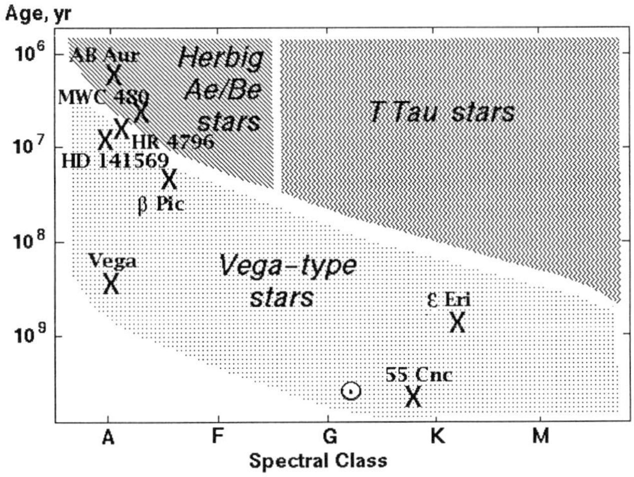

Figure 1. Classification of 'young solar systems'

the mid-IR, and PMS objects show strong emission already in the near-IR. Up till now thermal emission of dust has remained the main criterion for the selection of Vega-type candidates [39–41].

Spectroscopic observations in the IR region provide another good means of studying circumstellar dust. Whereas warmed dust emits mostly in the continuum, warm silicate particles produce also the well-known 10 μm emission feature which is observed in many dusty objects (see Sect. 3.2 for more details). The presence of silicates in crystalline form is also evidenced by emission features in the $20-70\,\mu$m region [42–45]. A family of emission bands (3.29 μm and associated $3.4-3.6\,\mu$m features, $6.2, 7.7, 8.6$ and $11.3\,\mu$m) from small carbon-rich particles, generally attributed to polycyclic aromatic hydrocarbons (PAHs), is also often observed [24,43–47].

It is well known that small dust particles are also effective absorbers and scatterers of stellar radiation. The amount of dust around Vega-type stars is, however, insufficient to cause observable circumstellar *extinction.* The optical depth of the discs is always much less than unity even in the visual. A crude measure of the optical depth is offered by the fractional luminosity of dust relative to the star L_{IR}/L_* (see e.g. [31]). For Vega-type stars, it is commonly less than 10^{-2} (see Table 1). Being dustier (which is well seen from comparison of the ages and values L_{IR}/L_*), younger (PMS) objects have, on average, more material along the line of sight, so that the circumstellar extinction is observable. Note, that the material around PMS stars may not be confined to a flat disc, but rather to a more extended envelope (see discussions by Pezzuto et al. [48], Waters and Waelkens [49] and Miroshnichenko et al. [50]).

Circumstellar dust also shows up in the observed *polarization* of stellar light. To cite an example, an optical polarimetrical monitoring of 18 Vega-type and post-T Tau stars undertaken by Oudmaijer et al. [51] has shown that the radiation of 15 of them was polarized. In Figure 2 the observed polarization of Vega-type stars and their precursors is plotted as a function of the observed fractional luminosity of dust. The tendency is clear: the integral polarization of star-disc systems is proportional to the disc optical depths. It is also proportional to the albedo of dust particles and depends on the inclination angle of the disc with respect to the line of sight [52,53]. This should and does introduce some dispersion into the observed values. For the oldest objects with tenuous discs, the integral polarization is, unfortunately, very low and close to the detection limits, which makes the study difficult. Further uncertainty is introduced by the interstellar component of the polarization. Fortunately, in the case of Vega-type systems this component is normally not large, as the objects are fairly close. Obtaining polarization maps instead of measuring the integral polarization may be more advantageous, as in this case, polarization in an optically thin disc is independent of the (typically poorly known) optical depth and is predominantly determined by the dust properties.

Imaging is, of course, a direct and the best way to probe circumstellar dust — both in thermal emission in the IR and in the scattered stellar light in the visual (including polarization maps). Indisputable leader among the systems with available 'portraits' is β Pic . Being one of the closest and with the favourable edge-on orientation, it is still dense enough to be imaged at different wavelengths from optical [3,63,65,67,70,96,97] through near- [68] and mid-IR [66,98] to the submillimetre region [23]. A number of other Vega-type sources have been recently imaged in different wavelength regions (see e.g. [30,100–102,136] and Table 1). The discs/shells of PMS stars are dustier, which should make them easier to detect. These objects are, however, more distant, so that it

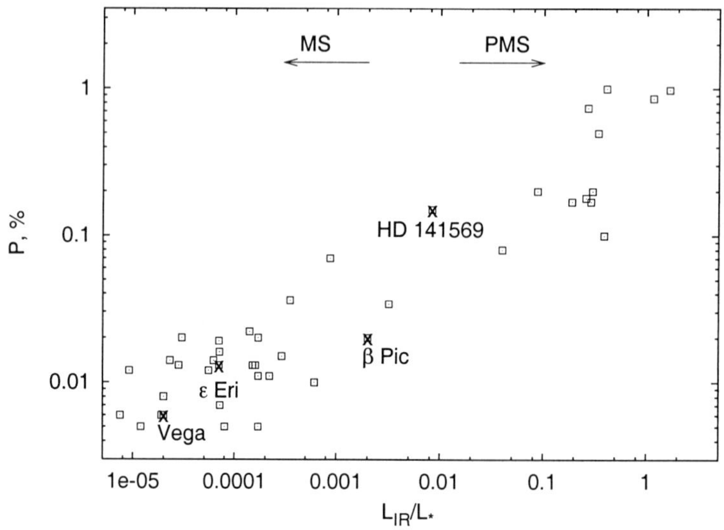

Figure 2. Polarization vs. L_{IR}/L_* for Vega-type and 'old' PMS stars

Table 1
Characteristics of discs around some PMS and MS stars

Star	Sp	Age(Myr)	L_{IR}/L_*	M_g/M_d	α^a	R_i^b(AU)	Images[c]	References
Sun	G2	4500	$\sim10^{-7}$	$\lesssim1$	~3–4?	~30	-	14,54,55
55 Cnc	G8	5000	7×10^{-5}		~4	13–35	nIR	56,57,29
ε Eri	K2	500–1000	10^{-4}	$\lesssim0.1$	ring	35	smm	31,25,58,27,20
Vega	A0	300±200	2×10^{-5}	$\lesssim0.1$			smm	31,59,25,23,20
α PsA	A3	200±100	8×10^{-5}	$\lesssim0.15$		60	fIR,mm smm	31,59,25,60,23,27,61,20
β Pic	A5	10–100	2×10^{-3}	$\lesssim0.1$	~3	~60	diff	62,25,63,64,23,65,66 68,28,69,70,71,3
HD 141569	B9.5	$\gtrsim10$	8×10^{-3}		ring	~150	n,mIR	72,73,48,74,75
HR 4796A	A0	8±2	5×10^{-3}	$\lesssim30$	ring	~50	n,mIR	75,76,77,78,79,80,22
HD 100546	B9	~10	$\gtrsim0.1$	~100	~3–4	<40	nIR	19,81,82,83,73
MWC 480	A2	6		100[d]	~7/4	(-)[e]	mm	84,85,82
WW Vul	A0-3	4±2	0.27	100[d]	~1.5	(-)[e]	-	86,87,88,82,89
BF Ori	A0-5	2.3±1	0.29	100[d]	~1.5	(-)[e]	-	90,88,82,89
AB Aur	A0	2±1	0.54	100[d]	~1	(-)[e] <150	diff	91,84,92,93,94,48,47 74,96

[a] α is the index of the power-law mid-plane density distribution outside the depletion zone;
[b] R_i is the size of the inner depletion zone;
[c] nIR, mIR, and fIR — near-, mid-, and far-IR, respectively; smm — submillimetre;
[d] no data available indicating non-standard ratio;
[e] the absence of the inner depletion zone is inferred from the spectral energy distribution.

is commonly not an easy matter to resolve an 'old' PMS system, either.

3. THEORETICAL INTERPRETATIONS

3.1. Structure of the discs

Yet before first images of Vega-type systems were obtained it had become clear that the observed excess emission in the IR stems from circumstellar dust grains rather than from other sources like free-free emission, background IS cirrus, cool companions, etc ([1]; see also [31]). Already first coronograph images of β Pic showed a clear disc-like circumstellar structure. Explanations alternative to the dust disc — equatorial mass loss or bipolar outflows from a pre- or post-main-sequence object — were quickly ruled out (see [31,103] for details). A set of subsequent images of β Pic resulted in a relatively good understanding of the global spatial distribution of dust in the disc. Not going into details (which can be found in other reviews centred on this intriguing object — e.g. [31,103–105]), we outline here the main ideas.

The disc seen nearly edge-on extends outward to more than 1000 AU [65,107]. The material is confined within the half-opening angle $\sim 7°$ [108] and the mid-plane number density of dust particles falls off approximately as r^{-3} (see [65] and references therein). At about 100 AU, however, the radial density distribution changes its profile [63–66,68].

Within this zone the number density increases much slower, so that the normal optical depth remains constant or even drops slightly. Two wings of the disc are substantially asymmetric. By analyzing their own and other observations, Kalas and Kewitt [65] identified in the disc of β Pic five types of asymmetry. These comprise a general brightness/radial extension asymmetry of two wings, their width asymmetry, and warp-like disc asymmetries, some of which are especially pronounced in the inner parts of the disc [64,109].

The *asymmetries* are also seen in the discs of some *other stars* (e.g. [23,58,80]) attracting considerable interest as a presumed manifestation of planets orbiting these stars. Whereas the warps and local asymmetries are, most probably, really the signatures of the alleged planets [64,109,110], the global brightness asymmetries may also be due to other mechanisms. The difference by about 20% observed in the brightness and polarization of the two wings of β Pic is well explained by slightly different size distributions [53]. Such a difference could, under plausible conditions, be put down to interactions of the circumstellar particles with interstellar dust grains [53,106,112]. On the other hand, the difference in the sizes may be attributed to the action of a planet, too. Lecavelier des Etangs et al. [113] suggests that, if the planetary orbit is slightly eccentric, it would cause alignment of periastrons of comets/planetesimals at larger distances from the star and therefore a non-axisymmetric dust production and distribution in the disc. Krivova et al. [114] point out another possible mechanism: as small dust particles are blown away by stellar radiation pressure, their enhanced collisional production in the regions where the disc parent bodies are trapped in planetary resonances would lead to an overabundance of hyperbolic grains in the adjacent outer parts of the disc.

The so-called *inner depletion zones* are another widely present feature of Vega-type systems, also attributed to the presumed planets. Apart from β Pic , the depletion zones are found in all cases when the disc around a MS or transitive-aged star is resolved — see Table 1. Many other MS stars with circumstellar discs have low near- and mid-IR fluxes, yet detectable emission at 60 and 100 μm. It is very difficult to explain such a distribution of energy unless the regions close to the stellar photospheres are devoid of dust particles. The gaps apparently occur more frequently in stars older than 10 Myr, indicating a clearing with age [115]. The inner parts of younger systems are not yet accessible to observations. But the spectral energy distribution suggests that the inner gaps are not yet formed and the inner boundary of the dust discs/shells is determined by dust grain sublimation (e.g. [88,116–118]). For a few T Tau stars, however, the spectral energy distribution shows an evidence of the gaps [119,120].

The *density profile* is not established in great detail for stars other than β Pic . It is clear, however, that the radial density slope for MS objects is typically much steeper than for young stars with discs/shells (cf. values of α in Table 1). In some cases, as for HD 141569 or HR 4796A [71,74,78,80], the discs have an abruptly bounded ring structure with intensity falling off fast both inwards and outwards.

3.2. Dust properties

In comparison with the dust distribution and structure of the discs, the properties of dust particles are much more difficult to find out. Nevertheless, some constraints can be placed, especially when a combination of different observational data, like in the case of

β Pic , is available.

Chemical composition. IR and optical data considered in combination allow an estimation of the mean albedo of dust particles. For β Pic it appears to be rather high — 0.5 ± 0.2 (see Artymowicz [105] for a discussion) thereby ruling out such materials as carbonaceous species, pure metals and iron-rich minerals. A similar value (about 0.4) is derived for HD 141569 [74]. Comparison of spectrophotometric data of β Pic with laboratory reflectance spectra together with the albedo restrictions led Artymowicz et al. [108] to the conclusion that the particles are likely to be composed of magnesium-dominated (i.e. iron-poor) olivines and pyroxenes or astronomical ices (H_2O, CO_2, NH_3, CH_4) with a small admixture of carbonaceous compounds. Spectroscopic observations of other Vega-type and younger systems imply essentially akin composition — silicates, typically magnesium-rich, small carbon particles (PAHs), iron oxide and crystalline water ice (e.g. [24,43–45,47,121,122]).

Spectrophotometry in the 10 μm spectral region has shown that silicates are indeed important constituents of dust in the disc of β Pic [124–126]. The silicate feature is quite similar to those in some Solar System comets with characteristic features (e.g. secondary peak at 11.3 μm) of crystalline silicates and differs significantly from those in very young stars. A similar profile of the feature is observed in some other cases, including intermediate-stage (post-Herbig Ae/Be) objects [44,43,47,127,128]. The presence of crystalline silicates in the discs is corroborated by emission features in the $20 - 70\,\mu$m region [42–45]. This suggests that the dust was reprocessed and supports the idea that the cometary bodies resupply the grains to the discs. For yet younger, PMS, objects, however, the spectral structure due to cristalline material is typically not seen, implying probably amorphous materials [121,122,129,130].

Size distribution. Another important characteristic of dust grains in the discs is their size distribution. The results obtained in the case of β Pic by different authors differ somewhat and at the first glance seem even somewhat controversial. First of all, the grey colours of the disc [67,70] are indicative of big particles — bigger than a few micron. Even larger sizes are suggested by far-IR and millimetre data [21,131]. To account for both IR and optical observations together, the major contribution to the cross section should be given by grains about $1 - 20\,\mu$m in size [108], though an additional population of hotter and smaller grains is required by some models (e.g. [98,132]). Micron-sized grains are also responsible for the observed silicate emission feature [125,126]. Though grain collisions should really supply the disc with small grains, particles smaller than $\sim 2\,\mu$m should be quickly swept out of the disc by stellar radiation pressure [133].

A detailed consideration of polarimetric data available for β Pic [53] shows, that there is, in fact, no contradiction in these data. The observed spatial distribution and the wavelength dependence of the polarization together with the disc colours can only be explained if particles in a wide size range are present in the disc with grains smaller than $\sim 3 - 10\,\mu$m in size being somewhat depleted, though still of importance. This general shape of the size distribution is confirmed by a model of the dynamical evolution of circumstellar dust grains — it is a result of the joint action of the particle collisions and stellar radiation pressure [53,114]. A detailed dynamical modelling [114] implies that the overall distribution has basically three different slopes. The slope is steeper, with the power-law exponent close to 3.5 typical of collision-governed systems, for grains smaller than the blowout limit and for big grains (tens of microns and bigger). For intermediate-sized, just above the blowout limit, particles prone not only to mutual collisions, but also to the destructive impacts of small grains streaming outwards, there is a gentle dip in the distribution. A similar shape of the distribution has long been known for the Zodiacal Cloud [134], although the distribution is set by different mechanisms.

As principally the same mechanisms of dust production and removal act in discs of other MS stars, the general shape of the size distribution is expected to be similar. Indeed, though the presence of large grains ($\gtrsim 5 - 20\,\mu$m) is inferred by observational data for many stars (see e.g. [21,31,57,58,135]), evidence of smaller hotter grains also exists (e.g. [71,72,80,99,102,123,127,136,137]). In younger and dustier systems, collisions should be very intensive, so that the depletion of smaller grains could be much less pronounced. On the other hand, the interaction of the dust particles with gas, which is not depleted in young systems, can dampen down the relative velocities of the grains, decreasing the probability of catastrophic collisions.

We are still in the dark about the *structure* of circumstellar dust grains. If, as is commonly believed, the particles in the discs are really of cometary origin (Sect. 3.3), then the aggregate structure proposed by Li and Greenberg [138] is a good probability. Fluffy aggregates in this model are made up of primitive interstellar dust particles consisting of a silicate core and an organic refractory mantle. An additional ice mantle could cover the grains in the outer region of the disc. Such a model provides a good fit to the $10\,\mu$m silicate feature and the continuum emission from near-IR to millimetre observed in the case of β Pic [138]. An extremely high porosity of the aggregates (consisting of more than 95% of vacuum) suggested by the authors seems, however, to be in conflict with the observed high level of the polarization ([53]; see also [111]).

3.3. Dust sources and sinks

It was quickly realized that dust particles in the discs around MS stars can not be left over after the star formation. Particles, the presence of which is indicated by different observational data, should be destroyed or removed from the discs by stellar radiation pressure, grain-grain collisions and sublimation of grains (e.g. [31]). Other possible, though less crucial, sinks of dust include sweeping up by larger bodies, destruction due to impacts of interstellar particles, sputtering due to the stellar wind. This implies that the particles must be continuously replenished. By analogy with the Solar System, Weissman [139] suggested that the dust material could be supplied by large bodies, like comets or planetesimals. Further observational evidence was found in support of this idea called Falling and Orbiting Evaporating Bodies (FEBs, OEBs) scenario [113,140–142].

1. Spectroscopic observation of β Pic in the visible and UV revealed strong variations of the circumstellar absorption lines Ca II, Mg II, Al III, C IV and others (see Lagrange et al. [143], Lagrange-Henri et al. [144], Beust et al. [141] and later works of this group for more details). The lines are almost always red-shifted and the variations are sporadic, which was interpreted as the result of vaporization of large bodies falling towards the star. The phenomenon was later observed for many other stars, including PMS ones (e.g. [145–149]). Moreover, spectroscopic monitoring by Grady [150] revealed red-shifted absorption lines in 17 of 18 programme MS and PMS stars.

2. Identification of the stable CO and C I circumstellar features in β Pic implies that the

disc is continuously replenished with these gaseous species [151]. A permanent or quasi-permanent source of gas is attested by stable circumstellar lines of many other elements [152]. The source should have standard abundances in refractory elements and the FEBs scenario provides a natural explanation of this.
3. A similarity of the profile of the 10 μm silicate emission feature in β Pic and some Solar Sytem comets discussed above indicates that the dust in the disc may be a product of the comet disruptions.
4. The hypothesis of planetesimals as the replenishing source of dust in circumstellar discs is implicitely supported by the existence of trans-Neptunian objects believed to be a primordial population left over after the formation of the planetary system [14,15].
5. The so-called UX Ori phenomenon — irregular brightness dimmings accompanied by an increase of the polarization and a blueing of the object observed for many PMS stars — is believed to be another manifestation of large comet-like bodies or planatesimals moving around the stars (e.g. [153,154]).

It should be mentioned that comet-like bodies or planetesimals produce mainly large dust particles [155,156] which then supply smaller grains to the disc through their collisional fragmentation (see [114]).

4. DUST IN MULTIPLE SYSTEMS

Observations show that at least half of all stars are members of multiple systems [157]. Let us, therefore, touch briefly on such systems (detailed reviews can be found in Lubow and Artymowicz [158] and Mathieu et al. [159]). The discs in binaries occur in two principal types: circumbinary (GG Tau, DQ Tau, UZ Tau E, UY Aur, HD 98800, BD+31 643) and circumstellar (HR 4796A, T Tau N, GG Tau, UZ Tau W) (see Jensen and Mathieu [160]; Mathieu et al. [159] and references therein). UZ Tau and GG Tau make intriguing examples of systems, in which both types of discs are found [161,162].

There are also planets found in binaries — 16 Cyg B, 55 Cnc, τ Boo [7,163]. It is well known that three-body stellar systems are unstable and tend to eject the lowest mass object (e.g. [164–166]). Benest [167,168] found that stable planetary orbits exist up to distances from each star of about half the binary's periastron separation. Stable circumbinary orbits also exist, when the semimajor axis, measured in units of the binary's semimajor axis, is larger than 2 [169]. Planets migrating to larger distances should be ejected from the system.

The gravitational field of close companion(s) perturbs also the orbits of dust particles. The orbits are unstable at circumstellar distances of the same order as the component separation in a multiple star system. Therefore discs in such systems are expected to be truncated (on the outside of circumstellar discs and on the inside of circumbinary discs) by tidal interactions [170,171]. A decrease in the disc mass is then expected in systems with a separation smaller than the size of a typical disc. Indeed, from millimetre observations discs in close binary systems (the separation is less than $\sim$ 100 AU) were found to be substantially less massive than the discs around single stars or in wide systems (e.g. [17,172–175]). In some cases evidence also exists of a tidal truncation of a circumstellar disc [76,78,162,176] and of a dynamically cleared central region in a circumbinary disc [100,160,177,178]. At the same time, for the survival times of the discs, no significant difference between singles and binaries was found [179].

5. CONCLUDING REMARK

Recent discoveries of (i) circumstellar discs around main-sequence, and particularly, around solar-type stars, (ii) extrasolar planets, particularly those in star-disc systems, and (iii) the EKB have shown that the 'solar system' phenomenon is not unique. We still do not know how typical our Solar System as it exists is. A detection of an Earth-mass planet outside the Solar System would undeniably be exciting, which makes planet-hunting a popular field of exploration. At the same time, information of no little significance can be gleaned from observations of circumstellar dust. The more so, as the dust is much easier to observe than the planets: whereas the masses of the observable discs in Vega-type systems are estimated to be less than the Earth's mass (e.g. [20,21,23–25]), planets of such masses are far beyond the detection limits. Of course, the physical processes and the dust properties in circumstellar discs are not easy to understand, and here our knowledge of the Solar System could be very helpful. In turn, the overall distribution of the material, particularly in the outer parts of the discs, is now much better known for other systems. Representing different evolutionary stages, other systems provide also a good test of the theory of the formation of the Solar System. So, to better understand the essence of the circumstellar discs, the extrasolar planets and the EKB in themselves and the planetary formation process as a whole, we must combine our efforts in these, as yet rather individual, areas.

Acknowledgements. I wish to thank D. Backman, V. B. Il'in, A. V. Krivov, A. Lecavelier des Etangs, and I. Mann for interesting and fruitful discussions and encouragement and J. A. M. McDonnell and I. P. Williams for the invitation that initiated this review.

REFERENCES

1. H.H. Aumann, C.A. Beichman, F.C. Gillett, T. de Jong, J.R. Houck, F.J. Low, G. Neugebauer, R.G. Walker and P.R. Wesselius. Astrophys. J. Lett. 278 (1984) L23.
2. F.C. Gillett. In Light on Dark Matter, Dordrecht, Noordwijk (1986) 61.
3. B.A. Smith R.J. Terrile. Science 226 (1984) 1421.
4. M. Mayor and D. Queloz. Nature 378 (1995) 355.
5. R.P. Butler and G.W. Marcy. Astrophys. J. Lett. 464 (1996) L153.
6. G.W. Marcy and R.P. Butler. Astrophys. J. Lett. 464 (1996) L147.
7. R.P. Butler, G.W. Marcy, E. Williams, H. Hauser and P. Shirts. Astrophys. J. Lett. 474 (1997) L115.
8. G.W. Marcy, R.P. Butler E. Williams, L. Bildsten, J.R. Graham, A.M. Ghez J.G. Jernigan. Astrophs. J. 481 (1997) 926.
9. G.W. Marcy and R.P. Butler. Publ. Aastron. Soc. Pacific 112 (2000) 137.
10. G.W. Marcy, W.D. Cochran and M. Mayor. In Protostars and Planets IV, (Eds. V. Mannings, A. P. Boss and S. S. Russell) Univ. Arizona Press), Tucson (2000) 1285.
11. J.J. Lissauer. ARA&A 31 (1993) 129.
12. S.V.W. Beckwith and A.I. Sargent. In Protostars and Planets III, (eds. E. H. Levy, J. I. Lunine and M.S. Matthews) Univ. Arizona Press, Tucson (1993) 521.

13. D. Jewitt and J. Luu. Nature 362 1993) 730.
14. D. Jewitt and J. Luu. Astrophs. Space Sci. 223 (1995) 164.
15. D.C. Jewitt and J.X. Luu. In Protostars and Planets IV, (eds. V. Mannings, A. P. Boss and S. S. Russell) Univ. Arizona Press, Tucson (2000) 1201.
16. B. Zuckerman. In Circumstellar Dust Disks and Planet Formation, (Eds. R. Ferlet and A. Vidal-Madjar) Frontières, Gif-sur-Yvette (1994) 131.
17. S.V.W. Beckwith, A.I. Sargent, R.S. Chini and R. Guesten. Astron. J. 99 (1990) 924.
18. V. Mannings. Mon. Not. R. Astr. Soc. 271 (1994) 587.
19. T. Henning, A. Burkert, R. Launhardt, C. Leinert and B. Stecklum. Astron. Astrophys. 336 (1998) 565.
20. T. Yamashita, T. Handa, T. Omodaka, Y. Kitamura, E. Kawazoe, S.S. Hayashi and N. Kaifu. Astrophys. J. Lett. 402 (1993) L65.
21. B. Zuckerman and E.E. Becklin. Astrophs. J. 414 (1993) 793.
22. B. Zuckerman, T. Forveille and J.H. Kastner. Nature 373 (1995) 494.
23. W.S. Holland, J.S. Greaves, B. Zuckerman, R.A. Webb, C. McCarthy, I.M. Coulson, D.M. Walther, W.R.F. Dent, W.K. Gear and I. Robson. Nature 392 (1998) 788.
24. H.J. Walker and I. Heinrichsen. Icarus 143 (2000) 147.
25. W.R.F. Dent, J.S. Greaves, V. Mannings, I.M. Coulson and D.M. Walther. Mon. Not. R. Astr. Soc. 277 (1995) L25.
26. H. Holweger, M. Hempel and L. Kamp. Astron. Astrophys. 350 (1999) 603.
27. R. Liseau. Astron. Astrophys. 348 (1999) 133.
28. J.S. Greaves, I.M. Coulson and W.S. Holland. Mon. Not. R. Astr. Soc. 312 (2000) L1.
29. D.E. Trilling and R.H. Brown. Nature 395 (1988) 775.
30. D.E. Trilling, R.H. Brown and A.S. Rivkin. Astrophs. J. 529 (2000) 499.
31. D.E. Backman and F. Paresce. In Protostars and Planets III, (eds. E. H. Levy, J. I. Lunine and M.S. Matthews) Univ. Arizona Press, Tucson (1993) 1253.
32. V.B. Il'in and N.V. Voshchinnikov. Astron. Rep. 37 (1993) 362.
33. H.H. Aumann. Publ. Aastron. Soc. Pacific 97 (1985) 885.
34. D.E. Backman and F.C. Gillett. In Cool Stars, Stellar Systems, and the Sun, (eds. J. L. Linsky and R. E. Stencel) Lect. Notes in Phys. 291, Springer, Berlin Springer (1987) 340.
35. H.J. Walker and R.D. Wolstencroft. Publ. Aastron. Soc. Pacific 100 (1988) 1509.
36. H.H. Aumann and R.G. Probst. Astrophs. J. 368 (1991) 264.
37. R.E. Stencel and D.E. Backman. Astrophys. J. Suppl. 75 (1991) 905.
38. K.P. Cheng, F.C. Bruhweiler, Y. Kondo and C.A. Grady. Astrophys. J. Lett. 396 (1992) L83.
39. V. Mannings and M.J. Barlow. Astrophs. J. 497 (1998) 330.
40. S.B. Fajardo-Acosta, R.E. Stencel, D.E. Backman and N. Thakur. Astrophs. J. 520 (1999) 215.
41. G. Decin, C. Dominik, K. Malfait, M. Mayor and C. Waelkens. Astron. Astrophys. 357 (2000) 533.
42. C. Waelkens, K. Malfait and L.B.F.M. Waters. Astrophs. Space Sci. 255 (1998) 25.
43. K. Malfait, C. Waelkens, L.B.F.M. Waters, B. Vandenbussche, E. Huygen and M.S. de Graauw. Astron. Astrophys. 332 (1998) L25.

44. C. Waelkens, L.B.F.M. Waters, M.S. De Graauw, E. Huygen, K. Malfait, H. Plets, B. Vandenbussche, D.A. Beintema, D.R. Boxhoorn, H.J. Habing, A.M. Heras, D.J.M. Kester, F. Lahuis, P.W. Morris, P.R. Roelfsema, A. Salama, R. Siebenmorgen, N.R. Trams, N.R. Van Der Bliek, E.A. Valentijn and P.R. Wesselius. Astron. Astrophys. 315 (1996) L245.
45. M.E. van den Ancker, J. Bouwman, P.R. Wesselius, L.B.F,M. Waters, S.M. Dougherty and E.F. van Dishoeck. Astron. Astrophys. 357 (2000) 325.
46. T. Brooke, A. Tokunaga and S. Strom. Astron. J. 106 (1993) 656.
47. R.J. Sylvester, C.J. Skinner, M.J. Barlow and V. Mannings. Mon. Not. R. Astr. Soc. 279 (1996) 915.
48. S. Pezzuto, F. Strafella and D. Lorenzetti. Astrophs. J. 485 (1997) 290.
49. L.B.F.M. Waters and C. Waelkens. ARA&A 36 (1998) 233.
51
50. A. Miroshnichenko, Ž. Ivezić, D. Vinković and M. Elitzur. Astrophys. J. Lett. 520 (1999) L115.
51. R. Oudmaijer, J.K. Davies, C. Eiroa, J. Palacios, F. Garzon,, D. de Winter, D.and the EXPORT-team. In Disks, Planetesimals and Planets, (eds. F. Garzon, C. Eiroa, D. de Winter and T. J. Mahoney) ASP Conf. Ser 219, (2000) 238.
52. N.A. Krivova, I. Mann and A.V. Krivov. In Meteoroids 1998, (Eds. W.J. Baggaley and V. Porubčan) Astron. Inst. Slovak Acad. Sci., Bratislava (1999) 291.
53. N.A. Krivova, A.V. Krivov and I. Mann. Astrophs. J. 539 (2000) 424.
54. H.F. Levison and M.J. Duncan. Icarus 127 (1997) 13.
55. S.A. Stern. Astron. Astrophys. 310 (1996) 999.
56. S.L. Baliunas, G.W. Henry, R.A. Donahue, F.C. Fekel and W.H. Soon. Astrophys. J. Lett. 474 (1997) L119.
57. R. Jayawardhana, W.S. Holland, J.S. Greaves, W.R.F. Dent, G.W. Marcy, L.W. Hartmann and G.G Fazio. Astrophs. J. 536 (2000) 425.
58. J.S. Greaves, W.S. Holland, G. Moriarty-Schieven, T. Jenness, W.R.F. Dent, B. Zuckerman, C. McCarthy, R.A. Webb, H.M. Butner, W.K. Gear and H.J. Walker. Astrophys. J. Lett. 506 (1998) L133.
59. D. Barrado y Navascues. Astrophs. Space Sci. 263 (1999) 235.
60. S.B. Fajardo-Acosta, R.E. Stencel and D.E. Backman. Astrophys. J. Lett. 487 (1997) L151.
61. S.A. Stern, M.C. Festou and D.A. Weintraub. Nature 368 (1994) 312.
62. D. Barrado y Navascues, J.R. Stauffer, I. Song and J.P. Caillault. Astrophys. J. Lett. 520 (1999) L123.
63. D.A. Golimowski, S.T. Durrance and M. Clampin. Astrophys. J. Lett. 411 (1993) L41.
64. S.R. Heap, D.J. Lindler, T.M. Lanz, R.H. Cornett, I. Hubeny, S.P. Maran and B. Woodgate. Astrophs. J. 539 (2000) 435.
65. P. Kalas and D. Jewitt. Astron. J. 110 (1995), 794.
66. P. Lagage and E. Pantin. Nature 369 (1994) 628.
67. A. Lecavelier des Etangs, G. Perrin, R. Ferlet, A. Vidal Madjar, F. Colas, C. Buil, F. Sevre, J. Arlot, H. Beust, A Lagrange Henri, J. Lecacheux, M. Deleuil and C. Gry. Astron. Astrophys. 274 (1993) 877.

68. D. Mouillet, A.-M. Lagrange, J.-L. Beuzit and N. Renaud. Astron. Astrophys. 324 (199 1083.
69. E. Pantin, P.O. Lagage and P. Artymowicz. Astron. Astrophys. 327 (1997) 1123.
70. F. Paresce and C. Burrows. Astrophys. J. Lett. 319 (1987) L23.
71. J.C.Augereau, A.-M. Lagrange, D. Mouilletand F. Ménard. Astron. Astrophys. 350 (1999) L51.
72. R.S. Fisher, C.M. Telesco, R.K. Piña, R.F. Knacke and M.C. Wyatt. Astrophys. J. Lett. 532 (2000) L141.
73. M.E. van den Ancker, D. de Winter and H.R.E. Tjin A Djie. Astron. Astrophys. 330 (1998) 145.
74. A.J. Weinberger, E.E. Becklin, G. Schneider, B.A. Smith, P.J. Lowrance, M.D. Silverstone, B. Zuckerman and R.J. Terrile. Astrophys. J. Lett. 525 (1999) L53.
75. J.S. Greaves, V. Mannings and W.S. Holland. Icarus 143 (2000) 155.
76. R. Jayawardhana, S. Fisher, L. Hartmann, C. Telesco, R. Pina and G. Fazio. Astrophys. J. Lett. 503 (1998) L79
77. D.W. Koerner, M.E. Ressler, M.W. Werner and D.E. Backman. Astrophys. J. Lett. 503 (1998) L83.
78. G. Schneider, B.A. Smith, E.E. Becklin, D.W. Koerner, R Meier, D.C. Hines, P.J. Lowrance, R.J. Terrile, R.I. Thompson and M. Rieke. Astrophys. J. Lett. 513 (1999) L127.
79. J.R. Stauffer, L.W. Hartmann and D. Barrado y Navascues. Astrophs. J. 454 (1995) 910.
80. C.M. Telesco, R.S. Fisher, R.K. Pinã, R.F. Knacke, S.F. Dermott, M.C. Wyatt, K. Grogan, E.K. Holmes, A.M. Ghez, L. Prato, L.W. Hartmann and R. Jayawardhana. Astrophs. J. 530 (2000) 329.
81. J.Y. Hu, P.S. Thé and D. De Winter. Astron. Astrophys. 208 (1989) 213.
82. A. Natta, V.P. Grinin, V. Mannings and H. Ungerechts. Astrophs. J. 491 (1997) 885.
83. E. Pantin, C. Waelkens, K. Malfait and P.O. Lagage. In Disks, Planetesimals and Planets, (eds F. Garzon, C. Firoa, D. de Winter and T. J. Mahoney) ASP Conf. Ser 219, (2000) 281.
84. V. Mannings and A.I. Sargent. Astrophs. J. 490 (1997) 792.
85. V. Mannings, D.W. Koerner and A.I. Sargent. Nature 388 (1997) 555.
86. A.V. Berdiugin, V.P. Grinin and N.H. Minikulov. Izv. Krym. Astrofiz. Obs. 86 (1992) 69.
87. C. Friedemann, H.G., Reimann, J. Gürtler and V. Toth. Astron. Astrophys. 277 (1993) 184.
88. N.A. Krivova and V.B. Il'in. Astron. Lett. 23 (1997) 791.
89. A.N. Rostopchina. Astron. Rep. 43 (1999) 113.
90. V.P. Grinin, N.N. Kiselev and N.K. Minikulov. Sov. Astron. Lett. 15 (1989) 448.
91. C.A. Grady, B. Woodgate, F.C. Bruhweiler A. Boggess, P. Plait, D.J. Lindler, M. Clampin and P. Kalas. Astrophys. J. Lett. 523 (1999) L151.
92. K.A. Marsh, J.E. van Cleve, M.J. Mahoney, T.L. Hayward and J.R. Houck. Astrophs. J. 451 (1995) 777.
93. R. Millan-Gabet, F.P. Schloerb, W.A. Traub,F. Malbet, J.P. Berger, And J.D. Bregman. Astrophys. J. Lett. 513 (1999) L131.

94. T. Nakajima and D.A. Golimowski. Astron. J. 109 (1995) 1181.
95. N.V. Voshchinnikov, F.J. Molster and P.S. Thé. Astron. Astrophys. 312 (1996) 243.
96. T.M. Gledhill, S.M. Scarrott and R.D. Wolstencroft. Mon. Not. R. Astr. Soc. 252 (1991) 50.
97. P. Kalas, J. Larwood, B.A. Smith and A. Schultz. Astrophys. J. Lett. 530 (2000) L133.
98. C.M. Telesco, R. Decher, E.E. Becklin and R.D. Wolstencroft. Nature 335 (1988) 51.
99. C.J. Skinner, M.J. Barlow and K. Justtanont. Mon. Not. R. Astr. Soc. 255 (1992) 31P.
100. P. Kalas and D. Jewitt. Nature 386 (1997) 52.
101. T.B. Guild, D.W. Koerner and A.I. Sargent. Bull. Amer. Astron. Soc. 195 (1999) 2503.
102. J. Silber, T. Gledhill, G. Duchêne and F. Ménard. Astrophys. J. Lett. 536 (2000) L89.
103. P. Artymowicz. In Circumstellar Dust Disks and Planet Formation (ed. R. Ferlet and A. Vidal-Madjar) Gif-sur-Yvette: Frontières (1994) 47.
104. A.M. Lagrange-Henri. Astrophs. Space Sci. 223 (1995) 19.
105. P. Artymowicz. Ann. Rev. Earth & Planet. Sci. 25 1997 175.
106. P. Artymowicz and M. Clampin. Astrophs. J. 490 (1997) 863.
107. B.A. Smith and R.J. Terrile. Bull. Amer. Astron. Soc. 19 (1987) 829.
108. P. Artymowicz, C. Burrows, and F. Paresce. Astrophs. J. 337 (1989) 494.
109. C.J. Burrows, J.E. Krist, K.R. Stapelfeldt and WFPC2 Investigation Definition Team. Bull. Amer. Astron. Soc. 187 (1995) 3205.
110. D. Mouillet, J.D. Larwood, J.C.B. Papaloizou and A.-M. Lagrange. Mon. Not. R. Astr. Soc. 292 (1997) 896.
111. N.A. Krivova, and V.B. Il'in. Icarus 143 (2000) 159.
112. J.J. Lissauer and C.A. Griffith. Astrophs. J. 340 (1989) 468.
113. A. Lecavelier des Etangs, A. Vidal-Madjar and R. Ferlet. Astron. Astrophys. 307 (1996) 542.
114. A.V. Krivov, I. Mann and N.A. Krivova. Astron. Astrophys. , 362 (2000) 1127.
115. M.F. Skrutskie, D. Dutkevitch, S.E. Strom, S. Edwards, K.M. Strom and M.A. Shure. Astron. J. 99 (1990) 1187.
116. C. Friedemann, J. Gürtler, and H.G. Reimann. Astron. Astrophys. 300 (1995) 269.
117. A. Miroshnichenko, Ž. Ivezić and M. Elitzur. Astrophys. J. Lett. 475 (1997) L41.
118. H.G. Reimann, J. Gürtler, C. Friedemann and H.W. Käufl. Astron. Astrophys. 326 (1997) 271.
119. K.A. Marsh and M.J. Mahoney Astrophys. J. Lett. 395 (1992) L115.
120. K.A. Marsh and M.J. Mahoney. Astrophys. J. Lett. 405 (1993) L71.
121. M.S. Hanner, T.Y. Brooke and A.T. Tokunaga. Astrophs. J. 438 (1995) 250.
122. M.S. Hanner, T.Y. Brooke and A.T. Tokunaga. Astrophs. J. 502 (1998) 871.
123. R.J. Sylvester and C.J. Skinner. Mon. Not. R. Astr. Soc. 283 (1996) 457.
124. C.M. Telesco and R.F. Knacke. Astrophys. J. Lett. 372 (1991) L29.
125. R.F. Knacke, S.B. Fajardo-Acosta, C.M. Telesco, J.A. Hackwell, D.K. Lynch and R.W. Russell. Astrophs. J. 418 (1993) 440.
126. D. Aitken, T. Moore, P. Roche, C. Smith and C. Wright. Mon. Not. R. Astr. Soc.

265 (1993) L41.
127. S.B. Fajardo-Acosta, C.M. Telesco and R.F. Knacke. Astrophys. J. Lett. 417 (1993) L33.
128. M.L. Sitko, C.A. Grady, D.K. Lynch, R.W. Russell and M.S. Hanner. Astrophs. J. 510 (1999) 408.
129. M.L. Sitko, C.A. Grady, M.S. Hanner, D.K. Lynch R.W. Russell. In Circumstellar Dust Disks and Planet Formation, (Eds. R. Ferlet and A. Vidal-Madjar) Frontières, Gif-sur-Yvette (1994) 389.
130. M.L. Sitko, D.K. Lynch, R.W. Russell and C.A. Grady. Bull. Amer. Astron. Soc. 195 (1999) 7806.
131. R. Chini, E. Krügel, E. Kreysa, B. Shustov and A. Tutukov. Astron. Astrophys. 252 (1991) 220.
132. D.E. Backman, F.C. Witteborn and F.C. Gillett. Astrophs. J. 385 (1992) 670.
133. P. Artymowicz. Astrophys. J. Lett. 335 (1988) L79.
134. E. Grün, H.A. Zook, H. Fechtig and R.H. Giese. Icarus 62 (1985) 244.
135. R. Chini, E. Krügel and E. Kreysa. Astron. Astrophys. 227 (1990) L5.
136. S. Skinner, A. Brown and W.R.F. Dent. Astrophs. Space Sci. 224 (1995) 555.
137. M.C. Wyatt, S.F. Dermott, C.M. Telesco, R.S. Fisher, K. Grogan, E.K. Holmes and R.K. Pinã. Astrophs. J. 527 (1999) 918.
138. A. Li and J.M. Greenberg. Astron. Astrophys. 331 (1998) 291.
139. P.R. Weissman. Science 224 (1984) 987.
140. R. Ferlet, A. Vidal-Madjar and L.M. Hobbs. Astron. Astrophys. 185 (1987) 267.
141. H. Beust, A.M. Lagrange-Henri, A. Vidal-Madjar and R. Ferlet. Astron. Astrophys. 223 (1989) 304.
142. H. Beust, A.M. Lagrange, I.A. Crawford, C. Goudard, J. Spyromilio and A. Vidal-Madjar. Astron. Astrophys. 338 (1998) 1015.
143. A.M. Lagrange, R. Ferlet and A. Vidal-Madjar. Astron. Astrophys. 173 (1987) 289.
144. A.M. Lagrange-Henri, H. Beust, R. Ferlet and A. Vidal-Madjar. Astron. Astrophys. 215 (1989) L5.
145. C.A. Grady, M.R. Perez, A. Talavera, K.S. Bjorkman, D. De Winter, P.S. Thé, F.J. Molster, M.E. van den Ancker, M.L. Sitko, N.D. Morrison, M.L. Beaver, B. McCollum and M.W. Castelaz. Astron. Astrophys. Suppl. Ser. 120 (1996) 157.
146. V.P. Grinin, O.V. Kozlova, P.S. Thé and A.N. Rostopchina. Astron. Astrophys. 309 (1996) 474.
147. K.P. Cheng, F.C. Bruhweiler and J.E. Neff. Astrophs. J. 481 (1997) 866.
148. A. Lecavelier des Etangs, M. Deleuil, A. Vidal-Madjar, A.M. Lagrange-Henri, D. Backman, J.J. Lissauer, R. Ferlet, H. Beust and D. Mouillet. Astron. Astrophys. 325 (1997) 228.
149. B.Y. Welsh, N. Craig, I.A. Crawford and R.J. Price. Astron. Astrophys. 338 (1998) 674.
150. C.A. GradyA. Mora, A. D. de Winter, D. and the Export team.. In Disks, Planetesimals and Planets, (eds. F. Garzon, C. Eiroa, D. de Winter and T. J. Mahoney) ASP Conf. Ser 219 (2000) 202.
151. A. Vidal-Madjar, A.-M Lagrange-Henri, P.D. Feldman, H. Beust, J.J. Lissauer, M. Deleuil, R. Ferlet, C. Gry, L.M. Hobbs, M.A. McGrath, J.B. McPhate and H.W.

Moos. Astron. Astrophys. 290 (1994) 245.
152. A.M. Lagrange, H. Beust, D. Mouillet, M. Deleuil, P.D. Feldman, R. Ferlet, L. Hobbs, A. Lecavelier Des Etangs, J.J. Lissauer, M.A. McGrath, J.B. McPhate, J. Spyromilio, W. Tobin and A. Vidal-Madjar. Astron. Astrophys. 330 (1998) 1091.
153. V.P. Grinin. Astron. Lett. 14 (1988) 27.
154. V.P. Grinin, A. Natta and L. Tambovtseva. Astron. Astrophys. 313 (1996) 857.
155. D.R. Davis and E.V. Ryan. Icarus 83 (1990) 156.
156. M.V. Sykes, L.A. Lebofsky, D.M. Hunten and F. Low. Science 232 (1986) 1115.
157. C. Leinert, H. Zinnecker, N. Weitzel, J. Christou, S.T. Ridgway, R. Jameson, M. Haas and R. Lenzen. Astron. Astrophys. 278 (1993) 129.
158. S.H. Lubow and P. Artymowicz. In Protostars and Planets IV, (Eds. V. Mannings, A. P. Boss and S. S. Russell) Univ. Arizona Press, Tucson (2000) 731.
159. R.D. Mathieu, A.M. Ghez, E.L.N. Jensen, and M. Simon. In Protostars and Planets IV, (Eds. V. Mannings, A. P. Boss and S. S. Russell) Univ. Arizona Press, Tucson (2000) 703.
160. E.L.N. Jensen and R.D. Mathieu. Astron. J. 114 (1997) 301.
161. E.L.N. Jensen, D.W. Koerner and R.D. Mathieu. Astron. J. 111 (1996) 2431.
162. S. Guilloteau, A. Dutrey and M. Simon. Astron. Astrophys. 348 (1999) 570.
163. W.D. Cochran, A.P. Hatzes, R.P. Butler and G.W. Marcy. Astrophs. J. 483 (1997) 457.
164. E.M. Standish. Astron. Astrophys. 21 (1972) 185.
165. J.J. Monaghan. Mon. Not. R. Astr. Soc. 176 (1976) 63.
166. S. Mikkola and M.J. Valtonen. Mon. Not. R. Astr. Soc. 223 (1986) 269.
167. D. Benest. Astron. Astrophys. 314 (1996) 983.
168. D. Benest. Astron. Astrophys. 332 (1998) 1147.
169. M.J. Holman and P.A. Wiegert. Astron. J. 117 (1999) 621.
170. D.N.C. Lin and J.C.B. Papaloizou. In Protostars and Planets III, (Eds. E. H. Levy, J. I. Lunine and M.S. Matthews) Univ. Arizona Press, Tucson (1993) 749.
171. P. Artymowicz and S.H Lubow. Astrophs. J. 421 (1994) 651.
172. E.L.N. Jensen, R.D. Mathieu and G.A. Fuller. Astrophys. J. Lett. 429 (1994) L29.
173. E.L.N. Jensen, R.D. Mathieu and G.A. Fuller. Astrophs. J. 458 (1996) 312.
174. M. Osterloh and S.V.W. Beckwith. Astrophs. J. 439 (1995) 288.
175. A. Dutrey, S. Guilloteau, G. Duvert, L. Prato, M. Simon, K. Schuster and F. Menard. Astron. Astrophys. 309 (1996) 493.
176. K.R Stapelfeldt, J.E. Krist, F. Menard, J. Bouvier, D.L. Padgett And C.J. Burrows. Astrophys. J. Lett. 502 1998 L65.
177. A. Dutrey, S. Guilloteau and M. Simon. Astron. Astrophys. 286 (1994) 149.
178. D.W. Koerner, E.L.N. Jensen, K.L. Cruz, T.B. Guild and K. Gultekin. Astrophys. J. Lett. 533 (2000) L37.
179. M. Simon and L. Prato. Astrophs. J. 450 (1995) 824.

Aperture synthesis observations of protoplanetary disks with the Nobeyama Millimeter Array

S. Yokogawa[a*] Y. Kitamura[b] M. Momose[c] and R. Kawabe[d]

[a]Graduate University for Advanced Studies, Japan

[b]Institute of Space and Astronautical Science, Sagamihara, Kanagawa, 229- 8510, Japan

[c]Institute of Astrophysics and Planetary Science, Ibaraki University, Mito, Ibaraki, 310-8512,Japan

[d]Nobeyama Radio Observatory, Nobeyama, Minamisaku, Nagano, 384-1305, Japan

We have made aperture synthesis observations of 2 mm continuum emission from the dust particles around pre-main-sequence stars with the Nobeyama Millimeter Array (NMA). We observed HL Tau, DO Tau, and GM Aur in the Taurus-Auriga dark clouds. By using the sparsest configuration of the NMA, we achieved a 1" resolution which corresponds to 140 AU at the region. As a result, we have resolved disklike structures toward all the objects. The flux densities and beam deconvolved sizes of the disks are 150 mJy and 1.12"×0.82" for HL Tau, 37 mJy and 0.79"×0.38" for DO Tau, and 31 mJy and 1.39"×0.76" for GM Aur. Using the disk sizes and the inclination angles which derived from our images and past studied SEDs, we determined the physical parameters of the disks.

1. INTRODUCTION AND OBSERVATIONS

The formation of low mass stars and planetary systems is one of the most interesting astronomical themes, because it is very closely related to the origin of the Solar System. Observationally, extensive flux surveys at mm and sub-mm wavelengths were performed by Beckworth et al. [1] and Osterloh and Beckwith [2]. They revealed that more than a half of young stellar objects (YSOs) in the Taurus Molecular Cloud (TMC) are accompanied with considerable millimeter excess emission. The excess can be explained well by the thermal emission from the compact dust disk whose radii are a few × 100 AU and masses are a few × 0.01 $M_\odot$ [1,3,2].

To investigate the disk properties more accurately, we have observed 141 GHz dust emission around 3 YSOs with the NMA from December 1998 to February 1999. The target sources are HL Tau, DO Tau, and GM Aur, which have strong continuum emission. HL Tau is now categorized as a protostar, whose stellar age is estimated to be 1 × 10^5

*Present address. Nobeyama Radio Observatory, Nobeyama, Minamimaki, Minamisaku, Nagano, 384-1305, Japan

yr. DO Tau and GM Aur are categorized as classical T Tauri star and their stellar ages are estimated to be 6×10^5 yr for DO Tau, and 2×10^6 yr for GM Aur [1,4].

2. RESULTS

Contour maps of 141 GHz continuum emission around 3 young stars are presented in Figure 1. The disklike structures are clearly resolved. The flux densities and beam deconvolved sizes of the disks are 150 mJy and 1.12" $\times$ 0.82" for HL Tau, 37 mJy and 0.79" $\times$ 0.38" for DO Tau, and 31 mJy and 1.39" $\times$ 0.76" for GM Aur. The peaks of the continuum emission are located at (α[1950],δ[1950]) = (04h 28m 44.39s, +18d 07m 35.0s) for HL Tau, (04h 35m 24.20s, +26d 04m 54.4s) for DO Tau, and (04h 51m 59.79s, +30d 17m 13.3s) for GM Aur.

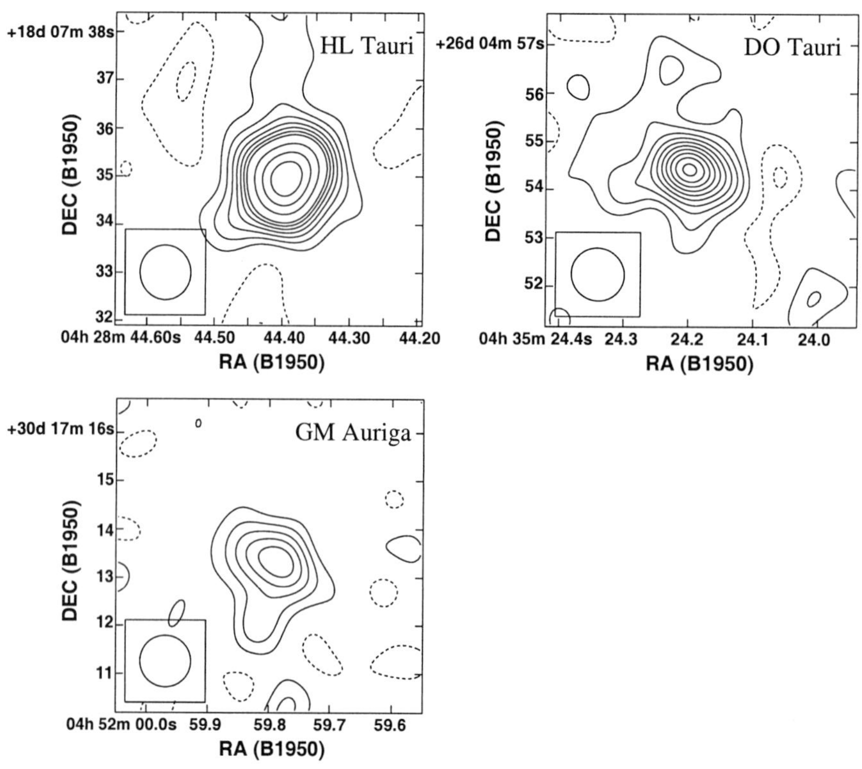

Figure 1. Contour maps of 141 GHz continuum emission from the dust particles around the 3 young stars, HL Tau, DO Tau and GM Aur. The contour levels are in steps of −3.0, −11.5, 1.5, 3.0, 4.5, 6.0, 7.5, 9.0, 10.5, 12.0, 15, 20, 25 and 30 times the rms level. The rms levels are 2.6 mJy $beam^{-1}$ for HL Tau, 1.8 mJy $beam^{-1}$ for DO Tau and 2.0 mJy $beam^{-1}$ for GM Aur.

In Figure 2, we show the best fit solutions to the spectral energy distributions (SEDs) of the 3 young stars. Applying a power-law model (discribed below) to the SEDs with the disk radii and the inclination angles which derived from our images, we determined the physical parameters of the disks. The best fit values are summarized in Table 1.

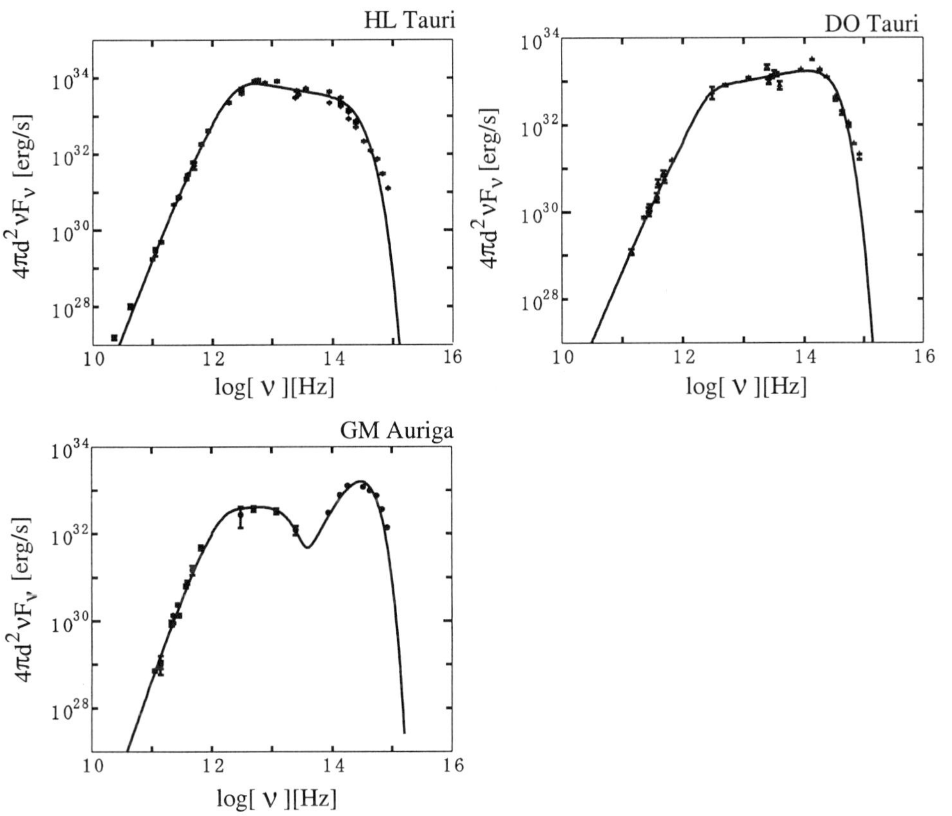

Figure 2. Spectral energy distributions of the 3 young stars. The 2 mm flux densities are from [1,3,6,9–14].

The flux density F_ν emitted from a disk at frequency ν can be expressed as

$$4\pi d^2 F_\nu = 4\pi \cos\theta \int_{r_{in}}^{r_{out}} B_\nu(T)(1-\exp(-\tau_\nu/\cos\theta))2\pi r dr, \quad (1)$$

where d is the distance to the objects ($\sim$ 140pc), $B_\nu(T)$ is the Planck function with temperature T, θ is an inclination angle of the disk (0° for edge-on), and τ_ν is the optical depth due to the dust opacity, ($\tau_\nu = \kappa_\nu \Sigma$). The dust opacity coefficient κ_ν at $\nu < 10^{12}$Hz is calculated from the following equation, $\kappa_\nu = 0.1(\nu/10^{12}[\mathrm{Hz}])^\beta$. At $\nu > 10^{12}$Hz, the coefficient from the interstellar extinction curve is adopted. The surface density and temperature distributions of the disk are assumed to have power-law forms of $\Sigma(r) = \Sigma_0(r/1\mathrm{AU})^{-p}$ and $T(r) = T_0(r/1\mathrm{AU})^{-q}$ respectively [1,3,5,6].

Table 1
Best fit parameters of the protoplanetary disks around the 3 young stars

Source	HL Tauri	DO Tauri	GM Auriga
disk radius [AU]	125 ± 2	147 ± 6	161 ± 4
inclination [deg.]	45 ± 5	62 ± 5	55 ± 6
disk mass [$M_\odot$]	0.069±0.003	0.004 ± 0.001	0.028 ± 0.002
Σ_0[g cm^{-2}]	1290(20-3500)	28(7-900)	200(70-1700)
p	$1.1^{+0.4}_{-1.1}$	$0.9^{+1.1}_{-0.3}$	$1.0^{+0.5}_{-0.3}$
T_0 [K]	320 ± 5	253 ± 5	200 ± 5
q	0.47 ± 0.01	0.54 ± 0.01	0.53 ± 0.01
β	1.19 ± 0.05	0.33 ± 0.11	1.02 ± 0.03
inner radius [AU]	0.02 ± 0.01	<0.01	0.83 ± 0.25

3. DISCUSSION

The power law index of the mass opacity coefficient, β, provides important infomation about the dust evolution in the protoplanetary disks. Theoretically, Miyake & Nakagawa [7] suggested that the index β decrease as the dust particles coagulate. Observationally, it is reported that β of protostar is larger than that of T Tauri stars [7,8].

Our results show no apparent correlation between β and the stellar ages. HL Tau and GM Aur show that $\beta \sim 1$, which is typical value of T Tauri stars. Whereas, DO Tau shows a significantly small β compared to other sources. In order to explain the small β of DO Tau, the mm emission would have to be either optically thick, or optically thin with $\beta \sim 0.3$. It is unnatural that the dust particles around DO Tau have already coagulate into significantly bigger sizes. So we are in favour of interpreting the observed continuum emission as optically thick emission from DO Tau circumstellar disk.

REFERENCES

1. S. Beckwith et al., Astrophys. J. 99 (1990) 924.
2. M. Osterloh and S. Beckwith, Astrophys. J. 439 (1995) 288.
3. S. Beckwith and A. Sargent, Astrophys. J. 381 1991) 250.
4. L.M. Close et al., Astrophys. J. 478 (1997) 766.
5. F. Adams, C. Lada, F. Shu, Astrophys. J. 326 (1998) 865.
6. F. Adams, J.P. Emerson G.A. Fuller, Astrophys. J. 357 (1990) 606.
7. K. Miyake and Y. Nakagawa, Icarus 106 (1993) 20.
8. G.H. Moriarty-Schieven et al., Astrophys. J. 436 (1994) 800.
9. A. Dutrey et al., Astro. Astrophys. 338 (1998) L63.
10. D.W. Koerner et al., Icarus 106 (1993) 2.
11. L.G. Mundy et al., Astrophys. J. Lett. 464 (1996) L169.
12. N. Ohashi et al., Astron. J. 102 (1991) 2054.
13. K.M. Strom et al., Astron. J. 97 (1989) 1451.
14. D.A. Weintraub et al., Astrophys. J. Lett. 340 (1989) L69.

DRVS and extrasolar planetary dust noise reduction

Rajendra VikramSingh

3611 Lupine Avenue, Palo Alto, CA 94303, USA.

Noise due to dust around extrasolar planets is not resolved by astrometry. Planetary-signal to dust-noise ratio is also not diminished by interferometry. DRVS (Differential Radial Velocity Spectroscopy) based on the difference in their Doppler shifts, allows the weak spectrum of an extrasolar planet to be filtered out from its parent star's spectrum [1,2,3,4]. The dust noise spectrum is essentially the same across the narrow spectral window within which a particular spectral line is Doppler shifted during each planetary orbit and Doppler filtering as such does not improve the signal/noise ratio. However, careful selection of strong and narrow spectral lines for inclusion in the matched filter, controlling their total number for processing gain while minimizing their combined bandwidth to reduce dust noise contribution, could optimize the planetary-signal/dust-noise ratio.

1. SOLAR SYSTEM DUST EMISSION

Interplanetary dust scatters the stellar emission and is also a source of intrinsic radiation. Dust emission is several hundred times stronger than emission from planet Earth [5]. Leinert states "The appearance of the solar absorption line spectrum in the zodiacal light proves that it is sunlight scattered by slowly moving particles. The motion of the scattering dust particles, apart from a slight broadening of the lines, leads to a Doppler shift of about 0.3 Angstroms" [6]. "Zodiacal light intensity (*I*) strongly depends on the heliocentric distance (*R*), the increase towards the Sun, being $I(R) \sim R\ e^{-2.3}$ [7]. Dumont and Levasseur-Regourd [8] write " To the first order, the zodiacal cloud is homogeneous and steady; it has a solar-like spectrum". Dumont [9] further declares "Doppler spectrometry can be used to retrieve the mean orbital velocity, v, of the interplanetary scatterers in the region of the terrestrial orbit". Dumont computes the difference of mean orbital velocity of the scatterers at 1 A.U. with respect to Earth from observational data to be $+9.1 \pm 5$ km s^{-1} [10] and $+7.6 \pm 5$ km s^{-1} [11].

2. EXTRASOLAR PLANETARY DUST

Precision radial velocity spectroscopy has been successfully applied to indirect detection of giant extrasolar planets in tight orbits around their parent stars by measuring the periodic Doppler shift of the stellar spectrum due to reflex motion of the parent star Orbital characteristics of most of the extrasolar planets discovered so far are surprisingly different from those of the planets in the solar system. The chemical composition, particle size distibution, and dynamics of dust in the environs of extrasolar planets may also be different from the solar system dust. Scattered starlight is likely to be the dominant feature of their dust spectra, similar to the appearance of the Sun's spectrum in the zodiacal light due to scattering by dust.

3. SPATIAL FILTERING OF EXTRASOLAR PLANETARY DUST NOISE

Precision astrometry will be more sensitive, due to its wideband collection of photons, in locating and tracking dust clouds in other stellar systems, but is unlikely to be able to identify a planet obscured by a dust cloud in a closely related orbit. Interferometers based beyond the solar system dust and providing a million times or even greater nulling of the signal from the parent star of an extrasolar planet may be operational someday. In the near future, Earth based interferometers are expected to provide a nulling factor >1000, thereby reducing the stellar infrared flux to within a hundred to a thousand times the extrasolar planet's flux. The nulled stellar flux will then be of almost the same order of magnitude as flux from extrasolar planetary dust. According to Beichman [12] "dust in target solar system ...is three hundred times brighter than planet". The dust surrounding an extrasolar planet obscures the line-of-sight to the extrasolar planet and cannot be nulled by interferometric spatial filtering. Emissions from dust in the vicinity of the extrasolar planet thus emerge as a major source of "noise" (undesired signal) against extrasolar planetary emissions (desired signal).

Noise from other interplanetary dust sources may however be minimized by observing the extrasolar planet in an appropriate epoch during which these dust sources do not lie along or near the line-of-sight to the extrasolar planet. It is also conceivable that short periods of direct observation of an extrasolar planet through relatively dust free holes in the extrasolar interplanetary dust may occur, and if recognized as such it may be possible to program further observations during these short epochs.

4. EXTRASOLAR PLANETARY SPECTRA

Probable detection of reflected starlight from the planet Peg 51 has been published [4]. Attempts to observe changes in the stellar spectrum during the near transit of the planet Peg 51 across the line of sight have not shown any conclusive results so far, but other transiting extrasolar planets are likely to be discovered in the near term and may prove more fruitful. Extraction of spectral signatures of specific molecules, e.g., methane, from extrasolar planets which may be buried in the glare of their parent stars has also been attempted but produced uncertain answers up until now [13]. A combination of interferometric spatial filtering and differential Doppler filtering using dynamic correlation filters tracking the planet's periodically changing radial velocity during its orbital motion is more likely to succeed. The extrasolar planetary signals are quite weak but large aperture interferometers coupled with high resolution spectrographs, namely, Keck (Hawaii), LBT (Arizona), and ESO (Chile) will be available soon.

5. EXTRASOLAR PLANETARY-DUST SPECTRA

The difference in the mean orbital velocity of the scattering dust with respect to the extrasolar planet leads to a periodically varying difference in their radial velocities, whose magnitude goes from zero when their orbits cross the line-of-sight to the star, to a maximum at the two orthogonal orbital positions. The stellar spectrum reflected from the extrasolar planet will have a Doppler shift different in shape and magnitude than the same stellar spectrum reflected from the orbiting dust cloud.

Correlation filters using very narrow windows and a large number of spectral lines may manage to separate these two differentially Doppler shifted copies of the stellar spectrum and confirm the separate existence of the extrasolar planet and the dust cloud envelope. If the scattered signal were strong enough, observing changes in spectral line shapes could be an alternative approach. The spectrum of a specific molecular species showing a similar differential Doppler separation would indicate its presence on the extrasolar planet as well as in the dust cloud. If no differential Doppler separation occurs the molecular species would be located only on either the extrasolar planet or in the dust cloud.

6. DOPPLER FILTERING OF EXTRASOLAR PLANETARY-DUST-NOISE

DRVS also faces new challenges in detecting and characterizing extrasolar planets embedded in dust clouds. "There is a significant excess of the orbital mean velocity of the scatterers surrounding the Earth's orbit, over the terrestrial velocity, of the order of +25 to +30 %" [9]. A similar difference in the mean orbital velocity of the extrasolar-planetary-dust with respect to the extrasolar planet would lead to a periodically varying difference in their radial velocities. For a circular orbit the two radial velocity curves would look like two sine waves with the same period and phase but different magnitudes. The differential radial velocity between the extrasolar planet and its dust envelope would vary during the orbit around the parent star from zero, when their orbits cross the line-of-sight from the observer to the star to several km s^{-1} at the two orthogonal positions in the orbit.

It may be recalled that radial velocity of the reflex motion of stars due to giant planets in close orbits varies over a range of tens of m s^{-1}, and for Earthlike planets only a few cm s^{-1}. Stellar "noise" (undesired signal) is a million (infrared) to a billion (visible) stronger than extrasolar planetary signal (desired signal). Precision radial velocity spectroscopy is required for measuring such small variations.

The periodic differential radial velocity between an extrasolar planet and its dust is several km s^{-1}, that is 3 to 5 orders of magnitude greater than the stellar planetary situation. Moreover, the dust is only a few hundred times brighter, that is 3 orders of magnitude lower, than the stellar planetary case. Precision radial velocity spectroscopy will not be needed. Lower signal-to-noise ratio and smaller processing gain in the Doppler spectral correlation filter will be good enough for separating the intrinsic emission spectra of the extrasolar planet and the dust obscuring it. There is a significant difference in the Doppler shift of the stellar spectrum scattered from the extrasolar planet and the extrasolar-planetary-dust. The reflex motion of the parent star is very small and its spectrum has minimal Doppler shift. (It could in fact be used as a local frame of reference.) The mixture of the direct signal from the star because of incomplete nulling, the two components of the scattered stellar signal, and the intrinsic emissions from the extrasolar planet and the dust cloud, each one with its unique Doppler shift during the orbital cycle, will require a fairly intricate dynamic correlation filter. The extraction of a signal whose spectral signature could be postulated, such as, a specific molecule from a very noisy background would be less difficult.

REFERENCES

1. R. Vikramsingh, in Progress in the Search for Extraterrestrial Life (rd. G. S. Shostak) ASP Conf. Ser. 74 (1995) 237.
2. R. Vikramsingh, Earth, Moon and Planets 70 (1996) 213.
3. R. VikramSingh, In Giant Planets to Cool Stars (Eds. C.A. Griffith and M.S. Marley) ASP Conference Series 212 (2000) 347.

4. A.C. Cameron, K. Horne, A. Penny and D. James. Nature 402 (1999) 751.
5. C.A. Beichman, N.J. Woolf and C.A. Lindensmith (eds), Terrestrial Planet Finder, Jet Propulsion Lab. (1999) 276
6. C. Leinert, Space Sci. Rev. 18 (1975) 281.
7. C. Leinert, I. Richter, E. Pitz and B. Planck, Astron. Astrophys. 103 (1981) 177.
8. R. Dumont and A.C. Levasseur-Regourd, Adv. Space. Res. 1 (1981) 8.
9. R. Dumont, Planet. Space Sci. 31 (1983) 12.
10. N.K. Reay and J. Ring, Nature 219 (1968) 710.
11. J.W. Fried, Astron. Astrophys. 68 (1978) 259.
12. C.A. Beichman, Pale Blue Dot Workshop 2, NASA Ames Research Center (1999).
13. D. Deming, G. Wiedeman and G. Bjoraker, in Giant Planets to Cool Stars (eds. C.A. Griffith and M.S. Marley) ASP Conference Series 212 (2000) 308.

Structure of the Edgeworth-Kuiper Belt (EKB) dust disk and implications for extrasolar planet(s) in ε Eridani

J.-C. Liou[a], H. A. Zook[b], J. S. Greaves[c], W. S. Holland[c], H. Boehnhardt[d] and J. M. Hahn[e]

[a]Lockheed Martin Space Operations, Houston, TX, USA.

[b]NASA Johnson Space Center, Houston, TX, USA.

[c]Joint Astronomy Center, Hilo, HI, USA.

[d]European Southern Observatory, Chile.

[e]Lunar and Planetary Institute, Houston, TX, USA.

Numerical simulations of the orbital evolution of dust particles from Edgeworth-Kuiper Belt (EKB) objects show that the three giant planets, Neptune, Jupiter, and Saturn impose distinct and dramatic signatures on the overall distribution of EKB dust particles. The features are very similar to those observed in the dust disk around the nearby star ε Eridani. Numerical simulations of dust particles in the ε Eridani system show that planetary perturbations may be responsible for the observed features.

1. INTRODUCTION

Over the past 10 years infrared surveys have discovered many stellar systems with dust rings or disks around the central stars. Small dust particles (micrometer to millimeter sized), once created from their large parent objects, spiral slowly toward the central star via Poynting-Robertson (PR) and stellar wind drag [1]. On their way toward the star, dust particles would interact with planets if planets exist in the system. The three major planetary perturbations on dust particles are resonance trapping, gravitational ejection, and secular perturbation. These perturbations may cause large-scale features on a dust disk. A good example is our Solar System's zodiacal cloud. Resonance trapping with the Earth triggers the formation of an enhanced dust ring around the Earth's orbit [2, 3] while secular perturbation causes the plane of symmetry of the cloud to be different from the invariable plane [4]. Similar features should exist in the EKB dust disk in the outer Solar System and, in fact, have been observed in several circumstellar dust disks. By analyzing the long-term evolution of dust particles under interactions with the planets in our Solar System, we may be able to decode some of the structures in other circumstellar dust disks and use the information to identify the existence of extrasolar planets.

2. STRUCTURE OF THE EKB DUST DISK

Since the discovery of the first EKB object in 1992, more than 280 trans-Neptunian objects have been discovered as at the end of July 2000. The dust production rate in the EKB region should be at least comparable to that in the main asteroid belt. Numerical simulations of dust particles between 1 and 100 μm from the EKB region show that giant planets induce large-scale structures on the EKB dust disk [5, 6]. Trapping into exterior mean motion resonances (MMRs) with Neptune dominates the orbital evolution of dust particles 5 μm and greater from the EKB region. The main resonances are the 2:1 and 3:2 exterior MMRs. Trapping into MMRs with Uranus is rare because gravitational perturbations from Neptune usually make the resonance trap highly unstable. Once particles escape MMRs with Neptune, they continue to spiral toward the Sun. About 80% of the particles are eventually ejected from the Solar System by Jupiter and Saturn. The signatures of giant planets imprinted on the EKB dust disk, due to these effects, include: (1) the deviation of radial spatial density distribution from that determined by PR drag alone; (2) an enhanced ring-like structure along the orbit of Neptune; (3) a brightness variation along the ring with an opening (a dark spot) located where Neptune is and with two bright arcs about 70° apart on either side of Neptune; (4) a seasonal variation of features in the ring that moves along with Neptune's orbital motion; and (5) a relative lack of particles inside about 10 AU.

3. EPSILON ERIDANI DUST DISK

Epsilon Eridani is a bright (V=3.7) K2V star at a distance of only 3.22 pc from the Sun. Its mass and luminosity are 0.8 M_{sun} and 0.33 L_{sun}, respectively. The age of the star is estimated to be less than 1 billion years. The dust disk around ε Eridani was revealed and imaged using the Submillimeter Common User Bolometer Array (SCUBA) at the James Clerk Maxwell Telescope (JCMT) on Mauna Kea Hawaii in 1997 [7]. At 850 μm wavelength, the observations clearly show a dust disk around the star with an azimuthal brightness variation along the disk. The disk has a 25° inclination with respect to the plane of sky. The inner and outer edges of the disk are approximately 30 and 90 AU, respectively, from the star with a peak brightness occurring around 60 AU. The brightness variation along the ring is about a factor of 2.5. The brightest spot/arc in the ring is a real feature. Other features (secondary bright spots and a dark gap) along the ring are less certain since they were identified in only half of the observational maps.

What makes the ε Eridani disk differ from other circumstellar dust disks is the azimuthal brightness variation along the ring and its almost face-on orientation. If one compares the pattern of the variation with the modeled structures of the EKB dust disk [6, 7], one finds global similarities between the two. Therefore, a simple explanation for the observed structures in ε Eridani is that they are caused by perturbations from objects (planets) orbiting the star.

To identify whether or not planets exist in ε Eridani we have performed 15 numerical simulations (with perturbations from planets) to see if we could produce dust disk patterns that match the observed features. The numerically produced dust disk patterns depend on several factors: (1) number of planets, (2) masses and orbital locations of the planets, (3) size distribution of the dust particles, and (4) the source region of the dust particles. As a first step, we have assumed there is only one planet in ε Eridani in our numerical simulations. In addition, we have placed the planet at either 30 or 40 AU from the star. It is based on (1) previous radial profiles of dust particles from the EKB dust simulations and the orbital location of the major perturber, i.e., Neptune [6], and (2) the radial profile of ε Eridani that shows a brightness peak at 60 AU [7].

Five different planetary masses are used in the simulations: (1) 5 Jupiter masses (the mass ratio between the planet and the star is five times that between Jupiter and the Sun), (2) 1 Jupiter mass, (3) 1 Neptune mass, (4) 5 Earth masses, and (5) 1 Earth mass. The eccentricity of the planet's orbit is assumed to be 0.01. Dust particles are assumed to be released from parent objects between 50 and 80 AU with eccentricities of 0.1 and inclinations of 10°. The effect of radiation pressure and PR drag are also included in the simulations. We simulate the orbital evolution of dust particles of 4 different sizes: 5, 10, 50, and 100 μm in diameter. In most cases (12 out of 15), there are 100 dust particles in each simulation. The remaining simulations include 50 dust particles. Since the age of ε Eridani is estimated to be between 500 million and 1 billion years, all simulations are carried out for at least 700 million years, unless all dust particles are ejected from the system or spiral into the star prior to that.

Results from each simulation are analyzed and the major MMRs that dominate the orbital evolution of dust particles are identified. To visualize a simulated dust disk (as viewed from a given direction) to compare with the actual observations, output positions of dust particles from each simulation are accumulated together and analyzed. For a massive planet, such as Jupiter, dust particles are trapped in various MMRs that are far away from the planet. However, as the mass of the perturbing planet decreases, those distant MMRs become weaker and are less likely to trap dust particles. Dust particles continue to spiral toward the star and get trapped in MMRs that are closer to the planet. The following examples are from 50 μm particles simulations. The major MMRs that trap particles in the four simulations are: (1) 1 Jupiter-mass planet: 5/2 (at 73.4 AU), 2/1 (63.3 AU), and 3/2 (52.2 AU), (2) 1 Neptune-mass planet: 5/2, 2/1, 5/3 (56.0 AU), and 3/2, (3) 5 Earth-mass planet: 2/1, 5/3, 3/2, and 4/3 (48.3 AU), (4) 1 Earth-mass planet: 3/2, 4/3, 5/4 (46.3 AU). The most dominant MMR in each simulation is: (1) 2/1, (2) 2/1, (3) 3/2, and (4) 3/2, respectively.

While trapped in an exterior MMR with a planet, a dust particle's trajectory, when viewed from a coordinate system co-rotating with the perturbing planet, forms simple ring-like geometric patterns around the star that consist of "loops" [2]. Different MMRs result in different numbers of loops. For example, 3:2 (2 loops), 2/1 (1), 5/3 (3), 4/3 (3). The loops are evenly spaced along the ring and in general are more stable when all loops remain far away from approaching the planet. The only exception is the 2/1 MMR where only one loop is formed and is kept about 90° from the planet in a stable trap. The 2/1 MMRs are also unusual in that there are two types of 2/1 MMRs. One is the more typical MMR where a dust particle gains orbital energy directly from the perturbing planet, by having the loop trailing behind the planet, to counterbalance the energy loss due to PR drag [2]. In the other 2/1 MMR, the loop leads the planet, and the dust particle gains orbital energy via the "indirect" perturbation that arises because the star is moving around the barycenter of the star and planet. An accumulation of dust particles trapped into an MMR for a long period of time will form a ring-like structure with an azimuthal brightness variation. The bright regions are associated with the loops while the region closest to the planet always appears as a brightness minimum in the ring since dust particles in stable MMRs tend not to pass near the perturbing planet.

When all simulations are analyzed, the following conclusions are obtained: (1) to have a ring-like structure with a strongly depleted inner region, a giant planet (≥ Jupiter) is needed to eject dust particles from the system, (2) to have an azimuthal variation along the ring, a planet (Earth-like to Jupiter-like) is needed to trap particles in MMRs, and (3) to have a single bright spot/arc in the ring, the 2:1 MMR with a giant planet has to be the dominant resonance. Figure 1 shows a simulated image of the ε Eridani dust disk with a Jupiter-like planet at 40 AU from the star. Figure 2 is the observed image published in 1998 [7]. The parts that match the key observation features include a ring-like structure around the star, the location of the bright peak in the ring and the adjacent bright arc-like structure. However, the faintest part of the ring as well as the width of part the ring are not well matched. Since the bright peak is the only reliable feature identified from the observations [7], it is somewhat premature to compare other features in the ring. Additional and improved observations will be needed to further characterize the ring features and, then, the property of the perturbing planet.

4. CONCLUSIONS

Large-scale dust structures should exist in the outer Solar System due to perturbations from Neptune. Similar interactions between dust disks and unseen planets should also exist in many extrasolar planetary systems. Using structures on a dust disk to detect and characterize planets provides a different way for extrasolar planet detection. In some cases, for example if a system has a close-to-face on orientation and/or the planet is very far away from the central star, this method may be more powerful than any other.

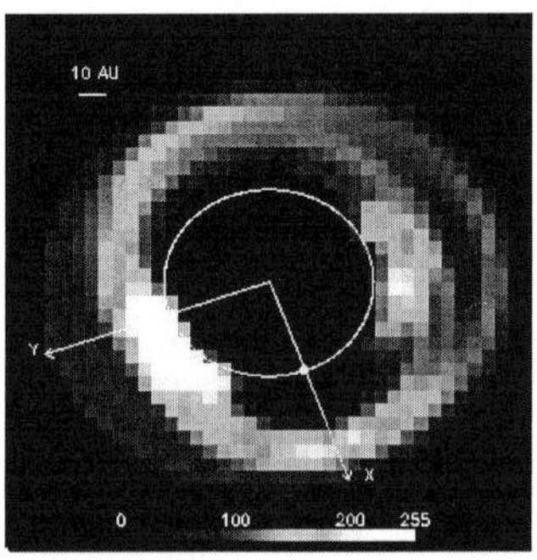

Figure 1. A simulated image of the ε Eridani disk with a Jupiter-like planet in the system.

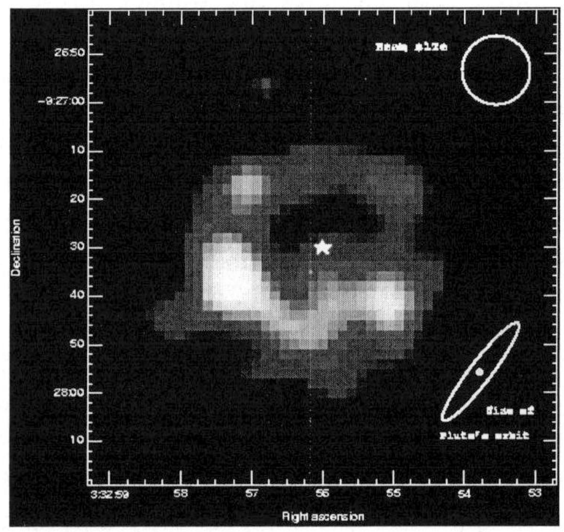

Figure 2. The observed ε Eridani disk, adapted from [7].

REFERENCES

1. J. A. Burns, P.L. Lamy and S. Soter, Icarus 40 (1979) 1.
2. A. A. Jackson and H. A. Zook, Nature 337 (1989) 629.
3. S. F. Dermott, S. Jayaraman, Y.L. Xu, B.A.S. Gustafson and J.C. Liou, Nature 369 (1994) 719.
4. S. F. Dermott, R.S. Gomes, D.D. Durda, B.A.S. Gustafson, S. Jayaraman, Y.L. Xu and P.D. Nicholson, In Astweroids, Comets Meteors 1993, (eds A. Milani, M. Di Martino and A. Cellino), Kluwer, Dordrecht IAU Symp. 152 (1992) 333.
5. J.-C. Liou, H.A. Zook and S.F. Dermott, Icarus 124 (1996) 429.
6. J.-C. Liou and H. A. Zook, Astron. J. 118 (1999) 580.
7. J. S. Greaves, W.S. Holland, G. Moriarty-Schieven, T. Jenness, W.R.F. Dent, B. Zuckerman, C. McCarthy, R.A. Webb, H.M. Burner and W.K. Gear, Astrophys. J. 506 (1998) L133.

Dust production in the Kuiper Belt and in Vega-like systems

C. Dominik[a]

[a]Sterrenkundig Instituut "Anton Pannekoek", Universiteit van Amsterdam, Kruislaan 403, NL-1098 SJ Amsterdam, The Netherlands

We discuss the observed 400 Myr lifetime of Vega-like disks in the frame of a simple analytical model of the collisional evolution of the cometary cloud responsible for the production of the dust. First results show that the 400 Myrs can be explained if the the comets move on highly eccentric and inclined orbits. However, the few old stars which also have a massive dust disk require either less excited orbits in order to allow the dust production to persist for a long time, or a specific event to turn on dust production at a later time.

1. INTRODUCTION

The Kuiper Belt in our own Solar System and the dust disks around Vega-like stars are both thought to be of similar origin. The dust in these systems is continuously produced by destruction of long-lived large bodies, probably in collisions. Replenishing the dust is necessary since the lifetime of dust particles in orbits around these stars is at most a few million years, and much shorter in systems with a lot of dust [1]. For an overview of the properties of Vega-like dust disks, see contributions of Backman and Krivova in this volume.

Recent observations with the Infrared Space Observatory (ISO) [2,3] combined with new age determinations of nearby main sequence stars [4] have revealed that most stars arrive on the main sequence with a strong infrared excess attributed to debris dust. The amount of dust is measured by the ratio of the infrared to the bolometric luminosity, $\tau_{\rm IR}$. Typical values for $\tau_{\rm IR}$ are around 10^{-4}. The main result of the ISO study was that most stars younger that 400 Myrs have a detectable dust disk. After 400 Myrs, most stars do not show a measurable IR excess which means that $\tau_{\rm IR}$ has dropped below 10^{-5}. This indicates that the cloud of comets responsible for the dust replenishment has a life time of about 400 Myrs. On the other hand, a few stars in the sample studied are several billion years old and still show a strong infrared excess. It has been suggested that these stars started out with a much more massive cometary disk so that they can still produce dust today. In this contribution we explore a simple model for the collisional evolution of the cometary cloud in order to discuss possible causes for the observed 400 Myr timescale.

2. A SIMPLE ANALYTICAL MODEL

We assume that the dust produced in the system is the tail of a collisional equilibrium size distribution which is ultimately due to the destruction of a group of large bodies which we will call *comets* for convenience. Let N_p be the initial number of such bodies with collisional cross section σ_p and mass M_p. Let these bodies occupy a volume V where they move around and collide with relative velocities $v_{\rm coll} = \nu v_K$ with v_K being the Kepler velocity in the regions where the comets exist. ν will be zero for circular orbits in a single plane and approach one for highly excited orbits. If all collisions are destructive, the time derivative of the number of comets in the system is given by

$$\dot{N}_p = -\frac{N_p^2}{t_s} \tag{1}$$

where t_s is the sweeping time, the time the by the cross section of a comet moving at velocity $v_{\rm coll}$ to sweep the entire volume V.

The solution of eq. (1) is obviously

$$N_p(t) = \frac{N_0}{1 + \frac{N_0}{t_s} t} \tag{2}$$

where N_0 is the initial number of comets. For times $t \gg \frac{t_s}{N_0}$, the solution becomes independent or N_0.

$$N_p(t) \rightarrow \frac{t_s}{t} \quad \text{for} \quad t \gg \frac{t_s}{N_0} \quad . \tag{3}$$

We may now proceed to compute the amount of dust grains produced by the collisional cascade of the comets. It has been show that collisional equilibrium produces a size distribution $f(m) \propto m^{-11/6}$ [5]. At the small end of the size distribution, particles are removed. The main removal processes are radiation pressure for grains which are small enough to receive an acceleration close to or higher than the gravitational acceleration of the star. Somewhat bigger grains can still be removed from the disk by Poynting Robertson drag. Which process dominates is dependent upon the dust density in the disk. If the density is high, collisional destruction dominates all the way down to small particles which are removed by radiation pressure. If the disk contains less mass, particles of a certain size have Poynting Robertson timescales which are smaller than the collisional time and are being pulled out of the cometary region towards the star (see contribution of Dermott, this volume). It turns out that in a system with as little mass as the solar Kuiper Belt (estimates indicate about 0.1 $M_\oplus$), Poynting Robertson drag dominates for 1μm sized grains. In the denser Vega-like systems collisions always dominate.

One can show that in a collisionally dominated disk the number of small grains is proportional to N_p while in a PR-drag dominated disk, the number of grains is proportional to N_p^2. The number of grains in the disk can then be calculated analytically and the infrared/bolometric luminosity $\tau_{\rm IR}$ be evaluated.

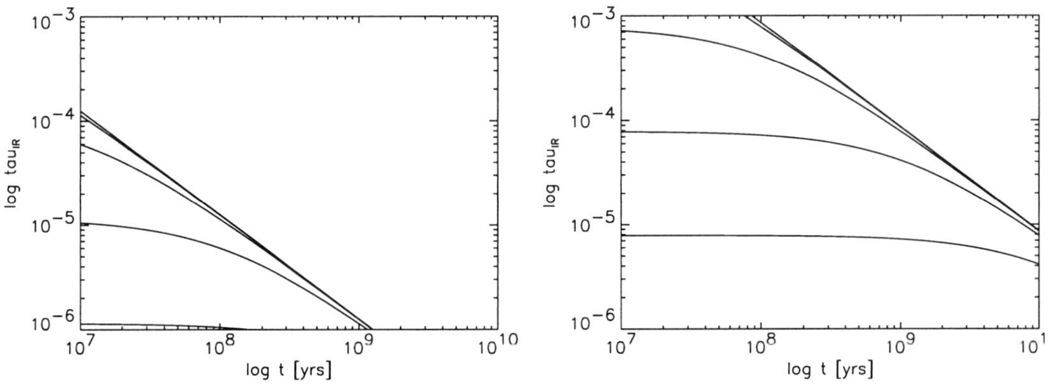

Figure 1. $\tau_{\rm IR}$ as a function of time. Left panel: a system with $\nu = 1$, i.e. highly eccentric orbits. Right panel: a system with $\nu = 0.1$, i.e. low-eccentricity orbits. Different curves are for different initial masses of the comet cloud. From bottom to top 0.1, 1, 10, 100, 1000 $M_\oplus$.

3. RESULTS AND DISCUSSION

Figure 1 shows the time evolution of $\tau_{\rm IR}$ for several different cases. We have used different initial disk masses (the total mass of comets in the disk). Also we have performed calculations with different collision velocities. High collision velocities indicate that the orbits of the comets are highly eccentric and/or inclined. Such orbits lead to a small sweeping time t_s and therefore to a short lifetime of the comets. The left pane in Figure 1 assumes collision velocities of the same order as the Kepler velocities at 50 AU from the star. The resulting decrease in $\tau_{\rm IR}$ agrees well with a lifetime of a few times 10^8 years. Note also that the different initial masses converge to the same solution after about 10^8 years. This result could already be seen from eq. (3). It is therefore clear that the old stars with Vega-like disks cannot be explained by a higher initial mass of the disk. The destruction time is the same for all masses.

However, the right panel in Figure 1 shows a similar calculation for collision velocities which are only one tenth of the Kepler velocities at 50 AU from the star. It is immediately evident that these disks last for much longer and that in fact *less* initial mass is required. Only $1 M_\oplus$ of comets is needed to reproduce the observed values of $\tau_{\rm IR} \approx 10^{-4}$. Higher initial masses at these low collisional velocities are in fact inconsistent with the observations.

3.1. Stellar Collisions?

Another possibility to explain a Vega-like excess of old stars is to assume that the star has a part of the disk which has been quiescent for most of its life, but only recently became active. It is known from dynamical calculations that any Kuiper Belt in the solar system outside 50 AU would have remained undisturbed by the gravitational influence of

Neptune and could have survived in a pristine state, with nearly circular and collision-free orbits.

How could such a disk be disturbed into a state where it produces copious amounts of dust? A possibility is that of a stellar encounter. Even though direct stellar collisions do not occur at the stellar densities in the solar neighbourhood, encounters within $d = 200$ AU can happen and have been suggested to explain the structure of the β Pictoris disk [6].

The time between collisions of two stellar systems is given by

$$t_e = \frac{1}{n_\star \sigma < v >} = 11\mathrm{Gyr}\frac{1\mathrm{pc}^{-3}}{n_\star}\left(\frac{200\mathrm{AU}}{d}\right)^2 \frac{30\mathrm{km/s}}{< v >} \tag{4}$$

where $n_\star$ is the stellar density in the solar neighbourhood, $< v >$ is the average relative velocity between stars and d is the maximum encounter distance at which the disturbing influence of the passing star is sufficient to stir up the cometary cloud.

Using eq. (4) at face value, we find with reasonable assumptions for the stellar density and the required encounter distance that about 5% of all 10 Gyr old stars have had such an encounter within the last 500 Myrs. The density of stellar objects in the solar neighbourhood is at least 0.1 pc^{-3}, but maybe higher due to an unknown number of brown dwarfs. It is, therefore, not entirely unreasonable to conclude that one or two of the old Vega-like stars discovered in the ISO study may be due to such close encounters.

4. CONCLUSIONS

The old Vega-like disks observed with ISO are physically different from the young disks. Either these disks started out with only moderately excited orbits, indicating less interaction with planets in the system. Or some of the systems have suffered a close encounter with another star or brown dwarf which has stirred up an until now quiescent part of the disk and induced a second Vega-like phase in the life of the star.

REFERENCES

1. P. Artymowicz and M. Clampin, Astrophys. J. 490 (1997) 863.
2. H. J. Habing, C. Dominik, M. Jourdain De Muizon, M. F. Kessler, R. J. Laureijs, K. Leech, L. Metcalfe, A. Salama, R. Siebenmorgen and N. Trams, Nature 401 (1999) 456.
3. H.J. Habing, C. Dominik, M. Jourdain de Muizon, R.J. Laureijs, M.F. Kessler, K. Leech, L. Metcalfe, A. Salama, R. Siebenmorgen, N. Trams and P. Bouchet, Astron. Astrophys. 365 (2001) 545.
4. R. Lachaume, C. Dominik, T. Lanz and H. J. Habing, Astron. Astrophys. 348 (1999) 897.
5. J.W. Dohnanyi, J. Geophys. Res. (1969) 2531.
6. P. Kalas, J. Larwood, B.Ã. Smith and A. Schultz, Astrophys. J. 530 (2000) L133.

Migration of matter from the Edgeworth–Kuiper, and main asteroid belts to the Earth

S.I. Ipatov * [a]

[a] Institute of Applied Mathematics (Moscow, Russia), NASA/GSFC (USA), ipatov@spp.keldysh.ru, siipatov@hotmail.com

A considerable proportion of near-Earth objects could come from the trans-Neptunian belt. Some of them have aphelia deep inside Jupiter's orbit for >1 Myr.

1. INTRODUCTION

The main asteroid belt (MAB), the Edgeworth–Kuiper belt (EKB), and comets, represent the main sources of dust in the Solar System. Most of the Jupiter-family comets come from the EKB, and comets can be disrupted due to close encounters with planets or the Sun, collisions with small bodies, or by internal forces. We support [1,2] Eneev's idea [3] that the largest objects in the EKB and MAB could be formed directly from the compression of rarefied dust condensations in the protoplanetary cloud rather than by the accretion of small (for example, 1 km) planetesimals. The total mass of planetesimals that entered the EKB from the feeding zone of the giant planets during their accumulation, could exceed tens of Earth's masses $M_\oplus$ [4,5]. These planetesimals increased eccentricities of 'local' trans-Neptunian objects (TNOs) and swept most of these TNOs. A small portion of such planetesimals could be left beyond Neptune's orbit in highly eccentric orbits. The results of previous investigations of migration and collisional evolution of minor bodies were summarized in [6,2]. Below we present mainly our recent results.

2. MIGRATION OF MATTER TO A NEAR-EARTH SPACE

Asteroids leave the MAB via regions corresponding to resonances with Jupiter, Saturn, and Mars. They enter these regions mainly due to collisions, although gravitational influence of the largest asteroids plays a smaller role. The number of resonances delivering bodies to the Earth is not small (more than 15) [7]. So even small changes in semimajor axis a, will allow some asteroids to enter the resonances. Thus the role of mutual gravitational influence of asteroids in their motion to the Earth may not be very small. Small bodies can also enter the resonant regions due to Yarkovsky orbital drift. For dust particles we also need to take into account the Poynting-Robertson effect, radiation pressure, and solar wind drag.

Objects leave the EKB mainly due to the gravitational influence of the planets [8]. However, during the last 4 Gyr, several percent of the TNO population could change a by

*This work was supported by INTAS (00-240), RFBR (01-02-17540), and NASA (NAG5-10776).

more than 1 AU due to gravitational interactions with other TNOs [2]. For most of the other TNOs, variations in a were less than 0.1 AU. The role of mutual gravitational influence of TNOs in the evolution of their orbits may be greater than that due to collisions. As even small variations in the orbital elements of TNOs can can lead to large variations in orbital elements due to the resulting gravitational influence of planets, TNOs could leave the EKB (and comets leave the Oort cloud) without collisions. Therefore, some cometary objects migrating into the Solar System could be large. The largest objects (with $d \geq 10$ km) that have collided with the Earth during last 4 Gyr, could be of mainly cometary origin.

We investigated the evolution for intervals $T_S \geq 5$ Myr, of 2500 Jupiter-crossing objects (JCOs) under the gravitational influence of all planets, except for Mercury and Pluto (without dissipative factors). In the first series we considered N=2000 orbits, near the orbits of 30 real Jupiter-family comets with period <10 yr. In the second series we took 500 orbits, close to the orbit of Comet 10P/Tempel 2 ($a \approx 3.1$ AU, $e \approx 0.53$, $i \approx 12°$). We calculated the probabilities of collision of objects with the terrestrial planets, using orbital elements obtained with a step equal to 500 yr, and then summarized the results for all time intervals and all bodies, obtaining the total probability P_Σ of collisions with a planet and the total time interval T_Σ during which the perihelion distance q was less than the semimajor axis of the planet. The values of $P_r = 10^6$, $P = 10^6$, P_Σ/N and $T = T_\Sigma/N$ are presented in the Table 1 together with the ratio r of the total time interval when orbits were of Apollo type (at $a>1$ AU, $q=a(1-e)<1.017$ AU, $e<0.999$) to that of Amor type ($1.017<q<1.33$ AU); r_2 is the same as r but for Apollo objects with eccentricity $e<0.9$. For observed near-Earth objects (NEOs) r is close to 1.

Table 1. Values of T (in kyr), $T_c = T/P$ (in Myr), P_r, r, r_2 for the terrestrial planets

		Venus	Venus	Earth	Earth	Earth	Mars	Mars	–	–
	N	T	P_r	T	P_r	T_c	T	P_r	r	r_2
JCOs	2000	9.3	6.62	14.0	6.65	2110	24.7	2.03	1.32	1.15
comet 10P	500	24.9	16.3	44.0	24.5	1800	96.2	5.92	1.49	1.34
3 : 1 reson.	144	739	529	1227	626	510	2139	116	2.05	1.78
5 : 2 reson.	144	109	54.5	223	92.0	416	516	19.4	1.28	1.15

For the integrations we used the Bulirsh-Stoer method (BULSTO) and a symplectic method. The calculated values of probability of collision of former JCOs with planets, were close to each other for both these methods, although bodies obtained resonant orbits more often in the case of BULSTO.

Besides JCOs, we considered asteroids with initial values of e and i equal to 0.15 and 10° respectively. For the asteroids initially located at the 3:1 resonance with Jupiter, we found that the ratio r_{hc} of the number of asteroids ejected into hyperbolic orbits to the number that collided with the Sun, was 5.6 for BULSTO, and 0.38 and 0.87 for a symplectic method using integration steps equal to 10 and 30 days respectively. So in some cases a symplectic method can give a large error. For the 5:2 resonance with Jupiter, r_{hc} equaled 20 and 10 for BULSTO and symplectic methods respectively. In Table 1, for asteroids we present only results obtained by the BULSTO code at T_S=50 Myr (at T_S=10 Myr the values of P and T are smaller by a factor less than 1.2 and 1.01 for the 3:1 and 5:2 resonances respectively) and for TNOs we present results obtained by both codes.

The total time during which 2000 former JCOs were in Apollo-type and Amor-type orbits was 28.7 and 21.75 Myr respectively, although 12.7 and 11.4 Myr of the above times were due to three objects. We found that several former TNOs spent more than 1 Myr in orbits with aphelion distance Q<4.7 AU. The time interval during which a body had Q less than 3.2 and 3.7 AU exceeded 0.1 and 2.6 Myr respectively.

Most of the collisions of former JCOs with the Earth were from orbits with aphelia inside Jupiter's orbit. The probability of collision with the Earth, for 3 former JCOs which had spent more than 1 Myr in Earth-crossing orbits (mainly with Q<4.7 AU) was 1.5 times greater than that for the other 1997 JCOs. About 1 in 300 JCOs collided with the Sun. In [9] we considered a much smaller number of objects, which did not get aphelia inside Jupiter's orbit, and the values of P_r and T were smaller than those in Table 1. For the 2000 JCOs we have considered, the mean probability of collision with Venus is about the same as with Earth, but 3 times smaller for Mars. These values are mainly due to a few bodies that spent more than 1 Myr in orbits with aphelia deep inside Jupiter's orbit (for such bodies usually more than 80% of collisions with planets were from orbits with Q<4.2 AU). If we consider 1000 JCOs, for which most of the collisions with planets were from orbits with Q>4.2 AU, then the mean probability for Venus and Mars is less by a factor of 1.6 and 3 respectively, than that for Earth. Therefore, the ratio of the total mass of icy planetesimals that migrated from the feeding zone of the giant planets and collided with an inner planet, to the mass of the inner planet, was greater for Mars than that for Earth and Venus.

The mean time during which an object crossed Jupiter's orbit was 0.13 Myr for 2500 JCOs. An object had a period P_a<10 yr usually only for about 12% of this time, so our consideration of initial objects with only P_a<10 yr does not influence much on the obtained results. At N=2000 for 10<P_a<20, 20<P_a<50 and 50<P_a<200 yr, we got 23%, 22% and 16% respectively. One former JCO spent some time in orbits with aphelia deep inside Jupiter's orbit, and then it moved for tens of Myr into the trans-Neptunian region, partly in low eccentricity and partly in high eccentricity orbits. This result shows that some bodies can get from the MAD into the trans-Neptunian region, and that typical TNOs can become scattered objects (with high eccentricities) and vice versa.

3. COLLISIONS WITH THE EARTH

The number of TNOs migrating to the inner regions of the Solar System can be evaluated on the basis of simple formulae and the results of numerical integration. Let $N_J = p_{JN} P_N N_{TNO}$ be the number of former TNOs with $d>D$ reaching Jupiter's orbit for the given time span T_{SS}, where N_{TNO} is the number of TNOs with $d>D$; P_N is the fraction of TNOs leaving the EKB and migrating to Neptune's orbit during T_{SS}; and p_{JN} is the fraction of Neptune-crossing objects which reach Jupiter's orbit for their lifetimes. Then the current number of Jupiter-crossers that originated in the zone with 30<a<50 AU equals $N_{Jn} = N_J \Delta t_J / T_{SS}$, where Δt_J is the average time during which the object crosses Jupiter's orbit. According to [8], the fraction P_N of TNOs that left this zone during T_{SS}=4 Gyr under the influence of the giant planets is 0.1–0.2 and p_{JN}=0.34. As mutual gravitational influence of TNOs also takes place [2], we take P_N=0.2. Hence, at Δt_J=0.13 Myr and $N_{TNO}=10^{10}$ (d>1 km), we have $N_{Jn} = 2 \cdot 10^4$. The number of former

TNOs now moving in Earth-crossing orbits equals $N_E = N_{Jn}T/\Delta t_J$. The characteristic time T_{cN} between collisions of former TNOs with the Earth is $T/(N_{Jn}P)$. For T=0.014 Myr and Δt_J=0.13 Myr, we have N_E=2150 and $T_{cN}\sim$0.1 Myr. N_E is larger than the estimated number N_{Ee} of Earth-crossers with d>1 km (750), and $T_c = T/P$ is larger than the characteristic time $T_{co}\approx$100 Myr ellapsed before a collision with the Earth is obtained for the fixed orbits of the observed NEOs. Such differences can be due to the fact that it is difficult to observe NEOs with high e and i and N_{Ee} doesn't include such NEOs. It may also be probable that the number of 1 km TNOs is smaller than 10^{10}. As comets can obtain NEO and asteroidal orbits, a considerable portion of dust produced by NEOs and even some dust produced in the MAB, can be of cometary origin.

The total mass of water delivered to the Earth during the formation of the giant planets is $M_w = M_J P_{JE} k_i$, where M_J is the total mass of planetesimals from the feeding zones of these planets that became Jupiter-crossers during their evolution, P_{JE} is a probability P of a collision of a former JCO with the Earth during its lifetime, and k_i is the proportion of water ice in the planetesimals. For $M_J = 100M_\oplus$, k_i=0.5, and $P_{JE} = 6.65 \cdot 10^{-6}$, we have $M_w = 3.3 \cdot 10^{-4} M_\oplus$. This value is greater by a factor of 1.5 than the mass of the Earth's oceans. The mass of water delivered to Venus can be of the same order of magnitude, and that delivered to Mars will be less by a factor of 3. Some TNOs with a>50 AU can also migrate to the orbits of Jupiter and Earth. Collisions of comets with small bodies and non-gravitational forces can decrease Q. Asher et al. [10] showed that the rate at which objects may be decoupled from Jupiter and attain NEO orbits is increased by a factor of four or five, if non-gravitational forces are included as impulsive effects. So the values of P_r and T can be larger than those in Table 1. Rickman et al. [11] also concluded that comets play an important role among all km-sized impactors. As it is easier to disrupt icy bodies than stone or metal bodies, the proportion of TNOs among NEOs for bodies with d<100 m may be greater than that for 1 km bodies. Perhaps in the future, when people make settlements on the Moon and other terrestrial planets, small icy comets could be moved by rockets into orbits around these celestial bodies in order to be sources of water.

REFERENCES

1. S.I. Ipatov, Celest. Mech. Dyn. Astron. 73 (1999) 107.
2. S.I. Ipatov, Adv. Space Res. 28 (2001) 1107.
3. T.M. Eneev, (in Russian) Sov. Astron. Letters 6 (1980) 295.
4. S.I. Ipatov, Earth, Moon, and Planets 39 (1987) 101.
5. S.I. Ipatov, Solar System Research 27 (1993) 65.
6. S.I. Ipatov, (in Russian) in Migration of celestial bodies in the Solar System (ed. URSS) Moscow (2000) 320.
7. A. Morbidelli and D. Nesvorny, Icarus 139 (1999) 295.
8. M.J. Duncan, H.F. Levison, and S.M. Budd, Astron. J. 110 (1995) 3073.
9. S.I. Ipatov and G.J. Hahn, Solar System Research 33 (1999) 487.
10. D.J. Asher, M.E. Bailey and D.I. Steel, in Collisional Processes in the Solar System (eds. M.Ya. Marov and H. Rickman) ASSL V. 261 (2001) 121.
11. H. Rickman, J.A. Fernandez, G. Tancredi and J. Licandro, in Collisional Processes in the Solar System (eds. M.Ya. Marov and H. Rickman) ASSL V. 261 (2001) 131.

V Cometary Dust

Comet Dust: The view after Hale-Bopp

M.S. Hanner[a]

[a]Jet Propulsion Laboratory, California Institute of Technology, Pasadena CA, 91109, USA

Extensive observations of comet Hale-Bopp from the ground and space have considerably expanded our knowledge of cometary dust. There is a clear correlation of stronger polarization, higher albedo, more prominent infrared silicate features, and higher infrared color temperature. The composition of the silicates includes both olivine and pyroxenes, in both crystalline and glassy or amorphous form, similar to the chondritic aggregate interplanetary dust particles.

1. INTRODUCTION

The apparition of comet Hale-Bopp (C/1995 O1) was a fortunate convergence of a bright comet observable for more than two years over a wide range in heliocentric distance and improved observational capabilities, enabling detailed measurements with high spatial resolution and broad wavelength coverage. ESA's Infrared Space Observatory (ISO) was in operation throughout the apparition, significantly extending the infrared observations in time and spectral range. As a result, considerable progress has been made in elucidating the properties of cometary dust.

Comet Hale-Bopp already displayed an extensive coma when discovered in July 1995 at heliocentric distance r = 7.2 AU. It soon became evident that the activity was driven by strong CO outgassing; the observed CO production rate at r = 6.6 AU was ~ 2 x 10^{28} mol/sec [1,2]. A strong infrared signal from warm dust was already detected at 4.9 AU [3]. The comet released a prodigious amount of dust; the dust area x albedo product near perihelion was 100 times that of comet Halley [4]. Complex patterns of jets and other structures were evident in the coma throughout the comet's apparition. Perihelion occurred on 1 April 1997 at 0.91 AU from the Sun. The comet was well observed during June to October 1996 (4.2–2.7 AU) and from February to May 1997 (1.25–0.91 AU).

This paper will highlight some of the new results from observations of the dust coma of Hale-Bopp. In particular, spatially resolved measurements of polarization and thermal emission permit us to correlate variations in several optical properties of the dust, while mid-infrared spectra have allowed us to resolve long-standing questions about the mineralogy of the cometary silicates.

2. POLARIZATION

2.1. Observations

Polarization carries important information about the properties of the scattering particles. The polarization $P(\theta)$ as a function of phase angle θ has been established by combining measurements from a number of comets. While the polarization from different comets is similar at small phase angles, the comets tend to divide into two classes at larger θ, having P_{max} ~15% and ~25% respectively near θ =90°–100° ([5,6] and references therein). Negative polarization of order –2% is observed in comets at θ < 20°; both classes of comets appear to have similar negative polarization at small θ.

Comet Hale-Bopp was observed at phase angles from 7° to 47° between June 1996 and May 1997 [7,8,9,10,11,12,13]. A phase curve for the average coma at λ~ 0.65 μm is shown in Figure 1. The polarization at all phase angles is distinctly higher than that measured for previous comets, even those in the high P_{max} class, including P/Halley. At θ >30° there is a clear trend towards higher polarization at longer wavelength; the higher the polarization, the stronger the wavelength dependence. A similar, but weaker, wavelength dependence was seen in comet P/Halley (e.g., [14]).

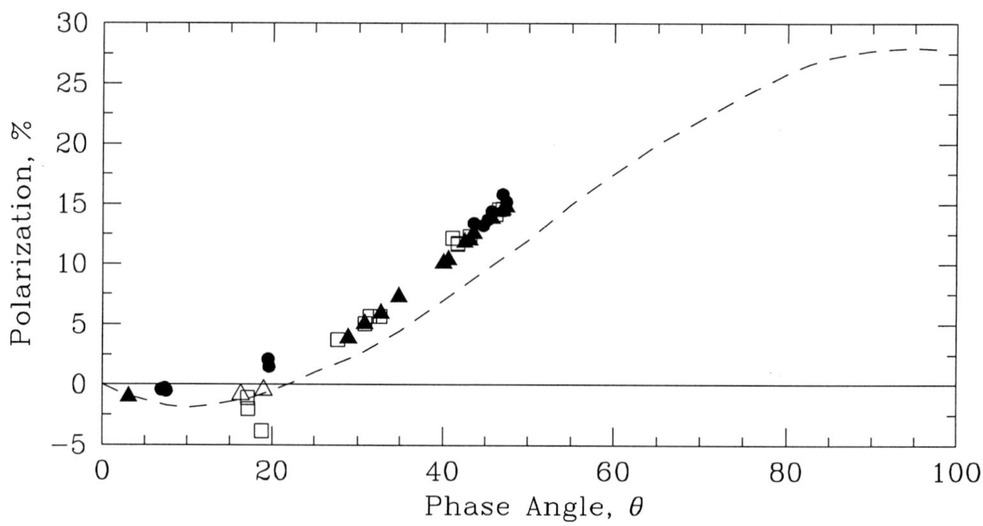

Figure 1. Hale-Bopp polarization (red filter λ ~ 0.67 μm). Filled circles: [9,15]. Squares: [8]. Filled triangles: [13] λ ~ 0.76 μm. Dashed curve: polarization for comets in high P_{max} group [6].

In single-aperture measurements, negative polarization of order -0.5% to –1% was recorded at θ < 20 °, similar to that in other comets at similar phase angles [8,13,12]. However, spatially resolved observations revealed differences in the negative polarization branch with position in the coma, with values as low as –5% near the photometric center at θ = 7° [15], while integrated measurements over the polarization image showed little or no negative polarization at all [15,9,16].

Widespread use of CCD arrays to construct polarization images for Hale-Bopp has facilitated the study of polarization variations within the coma [7,15,9,11,16]. Structure in the polarization images generally correlates with the structure seen in images of total intensity *I* x ρ, where ρ is the projected distance from the photometric center. The visible jets and arcs correspond to regions of higher polarization.

There is an inner region, extending a few thousand km in diameter, with lower polarization than the surrounding coma. This circumnuclear zone is evident in both 1996 and 1997, although the morphology of the inner coma changed considerably during that time. At small phase angles, this zone exhibits negative polarization down to –5% at 0.67 μm [15]. A similar circumnuclear region of lower polarization was detected in several other comets, including 1P/Halley [17], C/1990 K1 Levy [18], 47P/Ashbrook-Jackson [19] and 81P/Wild 2 [20]. However, C/1996 Q1 Tabur and C/1996 Hyakutake displayed higher polarization in the inner 2000 km [21].

2.2. Interpretation of Polarization

Interpreting the polarization in terms of particle properties is complex. The degree of polarization and its angular dependence, $P(\theta)$, depend on particle size, shape, composition, surface roughness, and aggregation. Trends, for example with composition, may differ in different particle size regimes. Predictions based on Mie theory will be inaccurate, even misleading in the general trends, for any particle shape except a smooth sphere.

Yanamandra-Fisher and Hanner [22] studied shape effects for particles with size parameter $X = 2\pi\, a/\lambda$ in the range 1–5. While silicate spheres with $X = 2.5$ exhibit negative polarization at all θ, other shapes show $P_{max} \sim +25\%$ and negative polarization at $\theta < \theta_c$, where θ_c ranges from 20 ° to 70°. In contrast, carbon particles with $X = 2.5$ display mainly positive polarization, $P_{max} \sim 50\%$; the phase angle of maximum polarization varies with particle shape. Mishchenko [23] computed the phase matrix for shape distributions of dirty silicate spheroids, $X = 3.5$. As the axial ratio of the spheroids increased, the polarization at intermediate phase angles changed from negative (characteristic of spheres) to positive, leaving a negative branch at 0°–30° phase angles, similar to the cometary $P(\theta)$.

In recent years, there has been encouraging progress in understanding the polarization by irregular particles, especially aggregates. Numerical methods have allowed exploration of particle shape and aggregate structure while direct scattering measurements have contributed data on particle structures too large to be handled easily by the numerical codes. West [24] and Zerull et al. [25] demonstrated via computation and laboratory measurements, respectively, that the polarization of a fluffy aggregate is determined primarily by the polarization properties of the constituent grains. Calculations by Kozasa et al. [26] confirmed that, indeed, fluffy aggregates composed of spheres much smaller than the wavelength produce Rayleigh polarization. Xing and Hanner [27] found that aggregates of intermediate porosity (~ 60%) generated polarization intermediate between that of the constituent grains and that of a compact particle with size parameter equal to that of the aggregate. They concluded that a mixture of silicate and carbonaceous aggregates of porosity ~ 60% and constituent grains of $X_i \sim 2.5$ could produce a polarization phase curve similar to that of comet dust.

Petrova et al. [28] computed $P(\theta)$ for aggregates of 15-33 constitutent grains having size parameter $X_i = 0.7$-2.5 and refractive index 1.65-*ik*, *k*=0.01-0.05. They find that aggregates of constituent grains with X_i = 1.5-1.65 can reproduce the cometary $P(\theta)$, including the small negative polarization branch at $\theta < 25°$. However, this small negative branch quickly disappears for other X_i.

Recently, two groups have published results of experiments designed to measure the polarization by fluffy aggregate structures:

The PROGRA[2] experiments make use of microgravity during parabolic airplane flights to levitate particles in the path of a laser and measure $P(\theta)$ [29,30,15,31]. In the recent experiments, the particles consist of submicron grains, radius < 0.1 μm ($X \leq 1$) forming fluffy aggregates. The aggregates can agglomerate into very loose structures up to millimeter size. Although the P_{max} is generally high, because of the small grain size, mixtures of silica and carbon aggregates show promise for reproducing the cometary $P(\theta)$, including negative polarization at small phase angles.

The microwave scattering laboratory at the University of Florida permits study of cometary analogue particles by scaling both the particle dimensions and the wavelength to the microwave domain [32]. Measurements can be made over a range of frequencies, to determine color and polarimetric color. Gustafson and Kolokolova [33] have presented color and polarization measurements for aggregates of silicate and absorbing grains, $0.5 < X < 20$ and have analyzed the trends of color and polarimetric color with grain size and composition. These results have led Kolokolova et al. [21] to interpret the change in polarization with distance from the nucleus in terms of sublimating organic refractory mantles on silicate grains.

Thus, both modeling and laboratory measurements show that irregular aggregates having constituent grains with X_I in a relatively narrow range of roughly 1.5 to 2 can exhibit $P(\theta)$ similar to that observed, including a small negative branch. However, scattering by comet dust generates a similar phase function, including the small negative branch, over a range of a factor 4 or more in wavelength, with only a modest increase in P_{max}.. Reproducing this behavior in the modeled particles remains a challenge. The cometary aggregate particles apparently have a range in structure and porosity such that they appear similar when viewed at different spatial scales.

3. CORRELATIONS IN OBSERVABLE QUANTITIES

In order to disentangle the effects of particle size, composition, and structure, it is useful to look at the correlations among observable quantities. The spatially resolved observations of the dust coma of Hale-Bopp and comparisons of Hale-Bopp with other comets allow us to summarize with some confidence the trends and correlations in the scattering properties of comet particles. Table 1a compares the average observed properties in the inner coma of Hale-Bopp with similar measurements of comet P/Halley. Table 1b makes a similar comparison for the jets and spirals in Hale-Bopp in February–April 1997 versus the background coma.

There is a clear correlation of higher polarization, redder polarimetric color, higher albedo, stronger silicate feature, higher infrared color temperature, and enhanced 3–5 μm thermal emission. (Although the color of the scattered light can also be diagnostic, (e.g., [34,21]), we do not have consistent color data for comparing Hale-Bopp and Halley.)

3.1. Polarization

As discussed in section 2.1, comets tend to form two groups with differing P_{max}. The comets with higher P_{max} generally exhibit a strong scattered light continuum and a conspicuous silicate feature. Hale-Bopp is consistent with this trend, displaying the highest polarization at a given phase angle and strongest silicate feature ever recorded. For grains of size comparable to the wavelength (X in the approximate range 2.5–6), polarization is higher for absorbing grains than for silicate grains and carbon grains lack the negative polarization produced by silicate grains. As X increases from 2.5 to 5, P_{max} decreases for silicate grains, but increases for carbon grains with regular, but non-spherical, shapes [22]. Thus, increased polarization could mean more absorbing grains relative to silicate grains in the coma; however, this would contradict the correlation between higher polarization and stronger silicate emission. Only if there is a significant contribution to the scattering from particles with $X < 2$, then a decrease in particle size would produce an increase in polarization for both silicate and absorbing grains, consistent with the higher silicate feature and higher 3–5 μm flux. However, for small grains, $X \leq 1.5$, polarization and albedo are anti-correlated, in contrast to the comet observations, and the negative polarization branch is lost.

Polarization by an aggregate particle depends on the porosity and the size of constituent grains (section 2.2).

Table 1.
Correlations among Observable Properties

a. Hale-Bopp versus Halley		Possible explanations
Polarization	higher	more small grains, $X < 2$ more absorbing grains, $a \sim \lambda$
Polarimetric Color	redder	more small grains, $X < 2$ more silicate grains, $a \sim \lambda$
Albedo	higher	more small grains, $a \sim 0.2$ μm more "clean" silicates
Continuum	stronger	more small grains more "clean" silicates higher dust/gas ratio
Silicate Feature	stronger	small silicate grains, $a < 1$ μm higher silicate/carbon abundance
Color Temperature	higher	warmer silicates more small absorbing grains
3-5 μm Flux	higher	more small absorbing grains, $a < 0.5$ μm
b. Jets versus Average Coma in Hale-Bopp		
Polarization	higher	more small grains, $X < 2$ more absorbing grains, $a \sim \lambda$
Polarimetric Color	redder	more small grains, $X < 2$ more silicate grains, $a \sim \lambda$
Silicate Feature	stronger	smaller silicate grains, $a < 1$ μm higher silicate/carbon abundance warmer silicates
Color Temperature	higher	more small absorbing grains
3-5 μm Flux	higher	more small absorbing grains, $a < 0.5$ μm

3.2 Polarimetric Color

Comets generally exhibit higher polarization at longer wavelength (red polarimetric color, $P(\lambda) > 0$) and $P(\lambda)$ was redder in Hale-Bopp than in Halley. Two studies of non-spherical particles are described here. Yanamandra-Fisher and Hanner [22] carried out calculations for compact, non-spherical shapes, $X = 1 - 5$. Gustafson and Kolokolova [33] measured the polarization by aggregate particles with porosity approximately 50% and 90%. For small, compact grains, $X < 2$, a decrease in grain size causes a redder polarimetric color, for both absorbing and silicate materials. In the limit $X << 1$, $P(\lambda) = 0$. An increase in polarization ($P(\lambda) > 0$) was measured for porous aggregates of silicate spheres as X for the spheres decreased from 1.75 to 1.32. For $X > 2$, compact absorbing grains with constant refractive index have higher polarization at larger X, thus blue polarimetric color, while silicate grains have decreasing polarization with increasing X, thus red polarimetric color. A porous aggregate of spheroids, refractive index $1.7 - 0.2i$ and $X \sim 4$ displayed slightly red polarimetic color. Compact particles with $X >> 1$ will tend toward neutral $P(\lambda) \sim 0$. The aggregates of silicate spheres with $X \sim 20$ showed essentially neutral $P(\lambda)$.

Correlations between color and polarimetric color are discussed by Gustafson and Kolokolova [33] and Kolokolova et al. [34,21].

3.3. Albedo

The average albedo, $A(\theta)$, at a given phase angle can be determined from nearly-simultaneous measurements of thermal emission and scattered light. (Note that $A(\theta)$ is neither the single-scattering albedo, which requires integration over all phase angles, nor the geometric albedo, defined for 0° phase [35]). Using the definition from Gehrz and Ney [36] we have

$$\frac{f\,(\mathrm{vis})}{f\,(\mathrm{IR})} = \frac{A(\theta)}{1-A(\theta)} \qquad (1)$$

where f (vis) and f (IR) are the integrated apparent intensities, measured at phase angle θ, of the scattered and thermal energy distributions, respectively. Mason et al. [37] find $0.25 < A(\theta) < 0.40$ for Hale-Bopp at $20^\circ < \theta < 40^\circ$, significantly higher than the typical values of $A(\theta) \sim 0.20$ in comet Halley [36].

There are no reliable measurements of the albedo in the jets and spiral structures. Such measurements would require accurate co-registration of simultaneous infrared and optical images having similar spatial resolution and point spread function. Strong jet activity was observed to be associated with higher albedo and bluer color in comet Halley [38,39].

For compact, slightly or moderately absorbing grains ($k \leq 0.25$), the single-scattering albedo reaches a maximum at $|m - 1|X \sim 1.5$, where $m = n-ik$ is the refractive index and X is the size parameter; at $\lambda = 0.5$ μm the maximum corresponds to grain radius ~ 0.2 μm. Consequently, decreasing the mean size of the grains in the coma from micron-sized to a few tenths of a micron would tend to increase the albedo, whereas decreasing the mean grain size below ~ 0.2 μm would tend to lower the albedo. (Since we measure the albedo at a specific phase angle, any change in the angular scattering function would also affect the observed albedo.) Loss of absorbing mantles on small silicate grains is another means of increasing the albedo. For fully absorbing grains ($k > 0.25$), the albedo increases monotonically with increasing grain size, so a decrease in grain size will lower the albedo.

Trends with porosity of an aggregate particle will depend on the size and absorptivity of the constituent grains. Hage and Greenberg [40] have shown that, for aggregates having constituent grains with size parameter $X \sim 0.2$, the albedo will decrease as the porosity increases, particularly for porosity > 75%. In their aggregate model of cometary dust, Greenberg and Hage [41] and Li and Greenberg [42] assume that the aggregates are composed of core-mantle interstellar grains with core radius 0.1 μm and overall radius ~ 0.14 μm; for this model the albedo will decrease as porosity increases. Only if the constituent grains would have radii > 0.2 μm and $k < 0.25$ would the albedo at $\lambda = 0.5$ μm increase with increasing aggregate porosity.

A very strong scattered light continuum, with respect to the gas emission, was observed in Hale-Bopp. The scattered light continuum was 100 times stronger than that of Halley at similar distances, while the gas production rate was ~ 20 times larger [4]. The higher albedo of the dust in Hale-Bopp can account for only a small part of the implied factor of 5 increase in the dust/gas ratio; the total cross-section of dust must have been higher relative to the gas production as well

3.4. Silicate Feature

The 10 μm silicate emission feature (as defined by the ratio of the total flux to the interpolated continuum at 10 μm) was stronger in Hale-Bopp than in any previous comet; total flux/continuum ratio was ~ 3 near perihelion. To show a strong, narrow 10 μm emission feature requires silicate grains with radius, $a < 1$ μm (e.g., [43]). The strength of the feature in a comet depends on its visibility above the continuum produced by featureless grains of all sizes. Therefore, a stronger silicate feature could be due to a higher cross section of small silicates, relative to other dust components, or to a higher temperature of the silicates, making their feature more visible. The effect of refractory organic mantles on the silicate emission feature depends on the refractive index of the mantle material at both optical and infrared wavelengths; higher absorption at $\lambda < 1$ μm will heat the mantled grains, increasing the visibility of the feature, whereas high opacity in the infrared will reduce the feature contrast (e.g. [41,42]).

Larger aggregate particles will generate a silicate feature only if the porosity is high enough for the radiation to interact primarily with the constituent submicron silicate grains. For their assumed size distributions, Greenberg and Hage [41] and Li and Greenberg [42] needed a porosity > 95% to produce a sufficiently strong silicate feature.

3.5. Thermal Emission

Comets typically have mid-infrared (5–13 μm) color temperatures, $T_{c,}$ 5–30% higher than that of an equilibrium blackbody; the ratio T_c/T_{bb} is often called the "superheat" (e.g., [36]). The high T_c can be explained by absorbing grains smaller than the infrared wavelengths. Such grains absorb sunlight at visual wavelengths more efficiently than they can radiate at $\lambda >$ 5 μm; they must heat up until they achieve a balance between absorbed and emitted energy. Because the grains do not radiate as blackbodies, T_c is not necessarily the physical temperature of the grains and T_c can be wavelength-dependent. The 8–13 μm superheat observed for Hale-Bopp was about 1.38 near 1 AU [44], compared with ~ 1.15 in P/Halley [45], while the T_c at 3–5 μm [44] and 5–8 μm [46] was ~ 1.8 times the blackbody temperature, far higher than that observed in any other comet [36]. The 3 5 μm thermal flux is a particularly sensitive indicator of hot sub-micron sized grains (grain radius $a < 0.5$ μm).

Gehrz and Ney plotted a clear correlation between the superheat and the strength of the silicate feature in their sample of comets, and Hale-Bopp fits with their plot [37].

Thermal images (5–18 μm) and spatially resolved 8–13 μm spectra were acquired by Hayward et al. [44]. The thermal images show the same jets and spiral structures as optical images from the same time intervals. Either the same particles were responsible for the optical and thermal excess radiation in the jets, or the two grain populations have the same dynamical properties and thus similar spatial distributions. The 8–13 μm T_c and the 3–5 μm fluxes are higher in the jets than in the background coma. The silicate feature was consistently stronger by 10–20% in the jets and spirals. In summary, there was always a correlation between the T_c, the 3–5 μm flux, and the strength of the silicate feature, both over time and with position in the coma.

Calculations of the motions of grains ejected from discrete active areas on the nucleus, as a function of their ejection velocities and $\beta = F_{rad}/F_{grav}$ indicate that the observed spiral patterns in these images are consistent with small grains having $\beta_{max} < 1$, but not small grains with $\beta_{max} > 2$ [44]. That is, the morphology is consistent with the excess radiation in jets and spirals arising from small clean or slightly absorbing silicate grains, but not with sub-micron sized absorbing grains. For submicron silicate spheres, β_{max} is < 1 when $k < 0.1$, where $m = n - ik$ is the complex refractive index.

A value of $k = 0.1$ would cause a silicate grain with $a \geq 0.5$ μm to be as warm as a carbon grain. A grain of 0.1 μm radius would be significantly warmer than a blackbody, but ~100 K cooler than a carbon grain at 1 AU; these grains probably could not produce the observed 3–5 μm flux in Hale Bopp. The temperature attained by a core/mantle particle depends on the refractive index and thickness of the mantle material (cf. [41] Figure 8).

3.6. Summary

No simple explanation is consistent with all of the parameters in Table 1. An increased abundance of small, slightly dirty ($k < 0.05$) silicate grains would cause a higher albedo, stronger silicate feature, and redder polarimetric color. However, their temperature would not be high enough to produce a high 3–5 μm flux nor $T_c > T_{bb}$. Their polarization would be higher only if the size parameter $X < 1.5$. A simultaneous increase of hot submicron absorbing grains would increase the 3–5 μm flux, but would tend to decrease the albedo Whether core/mantle particles could be hot enough to give a high 3–5 μm flux, while still causing an increase in albedo is not yet clear.

A synthesis of all these trends into a single, consistent model for the cometary dust is well beyond the scope of this paper. Indeed, such modeling is just beginning. To date, authors have considered only subsets of these parameters. Issues to be resolved include the correlation of higher albedo with higher polarization, correlation of higher albedo and silicate feature with higher 3–5 μm flux, possible effects of evaporating grain mantles, and the role of porous aggregates.

4. SILICATE MINERALOGY

Small silicate grains (submicron to micron size) exhibit spectral features in thermal emission at 8–13 μm and 16–33 μm. These silicate features were stronger – and more structured – in Hale-Bopp than in any previously observed comet, allowing us to infer the mineralogy of the silicate grains.

The 8–13 μm spectral region was well observed from the ground from $r > 4$ AU to 0.92 AU [47,48,49,44,50,51]. An example is displayed in Figure 2. One sees 3 maxima at 9.2, 10.0, and 11.2 μm and minor structure at 10.5 and 11.9 μm. The sharp peak at 11.2 μm is attributed to crystalline olivine, $(Mg,Fe)_2SiO_4$, based on the good spectral match with the measured spectral emissivity of Mg–rich olivine; it was first detected in comet P/Halley [52,53], and subsequently observed in several new and long-period comets [54,55]. The 11.9 μm shoulder is also due to crystalline olivine. The broad 10 μm maximum is characteristic of amorphous olivine [56]. Crystalline olivine has a secondary maximum at 10 μm as well.

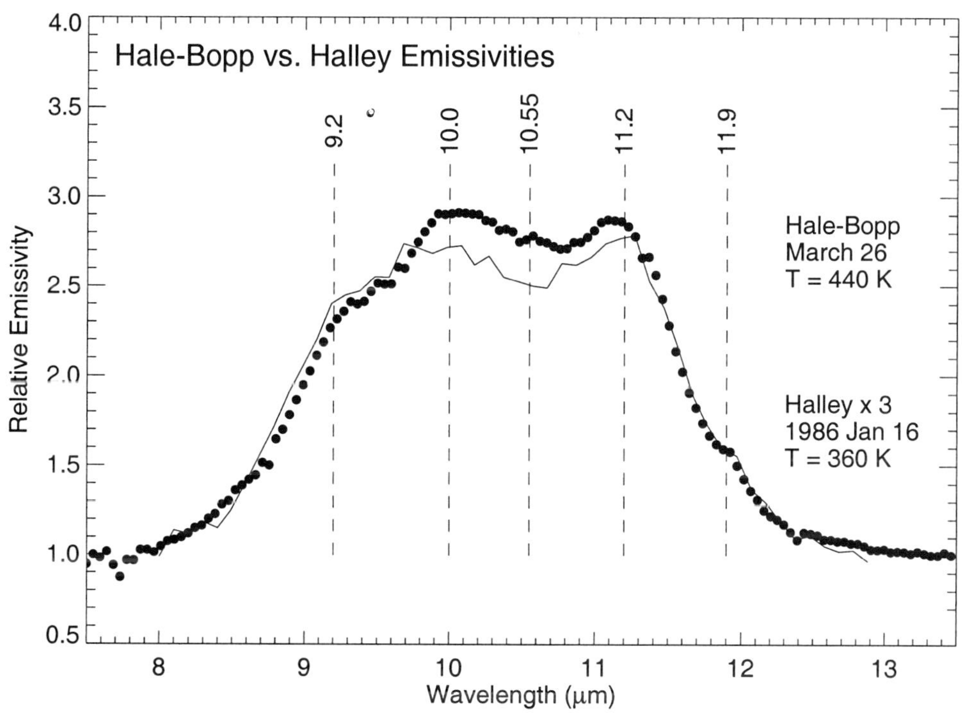

Figure 2. The silicate feature in Hale-Bopp (points) versus Halley (solid line), total flux/continuum. Spectral structure is marked [47].

The 9.2 μm maximum, first recognized in Hale-Bopp, is a signature of pyroxene, $(Mg,Fe)SiO_3$. Both amorphous and crystalline pyroxenes can exhibit a peak near 9.2 μm [56,57,58]. Crystalline pyroxenes have considerable variety in their spectral shape [59] and can contribute to the 10 μm maximum, the structure near 10.5 μm, and the overall width of the observed silicate feature. Thus, the major silicate minerals appear to be present in both crystalline and glassy or amorphous form.

The full 7–45 μm spectral region was observed with the ISO SWS spectrometer at $r = 2.9$ AU [60]. This spectrum shows several strong peaks that correspond to laboratory spectra of

Mg–rich crystalline olivine [58]. Minor structure is attributed to crystalline pyroxene [48]. In contrast, airborne spectra of P/Halley at 1.3 AU (the only other 16–30 μm spectra of a comet) show only weak olivine peaks at 28.4 and 23.8 μm [61,62]. That the silicates are Mg–rich is consistent with the elemental composition of the dust measured during the Halley spacecraft encounters [63].

The relative abundances of the various silicate components are difficult to determine from the spectra, because the strength of an observed feature will depend on the temperature of the emitting grains (as well as on their size and shape). The temperature of small silicate grains in a comet coma is determined by the amount of sunlight they absorb at visual wavelengths. Dorschner et al. [57] showed that the absorptivity depends on the Mg/Fe ratio. Pure Mg–pyroxene grains have extremely low absorption at visual wavelengths and, consequently, would be quite cold. For grain radius 0.5 μm, the difference between a pure Mg–pyroxene and an olivine grain with Mg/Fe = 1 can be hundreds of degrees at 1 AU [47]. However, the dust composition data from the Halley encounters showed that the silicates were usually associated with carbonaceous material [64]. Any absorbing material adhering to silicate grains will heat them.

Spectral models to match the Hale-Bopp spectra with a mixture of silicate minerals have been presented by Brucato et al. [65], Hanner et al. [47], Wooden et al. [48,66], and Hayward et al. (2000). Wooden et al. proposed that observed changes in spectral shape with heliocentric distance could be explained by temperature differences between more transparent (cooler) Mg–rich crystalline pyroxene and less transparent (warmer) olivine grains. In their model, the crystalline pyroxene grains produce the 9.2 μm maximum in 1997 and constitute ~ 95% of the small silicate grains in the coma, in order to produce observable emission at the cold temperatures.

Hayward et al. [44] fit their spectra with a composite silicate model plus a size distribution of absorbing grains; they fit the 9.2 μm maximum with amorphous Mg–pyroxene. The same silicate mixture matched all of the 1997 spectra of the central core, the jets, and the background coma. They concluded that, even assuming all silicate components have the same temperature, small pyroxene grains (amorphous and crystalline) must be more abundant than silicate grains of an olivine composition.

Thus, while there is not a single, unique model for the silicates in Hale-Bopp, one can say that pyroxenes (either glassy or crystalline) were more abundant than grains of olivine composition, with an abundance ratio of about 2:1 or higher. Because of its strong emissivity at resonances, olivine produces a strong peak at 11.2 μm with an abundance < 20% of the total small silicate grains.

Not all comets exhibit the strong, structured silicate emission features characteristic of comets Hale-Bopp, P/Halley, C/1990 K1 Levy, and C/1993 A1 Mueller (e.g., [54,67]). Whether the silicate grains in other comets are simply clumped into larger, optically thick particles or whether there is a real difference in the silicate mineralogy is not known.

5. OBSERVATIONS AT LONG WAVELENGTHS

Measurements of the thermal emission at far-infrared and submillimeter wavelengths can help us to assess the abundance of large dust particles in the coma, because the emissivity of small particles decreases as $\lambda^{-\alpha}$ $1 \leq \alpha \leq 2$, while the emissivity of large particles remains essentially constant.

The far-infrared spectral domain of Hale-Bopp was surveyed from the ISO satellite. The on-board photometer (PHOT) measured the thermal flux through filters at 7–160 μm, while the spectrometers (SWS and LWS) recorded the spectrum from 5–160 μm. To fit the slope of the spectral energy distribution with a size distribution of the form $n(a) \propto a^{-\alpha}$ requires α• 3.5, as shown in Figure 3 [3]. For an outflow velocity $v(a) \propto a^{-0.5}$, this result implies that the dust production size distribution from the nucleus has α • 4 and that the mass is concentrated in large particles.

Jewitt and Matthews [68] acquired submillimeter continuum images of Hale-Bopp in 1997. The observed submillimeter spectral index of 0.6 suggests that the emitting particles were millimeter sized. A dust production rate of 1–2 x 10^6 kg/s near 1 AU was derived, giving a dust/gas mass ratio of at least 5.

Thus, despite the obvious abundance of small grains giving rise to the silicate features and high color temperature at short infrared wavelengths, the mass of the dust emitted from the nucleus of Hale-Bopp appears to have been dominated by large particles. Li and Greenberg [42] argue that the small grains are incorporated into large, extremely fluffy aggregates with porosity > 97%. At such high porosity, these aggregates would be optically thin in the optical and mid-infrared and would be indistinguishable from a cloud of small grains.

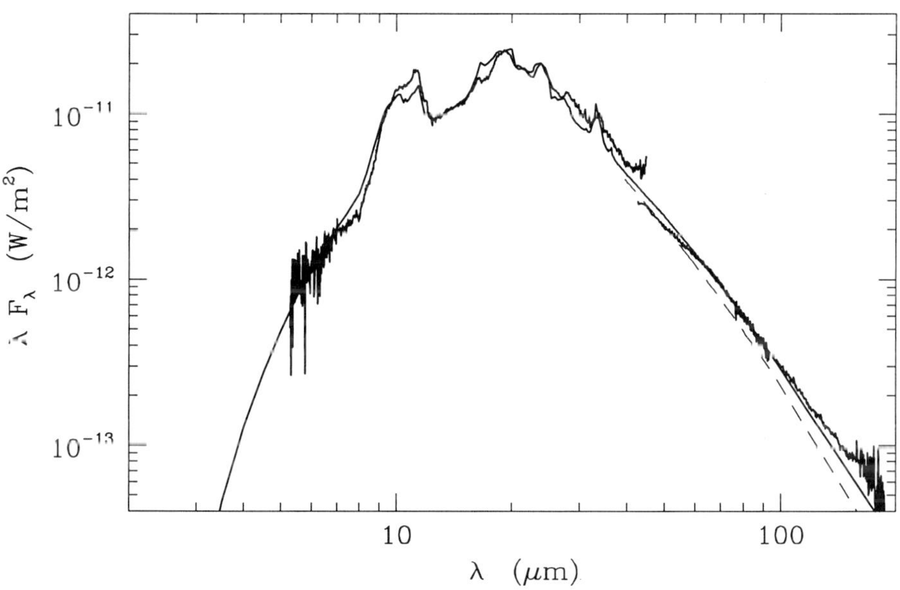

Figure 3. Comparison of ISO SWSand LWS spectra on 7 October 1996 (thick line) with dust models. This solid line: size distribution power law $\alpha = 3.5$; dashed line: $\alpha = 3.7$ [3].

6. RELATION TO IDPS AND INTERSTELLAR DUST

Small comet grains, such as those apparently dominating the observed optical and infrared radiation in Hale-Bopp, are rapidly removed from the solar system by radiation pressure (as are extremely porous aggregates of such grains). Larger cometary particles will remain in bound orbits and may be swept up by the Earth. The chondritic aggregate IDPs are of likely cometary origin because some of them have high atmospheric entry speeds [69]. These IDPs are heterogeneous aggregates of sub-micron sized silicates and other minerals in a carbonaceous matrix. The submicron grain size, high Mg/Fe abundance ratio, mix of crystalline and glassy olivine and pyroxenes, and high carbon content have no counterpart in any other known meteoritic material. These properties are a very good match with the inferred nature of cometary dust.

There are strong indications that at least one silicate component in the chondritic aggregate IDPs is of interstellar origin. Bradley [70] argued that the common 0.1–0.5 μm glassy silicate grains, or GEMS (Glass with Embedded Metal and Sulfides) are interstellar, because of their high radiation dosage, relict microcrystals, and similarity to observed properties of interstellar silicates. FeNi inclusions are sufficient for magnetic alignment of interstellar grains [71]. The 10 μm spectra of GEMS show a broad maximum that varies between 9.3 and 10.0 μm from sample to sample, as one would expect for a varying mix of glassy pyroxene and olivine [72]. Strong D/H and $^{15}N/^{14}N$ anomalies, approaching cold molecular cloud values, have been measured in the carbonaceous material in which the GEMS are embedded [73,74]. It seems plausible that the GEMS acquired these coatings in cold interstellar clouds (e.g. [75]).

The origin (or origins) of the crystalline silicates in comets is less clear. Heating in the coma or on the nucleus is not sufficient for annealing of glassy or amorphous grains, and the 11.2 μm peak in Hale-Bopp was just as prominent at 4.6 AU preperihelion [76] as at 0.92 AU. Crystalline grains can form by direct condensation from a hot gas at T=1200–1400 K. A few enstatite whiskers, ribbons, and platelets in probable cometary IDPs have growth patterns indicating such direct condensation [77]. Grain condensation could have occurred in the hot inner solar nebula or in presolar environments. If the crystalline silicates condensed in the inner solar nebula, then their presence in comets indicates large-scale mixing of high-temperature material to the cold outer regions where the comets formed.

Spectroscopy with ISO has given us a better picture of the distribution of crystalline silicates in astronomical sources. Mg–olivine was detected in the oxygen• rich outflows of some evolved stars [78] but is absent in spectra of the diffuse interstellar medium or molecular clouds such as the Trapezium. Nor are crystalline silicates detected in the circumstellar dust around most young stellar objects. Yet, crystalline olivine peaks are present in certain late-stage Herbig Ae/Be stars. The spectrum of HD100546 is very similar to that of Hale-Bopp [79]. These systems appear to have a population of sun-grazing comets, in order to explain transient gaseous emission spectral lines (e.g., [80]). The 11.2 μm peak is also present in the debris disk around β Pictoris [81]. Because the dynamical lifetime of the dust is shorter than the age of β Pictoris, comets are thought to be the replenishing source of the dust.

Non-solar isotopic ratios can be a signature of interstellar grains. The ability to analyze submicrometer sized volumes with nano SIMS ion microprobes is now possible. First results have identified oxygen isotopic anomalies in 0.1–1 μm silicate grains within IDPs, indicating a pre-solar origin for these grains [82]. Wider application of this technique in the future should lead to a better understanding of the prevalence of interstellar crystalline silicates.

7. CONCLUSIONS

Hale-Bopp has provided a wealth of new data for determining the physical properties of cometary dust. As with all remote observations, these have to be interpreted with the aid of models based on known scattering and emitting properties of small grains. We find a clear correlation of higher polarization, higher albedo, stronger silicate feature, higher 3–5 μm flux, and higher infrared color temperature. Much work remains to be done to synthesize all of the observed correlations into a dust model consistent also with what we know about interstellar dust and IDPs of likely cometary origin.

The origin of the crystalline cometary silicates remains puzzling. If they condensed in the inner solar nebula, then their presence in comets requires extensive mixing in the solar nebula. If they are circumstellar in origin, then one has to understand why their spectral features are not seen in interstellar dust.

Future progress will come from several directions. Continued work on the scattering by irregular and aggregate particles will allow better interpretation of the correlations outlined here. NASA's space infrared telescope, SIRTF, will extend long-wavelength spectroscopy and photometry to a number of comets. Although some very interesting results have come from analysis of a small number of likely cometary IDPs, the IDPs have not been exploited as fully as they could be to aid the interpretation of remote sensing data. For example, a survey of many chondritic porous IDPs should be conducted to quantify the relative abundance of crystalline olivine, crystalline pyroxene, and glassy silicate grains and their typical dimensions. Ultimately, we can look forward to ESA's Rosetta mission and to the return of cometary samples, beginning with NASA's STARDUST mission, now en route to encounter with comet P/Wild 2 in January 2004 [83].

ACKNOWLEDGEMENTS

This research was carried out at the Jet Propulsion Laboratory, California Institute of Technology, under contract with the National Aeronautics and Space Administration.

REFERENCES

1. D. Jewitt, M. Senay and H. Matthews, Science 271 (1996) 1110.
2. N. Biver, H. Rauer, D. Despois, R. Moreno, G. Paubert, D. Bockelee-Morvan, P. Colom, J. Crovisier, E. Gerard and L. Jorda, Nature 380 (1996) 137.
3. E. Grün, M.S. Hanner, S.B. Peschke, T. Muller, H. Boehnhardt, T.Y. Brooke, H. Campins, J. Crovisier, C. Delahodde, I. Heinrichsen, H.U. Keller, R.F. Knacke, H. Kruger, P. Lamy, C. Leinert, D. Lemke, C.M. Lisse, M. Muller, D.J. Osip, M. Solc, M. Stickel. M. Sykes, V. Vanysek and J. Zarnecki, Astron. Astrophys. 377, (2001) 1098.
4. D.G. Schleicher, S.M. Lederer, R.L. Millis and T.L. Farnham, Science 275 (1997) 1913.
5. G.P. Chernova, N. N. Kiselev and K. Jockers, Icarus 103 (1993) 144.
6. A.C. Levasseur-Regourd, E. Hadamcik and J.B. Renard, Astron. Astrophys. 313 (1996) 327.
7. R. Furusho, B. Suzuki, N. Yamamoto, H. Kawakita, T. Sasaki, Y. Shimizu and T. Kurakami, Publ. Astron. Soc. Japan 51 (1999) 367.
8. S. Ganesh, U.C. Joshi, K.S. Baliyan and M.R. Deshpande, Astron. Astrophys. Suppl. Ser. 129 (1998) 489.

9. E. Hadamcik, A.C. Levasseur-Regourd and J.B. Renard, Earth, Moon, Planets 78 (1999) 365.
10. H. Hasegawa, T. Ichikawa, S. Abe, S. Hamamura, K. Ohnishi and J. Watanabe, Earth, Moon, Planets 78 (1999) 353.
11. K. Jockers, V.K. Rosenbush, T. Bonev and T. Credner, Earth, Moon, Planets 78 (1999) 373.
12. N.N. Kiselev and F.P. Velichko, Earth, Moon, Planets 78 (1999) 347.
13. N. Manset and P. Bastien, Icarus 145 (2000) 203.
14. A. Dollfus, P. Bastien, J.F. Le Borgne, A.C. Levasseur-Regourd and T. Mukai, Astron. Astrophys. 206 (1988) 348.
15. E. Hadamcik, Ph.D. dissertation, University of Paris 6 (1999).
16. T.J. Jones and R.D. Gehrz, Icarus 143 (2000) 338.
17. A. Dollfus and J.L. Suchail, Astron. Astrophys. 187 (1987) 669.
18. J.B. Renard, A.C. Levasseur-Regourd and A. Dollfus, Ann. Geophys. 10 (1992) 288.
19. J.B. Renard, E. Hadamcik and A.C. Levasseur-Regourd, Astron. Astrophys. 316 (1996) 263.
20. E. Hadamcik and A.C. Levasseur-Regourd, C.R. Acad. Sci. Paris 2B (1999). C2-316.
21. L. Kolokolova, B.A.S. Gustafson and K. Jockers, This volume.
22. P. A. Yanamandra-Fisher and M.S. Hanner, Icarus 138 (1999) 107.
23. M.I. Mishchenko, Appl. Opt. 32 (1993) 4652.
24. R.A. West, Appl. Optics 30 (1991) 5316.
25. R.H. Zerull, B.A.S. Gustafson, K. Schultz and E. Thiele-Corbach, Appl. Opt. 32 (1993) 4088.
26. T. Kozasa, J. Blum, H. Okamoto and T. Mukai, Astron. Astrophys. 276 (1993) 278.
27. Z. Xing and M.S. Hanner, Astron. Astrophys. 324 (1997) 805.
28. E.V. Petrova, K. Jockers and N.N. Kiselev, Icarus 148 (2000) 526.
29. J.C. Worms, J.B. Renard, E. Hadamcik, A.C. Levasseur-Regourd and J.F. Gayet, Icarus 142 (1999) 281.
30. J.C. Worms, J.B. Renard, E. Hadamcik, A.C. Levasseur-Regourd and N. Brun-Huret, Presented at IAU Colloquium 181/COSPAR colloquium 11, Canterbury U.K. 10-14 April (2000).
31. E. Hadamcik, J.B. Renard, A.C. Levasseur-Regourdand J.C. Worms, This Volume..
32. B.A.S. Gustafson, in Light Scattering by Nonspherical Particles (eds. M.I. Mishchenko, J. W. Hovenier and L. D. Travis) Academic Press (2000) 367.
33. B.A.S. Gustafson and L. Kolokolova, J. Geophys. Res. 104 (1999) 31711.
34. L. Kolokolova, K. Jockers and G. Chernova, Icarus 126 (1997) 351.
35. M.S. Hanner, R.H. Giese, K. Weiss, and R. Zerull, Astron. Astrophys. 104 (1981) 42.
36. R.D. Gehrz and E.P. Ney, Icarus 100 (1992) 162.
37. C.G. Mason, R.D. Gehrz, T.J. Jones, M.S. Hanner, D.M. Williams and C.E. Woodward, Astrophys. J. 549 (2001) 635.
38. A.T. Tokunaga, W.F. Golisch, D.M. Griep, C.D. Kaminski and M.S. Hanner, Astron. J. 92 (1986) 1183.
39. C.S. Morris and M.S. Hanner, Astron. J. 105 (1993) 1537.
40. J.I. Hage and J.M. Greenberg, Astrophys. J. 361 (1990) 251.
41. J.M. Greenberg and J.I. Hage, Astrophys. J. 361 (1990) 260.
42. A. Li and J.M. Greenberg, Astrophys. J. 498 (1998) L83.

43. M.S. Hanner, A.T. Tokunaga, W.F. Golisch, D.M. Griep and C.D. Kaminski, Astron. Astrophys. 187 (1987) 653.
44. T.L. Hayward, M.S. Hanner and Z. Sekanina, Astrophys. J. 538 (2000) 428.
45. A.T. Tokunaga, W.F. Golisch, D.M. Griep, C.D. Kaminski and M.S. Hanner, Astron. J. 96 (1988) 1971.
46. D.M. Williams, C.G. Mason, R.D. Gehrz, T.J. Jones, C.E. Woodward, D.E. Harker, M.S. Hanner, D.H. Wooden, F.C. Witteborn, and H.M. Butner, Astrophys. J. 489 (1997) L91.
47. M.S. Hanner, R.D. Gehrz, D.E. Harker, T.L. Hayward, D.K. Lynch, C.G. Mason, R.W. Russell, D.M. Williams, D.H. Wooden and C.E. Woodward, Earth, Moon, Planets 79 (1999) 247.
48. D.H. Wooden, D.E. Harker, C.E. Woodward, H.M. Butner, C. Koike, F.C. Witteborn and C.W. McMurtry, Astrophys. J. 517 (1999) 1034.
49 P. Galdemard, P.O. Lagage, D. Dubreuil, R. Jouan and P. Masse, Earth, Moon, Planets 78 (1999) 271.
50. R.W. Russell, D.K. Lynch, A.L. Mazuk, G.S. Rossano, M.S. Hanner and M.L. Sitko, Bull. Amer. Astron. Soc. 29 (1997) 1041.
51. J.K. Davies, T.R. Geballe, M.S. Hanner, H.A. Weaver, J. Crovisier and D. Bockelee-Morvan, Earth, Moon, Planets 78 (1999) 293.
52. J.D. Bregman, H. Campins, F.C. Witteborn, D.H. Wooden, D.M. Rank, L.J. Allamandola, M. Cohen and A.G.G.M. Tielens, Astron. Astrophys. 187 (1987) 616.
53. H. Campins and E.V. Ryan, Astrophys. J. 341 (1989) 1059.
54. M.S. Hanner, D.K. Lynch and R.W. Russell, Astrophys. J. 425 (1994) 274.
55. M.S. Hanner, J.A. Hackwell, R.W. Russell and D.K. Lynch. Icarus 112 (1994) 490.
56. J.R. Stephens and R.W. Russell. Astrophys. J. 228 (1979) 780.
57. J. Dorschner, B. Begemann, T. Henning, C. Jager and H. Mutschke, Astron. Astrophys. 300 (1995) 503.
58. C. Koike, H. Shibai and A. Tuchiyama, Mon. Not. Roy. Astron. Soc. 264 (1993) 654.
59. S.A. Sandford and R.M. Walker, Astrophys. J. 291 (1985) 838.
60. J. Crovisier, T.Y. Brooke, K. Leech, D. Bockelée-Morvan, E. Lellouch, M.S. Hanner, B. Altieri, H.U. Keller, T. Lim, T. Encrenaz, A. Salama, M. Griffin, T. de Graauw, E. van Dishoeck and R.F. Knake, This volume.
61. T. Herter, H. Campins and G.E. Gull, Astron. Astrophys. 187 (1987) 629.
62. W. Glaccum, S.H. Moseley, H. Campins and R.F. Lowenstein, Astron. Astrophys. 187 (1987) 635.
63. H. Schulze, J. Kissel and E.K. Jessberger, ASP Conf. Ser. 122 (1997) 397.
64. J. Kissel, R.Z. Sagdeev, J.L. Bertaux, V.N Angarov, J. Audouze, J.E. Blamont, K. Buchler, E.N. Evlanov, H. Fechtig, M.N. Fomenkova, H. Vonhoerner, N.A. Inogamov, V.N. Khromov, W. Knabe, F.R. Krueger, Y. Langevin, V.B. Leonas, A.C. Levasseur-Regourd, G.G. Managadze, S.N. Podkolzin, V.D. Shapiro, S.R. Tabaldyev and B.V. Zubkov, Nature 321 (1986) 280.
65. J.R. Brucato, L. Colangeli, V. Mennella, P. Palumbo and E. Bussoletti, Planet. Space Sci. 47 (1999) 773.
66. D.H. Wooden, H.M. Butner, D.E. Harker and C.E. Woodward, Icarus 143 (2000) 126.
67. M.S. Hanner, D.K. Lynch, R.W. Russell, J.A. Hackwell and R. Kellogg, Icarus 124 (1996) 344.
68. D. Jewitt and H. Matthews, Astron. J. 117 (1999) 1056.

69. S.A. Sandford, in Origin and Evolution of Interplanetary Dust (eds. A.C. Levasseur-Regourd and H. Hasegawa) Kluwer Academic Publ. (1991) 397.
70. J.P. Bradley, Science 265 (1994) 925.
71. A.A. Goodman and D.C.B. Whittet, Astrophys. J. 455 (1995) L181.
72. J.P. Bradley, L.P. Keller, T.P. Snow, M.S. Hanner, G.J. Flynn, J.C. Gezo, S.J. Clemett, D.E. Brownlee and J.E. Bowey, Science 285 (1999) 1716.
73. L.P. Keller, S. Messenger and J.P. Bradley, J. Geophys. Res. 105 (2000) 10397.
74. S. Messenger, Nature 404 (2000) 968.
75. J.M. Greenberg, in Comets, (ed. L.L. Wilkening) Univ. Arizona Press, (1982) 131.
76. J. Crovisier, T.Y. Brooke, M.S. Hanner, H.U. Keller, P.L. Lamy, B. Altieri, D. Bockelee Morvan, L. Jorda, K. Leech and E. Lellouch, Astron. Astrophys. 315 (1996) L385 (1996).
77. J.P. Bradley, D.E. Brownlee and D.R. Veblen, Nature 301 (1983) 473.
78. L.B.F.M. Waters, F.J. Molster, T. de Jong, D.A. Beintema, C. Waelkens, A.C.A. Boogert, D.R. Boxhoorn, T. de Graauw, S. Drapatz, H. Feuchtgruber, R. Genzel, F.P. Helmich, A.M. Heras, R. Huygen, H. Izumiura, K. Justtanont, D.J.M. Kester, D. Kunze, F. Lahuis, H.J.G.L.M. Lamers, K.J. Leech, C. Loup, D. Lutz, P.W. Morris, S.D. Price, P.R. Roelfsema, A. Salama, S.G. Schaeidt, A.G.G.M. Tielens, N.R. Trams, E.A. Valentijn, V. Vandenbussche, M.E. van den Ancker, E.F. van Dishoeck, H. van Winckel, P.R. Wesselius and E.T. Young, Astron. Astrophys. 315 (1996) L361.
79. K. Malfait, C. Waelkens, L.B.F.M. Waters, B. Vandenbussche, E. Huygen and M.S. de Graauw, Astron. Astrophys. 332 (1998) L25.
80. C.A. Grady, M.L. Sitko, K.S. Bjorkman, M.R. Perez, D.K. Lynch, R.W. Russell, and M.S. Hanner, Astrophys. J. 483 (1997) 449.
81. R.F. Knacke, S.B. Farjado-Acosta, C.M. Telesco, J.A. Hackwell, D.K. Lynch and R.W. Russell, Astrophys. J. 418 (1993) 440.
82. S. Messenger, L.P. Keller and R.M. Walker, Lunar and Planetary Science Conf. XXXIII, in press (2002).
83. D.E. Brownlee, Burnett, D, Clark B., M.S. Hanner, F. Horz, J. Kissel, R. Newburn, S. Sandford, Z. Sekanina, P. Tsou, and M. Zolensky, in Physics, Chemistry, and Dynamics of Interplanetary Dust (eds. B.A.S. Gustafson and M.S. Hanner), ASP Conf. Series 104 (1996) 223.

Infrared spectroscopy of comets with ISO: what we learned on the composition of cometary dust

J. Crovisier[a], * T.Y. Brooke[b], K. Leech[c], D. Bockelée-Morvan[a], E. Lellouch[a], M.S. Hanner[b], B. Altieri[c], H.U. Keller[d], T. Lim[e], T. Encrenaz[a], A. Salama[c], M. Griffin[f], T. de Graauw[g], E. van Dishoeck[h] and R.F. Knacke[i]

[a]Observatoire de Paris-Meudon, F-92195 Meudon, France

[b]Jet Propulsion Laboratory, Pasadena, CA, USA

[c]ISO Data Centre, Astrophysics Division of ESA, Villafranca, Spain

[d]MPI fur Aeronomie, Katlenburg-Lindau, Germany

[e]Rutherford Appleton Laboratory, Didcot, Oxford, UK

[f]Queen Mary and Westfield College, Chilton, UK

[g]SRON, Groningen, The Netherlands

[h]Leiden Observatory, The Netherlands

[i]Penn State University, Erie, PA, USA

The Infrared Space Observatory (ISO) offered us the opportunity to observe the spectrum of celestial bodies from 2.4 to 196 μm. This spectral range is of particular interest for the study of cometary dust. We report here on the ISO observations of comets C/1995 O1 (Hale-Bopp) and 103P/Hartley 2, and discuss the nature of cometary dust.

1. COMET C/1995 O1 (HALE-BOPP)

The apparition of the long-period, exceptional comet C/1995 01 (Hale-Bopp) coincided with the ISO operation. The spectrum of this comet was observed as part of a target-of-opportunity programme by an international team. Observations took place pre-perihelion on April 1996 ($r_h = 4.6$ AU) and September-October 1996 ($r_h \approx 2.9$ AU), and post-perihelion on December 1997 ($r_h = 3.9$ AU) and April 1998 ($r_h = 4.9$ AU) [1,2-4]. Unfortunately, visibility constraints of the satellite precluded observations at smaller heliocentric distances.

High-resolution spectra with high signal-to-noise ratios in the thermal infrared region were obtained on September-October 1996 with the SWS instrument (Figure 1). These

*This work is based on observations with ISO, an ESA project with instruments funded by ESA Member States (especially the PI countries: France, Germany, the Netherlands and the United Kingdom) and with participation of ISAS and NASA.

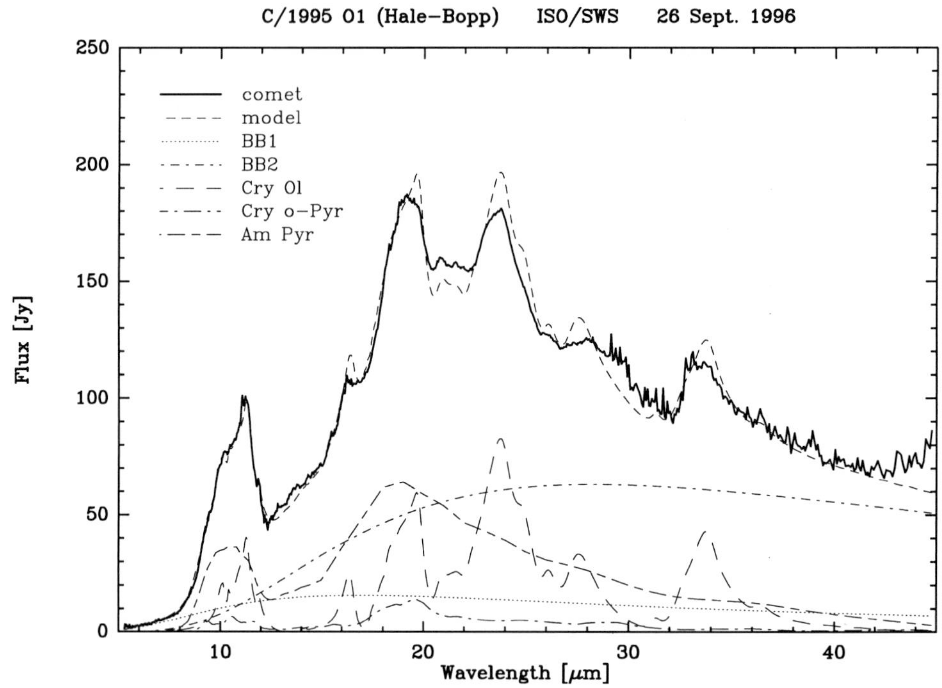

Figure 1. C/1995 O1 (Hale-Bopp) on 26 September 1996 at $r_h = 2.9$ AU. The 6 to 45 μm region of the SWS spectrum is represented here degraded to a resolution $\delta/\delta\lambda = 500$ (full dark line). The instrument apertures range from 14"×20" (short wavelengths) to 20"×33" (long wavelengths). The intensity scale has been normalized to a 14"×20" aperture assuming a l/ρ brightness distribution law and slight scale adjustments have been applied to make the spectrum continuous. As explained in [5], the observed spectrum has been decomposed into several emission components: blackbody 1 and 2, forsterite, ortho-pyroxene and amorphous silicates.

spectra show broad emission features superimposed over continuum thermal emission: the strongest appear at ≈10, ≈19.5, 23.5 and 33.5 μm, and weaker ones are present at 11.9, 16.25, 20.8, 21.6, 26.1 and 28.1 μm. The wavelengths of all these peaks correspond to crystalline, Mg-rich olivine (forsterite). A careful study of the spectra shows also signs of the presence of crystalline pyroxenes and amorphous silicates [1,5,2–4]. The contributions of these various components have been quantitatively evaluated [5] (see also [6–9]). There is no indication of change in the dust composition with heliocentric distance, or before and after perihelion.

Crystalline water ice is detected in spectra observed at 2.8 AU pre-perihelion, from emission features at 45 and 65 μm and possibly an absorption feature at 3.1 μm [10]. This suggests that grains of water ices are still present at this distance from the Sun, and that they could contribute significantly to the release of water in the coma. No sign of PAHs could be found in the spectrum of comet Hale-Bopp (especially in the 6–9 μm region, where they are expected to have strong bands not blended with silicate features, and where they are prominent in circumstellar dust discs).

2. COMET 103P/HARTLEY 2

The short-period comet 103P/Hartley 2 was observed in December 1997 to January 1998, close to its perihelion at 1 AU, as part of a guaranteed time programme [5,11] (see also [12]). Whereas Hale-Bopp was a long-period comet presumably coming from the Oort cloud, 103P/Hartley 2 is a Jupiter-family comet that supposedly originated from the Edgeworth-Kuiper belt. A low-resolution 5.5–17 μm spectrum obtained with the CVF of the ISOCAM instrument reveals a silicate feature much less intense relative to continuum than in comet Hale-Bopp. The presence of crystalline silicates is also tentatively indicated from the shape of its spectrum in the 8-13 μm region.

Cometary dust has thus striking similarities (composition, crystallinity) with Solar System interplanetary dust and circumstellar dust observed in both evolved and new stars [13,14]. The differences between comet Hale-Bopp and comet P/Hartley 2 may be ascribed to their belonging to the 'dust-rich' and 'dust-poor' comet classes, respectively: the dust-rich comet Hale-Bopp has very small, super-heated dust grains, whose silicate band emission is enhanced.

REFERENCES

1. J. Crovisier, T.Y. Brooke, M.S. Hanner, H.U. Keller, P.L. Lamy, B. Altieri, D. Bockelé-Morvan, L. Jorda, K. Leech and E. Lellouch, Astron. Astrophys. 315 (1996) L385.
2. J. Crovisier, K. Leech, D. Bockelée-Morvan, T.Y. Brooke, M.S. Hanner, B. Altieri, H.U. Keller and E. Lellouch, Science 275 (1997) 1904.
3. J. Crovisier, K. Leech, D. Bockelée-Morvan, T.Y. Brooke, M.S. Hanner, B. Altieri, H.U. Keller and E. Lellouch, in First ISO Workshop on Analytical Spectroscopy (eds. A.M. Herras, K. Leech, N.R. Trams and M. Perry) ESA SP-419 (1997) 137.
4. J. Crovisier, K. Leech, D. Bockelée-Morvan, E. Lellouch, T.Y. Brooke, M.S. Hanner, B. Altieri, H.U. Keller and T. Lim, in The Universe as seen by ISO (eds. P. Cox and M.F. Kessler) ESA SP-427 (1999) 137.
5. J. Crovisier, T.Y. Brooke, K. Leech,D. Bockelée-Morvan, E. Lellouch, M.S. Hanner, B. Altieri, H.U. Keller, T. Lim, T. Encrenaz, A. Salama, M. Griffin, T. de Graauw, E. van Dishoeck and R.F. Knacke, in Thermal Emission Spectroscopy and Analysis of Dust, Disks, and Regoliths (eds. M.L. Sitko, A.L. Sprague and D.K. Lynch) ASP Conf. Series 196 (2000) 109.
6. J.R. Brucato, L. Colangeli, V. Mennella, P. Palumbo and E. Bussoletti, Planet. Space Sci. 47 (1999) 773.
7. L. Colangeli, J.R. Brucato, L. Ferrini, V. Mennela, P. Palumbo and E. Bussoletti, Adv. Space Res. 23 (1999) 1243.
8. D.H. Wooden, D.E. Harker, C.E. Woodward, H.M. Butner, C. Koike, F.C. Witteborn, C.W. McMurtry, Astrophys. J. 517 (1999) 1034.

9. E. Grün, M.S. Hanner, S.B. Peschke, T. Muller, H. Boehnhardt, T.Y. Brooke, H. Campins, J. Crovisier, C. Delahodde, I. Heinrichsen, H.U. Keller, R.F. Knacke, H. Krüger, P. Lamy, C. Leinert, D. Lemke, C. Lisse, M. Müller, D.J. Osip, M. Solc, M. Stickel, M. Sykes, V. Vanysek and J. Zarnecki, Astron. Astrophys. 377 (2000) 1098.
10. E. Lellouch, J. Crovisier, T. Lim, D. Bockelée-Morvan, K. Leech, M.S. Hanner, B. Altieri, B. Schmitt, F. Trotta and H.U. Keller, Astron. Astrophys. 339 (1998) L9.
11. J. Crovisier, T. Encrenaz, E. Lellouch, D. Bockelée-Morvan, B. Altieri, K. Leech, A. Salama, M. Griffin, T. de Graauw, E. van Dishoeck, R. Knacke and T.Y. Brooke, in The Universe as seen by ISO (eds. P. Cox and M.F. Kessler) ESA SP-427, 161 (1999).
12. L. Colangeli, E. Epifani, J.R. Brucato, E. Bussoletti, C. de Sanctis, M. Fulle, V. Mennela, E. Palombo, O. Palumbo and A. Rotundi, Astron. Astrophys. 343 (1999) L87.
13. J. Crovisier, in IS0 beyond the Peaks, ESA SP-456 (2000) 23.
14. M.S. Hanner, this volume.

A search for trends in cometary dust emission

C.M. Lisse[a], M.F. A'Hearn[b], Y.R. Fernandez[c] and S.B. Peschke[d]

[a]Space Telescope Science Institute, Servicing Mission Office, 3700 San Martin Drive, Baltimore, MD 21218, USA.

[b] University of Maryland, Department of Astronomy, College Park, MD 20742, USA.

[c] Institute for Astronomy, University of Hawaii, 2680 Woodlawn Drive, Honolulu, HI 96822, USA.

[d]ISO Data Centre, Villafranca Del Castillo, ESA Satellite Tracking Station, P. O. Box (Apdo.) 50727, E-28080 Madrid, Spain.

We present the results of searching for statistical trends in the properties of dust emitted by 9 comets as determined using mid-IR observations. Trends in the aggregate dataset of dust PSD, total emission rate, and emission rate vs time were found versus dynamical class, with the short period comets slowly emitting mainly large, dark dust particles while the LP comets emitted most of their dust surface area (but not mass) rapidly in small, high albedo particles. These differences may be due to the effects of cometary evolution on the structure of the cometary surface and/or the depletion of cometary volatiles, although caveats concerning our currently small (but growing) dataset and the role of potentially important selection effects in our results. These results are consistent, however, with a larger sample of 46 infrared comet observations found in the literature.

1. INTRODUCTION

Studies of the physical properties of cometary dust are important for understanding the phenomena of comets as a class, and for understanding the formation and evolution of comets and the solar system. The quantities we can derive from infrared observations of the dust include the dust composition, mass, particle size distribution, and emission history. Given a large enough statistical population of observations, we can use the results of the measurements to understand the role of comets in the present day solar system, including, *e.g.*, the contribution of cometary dust to the interplanetary dust cloud, and the evolution and fate of highly aged comets. Since comets are among the most primitive bodies in the inner solar system, despite their evolution, we also want to use the results to understand the formation of the solar system from the proto-solar nebula, *e.g.*, the total and relative abundance of rock forming material, the formation and evolution of the accreting icy planetesimals (comets), and chemical variations in the proto- solar nebula. We hope to learn more about the aging process of comets, *e.g.*: the effect of weathering and mantling on the cometary surface and the effect of collisions on the cometary population.

Combined thermal infrared and optical observations are sensitive to the 0.1 - 100 um range of dust particle sizes [1, 2, 3]. The presence of a strong 8 - 13 μm silicate emission feature is also an indicator of relatively small dust grains ($\leq$ 10 μm), as is a spectral color temperature elevated more than 10% above the local equilibrium temperature, or a negative deviation from greybody behavior at 30 - 100 μm [4]. A total albedo for scattering, defined as the ratio of the scattered luminosity to the total luminosity observed at a given phase angle, is ~6% for large cometary particles (*e.g.*, comets Austin and Encke) and >12% for small particles (*e.g.* comets Levy, Hyakutake, and Hale-Bopp). Extensions of the dust tail morphology along the projected orbital velocity direction (*i.e.*, "trails") are due to large, heavy particle emission, while morphologies with only anti-solar tails are indicators of small particle emission [3]. Trends in long term light curves yield independent estimates of the particle size distribution and emission rate [1, 2].

There is growing evidence, from the Giotto encounter with the P/Halley nucleus [5], the IRAS observations of C/IRAS-Araki-Alcock 1983H1 [6], the COBE/DIRBE large angular scale observations of 4 comets [1], and the sub-mm observations of comet C/Hyakutake 1999 B2 and C/Hale-Bopp 1995 O1 [7, 8], that the majority (by mass) of grains emitted by comets are visibly dark and large, requiring the use of mid- to far-IR observations to probe the nature of cometary dust. This conclusion is very different from the current paradigm created mainly from large optical surveys of periodic comets [9, 10], which concludes that the majority of cometary dust mass is emitted by the optically bright comets.

2. OBSERVATIONS

Our study of the statistical properties of cometary dust emission began with the large-scale photometric imaging of 4 comets detected in the COBE/DIRBE all-sky survey of 1989 – 1990. These comets were selected only by our ability to detect them against the background in the COBE all sky survey, down to V ~ 12. As summarized [1], C/OLR and C/Austin were found to emit only large, dark dust grains not detected in the visible, while C/Levy was more Halley-like in its emission of many small dust grains. That the comet 73P/SW-3 was detected at all by COBE was quite surprising, as it was optically fainter than 4 other short period comets that were not detected; it was argued [1] that this was due to the enhanced release of large, dark particles due to the incipient breakup of the comet during its next perihelion passage.

The present paper adds 5 more comets to our database of comets well-characterized using multi-wavelength photometry in the thermal infrared (Table 1). Included are the extremely bright long-period comets of 1996 and 1997, C/Hyakutake 1996 B2 [11] and C/Hale-Bopp 1995 O1 [2, 12]; and the close-approaching periodic comets P/IRAS 1996, P/Encke 1997 [3], and P/Tempel-Tuttle 1998. Not included are comets like P/Wild2 1997, C/Utsunomiya 1997 T1, and C/LINEAR 1998 U5, which were observed by our group but only detected in broad-band N [13]. The reader is referred to the individual observational papers listed for each comet for the details of the observations and the photometric calibration.

3. MODELS

Results from the comet observations of the COBE mission [14, 1], have shown that simultaneous modeling of the shape and scattering/thermal emission spectrum of a comet's dust tail can constrain the silicate:carbonaceous composition ratio and particle size distribution of the dust. Our methods of spectral, dynamical, and temporal modeling of the thermal emission from the cometary dust have been extensively summarized in [1], where we demonstrate the ability to derive absolute dust production rates, dust-to-gas ratios, and rough compositional information for 4 comets using 1 - 300 μm multiwavelength broad-band infrared photometric imaging obtained by the COBE/DIRBE instrument in 1989 - 1990. [2] demonstrates our ability to use the same modeling techniques for comet C/Hale-Bopp from ground based 1 - 20 μm photometry.

We present here the aggregate results of modeling our IR observations of 9 different comets (Table 1, Figure 2). Our broad-band multi-wavelength photometry observations can be explained by dust with similar silicate:carbonaceous composition ratios of 2 to 3 : 1, but two very different dust particle size distributions (PSD's), the [15] and the $m^{-0.63}$ (or $1/\beta$; [1]) distributions. While both these PSD's are mass dominated by the largest (radius > 100 um)

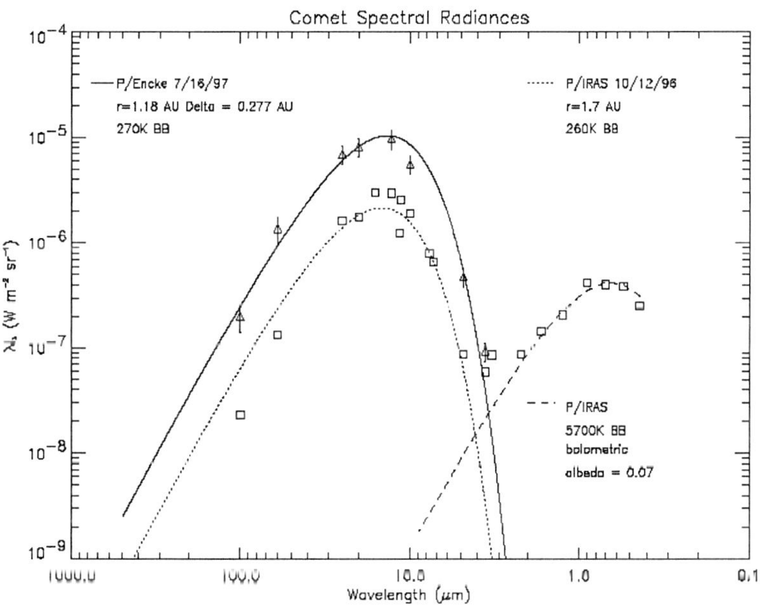

Figure 1. Example of small and large particle dominated dust emission infrared spectral observations. a) Spectral energy distribution of comet P/IRAS 1996 as taken by ISO from 3-100 μm, at the NASA/IRTF from 1-3 μm, and at the Lowell 42" on October 13, 1996 [3]. Squares - ISO observations with 20% error bars. Dotted Curve - best-fit greybody to the infrared data, 20% hotter than the local equilibrium temperature, due to the presence of small, hot grains. From 8 to 13 μm, there is a strong emission feature due to silicates in small dust grains. At 60 and 100 μm, there is a strong fall off from the blackbody curve due to decreasing emission efficiency from small grains at large wavelengths. Dashed Curve - solar spectrum normalized to ISO observation at 1.25 μm. There is good agreement between the observed scattered light spectrum and a blackbody of 5700 K, the average temperature of the Sun's emission. b) Spectral energy distribution of comet P/Encke 1997 as taken by ISO from 3-100 μm on July 16, 1997, and at 5-20 μm at the ESO 3.6m on July 18, 1997 [3]. Triangles - ISO observations with 20% error bars. Solid curve - best fit greybody to the emission. The best fit greybody temperature is very close to the local equilibrium temperature, a strong indication that only large grains are being observed. There is no evidence of a silicate emission feature or long wavelength emissivity falloff.

Table 1
Emitted dust properties for the 9 comets in the IR survey.

Comet	Time (UT Day)	r (AU)	Tc/Tbb	Dust Albedo (%)	Silicate Strength (%)	Q_{Dust} (kg/sec)	Q_{gas} (mol/sec)	Dust/Gas
C/OLR (New)*	1989 Dec 20	1.0	1.06	---	< 10	450	4×10^{28}	0.32
C/Austin (New)*	1990 May 15	0.94	1.08	6	~10	2400	1×10^{29}	0.79
73P/SW-3 (SP)*	1990 Apr 15	1.07	1.05	---	<20	~120	$<3 \times 10^{28}$	> 0.12
C/Levy (LP)†	1990 Sep 20	1.12	1.24	12	>30	6100	2×10^{29}	1.0
C/Hyakutake(LP)†	1996 Mar 25	1.05	1.20	13	>50	9000	2×10^{29}	1.7
126P/IRAS (H)†	1996 Oct 13	1.71	1.25	16	>35	300	3×10^{27}	3.3
C/Hale-Bopp(LP)†	1996 Oct 31	2.54	1.30	39	>200	140000	2×10^{31}	5.0
2P/Encke (SP)*	1997 Jul 15	1.18	1.03	5	<10	120	2×10^{27}	2.3
55P/Tempel-Tuttle (SP)*	1998 Jan 21	1.15	1.03	5?	< 20	210	3 x 1027	2.9

* - Large particle dominated comet. † - Small particle dominated comet. - q > 5 AU comet. N.B.: All Q_{dust}, D/G estimates are conservative, assuming minimum largest particle, with factor of +100% /-50% errors, with the dominant systematic uncertainty in the dust outflow velocity.

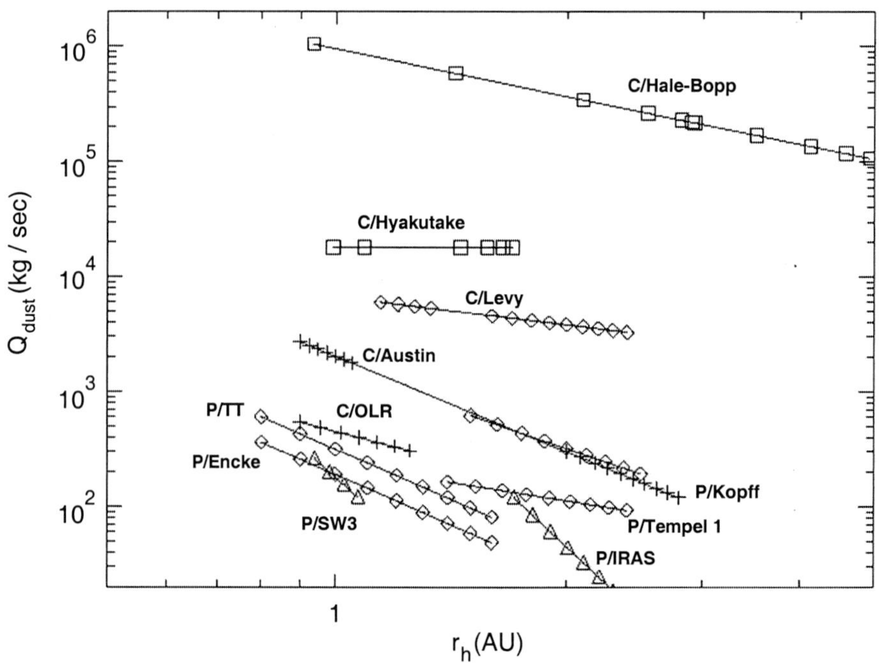

Figure 2. Derived mid-infrared dust emission rate vs heliocentric distance for the 9 comets in our survey. There is a general trend of steeper slope and lower absolute emission rate for periodic comets versus new or long period comets. A possible selection effect that may affect the results is our ability to detect and observe fainter periodic comets due to their known positions and orbits - although the 4 comets from the COBE dataset were detected in an unbiased all-sky survey.

particles, the emitted surface area (and optical brightness) is dominated by either 0.1 - 10 μm particles in "dusty comets" like 1P/Halley or > 10 um particles in "gassy comets" like 2P/Encke. The morphology of the coma and tail for the dusty comets is typically "normal", *i.e.* dominated by a bright, linear tail pointing roughly in the anti-solar direction; this tail is absent or weak and in addition to a curved tail pointing along the direction of orbital motion in the "gassy" comets.

4. RESULTS

Of the nine new well-measured comets in our infrared survey of cometary dust emission, the intermediate-aged long period comets C/Hyakutake 1996B2, P/IRAS 1996, and C/Hale-Bopp 1995O1 demonstrated high rates of dust emission dominated by small particles, including pronounced silicate features, long-wavelength emissivity falloffs, and large superheats, while the extremely processed short period comets P/Schwassmann-Wachmann 3, P/Encke, and P/Tempel-Tuttle emitted mainly large dust particles at relatively low rates. The dust from Comet Hale-Bopp was slightly anomalous in that the particles seemed especially dense and small [2], although of normal composition compared to other comets. Figure 3 presents the strong correlations between the temperature, albedo, and silicate strength of the emitted dust (brighter dust is hotter and has a larger silicate feature), between the dust production rate and the temperature of the dust (hotter dust is produced more rapidly), and the strength of the silicate feature and the Tisserand invariant (largest silicate feature and smallest dust particles are emitted by the LP comets) for the 9 comets in our survey.

Taken in aggregate, there appears to be a real difference in the particle size distributions for our 9 comet dataset of the dust emitted by the new and short period comets, which emit mainly large, dark dust particles, versus the small, bright dust emitted by the long period comets. The new comets also appear to have lower D/G ratios than the others.

Comparison of our results to the two classes of comets found in the optical polarimetric survey of Levasseur-Regourd [16] shows an excellent match of our large particle dominated emission with their "gassy, low polarization comets" and our small particle dominated emission with their "dusty, high polarization comets". This is consistent with the expectation that small particles have a higher albedo and polarization per unit dust mass [4, 12]. Hale-Bopp stands out as having unusually high polarization and albedo, which is consistent with our models showing it had unusually solid, small dust [2]. The strength of the 8 - 13 μm silicate emission feature versus the continuum is known to be inversely proportional to the dust particle size [17]; for the 5 of our comets also observed in the silicate feature at 8 – 13 μm by Hanner *et al.* [18], we find good agreement between our assignment of large or small particle dominated emission and their finding of large or small amplitude of the silicate feature.

Concern that our results may be preliminary due to small sample size and selection effects led us to search the literature for other infrared comet observations. (Note that while the 5 new comets were selected on the basis of their expected brightness and known orbits, the original 4 comets of the database, detected in the COBE all sky survey of 1989 – 1990, were selected only by their IR brightness against the background.) In Table 2 we present the results of searching for thermal IR observations of comets in the literature (*c.f.* [19] for references), ordered by whether small or large particle emission was evident via the superheat/silicate feature strength/long wavelength emissivity falloff criteria, and ordered within these two groups by perihelion distance. For the more than 40 literature comets observed, we again find

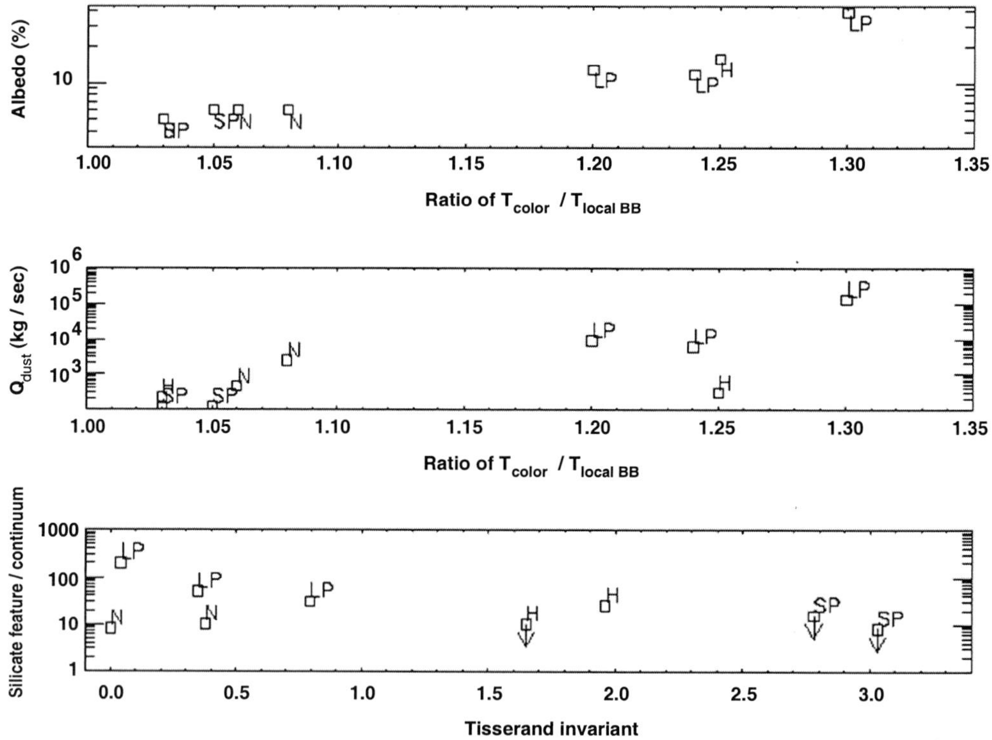

Figure 3. A search for correlations in the observations of emtted cometary dust by dust albedo, silicate feature ampltude/continuum ratio, dust color temperature, Qdust, and Tisserand invariant of the parent comet. a) Comparison of the dust temperature and albedo (not shown is a correlation of the dust albedo and silicate feature strength, which demonstrates a similarly strong correlation); b) of the dust production rate and the temperature of the dust; and c) of the strength of the silicate feature and the Tisserand. There are clear trends and distinct groups by dynamical class – the LP comets produce the most dust with the hottest color temperature, highest albedo, and largest silicate feature. The New and Short Period comets produce little dust and what they do produce is cold, low albedo, and without much silicate feature. The Halley comets are between the two extremes.

Table 2.
Cometary dust size distributions well observed to date in the mid-IR.

Proper Name	Tperi	q (AU)	e	P (yr)	i (deg)	Tisserand	Dynamical Type
Large Particle Dominated							
2P/Encke	1994 Feb 9	0.33	0.85	3.28	11.9	3.03	SP, old
23P/Brorsen-Metcalf	1989 Sept 11	0.48	0.97	70.5	19.3	1.12	H
73P/Schwassmann-Wachmann3	1995 Sept 22	0.93	0.69	5.34	11.4	2.78	SP, old
103P/Hartley 2	1991 Sept 12	0.95	0.72	6.26	9.25	2.64	SP, young
55P/Tempel-Tuttle	1965 Apr 30	0.98	0.90	32.9	163.	1.65	H
26P/Grigg-Skjellerup	1992 July 22	0.99	0.66	5.10	21.1	2.81	SP, mid
107P/Wilson-Harrington	1992 Aug 22	1.00	0.62	4.29	2.79	3.09	SP
21P/Giacobini-Zinner	1992 Apr 13	1.03	0.71	6.61	31.8	2.47	SP, mid
46P/Wirtanen	1991 Sept.20	1.08	0.65	5.50	11.7	2.82	SP, young

Table 2 continued

Proper Name	Tperi	q (AU)	e	P (yr)	i (deg)	Tisserand	Dynamical Type
24P/Schaumasse	1993 Mar 4	1.20	0.70	8.22	11.8	2.51	SP, old
7P/Pons-Winnecke	1989 Aug 20	1.26	0.63	6.38	22.3	2.69	SP,Middle
P/1983 V1 Hartley-IRAS	1984 Jan 9	1.28	0.83	21.5	95.7	0.809	SP
67P/Churyumov-Gerasimenko	1989 June 18	1.30	0.63	6.59	7.12	2.75	SP,young
10P/Tempel 2	1994 Mar 16	1.48	0.52	5.48	12.0	2.97	SP, old
9P/Tempel 1	1994 July 3	1.49	0.52	5.50	10.6	2.97	SP, old
28P/Neujmin 1	1984 Oct 8	1.55	0.78	18.2	14.2	1.55	H
81P/Wild 2	1990 Dec 16	1.58	0.54	6.37	3.24	2.88	SP, young
22P/Kopff	1996 July 2	1.58	0.54	6.45	4.72	2.87	SP, old
43P/Wolf-Harrington	1991 Apr 4	1.61	0.54	6.51	18.5	2.80	SP, mid
C/1989 X1 Austin	1990 Apr 10	0.35	1.000225		59.0	0.375	New
C/1975 N1 Kobayashi-Berger-Milon	1975 Sept 5	0.43	1.000097		80.8	0.128	New
C/1974 C1 Bradfield	1974 Mar 18	0.50	0.999697		61.3	0.425	New
C/1989 Q1 OLR	1989 Nov 12	0.64	1.000020		90.1	0.002	New
C/1983 H1 IRAS-Araki-Alcock	1983 May 21	0.99	0.990115		73.3	0.407	LP
C/Wilson C/1986 P1-A	1987 Apr 21	1.20	1.000312		147.	1.14	New
C/1983 J2 IRAS	1983 Jan 19	1.42	1.0		152.	9999	New?
C/1980 Y2 Panther	1981 Jan. 27	1.66	0.998974		82.6	0.207	LP?
C/1983 O2 IRAS	1983 Nov. 28	2.25	1.000196		121.	0.951	New
C/1980 E1 Bowell	1982 Mar. 12	3.36	1.057322		1.66	2.22	New
C/1983 O1 Cernis	1983 July 21	3.32	1.001968		135.	1.59	LP
Total : 16 SP, 3 H, 8 N, 3 LP							
Small Particle Dominated							
1P/Halley	1986 Feb 9	0.59	0.97	76.0	162	1.19	H
126P/1983 M1 IRAS	1983 Aug 2	1.70	0.70	13.2	46.2	1.96	H
19P/Borrelly	1994 Nov 1	1.37	0.62	6.88	30.3	2.57	SP,Young
38P/Stephan-Oterma	1980 Dec 5	1.57	0.86	37.7	18.0	1.89	H
4P/Faye	1991 Nov 16	1.59	0.58	7.34	9.09	2.75	SP,Young
D/1993 Shoemaker-Levy 9	1994 Mar 28	5.38	0.21	17.8	5.89	2.99	SP §
29P/Schwassmann Wachmann 1	1989 Oct 26	5.77	0.045	14.9	9.37	2.99	SP §
P/Chiron	1996 June 15	8.46					SP §
C/1973 E1 Kohoutek	1973 Dec 28	0.14	1.000008		14.3	0.453	New
C/1975 V1-A West	1976 Feb 25	0.20	0.999971		43.1	0.402	LP
C/1996 B2 Hyakutake	1996 May 1	0.23	0.999758		125.	0.346	LP
C/1969 Y1 Bennett	1970 Mar 20	0.54	0.996193		90.0	0.038	LP
C/1987 P1 Bradfield	1987 Nov 7	0.87	0.994736		34.1	0.987	LP
C/1995 O1 Hale-Bopp	1997 Apr 1	0.91	0.995069		89.4	0.040	LP
C/1990 K1 Levy	1990 Oct 25	0.94	1.000417		132.	0.795	LP
C/1993 A1 Mueller	1994 Jan 13	1.94	1.001779		125.	0.982	LP
C/1983 N1 IRAS	1983 May 3	2.42	0.999420		139.	1.45	LP
Total : 4 SP, 3 H, 1 N, 8 LP							

Well Observed = Photometry From at least > 5 - 20 μm, or 8-13 μm Using Spectroscopy
Large Particle Dominated = Silicate Emission/Continuum < 10% and/or $T/T_{bb} < 1.10$
Small Particle Dominated = Silicate Emission/Continuum > 10% and/or $T/T_{bb} > 1.10$
SP = Short period comet; LP = Long period comet, H = Halley Family, New – Dynamically New Comet
P = Photometric Observation, S = Spectroscopic Observation, I = Imaging Observation, § = q > 5 AU
Dynamical type - From [18]

the same distinction in the emitted dust between the SP + New and the LP comets. The Halley family comets seem to be split in their behavior.

There is no obvious trend in i, P, or e in Table 2. The only obvious trend in q is that all comets with $q > 5$ AU have small particle dominated emission. Using estimated nuclear radii from the literature [20, 13], we see no obvious trend with R_N. *E.g.*, LP comet Hale-Bopp, SP comets Chiron and SW-1 are all thought to be large, with varying levels of activity, while LP comets Levy and Hyakutake were as small as the average SP comet like Encke or Tempel-Tuttle. Thus any variability of the emitted dust due to nuclear structure differences dependent on size - like degree of melting and internal differentiation, and degree of heterogeneity - appear to be small.

5. DISCUSSION

What, then, leads to the very different dust emitted by the long-period group of comets? One possible answer is that the presence of small grains in a comet may be an indicator of the evolutionary state of a comet, with smaller grains originating mainly from comets with highly evolved mantles and jet activity on their surface created after numerous perihelion passages. (Whether the small particles are produced in the violent emission associated with jets or are due to chemical differences in the jets is not clear, however.) After many passages, though, the reservoir of volatile material runs out, and the jets turn off with age.

The similarity in the dust albedo for 73P/SW-3, 2P/Encke [3], 55P/Tempel-Tuttle, and that found for the 1P/Halley nuclear surface (p_V ~5% ; [5]), coupled with the preponderance of very large particles in the material emitted by the short period comets, suggest that the dust emitted is little modified from the that on the nuclear surface. I.e., large chunks of the comet's surface are being emitted slowly and relatively unchanged into the coma for the short period comets. In the other extreme of the dusty comets, we expect violent jet activity and/or the presence of ephemeral icy volatiles in the dust lead to changes in the emitted dust from that on the nuclear surface. Large increases in polarization and albedo in cometary jets, suggesting the production of excess small particles, has been found in the highly active comets Hale-Bopp [21] and Halley [22].

The evolutionary model also explains the increasing slope of the dust emission rate dependence with heliocentric distance for the more evolved short period comets (Figure 2), as the surface temperature has to be increased to higher values to drive off the denser mantle and/or scarcer volatiles in these comets. On the other hand, the large q short period comets, with surface temperatures unable to drive sublimation of the majority volatile H_2O, age very slowly and are locked into a state similar to the long-period comets.

A test of this paradigm can be found in the transition-case Halley family comets like P/IRAS, which seems to emit small dust typical of the long period comets. Since this comet has a relatively short period of 13.2 years, which will, if anything, most likely decrease along with the semi-major axis of the orbit as it becomes increasingly locked into the 2:1 resonance of the typical Jupiter family comet. Thus future monitoring of the dust emission from P/IRAS should show a change from small- to large-particle dominated emission as the comet ages. Another useful test will be an extended 1- 300 μm photometric imaging and 8 - 13 μm spectroscopic survey of a large number (> 50) of periodic and new comets using a single dedicated, well calibrated instrument, allowing us to improve the statistics and reduce the systematics of our search, such as will be possible with the upcoming SIRTF mission.

6. CONCLUSIONS

Our infrared observations of dust emission from 9 comets can be explained by particles with similar composition ratios but two very different PSD's. While both these PSD's are mass dominated by the largest (radius > 100 um) particles, the emitted surface area (and optical brightness) is dominated by either 0.1 - 10 um particles in "dusty comets" or > 10 um particles in "gassy comets". There is good agreement between the results of our small/large particle dominated classification and the high/low polarization comets of Levasseur-Regourd *et al.* [16] and the strong/weak silicate feature comets reported by Hanner *et al.* [18] and Williams *et al.* [12]. Trends in the aggregate dataset of dust PSD, total emission rate, and emission rate vs time were found versus dynamical class, with the short period comets slowly emitting mainly large, dark dust particles while the LP comets emitted most of their dust surface area (but not mass) rapidly in small, high albedo particles. These differences may be due to the effects of cometary evolution on the structure of the cometary surface and/or the depletion of cometary volatiles, although caveats concerning our currently small (but growing) dataset and the role of potentially important selection effects in our results. To partially address these caveats, we have found 46 infrared studies of cometary dust emission in the literature, which, when classified by cometary class, showed evidence of small or large particle dominated dust emission consistent with the trends found in our database.

ACKNOWLEDGEMENTS

This work was supported by NASA Grant NAG-53361 for ISO data analysis and by an ESO guest scientist grant in March 1997.

REFERENCES

1. C.M. Lisse, M.F. A'Hearn, M.G. Hauser, T. Kelsall, D.J. Lien, S.H. Moseley, W.T. Reach and R.F. Silverberg, Astrophys. J. 496 (1988) 971.
2. C.M. Lisse, C.M., Y. R. Fernandez, M. F. A'Hearn, T. Kostiuk, T. A. Livengood, H. U. Kaufl, W.F. Hoffmann, A. Dayal, M. E. Ressler, M. S. Hanner, G.G. Fazio, J.L. Hora, S. Peschke, E. Grün and L.K. Deutsch, Earth, Moon and Planets 78:1-3 (1999) 251.
3. C.M.Lisse, M.F. A'Hearn, Y.R. Fernandez, E. Grün, H.U. Kaufl, T.Kostiuk, D.J. Lien, D.J. Osip, S.B. Peschke and R.G. Walker, Icarus (in press)
4. C.G. Mason, R.D. Gehrz, E.P. Ney, D.M. Williams and C.E. Woodward, Astrophys. J. 507 (1998) 398.
5. H.U. Keller, W.A. Delamere, M.J. Reitsema, W.F. Huebner and H.U. Schmidt, Astron. Astrophys. 187 (1987) 807.
6. R.G. Walker, H.H. Aumann, J. Davies, S. Green, T. DeJung, J.R. Houck and B.T. Soifer, Astrophys. J. 278 (1984) L11.
7. D.C. Jewitt and H.E. Matthews, Astron. J. 113 (1997) 1145.
8. D.C. Jewitt, and H.E. Matthews, Astron. J. 117 (1997) 1056.
9. M.F. A'Hearn, R.L. Millis, D.G. Schleicher, D.J. Osip and P.V. Birch, Icarus 118 (1995) 223.
10. U. Fink and M.D. Hicks, Astrophys. J. 459 (1996) 729.
11. C.M. Lisse, Y. R. Fernández, A. Kundu, M. F. A'Hearn, A. Dayal, L.K. Deutsch, G.G. Fazio, W.F. Hoffmann and J.L. Hora, Icarus 140 (1999) 189.

12. D.M. Williams, C.G. Mason, R.D. Gehrz, T.J. Jones, C.E. Woodward, D.E. Harker, M.S. Hanner, D.H. Wooden, F.C. Witterborn and H. M. Butner. Astrophys, J. 489 (1997) L91.
13. Y.R. Fernandez, Physical Properties of the Nuclei of Comets, PhD Thesis (1999).
14. C.M. Lisse, H.T. Freudenreich, M.G. Hauser, T. Kelsall, S.H. Moseley, W.T.Reach and R.F. Silverberg, Astrophys. J. 432 (1994) L71.
15. J.A.M. McDonnell, P.L. Lamy and G.S. Pankiewicz, in Comets in the Post-Halley Era (eds. R.L. Newburn, M. Neugebauer and J. Rahe) Kluwer Academic, Dordecht (1991) 1043.
16. A.C. Levasseur-Regourd, E. Hadamcik and J.B. Renard, Astron. Astrophys. 313 (1996) 327
17. R.C. Gilman, Astrophys. J. 268 (1974) 397.
18. M.S. Hanner, D.K. Lynch and R.W. Russell, Astrophys. J. 425 (1994) 274.
19. C.M. Lisse, Infrared Observations of Cometary Dust by COBE, Phd Thesis (1992).
20. H.F. Levison and M.J. Duncan, in Origins of Solar Systems Workshop: The Origin, Evolution, and Detectability of Short period Comets, (SEE N94-11628 01-90) (1994) 1.
21. K.J. Meech, ACM 1996 Proceedings (in press)
22. T.J. Jones and R.D. Gehrz, Icarus 143 (2000) 338.
23. A.C. Levasseur-Regourd, E. Hadamcik and J.B. Renard, Astron. Astrophys. 348 (1999) 636

Evolution of cometary grains from studies of comet images

L. Kolokolova[a], B.Å.S. Gustafson[a], K. Jockers[b] and G. Lichtenberg[b]

[a]Dept. of Astronomy, University of Florida, Gainesville, FL, USA

[b]Max-Planck Institut für Aeronomie, Katlenburg-Lindau, D-37191, Germany

CCD colourimetric and polarimetric observations of comets, obtained using the two-channel focal reducer of MPAe, allowed to analyze change in colour and polarization with the distance from the nucleus that may indicate dust evolution in the coma. This paper concentrates on trends along the solar-antisolar direction for comets Hale-Bopp, Hyakutake, and Tabur in the innermost coma where the observed comets demonstrate a correlation between colour and polarization. We show that the observed correlation is most likely a result of evaporation/destruction of a dark (organic) material. The evolution and properties of the dust in the central coma of these comets are studied using a tentative model of cometary dust as an ensemble of core-mantle particles with sublimating mantles. Calculations for a variety of optical properties, core-mantle mass ratios and dust size distributions showed that this model could describe the observed change in colour and polarization in cometary comae at realistic ranges of input parameters. The best-fit particles have silicate cores of radius 0.05–0.2 μm with a porous organic-carbon mantle of the mantle/core mass ratio 0.5–1. Polarimetric colour can further restrict the range of satisfactory refractive indices of both the core and the mantle materials.

1. INTRODUCTION

Cometary grains evolve with the distance from the nucleus and are not the same in the regular coma as in jets, fans, shells, and in other specific cometary features. Non-imaging (aperture) observations refer to averages across the coma and can therefore only provide average characteristics for fresh, evolved, regular and peculiar particles mixed together. They cannot be used to obtain realistic characteristics of any specific comet dust populations. Although the importance of images in studying cometary dust has never been brought into question, studies usually concentrate on brightness images. These reveal the dust-filling factor and are therefore well suited to study dust distributions throughout the coma and yield at least relative dust concentrations. Here, we take advantage of the fact the dust concentration dependence can be removed through a simple normalization inherent in brightness ratios such as polarization or colour. We analyze colourimetric and polarimetric images of comets Hyakutake, Tabur and Hale-Bopp to reveal properties and evolution of cometary dust.

2. OBSERVATIONS

The observations of comets C/1996 Q1 (Tabur) and C/1995 O1 (Hale-Bopp) were made at the 2 meter RCC Zeiss telescope at the Pik Terskol Observatory (Northern Caucasus) using the Max-Planck-Institut für Aeronomie Two-channel Focal Reducer [1]. The red (643 nm) and blue (443 nm) narrow-band continuum filters were used. Comet C/1996 B2 (Hyakutake) was observed using the 1m Cassegrain reflector at the Hoher List Observatory near Bonn, Germany and a somewhat improvised focal reducer that employed in a single channel setup. Details of the observational conditions, processing of the images and the complete images of the comets are presented in [5] and [7]. In this paper we pay attention to the solar-antisolar trends (Figures 1 and 2) concentrating on the innermost coma ($R \leq 1000$ km for comets Hyakutake and Tabur and $R \leq 5000$ km for comet Hale-Bopp).

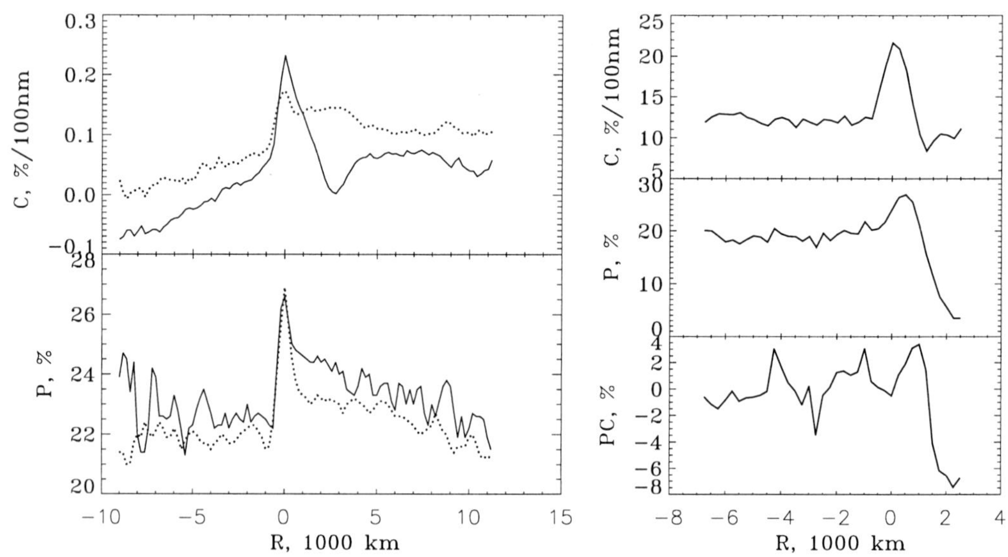

Figure 1. Profiles of colour, C, and polarization, P, for comet Hyakutake (left panel, the solid line shows the data for March 30, 1996 and the dotted lines are for April 1, 1996); colour, polarization, and polarimetric colour, PC (the difference in polarization observed at 642 and 443 nm), for comet Tabur (right panel, October 10, 1996) done along the solar(+)/antisolar(−) direction.

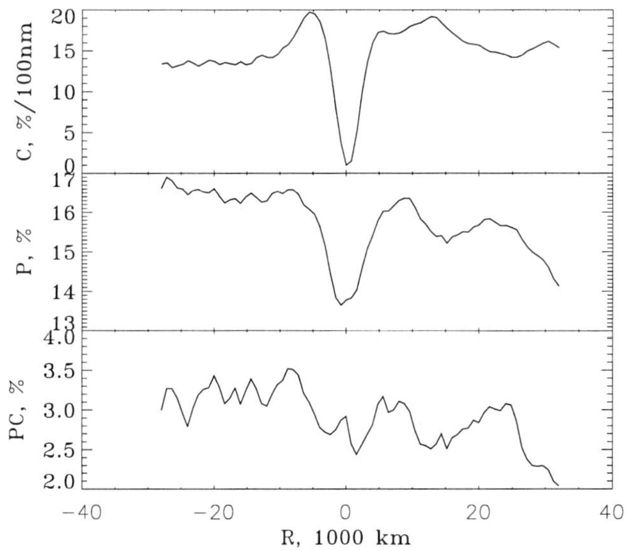

Figure 2. Profiles of colour, C, polarization, P, and polarimetric colour, PC (the difference in polarization observed at 642 and 443 nm), for comet Hale-Bopp (March 31, 1997) done along the solar(+)/antisolar(−) direction.

3. TENTATIVE EXPLANATION FOR THE OBSERVED TRENDS IN COLOUR AND POLARIZATION

From Figures 1 and 2 one can see that all the observed comets show a correlation between colour and polarization in the innermost coma. The correlation could be due to the following changes in the dust particles:

1. **Grain sizes decrease as a result of evaporation or particle break-up.** It is well known that particles much smaller than the wavelength (Rayleigh particles) have blue colour and show maximum polarization of $P = 100\%$ at 90° phase angle. As the particle dimensions grow further and no longer are negligible in comparison to the wavelength, the polarization decreases. Also, the colour starts to increase and becomes 'more red'. Thus, a consequence of a gradually changing particle size is an anti-correlation between colour and polarization.
2. **The composition changes as volatile materials evaporate or are otherwise destroyed.** Cometary dust composition changes, most likely, due to the sublimation of ice or the evaporation/destruction of some organics. Absorption increases with increasing wavelength for water ices and decreases for organics [6]. That translates in changing colour toward red as ice sublimates and toward blue during decomposition of organics. At the same time, polarization increases as water ice evaporates and decreases as organics disintegrates because the change in polarization is related to the change in absorption [6]

As a result, a positive correlation between colour and polarization should be observed.
3. The porosity increases as embedded volatiles evaporate. Using microwave analog measurements we studied light-scattering by aggregates of varying packing [4]. No significant influence of compactness on either colour or polarization in the range of packing factor (the ratio of the volume occupied by the material to the total volume) 10–50 % was found.

Thus, the observed correlation between colour and polarization can be only explained by changing particle composition due to sublimation of volatiles.

4. CORE-MANTLE MODEL OF COMETARY GRAINS

For a quantitative analysis we simulate cometary dust as aggregates of core-mantle particles packed so loosely that interactions between constituent grains can be neglected. This is reasonable since the expected packing factor should be near 10% (see [2], [3]). We suppose that there is some size distribution of the cores of the dust grains, which has maximum at a specific grain radius (we consider r_0=0.05, 0.1, 0.2, 0.5 and 1 μm) and is normally distributed around this maximum with some deviation σ (we consider σ=0.1, 0.5 and 1). To specify the mass ratio of organic to silicate material, we use the estimate based on the cosmic abundances [3] and consider that the mass ratio of organic to silicate can be $M_{o/s}$=0.5, 1 and 2. Input parameters include also the complex refractive index $m = n+i\kappa$ for silicates and for organics. We consider a range of refractive indices: from 1.5+i0 to 1.5+i0.5 for the silicate core and from 1.3+i0 to 1.7+i1 for the mantle, covering materials from ice to carbon for the mantle and silicates of different absorption for the core. Output parameters of the calculations are colour, C, polarization, P, and polarimetric colour, PC, defined at the phase angle 90° and at the blue (443 nm) and red (642 nm) filters. We calculated the initial mantle thickness based on the size distribution of the cores and on the organics to silicate mass ratio. We then obtained the presublimation values of C, P, and PC from the core-mantle light scattering calculations [9]. The calculations were repeated assuming a given mantle recession rate. We thus calculated the thickness of all mantles and new light-scattering properties of the core-mantle particles at each time step and obtained the change in colour, polarization and polarimetric colour. Table 1 presents the input parameters that provide the observed changes in the output parameters.

5. CONCLUSIONS

The trends in colour, polarization and polarimetric colour in the comae of comets Hale-Bopp, Hyakutake and Tabur have been studied. The main result is a correlation between colour and polarization in the innermost coma. The correlation can be due to changing the grain material during the sublimation of volatile constituents of the dust. Aggregated core-mantle particles whose mantles get thinner as the grains recede from the nucleus is a possible model of cometary grains. The observed correlations can be obtained from calculations of the light scattering by core-mantle particles at:

- the size of the cores peaks around 0.05–0.2 μm;
- the mantle material is absorptive and can be a porous organic-carbon material (more porous for comets Tabur and Hyakutake than for comet Hale-Bopp) with the

Table 1
Characteristics of near-nucleus dust particles, results of core-mantle calculations

Size distribution		Refractive index		Mass ratio,	Correlation between colour, C,
r_0	σ	core	mantle	$M_{o/s}$	and polarimetric colour, PC
Comets C/1996 B2 Hyakutake and C/1996 Q1 Tabur					
0.05	0.1	1.5 + i[0-0.1]	1.3 + i0.1	0.5-1	$C \sim 1/PC$
			1.5 + i0.5	0.5-1	$C \sim 1/PC$
		1.5 + i0.5	1.3 + i0.1	0.5	$C \sim PC$
0.1	0.1	1.5 + i[0-0.1]	1.3 + i0.1	0.5-1	$C \sim 1/PC$
			1.5+ i0.5	0.5	$C \sim 1/PC$
0.2	0.1	1.5 + i[0-0.1]	1.3+ i0.1	0.5	irregular change of PC
Comet C/1995 O1 Hale-Bopp					
0.05	0.1	1.5 + i0.5	1.7 + i0.5	0.5	$C \sim PC$
			1.7 + i0.1	1	irregular change of PC

imaginary part of the refractive index exceeding 0.1;

- the initial mass of the organic mantle is approximately equal or less than the mass of the silicate core.

The results are in a good agreement with the study [8] based on analysis of evolution of interstellar and circumstellar dust, origin of comets and cosmic element abundances. Both studies resulted in core-mantle silicate-organic cometary particles with the core sizes near 0.1 μm and an almost equal mass of organics and silicates.

REFERENCES

1. E.H. Geyer, K. Jockers, N. Kiselev and G. Chernova, Astrophys. Space Sci. 239 (1996) 259.
2. J.M. Greenberg and B.Å.S. Gustafson, Astron. Astrophys. 93 (1981) 35.
3. J.M. Greenberg and J.I. Hage, Astrophys. J. 361 (1990) 260.
4. B.Å.S. Gustafson and L. Kolokolova, J. Geophys. Res. 104 (1999) 31711.
5. K. Jockers, V.K. Rosenbush, T. Bonev, and T. Credner, Earth, Moon and Planets 78 (1997) 37.
6. L. Kolokolova and K. Jockers, Planet. Space Sci. 45 (1997) 1543.
7. L. Kolokolova, K. Jockers, B. Gustafson, G. Lichtenberg, J. Geophys. Res. 106 (2000) 10113.
8. A. Li and J.M. Greenberg, Astrophys. J. Lett. 498 (1998) L83.
9. O.B. Toon and T.P. Ackerman, Applied Optics 20 (1981) 3657.

High porosity for cometary dust: evidence from PROGRA^2 experiment

E. Hadamcik[a], A.C. Levasseur-Regourd[a], J.B. Renard[b], J.C. Worms[c]

[a]Aéronomie/CNRS/IPSL/Université PARIS VI, BP3, 91371 Verrières, France

[b]LPCE-CNRS, 3A av. de la Recherche Scientifique, 45071 Orléans-cedex, France

[c]ESSC-ESF, Parc d'Innovation, Boulevard S. Brandt, 67400 Illkirch, France

Laboratory measurements undertaken with the PROGRA^2 experiment on compact and fluffy particles floating in microgravity or lifted by an air draft are studied. The polarimetric phase curves of mixtures of aggregates of high porosity (95 to 99 %) built of submicron grains are consistent with the properties observed in cometary comae. The results suggest that the porosity of cometary particles is higher than 95 % and that their sizes may reach one millimetre.

1. INTRODUCTION

To deduce the physical properties of cometary dust, some diagnostics are often used: i.e. dynamical models for spatial evolution of brightness of the scattered light and spectral emission features. The linear polarization of the scattered light is another important diagnostic. Polarimetric phase curves have allowed a classification of comets in two classes, although comet C/1995 O1 Hale-Bopp has been found to have a higher polarization than the other comets [1,2,3]. Polarimetric maps emphasize variations in the physical properties inside the coma (e.g. [4,5,2,6]). In addition, the wavelength dependence for polarization or brightness provide a complementary diagnostic ([7,8]).

To tentatively interpret the observations some models of light scattering by the particles have been built. The particles are assumed to be compact grains (spheres or irregular particles) or aggregates with a relatively small number of constituent grains (due to computational limitations). For all these models, size is a main parameter.

This paper compares cometary polarimetric phase curves with results obtained for linear polarization of light scattered by large particles (compact and fluffy) studied with the PROGRA^2 instrument [9] which is a type of nephelometer. The compact particles levitate in microgravity, the fluffy ones are lifted in ground-based conditions by an air draft (Hadamcik, 1999). These particles are difficult to model due to their irregular shapes and the high number of grains (hundreds of thousands).

2. SAMPLES

The size distribution of the particles is in the 10 μm to 500 μm range depending on the sample. The irregular compact particles are mainly convex with sharp edges as described by Worms et al. [9]. Two different types of fluffy particles have been used, ground Orgueil meteorite (porosity 30 %; size 1 μm, see [11]); fluffy samples made of submicron grains (pyrogenic oxides, carbon black and mixtures of them, described in [10]). The smallest grain size (7 nm) is for a pyrogneic silica sample, the largest one for a carbon black sample (95 nm). These grains are linked together in chains giving aggregates. The size of the aggregates is of the order of ten micrometers and their porosity is higher than 95 %, with a specific surface area in the range of 50-300 $m^2 g^{-1}$. The densities of the grains are given by the company who provides the samples, the densities of the aggregates are estimated by measuring hardly packed powders. These aggregates stick together in agglomerates. When levitated, the particles in the field of view are such agglomerates. The densities of the agglomerates are estimated by measuring sifted powders.

3. RESULTS

The measurements are performed between 8° and 160° phase angles. Figure 1 gives a summary of typical phase curves in red light. Figure 1a is for compact particles, figure 1b for fluffy particles. We are mainly interested in two regions: the backscattering region at small phase angles and the maximum polarization (P_{max}) region.

3.1. Backscattering region

For phase angles smaller than 20°, three facts can be noticed. First, for compact particles (i.e. basalt), the polarization is negative or close to zero for different sizes of the particles [9]. For levitating fluffy particles (i.e. pyrogenic silica with 16 nm grains, pyrogenic silica with 40 nm grains), the polarization sharply increases near 10°-20° phase angle. This is not the case for mixtures of two fluffy samples; they produce a negative branch at phase angles smaller than 20° (the studied samples are: silica 16 nm and silica 40 nm; silica 40 nm and carbon 14 nm; silica 40 nm and carbon 95 nm; alumina 13 nm and carbon 14 nm).

3.2. Maximum polarization region

3.2.1. Size dependence

In figure 2a three different materials for compact particles are studied. P_{max} increases when the particles size (mean diameter of the single grain) increases, with a different maximum value for each material, as expected from Fresnel laws. In figure 2b two materials of fluffy particles (pyrognenic silica and carbon black) are considered. P_{max} decreases when the size of the constituent grains (mean diameter) increases. When the size is smaller than 15 nm (size parameter of about 0.07 in red light) Rayleigh scattering is observed. P_{max} for a mixture of two size distributions of pyrogenic silica is smaller than P_{max} for the constituent samples. More generally, a lower P_{max} than those obtained for the component materials is observed for all the studied mixtures of fluffy particles.

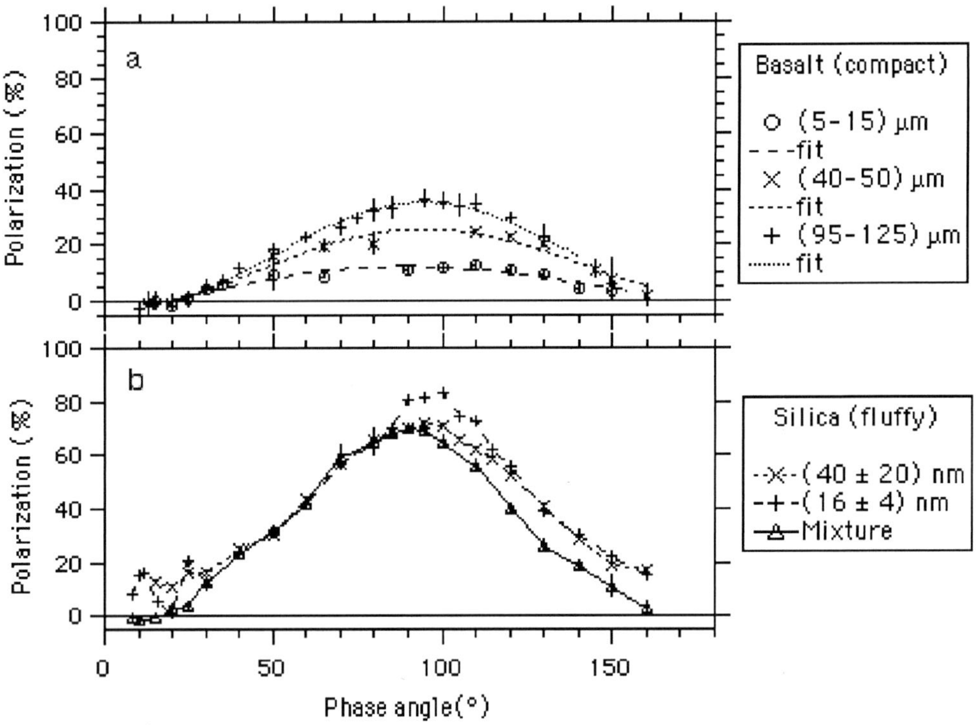

Figure 1. Polarimetric measurements (with error bars) and phase curves for different sizes of the grains. a) Compact particles in microgravity. b) Fluffy particles lifted in air.

3.2.2. Wavelength dependence

To study the wavelength dependence, a red and a green laser were used. For all the studied compact samples, the maximum value of polarization is found higher in green light than in red light. The same result was found for Orgueil meteorite [11]. For the other fluffy particles (highly porous and with submicronic grains), P_{max} is always found to be higher in red light than in green light [10].

4. DISCUSSION AND CONCLUSION

4.1. Comparison with microwave analog results

Gustafson and Kolokolova [12] have studied different compact particles and aggregates, with particular attention to the color dependence in brightness and polarization near P_{max}. Their work confirms that polarization of light scattered by a fluffy particle (aggregate) decreases with increasing size of the constituent grains and with increasing porosity of the particles. With large aggregates (ten times the wavelength or more), the polarization decreases and the color in brightness and polarization is redder [13]. These results are in complete agreement with our measurements.

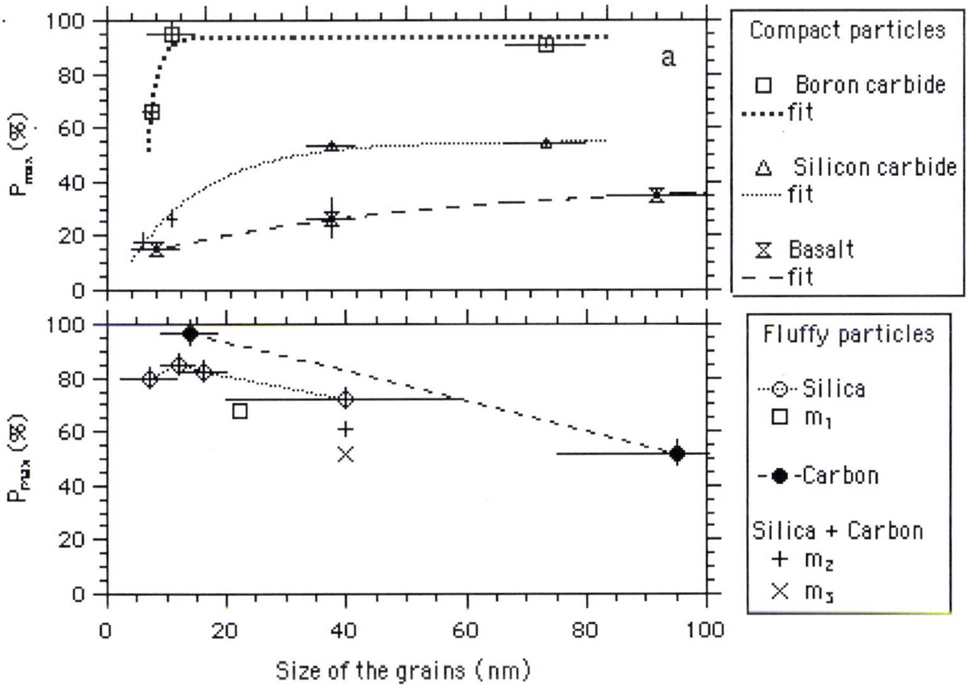

Figure 2: Size dependence. 3a: Compact particles (single grain): boron carbide (very dark), silicon carbide (dark), basalt (clear brown). 3b: Fluffy particles: silica and carbon with different grains sizes, mixtures with S=silica (40 nm): m_1= S + silica (16 nm), m_2 = S + carbon (14 nm), m_3 = S + carbon (95 nm).

4.2. Comparison with cometary observations

A nice agreement is found between the phase curves obtained for mixtures of fluffy particles and for those obtained for comets [1] with: a negative branch at phase angles smaller than 20°, a maximum value of polarization for a phase angle of about 100° with values smaller than 50 % when the grains size is larger than 50 nm, an increase of P_{max} with wavelength.

To better disentangle the physical properties of cometary dust, a comparison with other diagnostics is necessary. Infrared silicate features indicate the presence of submicron grains [14], but they also could be observed with high porosity (> 95 %) aggregates [15]. Interplanetary dust particles (IDPs) collected in the stratosphere, with pyroxene, olivine and high carbon contents, could have similar composition to cometary dust. The particles have porous structures of fine-grained heterogeneous aggregates [16]. The best fits by a dynamical model of the in-situ measured fluxes by Giotto/DIDSY and Giotto/OPE are obtained for densities of the order of 100 kg m^{-3} [17,18].

Mixtures of fluffy particles have physical properties similar to those expected for cometary particles with high porosity (95 to 99)% and submicron constituent grains. The polarimetric response of such mixtures would thus plead in favor of the presence, in cometary comae, of dust particles with similar physical properties.

REFERENCES

1. A.C. Levasseur-Regourd, E. Hadamcik and J.B. Renard, Astron. Astrophys. 313 (1996) 327.
2. E. Hadamcik, A.C. Levasseur-Regourd and J.B. Renard, Earth, Moon, Planets 78 (1997) 365.
3. N.N. Kiselev and F.P. Velichko, Earth, Moon, Planets 78 (1997) 347.
4. J.B. Renard, A.C. Levasseur-Regourd and A. Dollfus, Ann. Geophys. 10 (1992) 288.
5. N. Eaton, S.M. Scarrott and P.W. Draper, Mon. Not. R. Astr. Soc. 258 (1995) 384.
6. K. Jockers, Earth, Moon, Planets 79 (1997) 221.
7. E. Hadamcik, A.C. Levasseur-Regourd, J.B. Renard and J. Worms, in Physics, Chemistry and dynamics of Interplanetary dust (eds. B.A.S. Gustafson and M.S. Hanner), ASP Conf. Ser. 104 (1996) 391.
8. L. Kolokolova, K. Jockers, G. Chernova and N. Kiselev, Icarus 126 (1997) 351.
9. J.C. Worms, J.B. Renard, E. Hadamcik A.C. Levasseur-Regourd and J.F. Gayet, Icarus 142 (1999) 281.
10. E. Hadamcik, Thesis, Université Paris 6 (1999).
11. J.C. Worms, J.B. Renard, E. Hadamcik, N. Brun-Huret and A.C. Levasseur-Regourd, Planet. Space Sci. 48 (2000) 493.
12. B.A.S. Gustafson and L. Kolokolova, J. Geophys. Res. 104 (1999) 711.
13. B.A.S. Gustafson, in Light scattering by nonspherical particles (eds. Mishenko, Hovenier and Travis) Academic Press (2000) 367 (2000).
14. M.S. Hanner, this volume.
15. J.M. Greenberg and A. Li, Space Sci. Rev. 90 (1999) 149.
16. M.S. Hanner, Space Sci. Rev. 90 (1999) 99.
17. A.C. Levasseur-Regourd, N. McBride, E. Hadamcik and M. Fulle, Astron. Astrophys. 348 (199) 636.
18. M. Fulle, A.C. Levasseur-Regourd, N. McBride and E. Hadamcik, Astron. J.119 (2000) 1968.

VI Laboratory Studies

The nature of cosmic dust: laboratory data and space observations

L. Colangeli[a], J.R. Brucato[a], V. Mennella[a] and P. Palumbo[b]

[a]Osservatorio Astronomico di Capodimonte, Via Moiariello 16, 80131 Napoli, I.

[b]Università "Parthenope", Via A. De Gasperi 5, 80133 Napoli, I.

Astronomical observations and in situ measurements represent the tools to derive the properties of materials present in different space environments, from the interstellar medium (ISM) to the Solar System. By studying physics and chemistry of cosmic compounds it is possible to trace their evolution as a function of the environment in which they are observed and, thus, to deduce the characteristics of the ambient space. In this scenario, laboratory experiments represent a valid method to investigate how materials form in different space environments and to test how they evolve due to their interaction with the most efficient agents active in space (thermal, ion, UV processing, interaction with gas). The collection of results reported in this work demonstrates how relevant is the role played by laboratory data to interpret some of the most fascinating puzzles proposed by space exploration.

1. INTRODUCTION

The knowledge about the properties of cosmic materials has improved thanks to the advent of new ground-based observational instruments and of space-born telescopes, such as the Hubble Space Telescope and the Infrared Space Observatory. Thanks to such tools we have gained unprecedented new information about the spatial distribution and the spectral signatures of cosmic gas and dust components in the most relevant space environments outside (e.g., star forming regions, evolved stars, circumstellar and interstellar medium) and inside (e.g., planets, comets, asteroids) our Solar System. With specific concern to the solid component (dust), attention has been focused on the two main families of materials that are present in space: carbons and silicates. Interesting relations have been identified among the specific compounds of the two classes observed in different space conditions.

Infrared spectroscopic observations of carbon-based materials in the ISM [1] and in the Proto-Planetary Nebula CRL618 [2] present similar absorption features around 3.4 μm, typical of aliphatic solids. On the other hand, sharp emission bands around stellar sources rich in UV flux indicate the presence of aromatic compounds, such as Polycyclic Aromatic Hydrocarbons (PAHs) [3]. Emission spectra of comets present a convolution of bands probably due to volatile components (e.g., methane and ethane) and solid carbon [4].

As far as silicates are concerned, the information derived by ISO observations is quite innovative, as sharp emission bands in the 10 - 20 μm spectral range have been clearly identified as due to crystalline compounds, both in the circumstellar medium [5,6] and in comets [7]. On the other hand, both the diffuse and the dense ISM seems rich in amorphous silicates, as testified by the observation of broad absorption bands around 10 μm [8]. The main puzzle in this case is to find a reasonable evolution path from circumstellar to interstellar media and, then, to comets that might account for the observed structural variations.

The previous remarks evidence the complexity in interpreting the observations about cosmic materials. Actually, it is well understood that intimately different materials may provide similar features and that, on the other hand, the "intrinsic nature" of materials is closely related to their optical behaviour. Moreover, materials react to space environment conditions as different agents, such as thermal, UV and ion processes, are active with different efficiency depending on local conditions. Therefore, materials evolve in space and change in terms of physical, chemical and structural properties. A difficult task is to trace such evolution based on evidence coming from astronomical observations only.

In the case of comets, much help came from in situ analyses performed by the GIOTTO spacecraft during the fly-by of comet 1P/Halley. We know that comet grains present different chemical phases [9]. Organic (CHON) particles represent about 30 % in mass, while silicates are mainly Mg rich. Grains present densities between 2.5 g cm^{-3} (silicates) and 1 g cm^{-3} (mixture) and have a so-called core-mantle or fluffy structure. The overall composition (in mass) is [10]: 26 % silicates (Si, Mg, Fe, ...), 23 % organic material (carbon), 9 % small carbonaceous grains / PAH's, while the rest is ice (water and other condensed volatiles). Despite this direct evidence, also in the case of comets many questions are open about the pristine degree of the composing materials.

In this scenario, a fundamental contribution to the identification of physical and chemical characteristics of cosmic materials comes from laboratory investigation. In fact, the analysis of natural and synthetic compounds allows us to evidence relations between material properties and optical behaviour. This is a fundamental step for the quantitative interpretation of astronomical observations. Moreover, the new frontier of laboratory experiments is oriented to analyse the evolution of analogues due to the most important mechanisms which may alter materials in space (e.g., thermal annealing, UV irradiation, ion bombardment, hydrogenation). The experimental results obtained so far offer a solid background to interpret observations and to trace the changes induced in space. We are today tackling long standing astrophysical questions, which may find a definite clarification thanks to the contribution of deep experimental analyses.

In section 2 we will summarise the laboratory approach and some of the most recent and relevant results obtained in the laboratory concerning materials of interest in the view of applications to the interpretation of space observations. Section 3 will be devoted to demonstrate how the laboratory data can be fitted to observations to derive information about the actual status of cosmic materials. Finally, in section 4 we will indicate what are the most promising future steps both in laboratory and in space to achieve a deeper understanding on the main species populating key space environments.

2. THE LABORATORY APPROACH

Since the two main families of solids present in space are carbon- and silicon-based materials, laboratory experiments are focused on various forms of natural or laboratory synthesised pure and hydrogenated carbons and crystalline and amorphous silicates. In the latter case, main attention is devoted to olivine $(Mg_x,Fe_{1-x})_2SiO_4$ (Forsterite x = 1; Fayalite x = 0) and pyroxene $(Mg_x,Fe_{1-x})SiO_3$ (Enstatite x = 1; Ferrosilite x = 0), which appear promising candidates to be present in space.

2.1. Techniques

To produce micron/sub-micron sized grain analogues, vapour condensation, pyrolysis or grinding techniques are applied to pre-selected targets or natural pure minerals. The obtained samples can be subject to various kinds of processing which simulate mechanisms active in space. Thermal annealing is relevant in stellar outflows, as well as in the Oort cloud, while UV irradiation may be efficient to modify the structure and the chemical properties of dust during the permanence in the ISM. The ion bombardment may have a different impact during some critical evolution stages of materials: ISM, pre-comet phase, in comets and in the interplanetary medium. Moreover, it must be considered that materials are continuously in interaction with the gaseous components, so that at least the interaction with hydrogen must be accounted for properly.

At all different stages of material production/processing detailed analyses must be performed on the samples to identify the details of their status and to check the efficiency of the applied processes. The characterisation can rely on different complementary analytical techniques such as: scanning and transmission electron microscopy (useful to monitor morphology, surface status and structure), electron diffraction and energy loss spectroscopy (to identify the crystalline degree/structure), energy dispersive X-ray analysis (to determine the elemental abundance). Finally, absorption, emission, reflectance and Raman spectroscopic analyses in the widest spectral range (from the far UV to the mm) are applied to correlate composition, structure and morphology of samples with their spectral behaviour. Of course, those mentioned above are some of the most useful techniques applied so far for characterisation of materials. As long as new and more sophisticated methods are made available by the development of technologies, the investigation may become more accurate.

2.2. Results

The most common feature of grains produced by vapour condensation of various species is a fluffy structure formed by round-shaped grains in an amorphous status, as shown in Figure 1. The size distributions are skew, with average radii about 5 nm (for amorphous carbon grains) and about 30 nm (for amorphous silicate grains).

Carbon presents different structural properties and spectral features depending on the production conditions. In particular, amorphous carbon grains condensed by arc discharge in an inert argon atmosphere (ACAR grains) show a wide UV absorption peak around 240 nm. Grains condensed by the same technique, but in H_2 atmosphere (ACH2 grains), display no UV "bump". This spectral difference testifies a different internal structural organisation of the grains, due to the interaction of C atoms with hydrogen. In a hydrogen-free structure, C atoms can achieve a short-scale arrangement in (1-2 nm) aromatic networks, while in the presence of hydrogen, C-H bonds prevent their formation, giving rise to a less organised structure [13]. Interesting enough, ACH2 grains after thermal annealing up to 1050 K [14], UV irradiation up to 4×10^{22} eV cm^{-2} [15] or ion bombardment up to 660 eV/C-atom [16] always display the appearance of the UV feature, indicating an aromatisation process, also due to the hydrogen release. On the other hand, exposing hydrogen-free carbon to interaction with atomic H up to 7×10^{19} atoms cm^{-2} produces the appearance of a relevant 3.4 μm absorption feature clearly due to the formation of C-H aliphatic bonds [17].

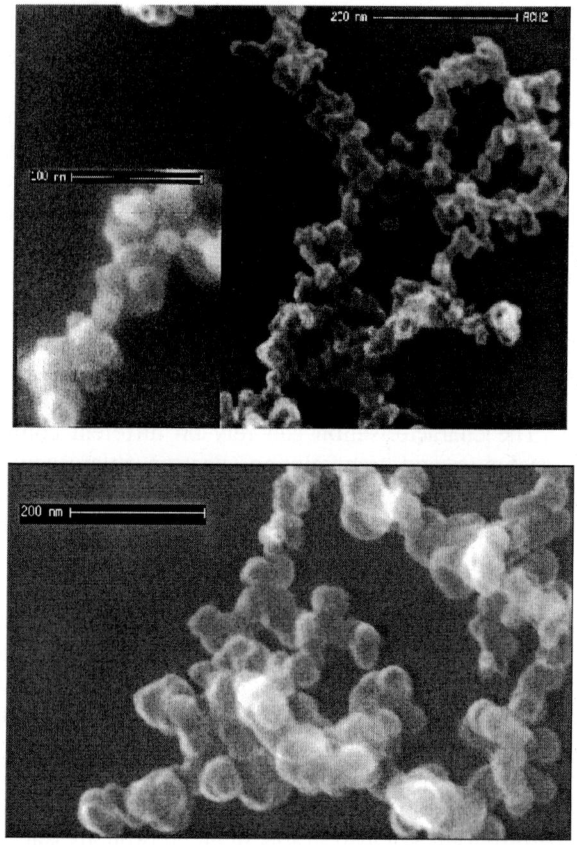

Figure 1. Transmission electron micro-graphs of laboratory dust analogues. Top: carbon grains produced by vapor condensation in an arc discharge between two carbon electrodes in an H_2 atmosphere at 10 mbar [11]. Bottom: amorphous silicate (forsterite) grains condensed after laser ablation of a natural forsterite target in O_2 atmosphere at 10 mbar [12].

The laboratory results mentioned above tell us that aromatic and aliphatic carbon forms compete to determine the arrangement of carbon nano-grains: the dominance of one of the two depends on the formation conditions (e.g., presence of hydrogen) and on the relative dominance of processing mechanisms, as well as interaction with hydrogen. All these phenomena are active, at different rates, in space environments. As a consequence the actual status of cosmic carbon is tightly related to the local conditions and it can be predicted on the basis of laboratory results.

For silicates, the bands falling around 10 and 20 μm, due to Si-O stretching and O-Si-O bending are diagnostic of chemical composition and crystalline degree. Sharp vs. broad features identify crystalline vs. amorphous silicates. The Fe/Mg ratio determines the detailed position of the band sub-structures (Figure 2).

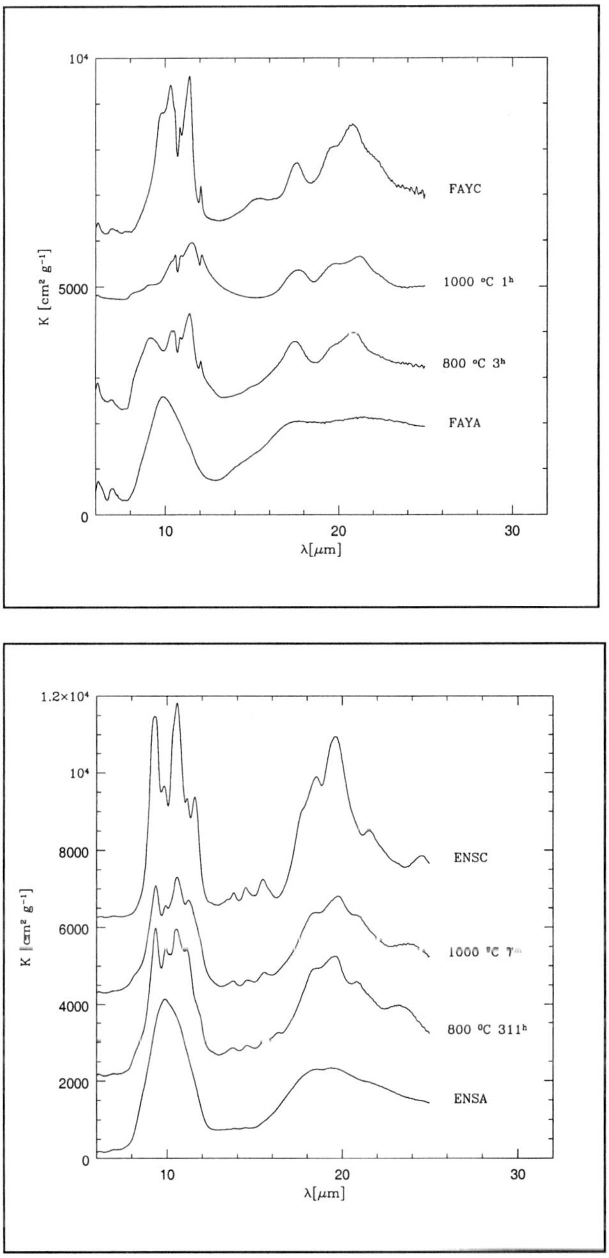

Figure 2. Infrared absorption spectra of crystalline, amorphous and thermally processed fayalite (top panel) and enstatite (bottom panel) [12,18].

Thermal annealing of amorphous silicates produces crystallisation (Figure 2). By analysing the progressive transformation of the broad into sharp features, it is possible to derive the so-called "activation energy" that relates time and temperature needed to perform a complete amorphous to crystalline transition. In general, the values obtained by various authors for different silicates [12,18,19], although not fully in agreement probably due to the use of slightly different samples, provide values of the activation energy of the order of some 10,000 K. This result may have important implications when applied to the astrophysical conditions, as it implies that about 10^6 yr are needed to crystallise silicates at ~ 1000 K. These conditions appear critical to achieve in space.

3. APPLICATIONS

The laboratory results reported in the previous section delineate a reference frame that can be applied to interpret the spectral features observed in various space conditions in terms of detailed chemical and structural composition of the carriers. Moreover, differences observed for various cosmic environments may find a justification in the processing of materials.

Important, long standing questions concerning the attribution of typical features observed in space seem to have recently found reasonable explanation thanks to data coming from laboratory experiments. This is the case, for example, of the so-called interstellar extinction "bump", falling at 217.5 nm. This is a broad band due to absorption of carbon-based materials dispersed in the ISM. Thanks to the use of laboratory data on carbon grains subject to various degrees of UV processing it has been possible to demonstrate that all the peculiarities observed for the bump (peak stability within 1 % and bandwidth variations within 25 %) are accounted for. Blends of amorphous carbon grains, which have suffered different degrees of UV treatment, provide a very satisfactory spectral fit to observations [20]. The solution to the puzzle is based on a material processing perfectly compatible with actual conditions in the ISM.

Similarly, the 3.4 μm aliphatic band observed in the ISM and in PPN's (see section 1) may find a clear attribution not only based on the good spectral match (see Figure 3) but also in a plausible justification of its genesis in terms of re-hydrogenation of amorphous carbon grains with the activation of C-H aliphatic bonds [17], as demonstrated by experiments (see section 2).

The complete infrared spectrum of comet Hale-Bopp obtained thanks to ISO [7] has been reproduced satisfactorily by a combination of silicates and carbon, where crystalline olivine plays a relevant role [21]. Once again, the good spectral match (Figure 4) is not enough to explain the observations, as questions arise about the occurrence of crystalline silicates in comets. As a matter of fact, comets should contain a record of the primordial material from which our Solar System originated about 4.5 x 10^9 years ago. Since observations on the ISM give no evidence of crystalline silicates, some specific mechanism must be invoked to account for the formation of this component during the early stages of the proto-planetary nebula evolution. It is unlike that grains participating to the comet formation in the external regions of the nebula have been thermally treated for a sufficient time at temperatures higher that 1000 K (see section 2.2). It must be, then, assumed that a significant mixing between internal and external zones of the pre-solar nebula has occurred during the solar system formation, allowing the involvement of inner processed materials also in the externally formed comets. A similar problem is faced to interpret ISO spectra of oxygen-rich stars [5,6] displaying

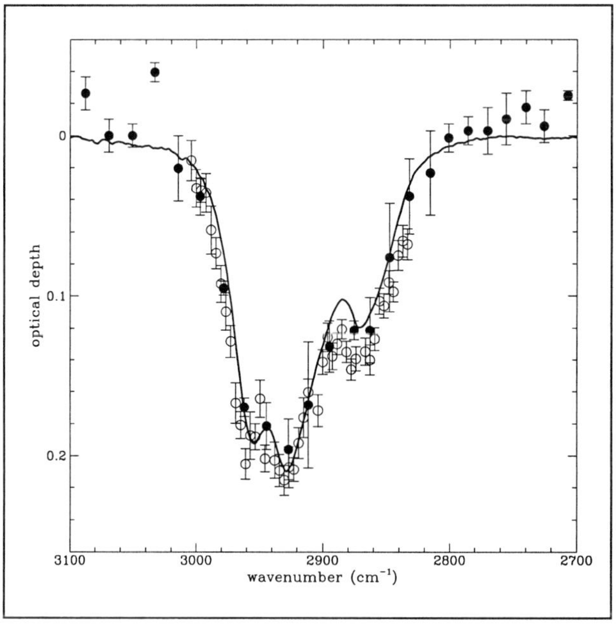

Figure 3. Comparison of the absorption spectrum towards GC IRS-6 E [1] (points) and laboratory data on re-hydrogenated carbon [17] (curve).

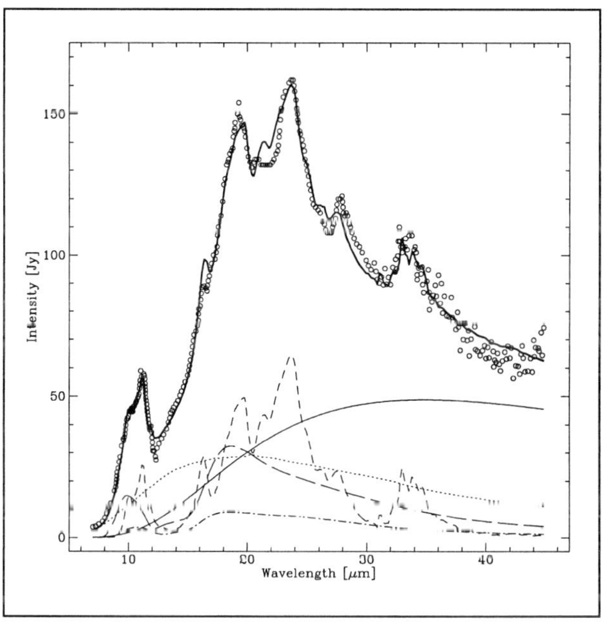

Figure 4. Comparison of the Hale-Bopp emission spectrum [7] with a simple model based on laboratory data for amorphous and crystalline silicates and amorphous carbon [21].

crystalline silicate features. In this case, a possible non-thermal crystallisation mechanism has been invoked.

4. CONCLUSIONS

The results reported in the present work give clear indication that the direct comparison of remote and in situ space measurements with laboratory data is a winning approach to interpret observations on cosmic materials. The new method is oriented at simulating formation and processing of cosmic materials in order to achieve the matching to observations by laboratory results complemented by a serious evaluation of the conditions and constraints that may (or may not) justify the presence of candidate compounds in various space environments. In this context, available elemental abundance seem to pose the most critical limitations [22].

Despite the many promising results, several astrophysical problems are still unsolved. The pathway for the amorphous-to-crystalline transition of silicates is still to be determined, accounting for the controversial indications coming from observations in comets, circumstellar and interstellar media. New experiments have been invoked to evaluate the capability that mechanisms, other then thermal annealing, may have in such a process, also depending on the chemical and structural status of the starting material.

The evolution of the aliphatic 3.4 μm band from diffuse to dense ISM is another open issue. Efficiency of mechanisms that may contribute to the destruction (e.g., UV and ion processing) or the formation (hydrogenation) of the relevant C-H bonds are strongly related to detailed environment characteristics, such as density and temperature. Again, new experiments must demonstrate how the various factors combine in such a complex scenario.

Last but not least, the future in laboratory is also oriented at a more and more detailed comparison of data obtained on actual extraterrestrial samples, such as interplanetary dust particles, that can tell us a story not easily traceable by remote observations only. The next ambitious steps include long standing in situ analyses on comets (e.g., the ESA Rosetta mission) and cometary sample return (e.g., the NASA Stardust mission) by space missions.

These considerations evidence how the basis for a deeper investigation of cosmic materials has been traced and that the future will reserve new fascinating results in such a field.

ACKNOWLEDGEMENTS

This work has been supported under ASI and MURST contracts. S. Inarta, N. Staiano and E. Zona are acknowledged for their technical assistance during experimental work.

REFERENCES

1. Y.J. Pendleton, S.A. Sandford, L.J. Allamandola, A.G.G.M. Tielens and K. Sellgren, Astrophys. J. 437 (1994) 683.
2. J.E. Chiar, Y.J. Pendleton, T.R. Geballe and A.G.G.M. Tielens. Astrophys, J. 507 (1998) 281.
3. T.R. Geballe, in From Stardust to Planetesimals (eds. Y.J. Pendleton and A.G.G.M. Tielens) ASP Conf. Series 122 (1997) 119.
4. T.Y. Brooke, R.F. Knacke, T.C. Owen and A.T. Tokunaga, Astrophys. J. 336 (1989) 971.
5. K. Malfait, C. Waelkens, L.B.F.M. Waters, B. Vandenbussche, E. Huygen and M.S. De Graauw, Astron Astrophys. 332 (1998) L25.

6. F.J. Molster, I. Yamamura, L.B.F.M. Waters. A.G.G.M. Tielens, T. de Graauw, T. de Jong, A. de Koter, K. Malfait, M.F., van den Ancker, H. van Winckel, R.H.M. Voors and C. Waelkens, Nature 401 (1999) 563.
7. J. Crovisier, K. Leech, D. Bockelie-Morvan, T.Y. Brooke, M.S. Hanner, B. Altieri, H.U. Keller and E. Lellouch, Science 275 (1997) 1904.
8. D.C.B. Whittet, A.C.A. Boogert, P.A. Gerakines, W. Schutte, A.G.G.M. Tielens, Th. de Graauw, T. Prusti, E.F. van Dishoeck, P.R. Wesselius and C.M. Wright, Astrophys. J. 490 (1997) 729.
9. H. Schulze, J. Kissel and E.K. Jessberger, in From Stardust to Planetesimals (eds. Y.J. Pendleton and A.G.G.M. Tielens) ASP Conf. Series 122 (1997) 397.
10. J.M. Greenberg, Astron. Astrophys. 330 (1998) 375.
11. A. Rotundi, F.J.M. Rietmeijer, L. Colangeli, V. Mennella, P. Palumbo and E. Bussoletti, Astron. Astrophys. 329 (1998) 1087.
12. J.R. Brucato, L. Colangeli, V. Mennella, P. Palumbo and E. Bussoletti, Astron. Astrophys.348 (1999) 1012.
13. V. Mennella, J.R. Brucato, E. Bussoletti, C. Cecchi Pestellini, L. Colangeli, E. Palomba, P. Palumbo, M. Robinson and A. Rotundi, in Cosmic Physics in the Year 2000 (eds. S. Aiello, N. Iucci, G. Sironi, Treves and U. Villante, Ed. Compositori) Bologna (1997) 225.
14. V. Mennella, L. Colangeli, A. Blanco, E. Bussoletti, S. Fonti, P. Palumbo and H.C. Mertins, Astrophys. J. 444 (1995) 288.
15. V. Mennella, L. Colangeli, P. Palumbo, A. Rotundi, W. Schutte and E. Bussoletti. Astrophys. J. 464 (1996) L191.
16. V. Mennella, G.A. Baratta, L. Colangeli, P. Palumbo, A. Rotundi, E. Bussoletti and G. Strazzulla, Astrophys. J. 481 (1997) 545.
17. V. Mennella, J.R. Brucato, L. Colangeli and P. Palumbo. Astrophys. J. 524 (1999) L71.
18. J.R. Brucato, L. Colangeli., V. Mennella, P. Palumbo, A. Rotundi and E. Bussoletti, ESA-SP 456 (2000) 335.
19. S.L. Hallenbeck, J.A. Nuth III and P.L. Daukantas, Icarus 131 (1998) 198.
20. V. Mennella, L. Colangeli, E. Bussoletti, P. Palumbo and A. Rotundi, Astrophys. J. 507 (1998) L177.
21. J.R. Brucato, L. Colangeli, V. Mennella, P. Palumbo and E. Bussoletti, Plant Space Sci. 47 (1999) 773.
22. T.P. Snow and A.N. Witt, Astrophys. J. 468 (1996) L65.

A new dust source for the Heidelberg dust accelerator

M. Stübig[a], G. Schäfer[a], T.-M. Ho[a], R. Srama[a] and E. Grün[a]

[a]Max-Planck-Institut für Kernphysik, Saupfercheckweg 1, D-69117 Heidelberg, Germany

The Max-Planck-Institut für Kernphysik in Heidelberg owns a dust accelerator, which is able to provide micron and submicron sized particles with speeds between 1 and 80 km s^{-1}. The type of dust materials and the speed and mass range of the accelerated particles is dependent on the design of the dust reservoir at the front end of the accelerator beam line. In order to improve the capabilities of this reservoir, a new dust source was developed. A simplification in the source design allows a more efficient particle extraction. Also new sorts of dust can now be accelerated at the Heidelberg facilities: Iron, silver, copper, carbon and latex. This paper describes the dust source and the properties of the first accelerated dust samples. The experiments were performed at both, the 20 kV test bench and the 2 MV accelerator.

1. INTRODUCTION

Nearly 40 years ago, the first experiments with dust accelerators started. The aim of these experiments were studies of hypervelocity impact physics. The first dust source was developed by Shelton in 1960 [1]. It was reconstructed from an ion source for ion beam accelerators. Micrometer sized particles are electrically charged with the dust source. The dust source is described in [2]. It runs best with a powder of spherical iron particles from 200 nm to 2 μm in diameter. After leaving the source, a high electrostatic potential will accelerate the particles to their final velocity. The accelerator works with a 2 MV Van-de-Graaff generator (Figure 1), which was developed by Friichtenicht [3]. Velocities from 2 to 10 km s^{-1} can be achieved. With particles less than 100 nm, even higher velocities were detected (80 km s^{-1}, [4]). A Particle Selection Unit (PSU) allows to select particles of a defined speed and charge. Particles not meeting these constraints are electrostatically deflected out of the beam line. Our dust accelerator is mainly used for studies of impact ionization, micro crater formation, calibration of impact detectors and time-of-flight mass spectrometers (TOF-MS). These detectors are used for cosmic dust space research [5–7]. In order to extend material types and the mass and speed range of the dust particles, a new dust source (Figure 2) was developed [8].

2. FUNCTION PRINCIPLE

The requirements of the source design are high: Potential differences higher than 20 kV over a few millimeters and a pressure difference of 18 bar between the interior and the environment. The most important change of the source from the previous system is the

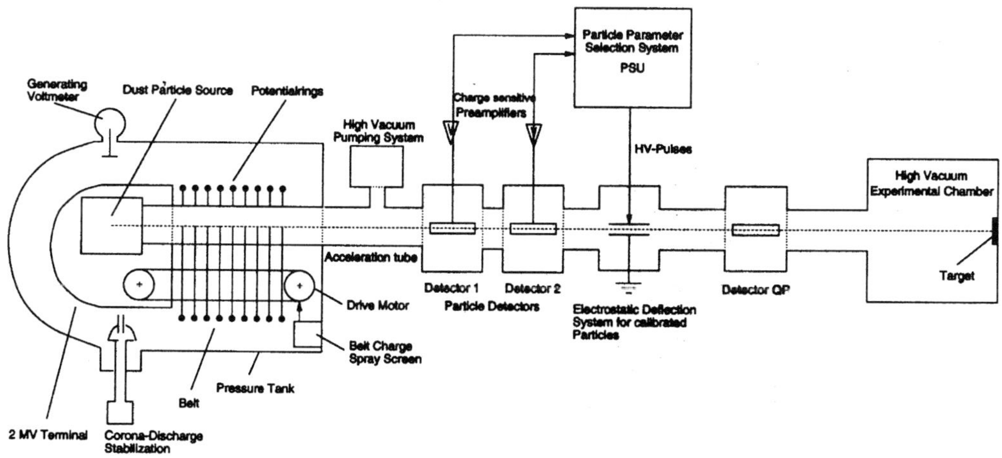

Figure 1. Schematic view of the Heidelberg 2 MV dust accelerator. On the left side the pressure tank with the belt generator and dust source. In the middle section of the beamline: Particle Selection Unit for preselection of desired particle parameters. On the right side: Vacuum chamber for experimental setups.

reservoir and the charging system [8]. The new reservoir is of cylindrical shape, rounded at the outlet side. A tungsten needle is centered in the reservoir and aligned to the accelerator beam line (Figure 2). In order to save weight, the housing of the new source is completely made of antimagnetic titan. The properties of the dust source are collected in Table 1. The dust reservoir itself lies on the same electric potential as the needle and is pulsed down frequently to blow up the dust powder. Electrically conductive particles carry surface charges and follow the time varying field conditions and swirl around the reservoir. If a particle hits the tip of the needle it gets its final charge and is accelerated by the field of the extraction plate. It has to pass through a small hollow in the extraction plate and a collimation system before it enters the acceleration section of 2 MV.

3. EXPERIMENTAL RESULTS

After a few experiments with different designs, we constructed a prototype and ran it on a test bench (Figure 3). The particles were only accelerated by the potential difference between the needle and the extraction plate. From the energy law one obtains:

$$v = \sqrt{2 \cdot \frac{q}{m} \cdot U_{acc}}, \tag{1}$$

where v is the final velocity of the particle, U_{acc} the acceleration voltage, q and m the particles charge and mass. An acceleration voltage of 20 kV results to a final velocity that is 1/10 of the velocity with $U_{acc} = 2$ MV.

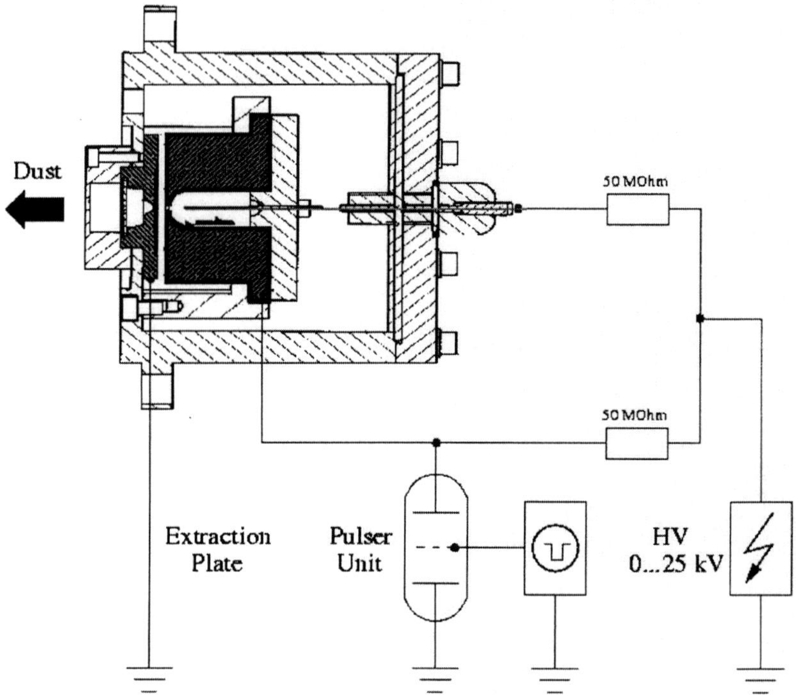

Figure 2. The function principle of the dust source. The needle is on a fixed potential while the potential of the dust reservoir is pulsed. Particles that hit the tip of the needle are extracted to the acceleration section (on the left side, not shown).

850 mm behind the dust source, a 100 mm long shielded influence tube with a TENNELEC *TC 161D* charge sensitive amplifier is located. Here, the flight time and charge of passing particles can be measured. The velocity of the particle is antiproportional to its flighttime. A conversion of equation (1) provides the particles mass. From a known density and the assumption of spherical particles (SEM-pictures support this assumption) one obtains the particle radius.

The first measurements at the test bench with several materials showed encouraging results. The tests were made with the following dust materials: Iron, carbon, copper, silver and latex spheres coated with a conductive layer [9]. All data were acquired with a needle potential of 20 kV, a repetition rate of 10 Hz and a pulse duration of 10 μs. The pulse amplitude was varied for each material. For the data in Table 2 100 particles of each material were selected.

Table 1
Properties of the new dust source.

Parameter				Value
Size	:	Length x Diameter	:	90 mm x 98 mm
		Reservoir	:	12 mm ⌀, 25 mm length
		Distance needle - extraction hole	:	3 mm
Weight	:	Total	:	1740 g
Material	:	Chassis	:	Titan
		Insulators	:	PEEK
		Needle	:	Tungsten (1 mm ⌀, tip $\leq$ 10 μm)
Pressure	:	Inside / Outside	:	10^{-10} bar / 18 bar ($N_2 + SF_6$)
Voltages	:	Needle	:	15 ... 25 kV, fixed
	:	Reservoir	:	same as needle, pulsed
Pulses	:	Duration / Repetition rate	:	10 μs, rectangular / 9 ... 35 s^{-1}
		Amplitude	:	variable, 0 kV ... Needle potential

Table 2
Results at the test bench. The pulse amplitude determines the voltage between needle and reservoir during the pulse. For better comparison, the surface fieldstrength E_{max} and the charge to mass ratio q/m is given for a 1 μm particle. The E- and q/m-values for Latex are calculated, assuming a r^{-1}-proportionality for E (constant potential) and a r^{-2}-proportionality for q/m.

Material	ρ [kg m^{-3}]	Grainsize [μm]	Pulse [kV]	E_{max} [GV/m]	q/m [C kg^{-1}]	Velocity [m s^{-1}]
Iron	7900	0.25...2.4	7.9	0.3	2.4	150...1300
			16.5	0.9	4.8	150...900
Carbon	2200	0.2...1.0	9.7	0.5	20.0	400...2800
Latex	1100	0.5	14.8	(0.3)	(14.5)	500...1250
			20.0	(0.5)	(24.1)	900...1550
Latex	1100	1.6	8.5	(0.4)	(19.3)	250...550
Latex	1100	2.1	7.4	(0.2)	(9.7)	150...300
Silver	10500	0.6...2.4	10.8	0.8	4.1	150...400
			20.0	1.3	5.6	200...700
Copper	8900	0.4...1.6	13.8	0.7	6.0	200...400
			16.1	1.1	6.4	300...500

4. DISCUSSION

The first step is to compare the old and new source design by taking iron particles. The data cover the same charge and velocity ranges, but with the new source even higher values were found [8]. Comparing the data at the test bench (Table 2), one can conclude that the average velocities rise with an increasing pulse amplitude. A higher potential difference between the tungsten needle and the reservoir causes a higher charge to mass

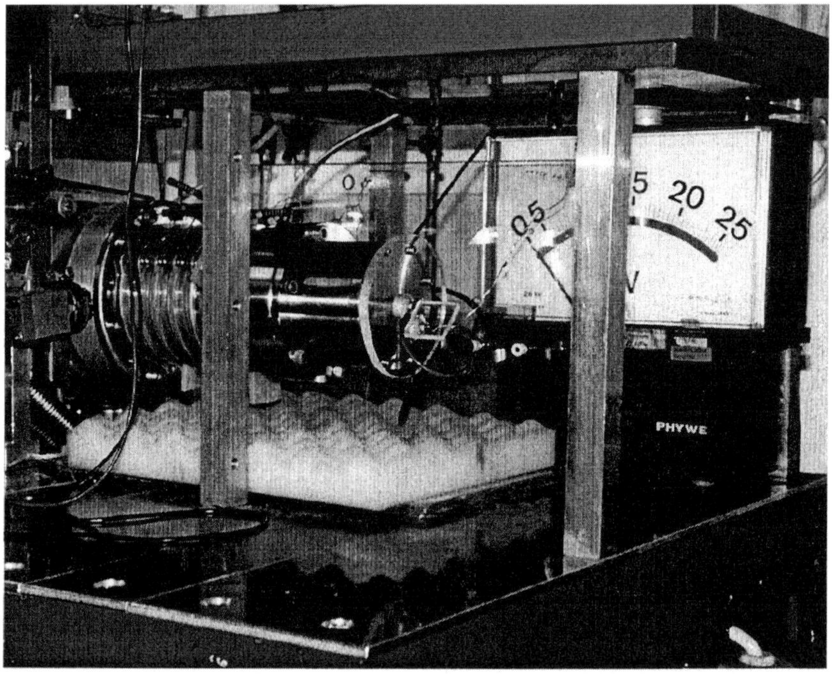

Figure 3. The new dust source at the test bench (in the middle of the picture). The high voltage ports are shown on the backside of the source. At the right: Voltmeter for the needle potential. At the left side from the source one can see the collimation section of the beamline and a high vacuum valve.

ratio. Since the total surface potentials of the particles are varying between 50 and 500 V, independent of their grainsize, smaller particles reach a higher charge to mass ratio and also higher velocities. Silver and copper are showing the highest fieldstrengths for 1 μm sized particles. Therefore they are the best chargeable materials followed by iron, carbon and latex. But due to their lower density latex and carbon obtain a higher charge to mass ratio and reach even higher velocities.

Extrapolating the test bench data of the latex samples to an acceleration voltage of 2 MV, and comparing these data with results from Canterbury [10] we obtain a lower top speed, for the 1.6 and the 2.1 μm particles ($\approx$ 7 and 4.5 km s^{-1} at Canterbury). The top speed of the 0.5 μm sample ($\approx$ 12km s^{-1} at Canterbury) is comparable to our extrapolated results. Using the maximum pulse amplitude even faster particles can be expected in Heidelberg.

5. SUMMARY

We developed a new dust source. Different dust samples have been accelerated at a test bench. A first data set with the prototype of the source was achieved at the 2 MV accelerator and met our expectations. We can now provide a wider range of projectile materials than before. E.g. the latex particles can simulate micrometeorites that have a low density and consist of organic materials. This will help to calibrate micrometeorite impact detectors and will improve our understanding of their time-of-flight spectra. It can also help to understand the impact physics better.

Acknowledgements. I like to thank M. Burchell (University of Kent, Canterbury, GB) for the latex samples and H. Shibata (University of Tokyo, Tokyo, JP) for the silver and copper powder.

REFERENCES

1. H. Shelton, C.D. Hendricks and R.F. Wuerker, J. Appl. Phys. 31/1 (1960) 1243.
2. H. Fechtig, E. Grün and J.Kissel, in Cosmic Dust (ed. J.A.M. McDonnell) Chichester, 1978.
3. J.F. Friichtenicht, Rev. Sci. Instr. 33/2 (1962) 209.
4. T.-M. Ho, Messung von schnellen und schwach aufgeladenen Teilchen mit einem hochsensiblen Staubdetektor, Masters Thesis, Max-Planck-Institut für Kernphysik, Heidelberg, 2000.
5. E. Grün, H. Fechtig, R.H. Giese, J. Kissel, D. Linkert, D. Maas, J.A.M. McDonnell, G.E. Morfill, G. Schwehm and H.A. Zook, Astron. Astrophys. Suppl. Ser. 92 (1992), 411.
6. E. Grün, H. Fechtig, M.S. Hanner, J. Kissel, B.A. Lindblad, D. Linkert, D. Maas, G.E. Morfill and H.A. Zook, Space Sci. Rev. 60 (1992), 317.
7. R. Srama et al., Space Sci. Rev. in press.
8. M. Stubig, Aufbereitung von Mineralen und Neukonstruktion einer Staubquelle für den Einsatz am Heidelberger Staubbeschleuniger, Masters Thesis, Max-Planck-Institut für Kernphysik, Heidelberg, 1999.
9. C. Barthet, S.P. Armes, S.F. Lascelles, S.Y. Luk and M.M.E. Stanley, Langmuir 14 (1998) 2032.
10. M.J. Burchell, M.J. Cole, S.F. Lascelles, M.A. Khan, C. Barthet, S.A. Wilson, D.B. Cairns and S.P. Armes, J. Phys. D: Appl. Phys. 32 (1999) 1719.

Development of low density dusts for impact ionization experiments

M.J. Burchell[a], M.J. Cole[a], M.J. Willis[a], S.P. Armes[b], A. Khan[b] and S.W. Bigger[c]

[a]Unit for Space Sciences and Astrophysics, University of Kent at Canterbury, Kent CT2 7NR, UK.

[b]School of Chemistry, Physics and Environmental Sciences, University of Sussex, Falmor, Brighton BN1 9QJ, UK.

[c]School of Life Sciences and Technology, Victoria University, Melbourne City, Australia

Hypervelocity impacts of dust on spacecraft are simulated in the laboratory using Van de Graaff dust accelerators. In this paper results are presented for acceleration of a new class of particle in such accelerators, namely particles coated by conducting polymers. The polymers can be made into discrete particles themselves, be used as a thin overcoating on a core particle (dimensions 100 to 1000 nm), or can be used to mix smaller grains of minerals (order 10 - 20 nm) into a composite particle. Results for all three types of particle are presented.

1. INTRODUCTION

Impact ionization experiments have had a distinguished role in helping to explore and better understand our Solar System. They rely on the impact of micron and sub-micron sized particles at high speed onto spacecraft. Because the typical relative speeds of spacecraft and dust particle are in the many km s^{-1} range the resulting impact is in the hypervelocity regime. Such impacts involve impact speeds above the speed of the resulting compression waves in both projectile and target. The result is an impact shock which develops extreme high pressures and temperatures in the projectile and the impact site in the target as the shock and subsequent release waves propagate. These temperatures and pressures are sufficient to not only melt but also vaporize the projectile and part of the target. The bulk target material also undergoes motion away from the impact site, excavating an impact crater. It is this crater that is the surviving legacy of the impact.

This impact process is perfectly natural in space. Due to the flux of small (dust) particles in space it occurs regularly on spacecraft. Measuring the dust flux throughout the Solar System has been a long term aim of space science since its earliest days. Initially, this was to determine the degree of damage to spacecraft, and it is now a science in its own right. This series of Dust Colloquia, stretching back many years, is an indication of the importance attached to these studies by the academic community. A good spacecraft deployed measurement system for dust is one that can generate a real time electronic signal characterizing the impact. This would determine the mass, velocity and composition of the particle. The information has to be real time, so a time dependent flux can be calculated, and electronic as the spacecraft may not be retrieved from space and will have to transmit the data to Earth. Fortunately, since the 1960's, it has been realized that the phenomena of impact ionization provides such a measurement opportunity.

During a hypervelocity impact as described earlier, the particle is vaporized and forms an ionic plasma. If a clean metal target is used, with a grid above it, with an electric field between target and grid, the presence of this impact generated plasma will allow a brief current flow and hence a pulse in external circuitry. Under the right conditions, the rise time of this pulse is related to the impact speed. Further, the magnitude of the ionization charge (I) measured can be related to the particle mass (m) and impact velocity (v) by a relationship of the form:

$$I/m = \alpha v^{\beta} \tag{1}$$

The determination of the relationship between rise time and impact speed and the measurement of the coefficients α and β in eqn. (1) represents the laboratory calibration of a particular dust detector. Once these are known, then for an impact (where I and V are measured), the mass m can be determined. Many such detectors have been deployed in space and several are still operational. For example, one family of similar instruments is widely deployed; in Earth Orbit the Gorid detector [1] is on the Russian Express2 telecommunications spacecraft, the Ulysses spacecraft carries one [2] in a Solar polar orbit and Galileo carries one [3] on its mission to the Jovian system. All of these map the dust flux in various regions of the Solar System.

However, this does not assist in determining particle composition. To do this another feature of the ionic plasma can be exploited. If the ionization detector has an intense electric field for a short height above the target the ions will undergo a rapid acceleration. If they are singly ionized, then their resultant velocity is dependent on the ion mass. If a further, weak, field is then applied to guide the ions to a detector a distance away, then the ion arrival time at the detector is related to the square root of the ion mass. Thus, in effect, a time of flight (mass) spectrometer can be manufactured. Two versions of such detectors are currently deployed in space. The first is the Cosmic Dust Analyser (CDA) [4] on the Cassini spacecraft to Saturn. This is essentially as described, and is a modification of the instruments on board Ulysses, Galileo etc. The second type is CIDA [5], deployed onboard the Stardust spacecraft travelling to Comet-P-Wild-2. In this type of instrument during the drift to the detector, the ions are focussed by an electric field. This removes the spread on the arrival time of an ion species due to slight differences in initial energy in their initial production. The result is a better mass resolution. Just as before, both types of instruments have to be calibrated in the laboratory.

Laboratory calibrations are usually carried out using Van de Graaff (VdG) dust accelerators. These are traditional style VdG machines, into which a dust source is placed at the top terminal. These sources charge the dust particles (mass m) with charge q and eject them into the full potential difference (V) of the VdG. The beam lines are maintained at a vacuum of typically 10^{-6} mbar, and the result is that the charged particle is electrostatically accelerated to a velocity v given by:

$$v = \sqrt{(2qV/m)} \tag{2}$$

The charge on each accelerated particle can be measured in flight by its passage through a conducting ring positioned along the axis of flight. The time difference in the signals from passage through two such rings separated along the line of flight by a known distance gives the velocity. The VdG voltage is monitored continually, so for each particle accelerated the mass m can be found from eqn (2). A typical such VdG is described in detail in [6].

The type of dust used in such VdG dust accelerators is limited to those that are easily charged. Iron has been the dust of choice for many years. However, iron is limited in its suitability as an analogue for Solar System dust. Attempts have been made to use other materials. In the 1970's use was made of an ion beam to charge particles and then accelerate them in a VdG [7], but this did not gain general use. Later the coating of particles with conducting metals was used [8], but again has not found general use. More recently a new material has been used in dust accelerators, namely conducting polymers. The first report [9] accelerated pure particles of the polymer polypyrrole to 5 km s^{-1}, although no data for ionization resulting from such impacts was given. The technique was then developed further to use the conducting polymer as a thin coating on the outside of a core particle. This allowed the acceleration of materials which are non-conducting. This is reported in [10], where use of three different conducting polymers as coatings was demonstrated, polypyrrole (PPY), polyanilane (PANi) and poly(3,4-ethylenedioxythiphene) (PEDOT). The core particle in each case was polystyrene latex which could be chosen in the size range 0.1 to 5 microns diameter.

This represents a significant advance in laboratory impact ionization experiments. It is relatively straightforward to synthesize the materials involved and carry out the coating. The conducting polymers are low density, as is the latex, with typical coated particle densities of 1080 to 1160 kg m^{-3}. Further, the size range of latex used allowed acceleration to different velocities, up to 20 km s^{-1} for the coated particles. In addition, the materials used are organic, allowing a controlled experiment where the resulting ionization signals and time of flight (mass) spectra can be studied. For the latter this is important, as the thermal decomposition of latex has been well studied in the past. Thus the effects of hypervelocity impact ionic plasma production can be compared to the properties of a well understood organic material.

In this present work we report on three further developments in this field, namely the acceleration of pure PPY particles to a velocity of 35 km s^{-1}, the use of PPY coated latex where the core latex particle has been doped with bromine, and the acceleration of nano-composite particles where 20 nm grains of silica were mixed with PPY to form mixed particle.

2. RESULTS

The first sample were pure PPY particles, 100 nm in diameter with density of 1500 kg m^{-3}. They were charged successfully and the resulting mass:velocity distribution is shown in Fig. 1a. The cluster of data at mass 1.1×10^{-18} kg represents the individual particles. Higher mass data occurs when individual particles clump together. The data for the individual particles covers the velocity range 26 - 37 km s^{-1}. This is significantly higher than previously [10] reported for these low density particles.

The second type of particle used was a PPY coated bromostyrene latex of diameter 590 nm. In the latex synthesis a chemical substitution had been made, such that bromine accounted for 40% of the mass of the particle. This changed the mean density to 1603 kg m^{-3}, compared to the typical 1100 kg m^{-3} for the previous coated latex particles [10]. The particles charged well and the mass:velocity distribution measured in flight is shown in Fig. 1b. The average particle mass was 1.9×10^{-16} kg. The velocity range covered was 4 to 10 km s^{-1}. There were a few faster, smaller particles recorded. However, these are probably fragments of the PPY coating which have detached themselves during acceleration.

The final type of particle tested was a PPY-silica mixture. Here the small grains of PPY (10 nm or less) are mixed with 20 nm grains of silica. The resulting composite particles were nominally 270 nm in diameter, with density 1637 kg m^{-3}. Given the bulk densities of PPY and the silica, this suggests the particles are typically 21% silica by mass. They charged well and

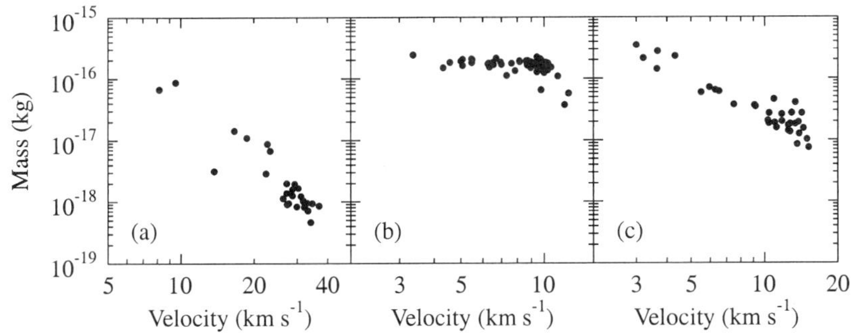

Figure 1. Mass velocity distributions for (a) pure PPY, (b) bromine doped latex, (c) PPY-silica mixed particles.

were accelerated, giving the mass:velocity distribution shown in Fig 1c. The single particles have an average mass of 2.3×10^{-17} kg and cover a velocity range 7 to 15 km s^{-1}. Again, some larger mass clumps of particles were observed during the measurements.

3. CONCLUSIONS

In this present paper the use of conducting polymers in VdG dust accelerators has been improved to show how such particles can extend the range of materials used in impact ionization studies. The samples used were: Pure PPY, achieving a maximum velocity of 37 km s^{-1}, excellent for such work. A bromine doped latex, showing how the density of the particles can be controlled and how specific elements can be added to them for delivery to the target in experiments. And finally a new PPY-silica nano-composite mixed particle. This last type of particle is of particular interest as it shows than one can readily make mineral grains available in impacts. These dusts are now all being used at the University of Kent for impact ionization experiments, including time of flight (mass) spectrometry investigations.

REFERENCES

1. H. Svedhem, G. Drolshagen, E. Grün, O. Grafodatsky and U. Prokopiev, Adv. Space Res. 25 (1999) 309.
2. E. Grün, Fechtig, R.H. Giese, J.Kissel, D.Linkert, D.Maas, J.A.M. McDonnell, G.E. Morfill, G. Schwehm and H.A. Zook, Astron. Astrophys. Suppl. Series 92 (1992) 411.
3. E. Grün, H. Fechtig, M.S. Hanner, J.Kissel, B.A. Lindblad, D.Linkert, D.Maas, G.E. Morfill and H.A. Zook, Space Sci. Rev., 60 (1992) 317.
4. R. Srama and E. Grün, Adv. Space Res. 20 (1997) 1467.
5. M. Landgraf, M. Muller and E. Grün, Planet. Space Science 47 (1999) 1029.
6. M. Burchell, MJ. Cole, J.A.M. McDonnell and J.C. Zarnecki, Meas. Sci. Technol. 10 (1999) 41.
7. B.-K. Dalmann, E. Grün and J. Kissel, Planet. Space Sci. 25 (1977) 135.
8. J.R. Göller and E. Grün, Planet. Space Sci. 37 (1989) 1197.
9. S. P. Armes, M. Aldissi, G.C. Idzorek, P.W. Keaton, L.J. Rowton, G.L. Stradling, M.T. Collopy and D.B. McColl, J. of Colloid and Interface Science 141 (1991) 119.
10. M. Burchell, M.J. Cole, S.F. Lascelles, M.A. Khan, C. Barthet, S.A. Wilson, D.B. Cairns and S.P. Armes, J. Phys. D: Applied Phys. 32 (1999) 1719.

Application of new, low density projectiles to the laboratory calibration of the Cassini Cosmic Dust Analyser (CDA)

B.J. Goldsworthy[a], M.J. Burchell[a], M.J. Cole[a], S.F. Green[a,d], M.R. Leese[a,d], N. McBride[a,d], J.A.M. McDonnell[a,d], M. Müller[a,d], E. Grun[b], R. Srama[b], S.P. Armes[c], and M.A. Khan[c]

[a]Unit for Space Science and Astrophysics, University of Kent at Canterbury, CT2 7NR, UK

[b]Max-Planck-Institut für Kernphysik, Postf. 103980, 69029 Heidelberg, Germany

[c]School of Chemistry, Physics & Environmental Science, University of Sussex, Brighton, BN1 9QJ, UK

[d]Now at: Planetary and Space Science Research Institute, The Open University, Walton Hall, Milton Keynes, MK7 6AA, UK

The Cassini Dust Analyser (CDA) has the capacity to provide composition information from impacting micrometeoroids through impact ionization, time-of-flight mass spectrometry. The instrument, aboard the Cassini-Huygens mission, has been collecting data since March 1999 during its journey to the Saturnian system (2004) via Jupiter (December 2000).

The University of Kent at Canterbury is using A 2-MV Van de Graaff electrostatic accelerator and laboratory CDA model to perform a calibration programme primarily focused upon the Chemical Analyser. The particles used to simulate impacts and characterise the instrument responses are of the same velocity (up to 37 km s^{-1}) and mass (10^{-19}-10^{-15} kg) as those expected from interplanetary space. A variety of materials are used although of particular interest are revolutionary low-density, coated-latex particles. They enable complex organic spectra to be obtained under laboratory conditions.

The suitability of these particles for the calibration of CDA type instruments and the current status of the ongoing programme to investigate the system's response are reported.

1. INTRODUCTION

Onboard the Cassini-Huygens mission destined for the Saturn system, the Cassini Cosmic Dust Analyser (CDA) instrument will provide extended in-situ observations of the dust environment of particle mass 10^{-15} - 10^{-9}g. The seven year cruise enables an extended period for measurement of the interplanetary environment and of the Saturnian system on arrival. Additionally, during the Jupiter fly-by in December 2000 there arises the unique opportunity to take simultaneously measurements with the CDA instrument and its forefather the Galileo dust flux instrument [1]. CDA is the next generation descendant of Galileo, with the inclusion of a chemical analyser capable of providing compositional information. The principle of a time-of-flight mass spectrometer relying on plasma production by hypervelocity impact is first reported in [2]. Impact ionization is produced when a hypervelocity impact occurs; the velocity of an impact is greater than the speed of the compression wave produced in both the projectile and target. This results in immense shock waves within the projectile and target material causing intense heating and vaporisation hence producing an ionic plasma.

On CDA the curved chemical analyzer target has a potential of 1 kV with an earthed grid 3 mm above. This accelerates the positively charged constituent of the ionic plasma towards a centrally located detector beyond a short drift region, resulting in a time-of-flight spectrum detailing the anionic components of the impacting projectile and the target material [3].

In addition to CDA, the Cometary and Interstellar Dust Analyzer (CIDA) aboard the Stardust mission has collected data on several occasions since February 1999. Some CDA and most of the spectra measured by CIDA have shown features quite different from the laboratory calibrations, characterised by broad features with masses up to several hundred amu. The higher mass resolution of CIDA has led to interpretation of these features as due to complex organics containing nitrogen and/or oxygen [4]. Interpretation of the CDA spectra to glean impactor compositional information is nontrivial with many perturbing influences. Such interpretation must rely upon extensive laboratory calibration to characterise the instrument and constrain the processes involved.

The University of Kent at Canterbury 2 MV Van de Graaff electrostatic accelerator [5] accelerates suitable projectiles to simulate impacts on a CDA representative model. Traditional projectiles (e.g. iron) are used to characterise instrument responses, although the resulting time-of-flight spectra contain no molecularly bonded high-mass components. Revolutionary low-density projectiles are enabling the investigation of complex organic spectral responses within a laboratory environment, aiding the interpretation of flight data.

2. EXPERIMENTAL PROCEDURE

Advances in particle coating processes have enabled the coating of low-density polystyrene latex spheres with a number of different conducting polymers suitable for acceleration [6]. The projectiles are synthesised at the University of Sussex [7,8]. The conducting polymer is deposited as an ultrathin overlayer (less than 10% by mass) onto the surface of a polystyrene latex sphere, producing a well-defined core-shell morphology. Each sample has a fairly monodisperse size distribution clearly apparent in Figure 1, an electron micrograph of a PEDOT coated polystyrene latex of size 1.8 μm diameter. The non-conducting core latex is readily doped with different elements and the versatility of core size and coating thickness enables a range of projectile diameters and compositions to be synthesised.

The particles' ability to retain surface charge makes them suitable for use in an electrostatic accelerator, and they represent better substitutes for carbonaceous cosmic particles than traditional projectiles. They mimic the carbon-rich composition of cosmic dust grains and their densities are comparable, although cosmic grains are believed to contain more dense elements they are also believed to be micro-porous. The controllable size enables a range of velocities to be investigated and a mean projectile mass to be accurately obtained for each sample. The method of acceleration with the 2MV electrostatic accelerator is discussed in [5]. The laboratory CDA model was dimensionally identical to the flight instrument and had a representative rhodium target; its operation is given in [9]. The accelerated particles were within the same velocity and mass ranges as those expected from interplanetary space (e.g. up to ~37 km s^{1} and 10^{19} – 10^{15} kg.).

In the Van de Graaff the velocity (v) and particle charge (q) is measured for each event through detection along the beam line. The mass (m) of the impacts can be determined from the accelerating voltage (V):

$$m = (2\,q\,V) / v^2 \qquad (1)$$

For flight events the velocity is obtained from empirically calibrated rise times of the charge signals on the instrument.

The mass spectra are obtained in terms of time-of-flight and must have an atomic mass scale applied. For this work we assumed that only single ionization has occurred. The method adopted has been to identify two known features within each spectrum to "anchor" the mass scale. Sodium and potassium, as contaminants within the system were always present and clearly identifiable within the complex spectra of these low-density projectiles. The resulting calibration was verified for accuracy through comparison with the impact time as recorded by the target signal.

3. RESULTS AND ANALYSIS

We present three successfully accelerated low-density projectile samples in addition to iron used as a control. Two different conducting coatings - poly(3,4-ethylenedioxythiophene) (PEDOT) and polypyrrole (PPY) - were used over polystyrene latex (PS) as well as a sample of pure polypyrrole (PPY). Table 1 shows sample characteristics and responses. The mean mass has been calculated from the mean sample diameter and average density; note this hasn't been determined for iron as its size isn't monodispersive.

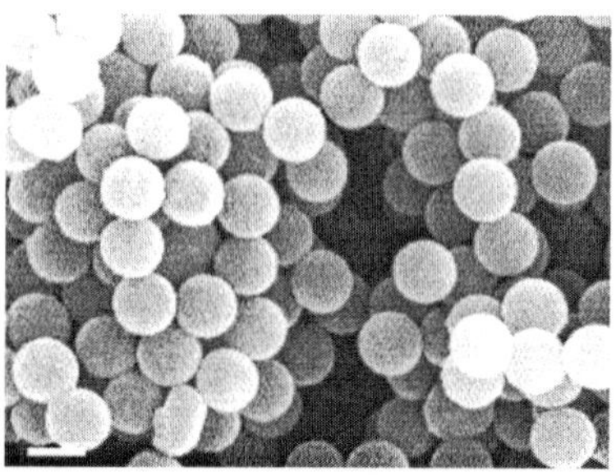

Figure 1. Scanning electron micrograph of 1.8 μm PEDOT coated (5.8% by mass) polystyrene latex. The scale bar (lower left) represents 2 μm.

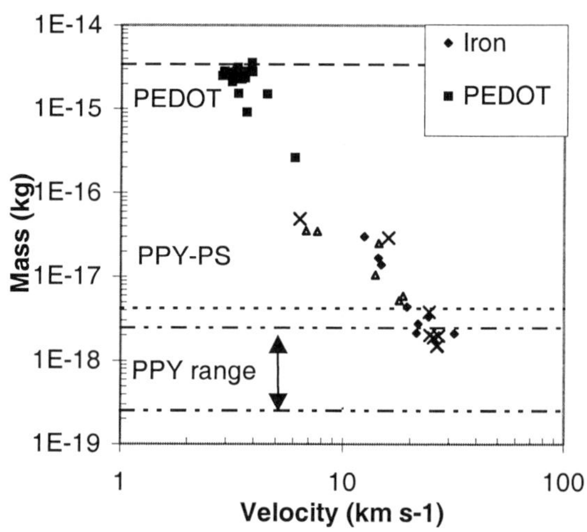

Figure 2. Mass velocity plot. (the mass is obtained from equation 1). Horizontal lines indicate the measured sample grain mass with maximum and minimum limits for pure PPY.

Figure 2 shows the mass distribution for each sample with respect to velocity. The horizontal lines represent the mean sample mass, where the range of values for pure PPY is indicated with a maximum and minimum line. Iron has no clearly defined mean mass and this is reflected in the plot with the mass ranging 2.5 - 24.7x10^{-18} kg. The PEDOT latex sample predominately falls close to the measured sample grain mass. However, the PPY latex and pure PPY samples appear to give a range of values with typically a larger mass than expected. It is possible these are multiple grains clustered together.

The spectra obtained from the low-density particles display a complex response, completely contrasting with the clean elemental spectra obtained from iron. There are many features of mass ranging 1 - 200amu consisting of both elemental components and molecularly bonded fragments. The spectra obtained within each sample type display consistent peak responses for similar velocities. However, there are variations in peak intensities. This is possibly due to slight variations in event conditions, e.g. target morphology. Also there seem to be different peak formations with different velocities, which needs further investigation. The spectra from those events with a high speed of impact reflect a greater abundance of peaks with lower mass and lower abundance of large molecular groups. The sample with most events, PEDOT, has a strictly constrained velocity response.

Figure 3 shows a representative spectrum. System contaminants sodium (Na, 23 amu) and potassium (K, 39 amu) are clearly identifiable and labelled, and there are many higher peaks, one of which may be the target material, rhodium (although this is not expected at such low velocities). A very strong signal is evident at 91 amu. The core material, polystyrene latex, produces a strong response at 91 amu under thermal decomposition [10] and interestingly theis isn't evident in the pure PPY spectra, although the different velocity regimes could account for this.

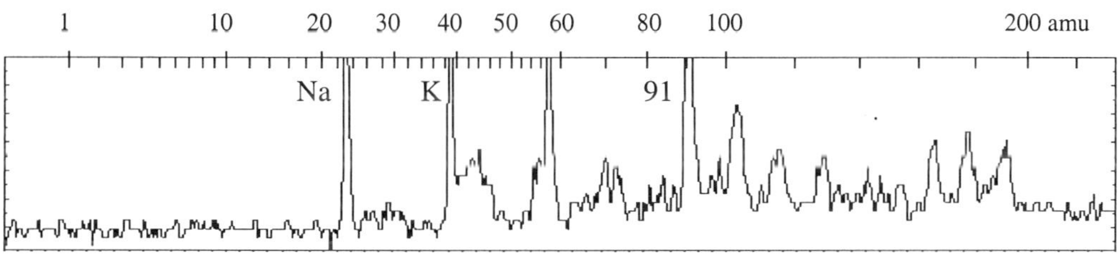

Figure 3. Typical mass spectrum for PEDOT-PS event, velocity 3.25km s^{-1}. Mass scale calibrated using Sodium (23amu) and Potassium (39amu)

Table 1. Summary of Sample Characteristics

	Name	Mean Diam. (μm)	Mean Mass (kg)	No. Events	Velocity (km s^{-1})
1	Iron	N/A	N/A	19	12.40-31.70
2	PEDOT - PS	1.81	3.42 x 10^{-15}	40	2.86-6.05
3	PPY - PS	0.19	4.17 x 10^{-18}	7	6.80-18.50
4	Pure PPY	0.07-0.15	(0.25-2.47) x 10^{-18}	8	6.37-26.80

4. CONCLUSIONS

The first complex organic spectra have been obtained in the laboratory from hypervelocity impacts using a variety of low density carbon-rich projectiles on a CDA representative instrument. There is evidence for molecularly bonded components within the impact produced ionic plasma even at low velocities, ~3 km s^{-1}. Spectral responses are consistent although possible velocity dependencies have been identified requiring further investigation. Despite statistically small sample sizes, differences in spectral response due to composition are apparent and will be further characterised in the future.

The suitability of these projectiles to the calibration of CDA type instruments has been demonstrated. The well known nature of the organic projectiles will permit an improved understanding of the physics of ionic plasma formation in hypervelocity impact.

ACKNOWLEDGMENTS

We are grateful to PPARC for their continued support of this research

REFERENCES

1. E. Grün, H. Fechtig, M. Hanner, J. Kissel, B. Lindblad, D. Linkert, D. Maas, G. Morfill and H.A.. Zook, Space Sci. Rev. 60 (1992) 317.
2. J.F. Friichtenicht and J.C. Slattery, NASA Technical Note D-2091, (1963).
3. R. Srama, E. Grün and The CASSINI-Dust-Science Team. ASP Conference Series, 104 (1996) 227.
4. J. Kissel, personal communication.
5. M.J. Burchell, M.J. Cole, J.A.M. McDonnell and J.C. Zarnecki, Meas. Sci. Technol. 10 (1999) 41.
6. M.J. Burchell, M.J. Cole, S.F. Lascelles, M.A. Khan, C. Barthet, S.A. Wilson, D.B. Cairns and S.P. Armes, J. Phys. D: Appl. Phys. 32 (1999) 1719.
7. S.F. Lascelles and S.P. Armes, J. Matter. Chem. 7(8) (1997) 1339.
8. M.A. Khan and S.P. Armes, Langmuir 15 (1999) 3469.
9. P.R. Ratcliff, J.A.M. McDonnell, J.G. Firth and E. Grün, J. Brit. Interplan. Soc. 45 (1992) 355.
10. E.M. Silverstein and G.C. Bassler, Spectrometric Identification of Organic Compounds, John Wiley and Sons, (1964).

Analysis of micro-craters on metal targets formed by hyper velocity impacts

Y. Hamabe[a], S. Sasaki[a], H. Ohashi[b], T. Kawamura[c], K. Nogami[c], H. Yano[d], S. Hasegawa[d], and H. Shibata[e]

[a]Earth and Planetary Science, University of Tokyo, 3-7-1 Hongo, Bunkyo, Tokyo 113-0033, JAPAN.

[b]Department of Ocean Science, Tokyo University of Fisheries, Tokyo 108-8477, JAPAN.

[c]Departiment of Physics, Dokkyo University School of Medicine, Tochigi 321-0293, JAPAN

[d]Institute of Space and Astronautical Science, Kanagawa 229-0022, JAPAN

[e]Research Center for Nuclear Science and Technology, University of Tokyo, Tokyo 113-0032, JAPAN

This paper reports the impact craters formed by Ag microparticles impacts, comparing with the TOF-MS (Time-Of-Flight Mass Spectrometer) spectra obtained simultaneously. The TOF spectra suggested that ions were hardly produced on Al target compared to Mo and Au targets. To determine the reason for this and to select the best material that ionizes the impacting dust particles, we observed impact craters on these targets with SEM and STM. Our results suggest that the ionization depends on the materials and that the combination of target and projectile materials causes the differences of crater shape and the Ag residue distribution.

1. INTRODUCTION

The production of plasma by hypervelocity impacts has been studied with the experimental investigations. Previous works suggested that the ionization required higher impact velocity above 10km/s [1,2], and that the threshold velocity depended on the characteristics of projectile or target materials such as the ionization energy and the temperature [3]. However, ion formation is not explained only by the simple physical processes because of the complicated energy division. The division ratio of the impacting energy is related to some factors such as the differences in their density, melting point, ductility, and so on.

2. IMPACT EXPERIMENTS OF MICROPARTICLES AND TOF-MS SPECTRA

Impact experiments of microparticles were performed with a linear type TOF-MS at HIT (High Fluence Irradiation Facility, University of Tokyo) in Japan [4]. Silver particles around 0.5-2.0μm in diameter were accelerated up to 2-7km/s, and they impacted on a metal target (Al, Mo and Au) with the incident angle of 45 degrees. Ions are produced by the hypervelocity impacts of Ag projectiles and accelerated in the accelerating region between the target surface and the front Cu mesh. After ions are extracted from the target region, they fly in the drift region of length 141mm and achieved at the detector (MSP; Micro Sphere Plate).

It is expected that ions both from the projectile and the target are detected. However, we could hardly obtain Ag^+ and Al^+ for the impacts of Ag projectiles on Al target. On the contrary, Mo^+, Au^+ and Ag^+ were sometimes detected for Mo and Au targets, especially when the impact velocity exceeded above 3km/s. The ion peaks from projectile (Ag^+) and target (Al^+, Mo^+, and Au^+) were not obtained repeatedly. The peaks of Na^+ and K^+, the contaminants in the target or projectile materials, were detected in the most spectra.

3. OBSERVATIONS OF MICRO-SIZED CRATERS ON METAL TARGETS

Does the difference of target materials cause the difference in ion production from a projectile and a target? To find a clue to elucidate what caused such difference in detecting ions, we analyzed impact craters on these metal targets with SEM and STM. The representative three craters on a target are shown in Figure 1, where the projectiles were impacted from the upper side with the incident angle of 45 degrees. The distributions of Ag residues are shown below the SEM images of the craters.

Impact craters are roughly divided into two types in shape. One is a ripple-shaped crater, which is created when an Ag projectile bombarded onto the target with higher velocity and the projectile was evaporated completely. The other is the case that an Ag projectile did not have enough velocity to vaporize itself and it remained in/around the crater as residues. The distributions and the amounts of Ag residues are different on the respective targets.

The craters on Al target have two characteristics. One is the quantity of Ag residues; the craters have larger quantities of Ag residues inside the craters but not around the craters (a-1 and a-2). Another characteristic is that more than 80% of craters on Al target have a widespread melting trace around the crater (a-3). The spread of melting trace is more than a few times as large as the diameter of crater. The craters with the melting trace are larger than the other craters without it.

On Mo target, more ripple-shaped craters (b-1) are observed compared to the craters with Ag residues inside the crater or/and on the rim (b-2 and b-3). Among 217 craters we analyzed, the ripple-shaped craters and the craters with Ag residues are about 9.7% and 84%, respectively. As can be seen in Table 1, the mean diameter of ripple-shaped craters is larger than that of craters with Ag residues. The SEM images show that the craters on Mo target are shallower than ones on other targets.

The craters on Au target have the largest mean diameter of 3.05μm (Figure 1c, Table 1). However, there is a clear distinction between the mean diameters of the ripple-shaped crater and the craters with Ag residues. Most of Ag residues distribute outside the craters in all directions, and they distribute broader than other targets. It is also noticed that the existence probability of ripple-shaped crater of 43% is higher than other targets.

Crater depth is one of the most important factors when we discuss the relation between the crater shape and the impact velocity. According to the SEM images, the crater depth of Al target is deeper than that of other targets, which are confirmed by stereo pairs of SEM images. Since the absolute value of crater depth cannot be measured with SEM, we observed some craters on Al and Au targets with a laser microscope and STM. The crater depth is about 5-6μm on Al target which is measured with a laser microscope, and the STM image shows that the depth is about 0.8μm on Au target.

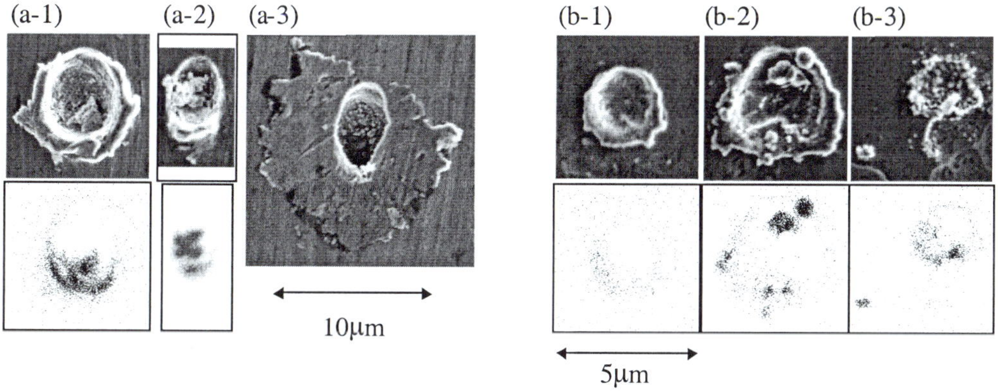

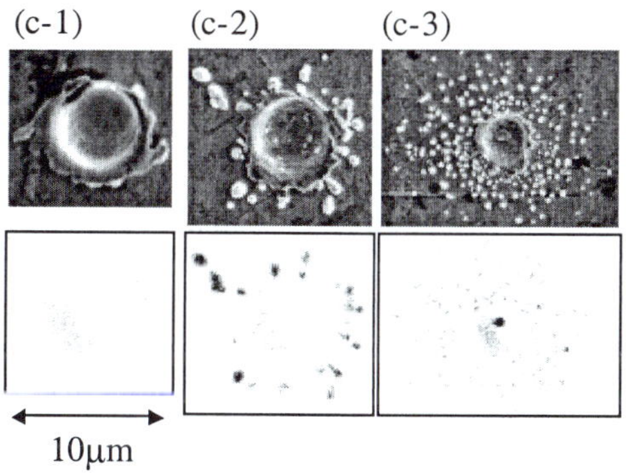

Figure 1. The SEM images of impact craters.
Silver projectiles impacted from the upper direction with the incident angle of 45 degrees. The distributions of Ag residues are shown below the crater images.
(a) Al target
(b) Mo target
(c) Au target

4. DISCUSSION

The size and shape of impact craters mainly depend on the impact velocity, and the distribution of Ag residues is related to the characteristics of target materials such as the melting point and the evaporation energy. The melting point of aluminum is the lowest of all target materials used in these experiments; Al (660° C)< Au (1064° C)< Mo (2620° C) Silver has a higher melting point than aluminum, which is lower than gold and molybdenum. This leads to the distribution of Ag residues; the observation that Ag residues remain inside the craters on Al target is due to the nature that the melting point of aluminum is lower than that of silver, Ag residues distribute inside and around the craters on the other targets because silver has lower melting point than gold and molybdenum. The Ag projectiles buried into the Al target deeply, while they scattered around the craters on Au and Mo targets.

The crater depth on Mo target is relatively the shallowest of all targets. The shallow crater may be also related to the highest melting point of molybdenum and its largest evaporation energy. This means that much energy is required to dig the Mo target, and that Al target is dug deeply because a little energy is required to evaporate much volume of aluminum. This is supported by the observations of craters with the SEM. Evaporation energy has a great influence on excavations of the metal targets.

The ionization of impacting projectiles depends on the impact velocity, the melting point and the density difference of the projectile and target materials. The SEM images show that Ag residues remained inside and around the craters on all targets. Such a large quantity of Ag residue is mainly due to the low velocity of Ag projectile. In other words, Ag projectiles could not be vaporized sufficiently at these low impact velocities because the impact energy is divided into the energies to vaporize the projectile and to dig the target. The division may be related to the density difference, which is supported by the TOF spectra; Mo, Au and Ag ions were often detected for the impacts of Ag projectile on Mo and Au targets, while Al and Ag ions were not obtained for Al target.

5. CONCLUSION

The results of our experiments showed that ion production depends on several factors such as the density difference between the projectile and the target and evaporation energy as well as the impact velocity of Ag projectile. These dependences are confirmed by analyses of impact craters on different targets (Al, Mo and Au). The existence of Ag residues means the impact velocity is not enough to vaporize the Ag projectile, and the crater depth shows the relative difference of degree in the evaporation of the target and the .projectile. The best material for the target of a dust analyzer is one capable of ionizing dust particles effectively. Our results suggest that molybdenum is the best target material in the three materials only when silver is used as the projectile; molybdenum can vaporize a projectile much efficiently because molybdenum has larger evaporation energy than other materials.

Table 1. Classification of craters by shape and size.
The numbers in parentheses indicate the mean diameters of each type of crater.

	Number	Ripple-Shaped [%]	Ag Residues [%]			Unknown [%]	Melting Trace	
			Inside	Outside	Both		O	X
Al	1239 (2.79)	5.2 (2.67)	89.1 (2.80)	0	5.2 (2.84)	50.0	82.2 (2.80)	17.3 (2.75)
Mo	217 (2.85)	9.7 (2.85)	77.0 (2.59)	0.5 (2.61)	6.8 (2.60)	6.0		
Au	669 (2.93)	43.2 (2.93)	35.4 (3.14)	9.1 (3.05)	11.4 (3.13)	0.9		

REFERENCES

1. J. Kissel and F. R. Krueger, Appl. Phys. A 42 (1986) 69.
2. F. R. Krueger and J. Kissel, ESA SP 224 (1984) 43.
3. P. R. Ratcliff, M. J. Burchell, M. J. Cole, T. W. Murphy and F. Allahdadi, Int. J. Impact Engng. 20 (1997) 663.
4. S. Hasegawa, A. Fujiwara, H. Yano, T. Nishimura, S. Sasaki, H. Ohashi, T. Iwai, K. Kobayashi and H. Shibata, Adv. Space Res. 23 (1999) 119.

Grain-target collision experiments and astrophysical implications

T. Poppe and Th. Henning [a]

[a]Astrophysical Institute and University Observatory,
Schillergäßchen 3, 07745 Jena, Germany

The collisional behaviour of micron-sized grains in the solar nebula is regarded to be important for the preplanetary dust aggregation. Collision experiments with analogous materials and velocities between about 0.1 and 100 m s^{-1} demonstrated that both the sticking efficiency and the collisional grain charging is higher than previously assumed. We summarize the results and we discuss some implications in the frame of models of preplanetary dust aggregation. Based on the experimental results, we also point out the possible role of magnetic forces on charged particles and make conclusions concerning the possible chondrule formation by nebular lightning.

1. INTRODUCTION

It is generally accepted that the first step of the planet formation process is the dust aggregation of initially submicron-sized and/or micron-sized grains in the dilute gas of the early solar nebula. Among the small grains, the tiny gravitational interaction played no significant role, but the grains formed planetesimals due to other types of interactions involving surface forces, electrostatic forces, and aerodynamic forces. Any modelling of the dust aggragtion process must be based on assumptions about their collisional behaviour. Therefore, we conducted laboratory experiments on dust grain collisions whose results are useful to draw conclusions on the dust aggregation process and on other physical conditions in the forming solar system.

2. SUMMARY OF EXPERIMENTAL RESULTS

We directly observed individual grain-target collisions in vacuum by imaging particle trajectories close to the target. A detailed description of the experimental work can be found in [1] and [2]. Here, we will give a short summary.

2.1. Sticking efficiency

We determined the sticking probability as a function of impact velocity, the energy loss in bouncing collisions and the collisional grain charging of micron-sized (0.14...1.4 μm diameter) grains consisting of materials (SiO_2, $MgSiO_3$, diamond, SiC) which are either astrophysically relevant themselves or have properties typical for certain relevant materials. We found that spheres impacting flat surfaces have a capture velocity, which is of the order of 1 m s^{-1} for particles of 1 micron diameter and which is both higher and less well defined for smaller particles. The capture velocity is determined by an effective dissipation of kinetic energy in the bulk material, whereas the deformation of surface roughnesses and the surface energy are less important for the analogous materials we used. Irregularly-shaped grains stick better due to their shape which supports multiple contacts and, thereby, they more effectively dissipate the impact energy. This leads either to a many-times higher capture velocity or, even more often, to a constant or slightly decreasing sticking probability in the investigated velocity interval. A fraction of several 10 percent of the submicron-sized grains stuck even at velocities exceeding 50 m s^{-1}. The influence of shape, particle size, surface roughness, and primarily the difference between spherical and irregular grain shape, turned out to be important, whereas an influence of material was not found among the grains used. Generally, in a bouncing collision, most of the kinetic energy is lost.

2.2. Collisional grain charging

Silica spheres of 1.2 μm diameter impacting silica targets acquire several 100 elementary charges in collisions with impact velocities of up to 50 m s^{-1}. The charge density at the point and time of contact is of the order of 10^{-4}C m^{-2}, and there is a maximum overall charge density of 10^{-5}C m^{-2} on silica targets and silica spheres which cannot be increased by further collisions. Temporarily higher charge densities at the point of contact decrease as charges spread across the surface. The charging in a collision with a certain impact velocity varies over 1 order of magnitude. The translational impact energies investigated ranged over 3 orders of magnitude from $E_{kin} = 10^{-15}$J to 10^{-12}J. Within this interval, the number n of net acquired elementary charges was found to be between 0 and a few hundred and scaled roughly linearly with impact energy. The sign of particle charging was mostly, but not exclusively, negative. There is no evidence that even highly precharged particles consisting of insulating material ever discharged upon a bouncing contact with a target carrying charge of the opposite sign. Collision-induced charging often resulted in attracting the particles which mechanically rebounded. Then the particles often returned to the target and, after literally 'jumping' on the target, eventually stuck.

The strong collisional grain charging must not be regarded as an exceptional effect of the materials used. It occured with all dust sample materials. Moreover, further experiments showed that collisional grain charging is not an effect restricted to insulators only: We observed 'jumping' SiO_2 spheres also on a metal-coated silica plate [3] and on aluminium targets (not published elsewhere) which were marked by the same altitude of a few hundred microns. This experimental basis demonstrates (1) that the charging in collisions between insulating and metallic materials is of the same order of magnitude as between two insulating materials and (2) that collisional grain charging between targets and particles made of the same material is not fundamentally different from the charging between targets and particles made of different materials.

2.3. Comparison of results to assumptions on grain collisions in the literature

The theoretically calculated capture velocity of micron-sized spherical grains [4] was too low by more than one order of magnitude. Furthermore, the even higher sticking efficiency of irregular grains was not treated and the lack of a capture velocity for many irregular grains was not discussed. Also, collisional grain charging is by one to two orders of magnitude stronger than previously assumed in astrophysical literature (e.g. [5] and [6]). Only recently, Desch and Cuzzi [7] pointed out that collisional charging could be much stronger. However, they assume a grain charging in excess of our measurements and they address a charging mechanism which did not cause the charging we observed. Based on [8], Desch and Cuzzi [7] assumed that, due to a different contact potential, electrons migrate during collision. This necessarily requires (1) the contact of different materials to allow charging, (2) an influence of the material combination on the charging strength, and (3) a uniform charge sign in a given particle-target combination which all is not in agreement with our results.

3. ASTROPHYSICAL IMPLICATIONS

3.1. Dust aggregation proccss in the early solar nebula

Weidenschilling and Cuzzi [9] developed a model of dust aggregation in which the relative velocities between grains embedded in the solar nebula gas are determined by their sizes. A crucial point in this model is the question under which conditions a collision will lead to sticking or not. Based on the collision velocities given in [9] for 1 AU solar distance, our results indicate that micron-sized grains always stick to objects up to cm to dm size and can stick to objects of dm to m size due to effects of irregular grain shape or collisional grain charging. The latter effect can directly contribute to the efficiency of the preplanetary dust aggregation because rebounding grains can be trapped in the collision-induced electrostatic fields. Particle sticking in the velocity regime of several 10 m s^{-1} due to effects of irregular grain shape or collisional charging is important in the frame of the model because the authors identified the growth in the cm to m range as the "crucial gap". Both for smaller and for larger bodies, the relative velocities in collisions are smaller than for this size range and can thus more easily explain particle growth. Also for the outcome of a more recent modelling by Weidenschilling [10] which is more related to comct formation but still based on the same model as used before in [9], the stickiness in the velocity regime of several 10 m s^{-1} is critical.

Electrostatic forces are not the only long-range interactions which support the dust aggregation process. Nübold and Glassmeier [11] theoretically investigated the aggregation of magnetic particles in an astrophysical context and additional experimental [12] and numerical work on this topic is under way.

3.2. Influence of magnetic fields on charged particle motion?

Weidenschilling and Cuzzi [9] assumed that the motion of particles in the solar nebula was exclusively caused by gravitation and aerodynamical drag forces. However, our results show that the particles or a fraction of them could be highly charged which rises the question about influences of magnetic fields. In [2] we demonstrated that, according to presently discussed magnetic conditions in the solar nebula, magnetic forces on micron-sized particles could dominate aerodynamical ones. However, we also pointed out that assumptions about the magnetic field and its motion are highly uncertain today. Further work will be necessary to investigate the possible action of magnetic forces on particles which could influence the spatial dust distribution in the solar nebula, the relative velocities in collisions. Concerning the idea that, in case that magnetic forces cause fast enough collisions, further collisional charging could lead to a self-sustained collisional grain charging process.

3.3. Chondrule formation by lightning

Meteoritic chondrules are stony spheroids of mm-size which underwent a melting by a short-duration heating in the time of the solar system formation. Chondrules are common in meteorites suggesting that they are important for understanding physical conditions in the solar nebula. Collisional dust grain electrification, subsequent large-scale electric field generation by separation and accumulation of oppositely charged objects, and, finally, electrical breakdown in the neutral gas of the early solar nebula has long been discussed to lead to lightning capable of melting meteoritic chondrule precursors. Although this explanation is, compared to others, not favored today [13], it is still seriously under discussion. Three points of recent work show that a detailed treatment of this topic might lead to new insights which support the idea of chondrule formation by lightning. (1) Grain charging is, according to our experimental results, by orders of magnitude stronger than in former models [5] [6]. The work of Desch and Cuzzi [7] showed that a numerical model involving stronger charge transfers in grain-grain collisions could explain chondrule formation by lightning. (2) Based on experimental work, it was pointed out that the dust aggregates are less compact than assumed in former work resulting in larger surface-to-mass ratios [14]. With the larger surface, a larger surface charge can be carried per particle mass. This effect is comparable to an effect investigated in [6] who treated the role of a fine dust component. In the frame of this model, the efficient transport of electric charges due to a large surface-to-mass ratio would support the the chondrule formation by lightning if such a dust component existed. Our experiments help to quantify the charge transport because surface charge densities resulting from collisional charging were determined, and they indicate a high charge transport efficiency. (3) The awareness of and the basis for treating the large-scale separation of oppositely charged objects due to interactions with the turbulent nebular gas has improved. E.g. Cuzzi et al. [15] reviewed this topic with respect to chondrule formation, and Klahr and Henning [16] investigated the role of gas eddies in trapping and concentrating solid particles depending on their sizes and aerodynamical properties. Based on the results by Poppe et al. [2] summarized in this contribution, we propose to focus the interest on magnetic interactions with charged particles which could play an important role in determining the particle motion and the spatial distribution of the particles. This suggests that an additional separation mechanism directly sensitive to the sign of charged particles could be viable.

REFERENCES

1. T. Poppe, J. Blum and Th. Henning, Astrophys. J. 533 (2000) 454.
2. T. Poppe, J. Blum and Th. Henning, Astrophys. J. 533 (2000) 472.
3. E. Palomba, T. Poppe, L. Colangeli, P. Palumbo, J.M. Perrin, E. Bussoletti and Th. Henning, Planet. Space Sci. 49 (2001) 919.
4. A. Chokshi, A.G.G.M. Tielens and D. Hollenbach, Astrophys. J. 407 (1993) 806.
5. S.G. Gibbard, E.H. Levy and G.E. Morfill, Icarus 130 (1997) 517.
6. W. Pilipp, T.W. Hartquist, G.E. Morfill and E.H. Levy, Astron. Astrophys. 331 (1998) 121.
7. S.J. Desch and J.N. Cuzzi, Icarus 143 (2000) 87.
8. W.R. Harper, Contact and Frictional Electrification, Clarendon Press, Oxford, 1967.

9. S.J. Weidenschilling and J.N. Cuzzi, in Protostars and Planets III (eds. E.H. Levy and J.I. Lunine) University of Arizona Press, Tucson (1993) 1031.
10. S.J. Weidenschilling, Icarus 127 (1997) 290.
11. H. Nübold and K.-H. Glassmeier, Icarus 144 (2000) 149.
12. H. Nübold, T. Poppe and K.-H. Glassmeier, Adv. Space Res. (2000) submitted.
13. A.P. Boss, in Chondrules and the Protoplanetary Disk (eds. R.H. Hewins, R.H. Jones and E.R.D. Scott) (1996) 257.
14. J. Blum, Planetesimalentstehung im frühen Sonnensystem – Beiträge der Laborastrophysik, Habilitationsschrift (2nd Thesis), University of Jena, 1999.
15. J.N. Cuzzi, A.R. Dobrvolskis and R.C. Hogan, in Chondrules and the Protoplanetary Disk (eds. R.H. Hewins, R.H. Jones and E.R.D. Scott) (1996) 35.
16. H. Klahr Th. Henning, Icarus 128 (1997) 13.

Space weathering: spectral change and formation of nanophase iron due to pulse laser irradiation simulating impact heating of interplantary dust flux

S. Sasaki[a], T. Hiroi[b], K. Nakamura[c], Y. Hamabe[a], E. Kurahashi[a], and M. Yamada[a]

[a]Department of Earth and Planetary Science, University of Tokyo, Tokyo 113-0033, Japan

[b]Department of Geological Sciences, Brown University, Providence, RI 02912, USA

[c]Department of Earth and Planetary Sciences, Kobe University, Kobe 657-0013, Japan

The spectral mismatch between S-type asteroids and ordinary chondrites, is explained by the so-called "space weathering" process where impacts of interplanetary dust should change the optical properties of asteroid surfaces. To simulate the heating by dust impacts, comminuted olivine and pyroxene samples were irradiated by a nanosecond pulse laser. Laser-irradiated samples show significant depletion and reddening of the reflectance. Some asteroid spectra such as 349 Dembowska and 446 Aeternitas are reproduced by mixing of irradiated spectra. Changes of pyroxene spectra are much smaller than those of olivine. This is consistent with the statistical data of asteroids where olivine-rich asteroids have more reddened spectra. If energy efficiencies for changing the optical properties are the same between dust impact and laser irradiation, olivine-rich surfaces should be darkened and reddened in 10^8 yrs.

According to TEM analysis, nanophase particles (several to 30 nm in size) probably of iron are widely spread in the rim region of irradiated olivine grains. In contrast, no trace of structural change was observed on irradiated olivine crystal samples. The presence of a regolith-like surface would be essential for effective space weathering on asteroids.

1. INTRODUCTION

A large number of asteroids are categorized into S-type based on their reflectance spectra. Those S-type asteroids are believed to be parent bodies of ordinary chondrites [1]. However, the steep reddened slope of their reflectance spectra and the derived mineralogies of S-type asteroids are different from those of ordinary chondrites. The mismatch between spectra of ordinary chondrites and S-type asteroids is considered to be caused by so-called"space weathering", where high-velocity impacts of interplantary dust should change the optical properties of the uppermost surfaces of asteroids possibly covered with regolith. To explain optical property changes such as overall reflectance depletion, reddening, and weakened absorption bands, Hapke et al. [2] proposed formation of submicroscopic metallic (SMFe) iron particles (10 nm in size) within coatings produced by the recondensation of ferrous silicate vapor. Impacts of interplanetary dust particles are a plausible process to produce the vapor. Such nanophase iron particles were found on the vapor-deposited rims of lunar soils [3, 4]. Recently it was shown theoretically that nanophase iron particles on the grain surface rim should decrease and redden reflectance spectra of ordinary chondrites and lunar soils [5]. However, realistic simulation of the space weathering is not easy: at present hyper-velocity (>10 km s^{-1}) dust accelerators can hardly produce sufficient dust flux to simulate space weathering.

2. PULSE LASER IRRADIATION SIMULATING DUST IMPACT HEATING

Pulse laser beams have been used for small-scale short-duration heating of samples. Moroz et al. [6] observed changes of reflectance spectra of silicate samples after pulse laser heating. However, the pulse duration they used was 0.5-1 μs, which is 1000 times the real timescale of micrometeorite (1-10 μm size) impacts. To simulate space weathering by impact heating of dust particles as small as 1 μm and impact velocity higher than 10 km s^{-1}, nanosecond pulse duration is necessary. We irradiated powder and pellet samples of olivine and pyroxene by a pulse laser beam (1064 nm) with pulse duration 6-8 ns, which is comparable with a real dust impact [7, 8]. To perform uniform scanning of pulse laser irradiation, we made a small chamber on an X-Y stage [8]. The pulse energies, 1 and 30 mJ, can supply the total irradiation energy in unit area of 8×10^3 and 2.4×10^5 J m^{-2}, respectively. Since the footprint diameter of the beam is focused to be as small as 500 μm, the energy deposition rate under 30 mJ reached 10^{10} W cm^{-2}, where evaporation and ion formation is strongly expected [9]. After the laser irradiation, bidirectional reflectance spectra of samples were measured; a spectral range 250 – 2600 nm was recorded at every 10 nm.

3. CHANGE OF REFLECTANCE SPECTRA

In the present experiments, olivine is San Carlos Olivine with 8.97 wt% FeO and pyroxene is an enstatite from Bamble, Norway with 9.88 wt% FeO. Figure 1 shows a pellet sample of olivine after the pulse laser irradiation. Reflectance of olivine is apparently decreased. Measured bidirectional reflectances of olivine and enstatite are shown in Figure 2. The reflectance of the mm-size region in the center of the sample (Figure 1) was measured. Laser-irradiated samples show significant reddening: reduction of spectra is much larger in the visible region than in near infrared region. Measured from the continuum slope, relative depth of the absorption at 1 μm (and 2 μm for pyroxenes) is not changed largely. Figure 3 shows comparison of two asteroid spectra with best-fit mixings of irradiated spectra [11]. Treatment with pulse laser can reproduce those asteroid spectra very well. 446 Aeternitas is well explained by the irradiated olivine. 349 Dembowska, which is pyroxene-rich with distinct 1 and 2 μm absorption bands, is also explained.

In Figure 2, the enstatite spectrum is changed only after repetitive irradiation. Reflectance of olivine should be more easily changed than that of pyroxenes: changes of hypersthene and diogenite are also smaller than that of olivine [8]. As seen in the raw reflectance samples in Figure 2, reflectance at 1064 nm is nearly the same (60%); the difference cannot be ascribed to the difference of absorbed energy. The degree of optical change on simulated space weathering by pulse laser heating depends on the mineral composition. We also compiled asteroid reflectance data using areas of 1 μm and 2μm absorption bands [11, 12]. Figure 4 shows that S-type asteroids with a weaker 2 μm absorption band (relatively olivine-rich) show redder spectra than those with a distinct 2 μm band (relatively pyroxene-rich). The difference of weathering degree according to mineral composition is also observed in the space.

Figure 1. An olivine pellet sample after pulse laser (30 mJ) irradiation. Each beam spot diameter is about 500 μm and the irradiated area is approximately 10 mm x 10 mm. Size of pellet with an aluminum ring is 20 mm. The samples are formed of particles smaller than 75 μm.

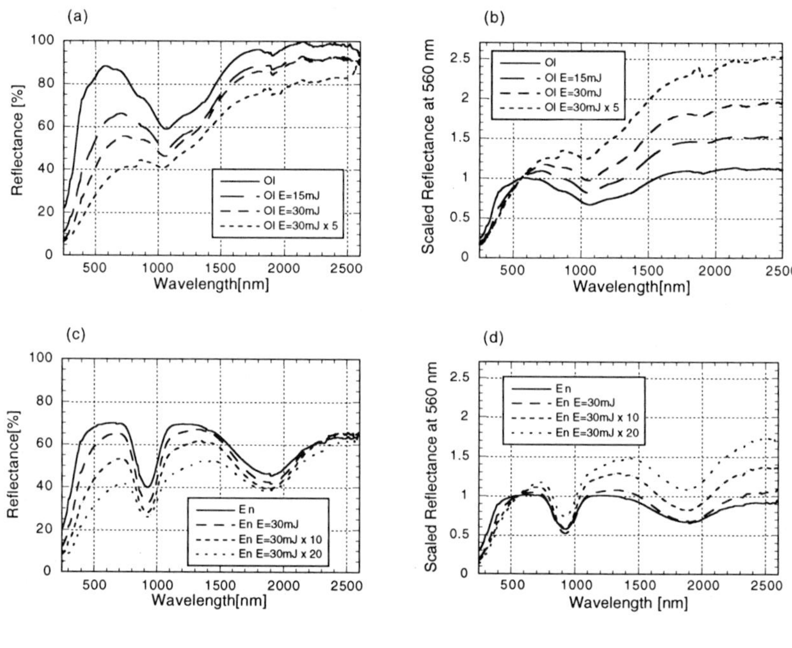

Figure 2. Reflectance spectra of laser-irradiated olivine and enstatite pellet samples. (a) Absolute olivine spectra. (b) Olivine spectra scaled at 560 nm. There are four spectra (raw, 15 mJ, 30 mJ, and 5 times 30 mJ). (c) Absolute enstatite spectra. (d) Enstatite spectra scaled at 560 nm. There are four spectra (raw, 30 mJ, 10 times 30 mJ pulse, and 20 times 30 mJ).

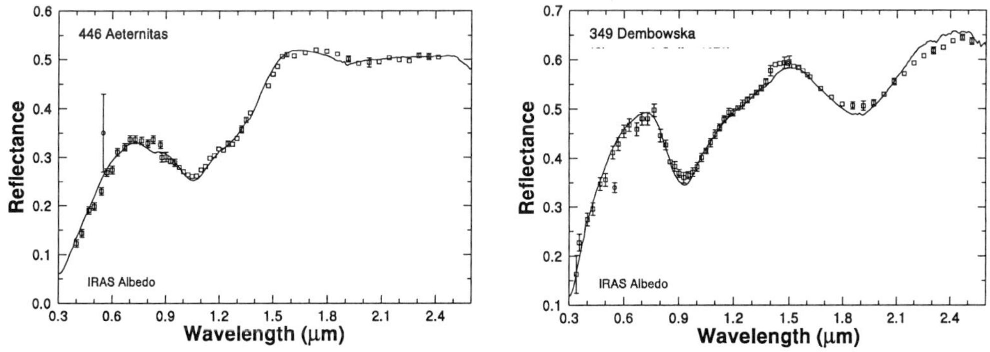

Figure 3. Asteroid spectra and best match mixing of pulse-laser irradiated samples [11]. (a) 446 Aeternitas. Mixing ratio: Ol (Non-irradiated) 2.0%; Ol (15 mJ) 19.4%; Ol (30 mJ) 73.7%; En (30 mJ x 10) 0.6%; En (30 mJ x 20) 0.3%; Chromite 4.0%. (b) 349 Dembowska. Mixing ratio: En (30 mJ) 7.0%; En (30 mJ x 20) 37.3%; Ol (raw) 0.7%; Ol (15 mJ) 55.0%.

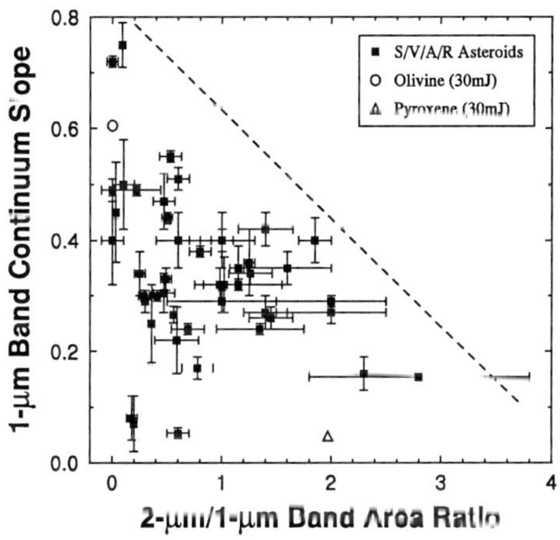

Figure 4. As for S/V/A/R asteroids, the continuum slope of the 1-μm band is compared with the area ratio between the 1-μm and 2-μm bands [13]. Data from laser-irradiated olivine and pyroxene pellet samples (Figure 2) are also shown for comparison.

In space, the impact rate of dust particles of around 10^{-12} g (1 μm size) is about a few 10^{-4} m^{-2} s^{-1} at 1 AU [10]. Relatively high-velocity beta meteoroids would be effective in producing the space weathering. One dust particle loses energy 2×10^{-7} J at the impact with velocity 20 km s^{-1}. Then, the total energy deposition rate by dust impacts is about 10^{-3} J m^{-2} yr^{-1}. Supposing a simple comparison between the irradiation energy by pulse laser and released energy at dust impacts, irradiation with a 30 mJ pulse laser in our experiments corresponds to a few 10^{8} yr in the space. If energy efficiencies for changing the optical properties are the same between dust impact and laser irradiation, olivine-rich surfaces should be darkened and reddened in 10^{8} yrs, whereas pyroxene-rich surface should need more time to be optically reddened.

4. TEM OBSERVATION

During pulse laser irradiation of samples in a vacuum chamber, evaporation should proceed on the grain surface since frequent changes of pressure were observed. But it was not clear whether evaporation or recondensation is responsible for the surface alteration. In order to clarify microscopic processes and the cause of reflectance change, we observed the irradiated samples by a transmission electron microscope (TEM).

From the irradiated samples, altered surface grains were picked up. Their thin sections (about 80-100 nm thickness) were produced using an ultramicrotome and observed by 200 keV High-Resolution TEM JEOL2010 at Kobe University equipped with EDS. Some sections of the rim region of irradiated olivine grains contain a large number of nanophase particles as shown in Figure 5. The size of those particles, which have never been observed in raw olivine samples, is between several to 30 nm. According to EDS analysis, the rim region with the nanophase particles is rich in iron. Moreover, electron diffraction patterns and high resolution TEM image showed that those nanophase particles are body-centered cubit alpha-Fe [14]. These nanophase particles should be responsible for the change of optical properties such as darkening and reddening as proposed [2] and calculated [5]. These particles may be also related to SMFe in lunar soils [3, 4]. We have also confirmed that nanophase iron particles are also found in irradiated pyroxene samples (Nakamura et al., in preparation).

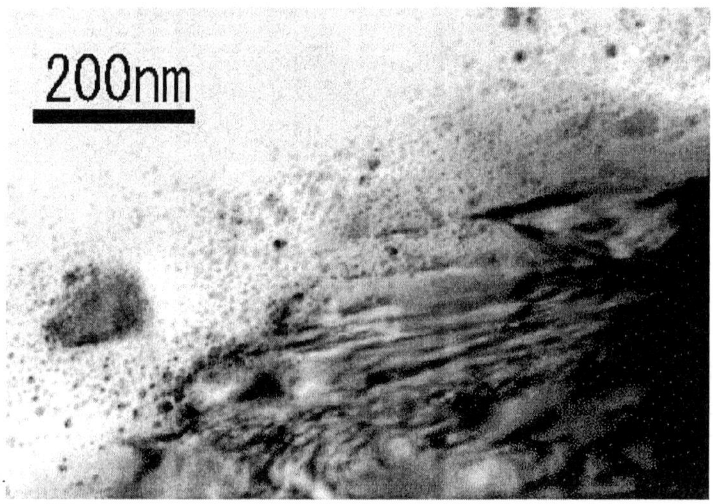

Figure 5. A bright-field TEM image of the rim of an altered olivine grain from the pulse-laser irradiated pellet sample (30 mJ x 5). There are many nanophase particles whose size is several to 20 nm. The rim is probably deposited from recondensation.

We also irradiated pulse laser on the flat surface of olivine crystal. Olivine crystal samples hardly changed optically, although surface evaporation was confirmed during laser irradiation. Irradiated crystal samples were investigated by TEM but no structural change on the crystal surface was observed. Evaporated materials should escape quickly from a flat crystal surface, whereas on pellet samples the evaporated phase could recondense and deposit on nearby grain surfaces. The presence of a regoligh-like surface would be essential for effective space weathering on asteroids. If very small asteroids are free from regolith, their surface could not be weathered easily.

ACKNOWLEDGEMENTS

The authors thank S. Hasegawa, H. Nagahara, H. Ohashi, H. Otake, and K. Tomeoka for their help with the experiments and discussions. The authors acknowledge T. Kogure for TEM measurements of olivine crystal samples.

REFERENCES

1. C. R. Chapman, Meteoritics and Planet. Sci. 31, (1996) 699.
2. B. Hapke, W. Cassidy and E. Wells, Moon 13 (1975) 339.
3. L. P. Keller and D. S. McKay, Science 261 (1993) 1305.
4. L. P. Keller and D. S. McKay, Geochim. Cosmochim. Acta 61 (1997) 2331.
5. B. Hapke, J. Geophys. Res. 106 (2001) 10039.
6. L. V. Moroz, A.V. Fisenko, L.F. Semjonova, C.M. Pieters and N.N. Korotaeva, Icarus 122 (1996) 366.
7. M. Yamada et al., Antarctic Meteorites XXIII (1998) 173.
8. M. Yamada, S. Sasaki, H. Nagahara, A. Fujiwara, S. Hasegawa, H. Yano, T. Hiroi, H. Ohashi and H. Otake, Earth, Planet Space 51 (1999) 1255.
9. J. Kissel and F. R. Krueger, Appl. Phys. A 42 (1987) 69.
10. E. Grün, H. Fechtig, M. S. Hanner, J. Kissel, B. A. Lindblad, D. Linkert, G. Morfill and H. A. Zook, in The Origin and Evolution of Interplanetary (eds. A.-C. Levasseur-Regourd and H. Hasegawa), Kluwer, Dordrecht, IAU Colloquium 126 (1991) 21.
11. T. Hiroi, T. and S. Sasaki, Meteoritics Planetary Sci. 36, (2000) 1587
12. T. Hiroi, T. and S. Sasaki, Lunar Planet. Sci. XXX (1999) #1444.
13. M. J. Gaffey, J.F. Bell, R.H. Brown and T.H. Burbine, Icarus 106 (1993) 573.
14. S. Sasaki, K. Nakamura,Y. Hamabe, E. Kurahashi and T Hiroi, Nature 410 (2001) 555.

Light scattering by flakes

K. W. T. Waldemarsson and B. Å. S. Gustafson

Department of Astronomy, University of Florida, Gainesville, FL 32611,USA.

This work examines how thin flakes in for example a cometary atmosphere can be recognized by the way they scatter light. The investigation includes both theoretical modeling and microwave analogue measurements of light scattered by flakes with a size of the order of a few wavelengths and larger. The theoretical modeling adds the transmitted field found using Babinet's principle and Fresnel's equations, to an existing model for dielectric edges within the framework of geometrical theory of diffraction. This gives an approximate solution to the scattering by thin 2-dimensional dielectric flakes. The laboratory measurements simulate light scattering by 0.25 μm thick randomly orientated circular silicate flakes of 8 μm diameter.

1. INTRODUCTION

The modeling of light scattering by flakes is motivated by the desire to recognize flake-like particles in a cometary atmosphere through their light scattering properties. The dynamics of Geminid meteoroids during atmospheric flight and of the Geminid meteor stream in space indicate that flakes may have been produced during cometary activity on asteroid 3200 Phaethon [1,2]. Further, laboratory simulation of sublimating particle/ice mixtures representing "dirty" ice surfaces on atmosphereless Solar System bodies has shown that thin flakes may indeed form [3].

The geometric optics approximation is applicable when the dimensions of the particle are much larger than the wavelength. The T-matrix method has been used for scattering by flakes up to 15 wavelengths in extent [4]. In this work, we model the light scattering by flakes within the premises of geometric theory of diffraction (GTD), in order to obtain a computationally fast yet accurate method for the scattering by thin flakes in the size range of a few wavelengths and larger.

Parallel to the theoretical modeling we have done an experimental investigation, in which we take advantage of the unique microwave scattering facility at the University of Florida [5]. Here, based on the principle of electromagnetic similitude, micron sized particles in the visual region are scaled to centimeter sized particles in the microwave region, which allow precise control of the particle and its orientation. The broadband capability of this facility allows us to measure the scattering in the wavelength interval 2.7-4 mm, which we use to simulate the 0.44-0.65 μm visual range. This enables us to obtain spectral properties such as color and polarimetric color experimentally in addition to intensity and polarization.

2. THEORETICAL MODELING

To illustrate the importance of diffraction when modeling the scattering by a flake, the scattering by a perfectly conducting thin flake with parallel edges was calculated using the

GTD solution in [6]. Figure 1 shows the intensity (given in units of the scattering matrix following the notation in [7]) and polarization of the scattering by an infinitely long 5.5 λ wide flake averaged over rotation is shown and compared with microwave test measurements. The experimentally obtained intensity (but not the polarization) has been shifted to best fit the intensity calculation since the laboratory scattering measurements necessarily are for a flake of finite length. Scattering averaged over rotation here means the average of the scattered light as the flake is rotated about its long dimension, which is kept perpendicular to the scattering plane. In this case there is no transmitted radiation and the comparison with the microwave data is very good. Note that using geometric optics only, or geometric optics with scalar Fraunhofer diffraction to model the scattering would fail in this case since these approximations yield no polarization.

To extend the modeling to the light scattering by a 2-dimensional dielectric thin flake, we first apply an approximate solution for scattering by impedance wedges [8] to treat the two parallel edges of the flake. However, as the transmitted radiation is not taken into account, this yields only the externally diffracted field. Using Babinet's principle and Fresnel's equations we find a solution also for the transmitted field. By adding the external and transmitted diffracted field vectors we arrive at an approximate model for the total scattering by a 2-dimensional thin dielectric flake. The result from this approach to describe the light scattered by a 6.3 λ wide dielectric (complex refractive index m = 2.5 + i0.02) thin flake averaged over rotation, is shown in Figure 2. An overall good agreement with test measurements is found in intensity and the main structure in the polarization is also reproduced.

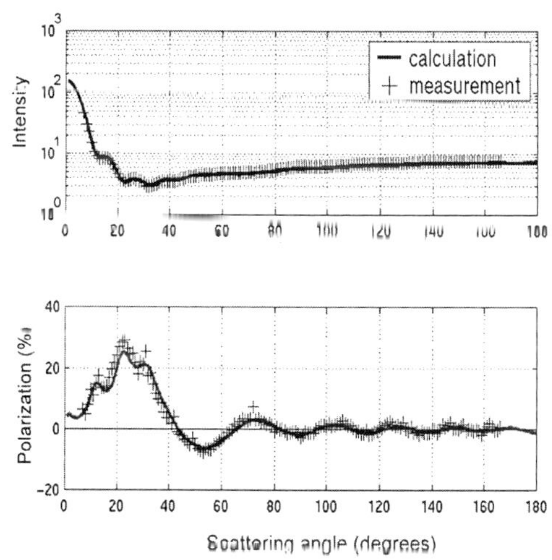

Figure 1. Scattering by a 5.5 λ wide thin conducting flake averaged over rotation.

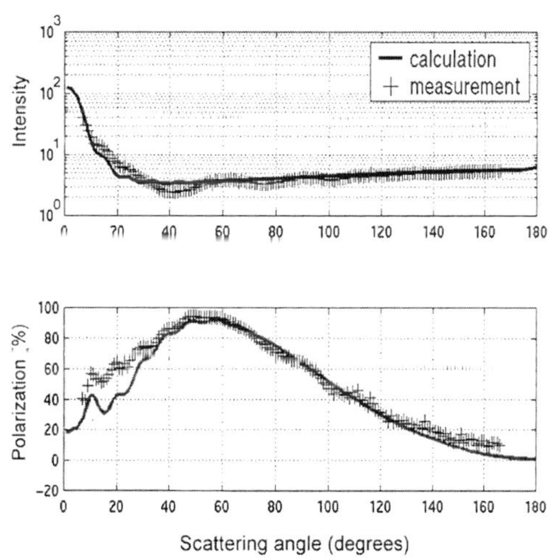

Figure 2. Scattering by a 6.35 λ wide dielectric thin flake averaged over rotation. Complex refractive index m = 2.5 + i0.02.

3. MICROWAVE ANALOG EXPERIMENTS

The refractive index of acrylic at microwave frequencies is in the range typical of silicates in the visual. An acrylic model was therefore used in the microwave scattering facility to simulate the scattering by an 8 μm diameter and 0.25 μm thick silicate flake in the visual part of the spectrum.

Figure 3 shows the intensity and polarization of the scattered light averaged over a uniform distribution of orientations to simulate the scattering from a thin cloud of flakes in random orientations. The total number of orientations used in the average is 1620 and the tilt angle between the normal to the surface of the flake and the spin axis is varied in steps of 10 degrees. Again, the spin axis is perpendicular to the scattering plane. The data has been

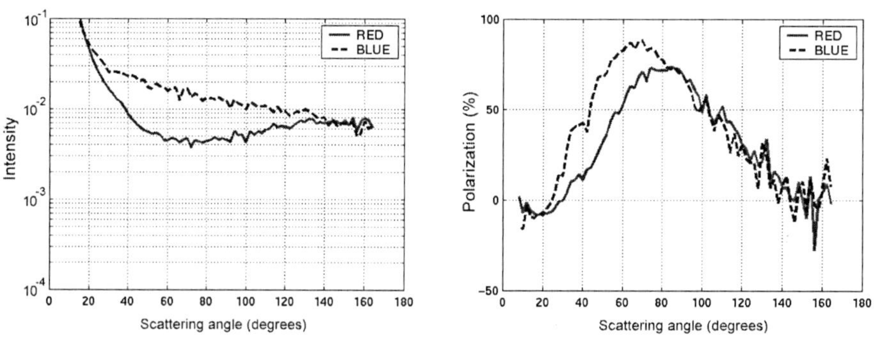

Figure 3. The dependence of intensity and polarization on scattering angle averaged over a uniform distribution of orientations to simulate randomness. Solid and dashed lines represent red and blue wavebands, respectively.

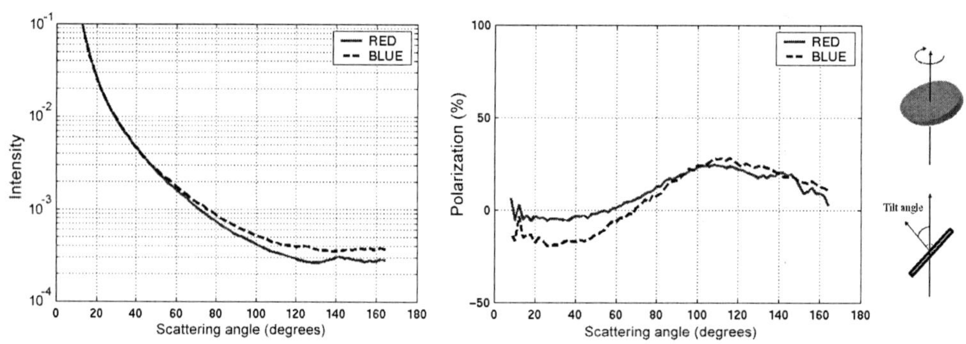

Figure 4. The dependence of intensity and polarization on scattering angle when averaged excluding specular reflection produced at tilt angle of 90 degrees. Solid and dashed lines represent red and blue wavebands, respectively.

reduced in two wavelength intervals, 2.7-3 mm and 3.5-4 mm corresponding to blue and red in the visual. The fine structure in the plots is likely due to the finite number of orientations used in the averaging. The main trends are that the scattered light in the blue waveband has a higher intensity than in the red, thus indicating blue color, and for scattering angles smaller than 90 degrees the polarimetric color is blue, i.e. the polarization is greater in the blue waveband than in the red. This is interesting as blue color is often an indication of very small particles, which usually have red polarimetric color. This may make flakes distinguishable from other common particle shapes.

Figure 4 shows the intensity and polarization excluding the most dominant set of orientations, the average over rotation when the tilt angle is 90 degrees. The scattering is much different from the case simulating average over random orientation. The intensity is of course much lower but the color is still blue, and the polarimetric color at smaller scattering angles has changed to red. This illustrates, that randomly orientated flakes and flakes that are aligned in some preferred orientation(s), might be distinguishable based on their very different scattering properties.

4. SUMMARY

We have obtained an approximate computational model for light scattering by a 2-dimensional thin dielectric flake by applying a solution for scattering by an impedance wedge to the edges of the flake and adding the transmitted diffracted field acquired using Babinet's principle and Fresnel's equations. Results from our model have been shown to agree well with microwave test measurements. Analog microwave measurements simulating the light scattering by a silicate flake averaged over random orientation, show over all blue color, as well as blue polarimetric color at scattering angles smaller than 90 degrees. Blue color is thus not necessarily an indication of very small particles but can also indicate flakes. We suggest that it could be possible to use polarimetric color data to distinguish these two types of particle geometries. By looking at subsets of the measurements, it was shown that aligned flakes and flakes which are arbitrarily orientated, can show very different light scattering characteristics.

REFERENCES

1. B.Å.S Gustafson, in Asteroids, Comets, Meteors III (eds. C.-I. Lagerkvist, P. Magnusson, and H. Rickman) Uppsala Univ. Press, (1990) 523.
2. B.Å.S Gustafson. Astron. Astrophys, 225 (1989) 533.
3. J.R. Stephens and B.Å.S Gustafson, Icarus 94, (1991) 209.
4. M.I. Mishchenko, D.J. Wielaard and B.E. Carlson. Geophys. Res. Lett. 24 (1997) 771.
5. B.Å.S Gustafson, J. Quantitative Spectroscopy and Radiative Transfer 55 (1996) 663.
6. R.G. Kouyoumjian and P.H. Pathak, Proc. IEEE 62 (1974) 1448.
7. H.C. van de Hulst, Light Scattering by Small Particles, Dover Publications Inc., New York, 1957.
8. H.H. Syed and J.L. Volakis, Radio Sci. 30 (1995) 505.

Aggregation experiments with magnetized dust grains

H. Nübold[a*], T. Poppe[b] and K.-H. Glassmeier[a]

[a]Institut für Geophysik, TU Braunschweig, Mendelssohnstr. 3, D-38106 Braunschweig, Germany, Email: h.nuebold@tu-bs.de

[b]Astrophysikalisches Institut der Universiät Jena, Schillergäßchen 2-3, D-07745 Jena, Germany

The growth of fractal aggregates from small dust grains is believed to be the first step in the formation process of planetesimals and cometesimals. The grain-grain interactions treated up to now were mainly of mechanical and electrostatic nature. However, there is evidence that magnetic interactions between permanently magnetized dust grains could have played an important role in the aggregation process. We prepared aggregation experiments with micron-sized magnetized dust analogues in a dilute gas. In this paper, we describe a feasibility study on the use of the magnetized dust in existing experimental facilities. First results are presented.

1. INTRODUCTION

The formation of macroscopic objects in the early Solar System is generally believed to have been initiated by collisional growth of dust aggregates from micron-sized grains. This mechanism has replaced gravitational instability as the 'standard' way of assembling planetesimals in the solar nebula, since it has been shown, that turbulence should inhibit the formation of a dust layer dense enough to become gravitationally unstable [1]. On the other hand, today's picture of preplanetary dust aggregation is far from being complete. The problem of how collisional grain growth could proceed fast enough to create the building blocks of our Solar System within the relatively short period of $10^6 - 10^7$ years [2] remains to be solved. The very first stage of the aggregation process produces fluffy clusters of fractal nature. How and when does compaction toward more massive objects occur [3]? These are just a few considerations showing that preplanetary dust aggregation deserves further study.

Up to now, preplanetary dust aggregation scenarios mainly included mechanical (surface) interactions between dust particles [4,5]. Few authors considered long ranged interactions: Ossenkopf [6] as well as Horanyi and Goertz [7] discussed electrostatic interactions between charged grains, Nuth et al. [9] performed experiments with small magnetic iron particles. We believe that the latter deserves additional attention since magnetic interactions between magnetized grains add a some interesting features to the dust aggregation process. To this end, we assume that a magnetizable dust population was present in the

*supported by the *Studienstiftung des deutschen Volkes* through their PhD-program.

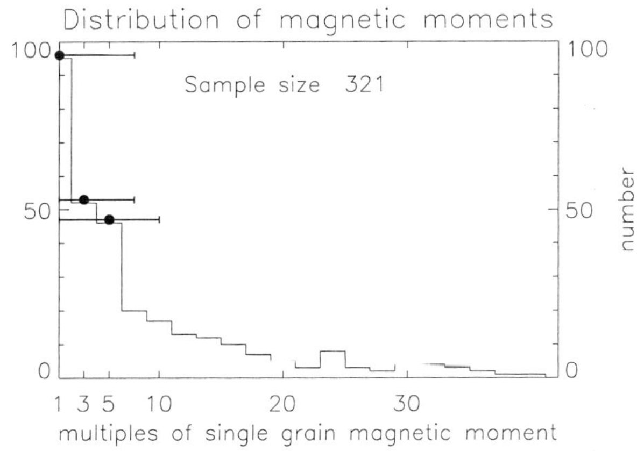

Figure 1. Measured magnetic dipole moments of individual grains. The data has been normalized with the nominal value $|\boldsymbol{\mu}| \approx 3 \cdot 10^{-13}$ Am^2. Errors apply to all the data.

solar nebula as a result of various melting processes [9,10]. Potentially magnetic minerals such as magnetite, phyrrotite, and iron-nickel have been detected in meteoritic samples [11] and in comet P/Halley's dust [12]. Together with the strong nebular magnetic fields that are inferred from the paleomagnetic record in meteorites [11], this should have produced a population of magnetized dust grains.

The long ranged interactions between magnetized dust grains are responsible for an enhanced collisional cross section σ for these particles, thus inducing a selective growth mechanism together with an acceleration of the coagulation process itself. Numerical simulations by Nübold and Glassmeier [13] and Nübold, Dominik, and Glassmeier [14] have been able to quantify this enhancement. As has been shown experimentally by Nuth et al. [9], magnetic dust aggregation creates web-like structures that could act as 'fishing nets' for nonmagnetic particles. This could serve as a starting point for rapid grain growth and new compositional features [13]. Ideally dipolar interactions between magnetized grains lead to elongated structures that could explain interstellar polarization of starlight (e.g. [15]) and may have been found in meteoritic samples [8]. As an experimental test for these considerations we performed analogous experiments with magnetized micron-sized dust grains. In this paper, we will report on first results and future perspectives.

2. MAGNETIC DUST AGGREGATION EXPERIMENTS

Magnetic dust aggregation has been studied numerically [16,13,14]. In order to have a means for calibrating and validating these simulations, we investigated the effects of magnetic interactions on the preplanetary grain growth process in analogous laboratory experiments. To this end, we used single domain (SD) ferritic (Ba,Sr) $Fe_{12}O_{19}$ powder in the levitation drum setup described by Blum et al. [17]. SD ferritic grains are used as a raw material for the industrial production of sintered permanent magnets [18]. The

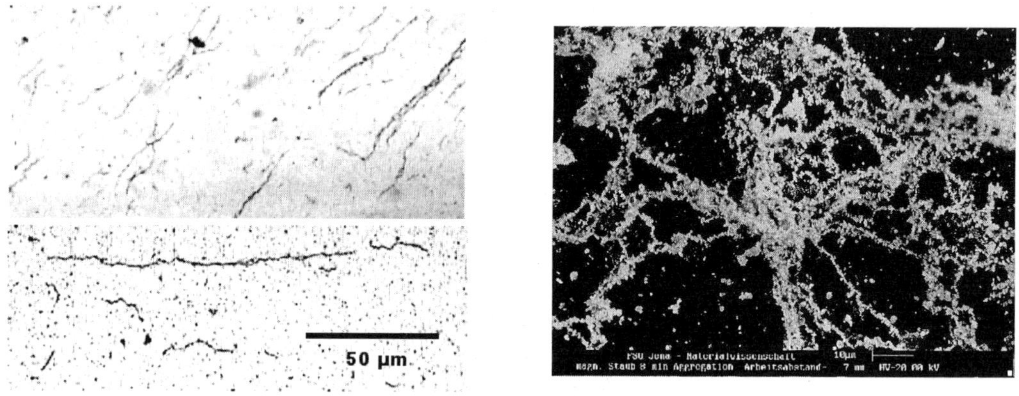

Figure 2. Optical and electron microscopic images of magnetic dust aggregates that formed inside the levitation drum. Particles coagulate rapidly and the resulting aggregates are visible to the unaided eye. Left image: chain-like aggregates from the levitation drum. Top; the picture width is approximately 1 mm. The alignment may be due to aerodynamic effects. Bottom; close-up. Right image: a web-like structure containing several elongated aggregates in a spoke-like pattern around a central cluster. These aggregates could act as a "fishing net" for nonmagnetic grains.

grains in our dust sample are micron-sized and strongly magnetized due to the existence of only one magnetic domain. In an approximate way, they can be regarded as magnetic dipoles with a dipole moment $|\boldsymbol{\mu}| \approx 3 \cdot 10^{-13}$ Am^2. This value can be obtained from direct calculations using the mean particle size and the material parameters. In order to verify this result, we have measured the magnetic moment of individual grains directly by means of a specifically designed magnetic coil target [19]. The magnetic grains are deagglomerated by a rotating cogwheel [20], enter the target and are deflected by a strong gradient of the magnetic field generated by the target coils. The particle motion is imaged by means of a pulsed laser and a CCD-camera, and can be used to calculate the magnetic moment of the deflected particle. The measured distribution of magnetic dipole moments is shown in Figure 1. It depicts a strong maximum around the calculated value. There are rather huge systematic errors due to the loss of 3D information and the finite resolution of the CCD image. In some cases, the deagglomeration may not be perfect. This will produce grain clusters that carry more than one nominal dipole moment and would explain the long tail of the distribution in Figure 1.

3. FIRST RESULTS AND FUTURE PROSPECTS

Aggregation experiments were conducted inside a rotating levitation drum. The dust grains are deagglomerated by means of a pyrotechnical dispersion device and are supported against Earth's gravity by the aerodynamic drag forces of the explosion gas which rotates with the drum [17]. A particle component with number density n has collision times

$$\tau = \frac{1}{n \cdot \sigma \cdot v_r}, \tag{1}$$

that depend on the individual particle's cross section σ and the relative velocity v_r between grains. Adopting the numerical results obtained by Nübold and Glassmeier [13], the collision times for nonmagnetic grains are by several orders of magnitude longer than for magnetic particles. Therefore, magnetic dust aggregation should proceed rapidly, producing huge elongated aggregate structures.

This could indeed be seen in our experiments: on time scales of the order of seconds only, we grew long chain-like aggregates and more complex, web-like structures (Figure 2). The biggest aggregates were approximately 0.5 cm long and could easily be seen with the naked eye. The structural differences relative to growth experiments with magnetically neutral silicate grains conducted by Heim [21] are easily noticeable. Since those were formed inside the same experimental setup using nonmagnetic silica (SiO_2) spheres, we conclude that the structural differences are due to the magnetic interactions between the ferrite particles.

The aggregates were collected for further study using optical and electron microscopy. The microscopic images also show the more complex web like structures (Figure 2; right image) described by Nuth et al. [9]. The latter contain chain-like aggregates similar to that in Figure 2 as 'spokes' around central clusters. We carried out a first analysis in order to establish the fractal dimensions of these aggreagtes using the density-density correlation function. The simple chain-like structures have $D = 1$, whereas $D = 1.2 \pm 0.1$ for the more complex aggregates.

The microphysics of the coagulation process will be subject to accompanying theoretical and numerical studies [14]. The highly irregular shape of our ferrite grains makes the comparison with numerical models, that usually rely on spherically shaped particles, rather difficult. Repeating these experiments with regularly shaped dust samples is one of our future goals. Furthermore, a microgravity experiment on the same subject will take part in this year's ESA student parabolic flight campaign (www.tu-bs.de/institute/geophysik/forschung/projekte/hotzenplotz/index.html). As can be shown using Equation 1, only magnetic grains have time to coagulate in a parabolic flight aggregation experiment. On the other hand, the neutral dust component might be trapped by magnetic aggregates of the type shown in Figure 2 (right image). Using samples of magnetic and nonmagnetic dust, the 'fishing net' hypothesis will thus receive special attention.

ACKNOWLEDGEMENTS

The ferrite dust sample was provided by Magnetfabrik Bonn/Germany. H. Nübold and K.-H. Glassmeier gratefully acknowledge the permission to use the Jena dust experimentation facilities for the present studies.

REFERENCES

1. Weidenschilling, S. J., *Icarus* **116** (1995) 433

2. Sahijpal, S., J. Goswami, A. Davis, L. Grossman, and R. Lewis, *Nature*, **391-6667** (1998) 559
3. Wurm, G., PhD-thesis, University of Jena, 1997
4. Chokshi, A., A. G. G. M. Tielens, and D. Hollenbach, *Astrophys. J.* **407** (1993) 806
5. Dominik, C. and A. G. G. M. Tielens, *Astrophys. J.* **480** (1997) 647
6. Ossenkopf, V., *Astron. Astrophys.* **280** (1993) 617
7. Horanyi, M. and C. K. Goertz, *Astrophys. J.* **361** (1990) 155
8. Kuebler, K. E., H. Y. McSween Jr., W. D. Carlson, and D. Hirsch, *Icarus* **141** (1999) 96
9. Nuth, J. A. III, O. Berg, J. Faris, and P. Wasilewski, *Icarus* **107** (1994) 155
10. Rietmeijer, F. J. M., J. A. Nuth III, and J. M. Karner, *Astrophys. J.* **527** (1999) 395
11. Sugiura, N. and D. W. Strangway, in: *Meteorites and the Early Solar System*, edited by J. F. Kerridge and M. S. Matthews, 595, 1988
12. Schulze, H., J. Kissel, and E. K. Jessberger, *ASP Conference Series* **122** (1997) 397
13. Nübold, H., K.-H. Glassmeier, *Icarus* **144** (2000) 149
14. Nübold, H., C. Dominik, and K.-H. Glassmeier, in preparation.
15. Roberge, W. G. and D. C. Whittet (eds.), *Polarimetry of the interstellar medium*, ASP Conference Series **97**, 1996
16. Nübold, H., K.-H. Glassmeier, *Adv. Space Res.* **24-9** (1999) 1167
17. Blum, J., G. Wurm, T. Poppe, and L.-O. Heim, *Earth, Moon, and Planets* **80** (1999) 285
18. Hadfield, D. (ed.), *Permanent Magnets and Magnetism*, Ilife Books Ltd., London, UK, 1962
19. Nübold, H., T. Poppe, and K.-H. Glassmeier, *Adv. Space Res.* (2000) submitted
20. Poppe, T., J. Blum, and Th. Henning, *Astrophys. J.* **533** (2000) 472
21. Heim, L.-O., PhD-Thesis, University of Jena (in preparation), 2000

Crystallization processes in amorphous $MgSiO_3$

S.P. Thompson and C.C. Tang

Daresbury Laboratory, Warrington, Cheshire, WA4 4AD, UK.

Structural changes during crystallization of annealed $MgSiO_3$ are investigated. The most important changes occur in the initial stages of annealing. Significant amounts of amorphous structure survive up to high temperatures limiting crystal development. Grains initially processed at lower temperatures will have had their amorphous component strengthened by the annealing process and could retain some structural signature of their pre-processed form.

1. INTRODUCTION

Many main sequence stars show the Vega phenomenon whereby large amounts of refractory materials persist long after star formation. The disk once possessed by our Sun survives today as interplanetary debris (comets, dust, etc.), while the best known extrasolar debris disk is β Pic, whose 10μm band is similar to certain solar system comets in that both show a sub-feature at 11.2μm attributed to crystalline olivine [1]. This feature is also seen in a subclass of IDP's believed to have suffered extensive cosmic ray exposure in the ISM or proto-stellar nebula [2]. Crystalline olivine is perplexing as it is seen only in some comets and in the outflows of certain evolved stars where dust is condensing [3] but not in star forming regions or diffuse interstellar clouds [4]. Crystalline cometary dust is also troublesome as comets are believed to be frozen reservoirs of the most primitive pre-solar materials. As crystalline olivine condenses at >1200K (typical of the inner solar nebula) it is possible that it could have been transported to the outer comet forming zones and mixed with ice grains. How this could have been accomplished is not known as crystalline features are not seen in star forming regions where inner nebula condensation would occur. Observations of evolved AGB stars show crystallites to be confined to systems with high mass loss rates, only constituting ~10% of the total dust outflow [5], contrary to the expectation that more crystalline material should be observed in higher temperature shells. Alternatively, post-condensation annealing of amorphous grains could have occurred. However the annealing properties of silicates are poorly defined, with few laboratory results available [6,7]. We present here a basic experimental study of annealing-induced crystallization. We have concentrated on Mg silicate because: no evidence for Fe rich dust around AGB stars exists [8]; laboratory spectra for Mg-rich silicates give good fits to cometary 11.2μm features [6,9] and in condensation experiments mixed Fe-Mg solids do not readily form [8].

2. EXPERIMENTS

An amorphous sample of $MgSiO_3$ was continuously annealed at various increasing temperatures (see §3 below). Simultaneous to this, X-ray powder diffraction (XRD) data were collected. As a long-range structural probe XRD is ideally suited to measuring structural modification during crystallization. The annealing furnace was mounted at the centre of rotation of the 2.3 diffractometer at Daresbury Laboratory's Synchrotron Radiation Source.

The furnace enclosure has kapton entrance/exit windows allowing data collection during annealing. Heating is via 2kW RF Cu coil and temperature is measured by thermocouple at the sample crucible. The operational range is 290K – 2000K, with heating response times of ~30s, while thermal stability (±1K) is achieved in under a few minutes (see [10,11]). Standard θ - 2θ XRD geometry was used, but the need to cover a wide angular range (~55000 points, 1s per point) limits temporal resolution to ~1.6hrs per scan.

3. RESULTS

The sample was scanned at room temperature, then raised to 1000K and repeatedly scanned for 19.5hrs. During this time the XRD patterns showed the evolution of a crystallite. At the end of this period a developed crystalline diffraction pattern was visible, superposed upon an amorphous background that had changed little throughout. The crystalline component had formed after only ~2hrs and after ~5hrs had developed to its full extent. Neither component changed from then on. The temperature was increased to 1031K and the sample annealed for 24.5hrs. During this time no major changes in either phase occurred. Further processing for 24hrs at 1043K again produced no major change. The temperature was then increased in steps through 1083K, 1113K and 1143K over 5hrs. Again, no changes were observed. Finally raising the temperature to 1173K, annealing continued for a further 14.5hrs. After 5-6hrs at this temperature, the crystalline phase underwent further growth while the amorphous features diminished. Figure 1 shows XRD data for selected temperatures and times.

4. DISCUSSION

The evolution of our sample matches published results for the 10μm band [6] where initial annealing gave a band profile characteristic of astronomical silicate, followed by the development of a 11μm sub-feature. After this is a stall period with no further development. On exiting the stall full crystalline features develop. Changes in the 10μm band originate in structural changes induced by annealing and are mirrored in our structural data. Our sample quickly entered a structural stall, only exiting after substantial annealing. The crystalline XRD pattern is matched by forsterite (Mg_2SiO_4) [10]. If all available Mg is in the crystal phase then, given an $MgSiO_3$ composition, the amorphous phase will be SiO_2-rich. The permanency of this component is shown by the widths of the broad features at ~30°and ~50° 2θ. At room temperature these were ~8.5° and ~6° 2θ. After ~2hrs at 1000K they shrank to ~4° each and remained constant until 1173K when the sample began to change. XRD feature widths reflect the disorder in the reflecting lattice layers. Some alteration to the amorphous phase thus occurred as the crystallite formed, but then showed strong resilience to higher temperatures.

A likely explanation is that, as well as promoting crystallization, annealing strengthens the amorphous phase. Laboratory and cosmic silicates will have significant hydroxyl content [12,13,14]: molecular hydrogen dissolves into silicate structures over a wide temperature range, reacting with SiO_2 to form hydroxyl, or hydroxyl/hydride pairs [15,16]. Network bonds can also break by ionization or direct displacement and react with void space H to form Si-OH or Si-H [17]. Hydroxyl and hydride are removed from amorphous SiO_2 at >973K and completely removed over a period of hours [18]. Plotting the percentage of hydroxyl/hydride removed as a function of time, the data for each fall on the same curve implying that the effective diffusion coefficients are the same for both. The recombination process is [18]

$$SiOH + SiH \rightarrow Si - O - Si + H_2, \tag{1}$$

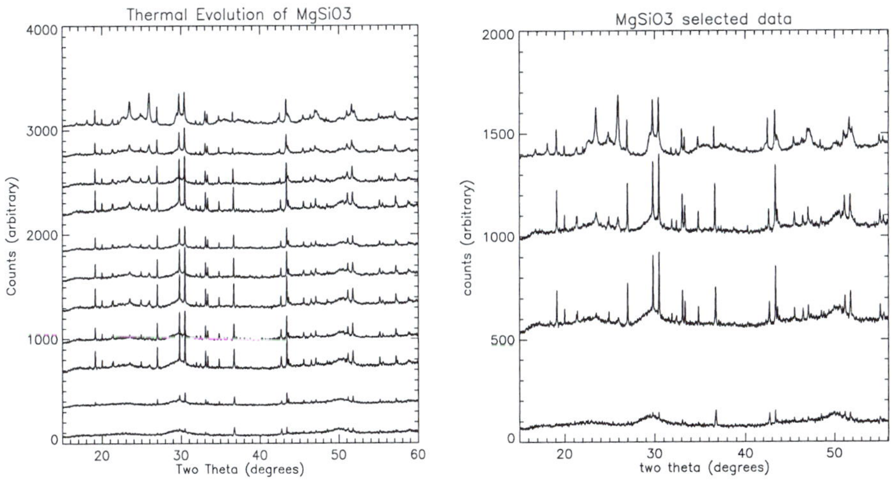

Figure 1. Left: crystallization of $MgSiO_3$, bottom to top: initial 1000K; after 3 hrs; after 19.5 hrs; initial 1031K; after 24.5 hrs; initial 1043K; after 24 hrs; at 1083K; at 1143K; initial 1173K; after 14.5 hrs. Right: selected data, bottom to top: Initial 1000K; after 19.5hrs; 1143K; after 14.5hrs at 1173K

and is confirmed experimentally [19]. Below ~70% removal, the removal curves approximate those of simple diffusion with an effective diffusion coefficient of 1.4×10^{-7} cm^2 s^{-1}, which is much less than the expected value of 6×10^{-6} cm^2 s^{-1} for simple diffusion [20]. Taken in conjunction with deviations of the removal curves from simple diffusion above 70% removal, this implies that H_2 outflow is controlled by the reaction rate and not the diffusion coefficient [19] and is supported by an activation energy of 260-266 kJ mol^{-1} [18] which matches the SiO–H bond energy of 264 kJ mol^{-1} [21]. Dehydroxylation is thus controlled by the breaking of this bond. X-ray absorption spectroscopy at the Si K shell absorption edge in annealed Mg silicates [22] revealed shifts in the energy of the absorption features, which were greatest below the crystallization threshold. Such shifts are characteristic of changes in the oxidation state of the absorbing species (i.e. Si polymerization) [23], confirming that most alterations to the local atomic environment of Si occur early on. The closeness in threshold temperature for (1) and the initial temperature in our experiment where early alteration to the amorphous component occurred, strongly suggests that dust grains annealed at low-end temperatures will be likely to retain an amorphous component at higher temperatures.

The net effect of (1) is to increase the overall SiO_2 network by reconnecting intertetrahedral Si–O–Si bridging bonds, which raises the average number of bridging oxygen atoms per tetrahedral unit (*BO/T*). Since forsterite comprises units with $BO/T=0$, the extent of its development is limited by the need to break bridging bonds on units with $BO/T\geq1$. Amorphous silicates however will contain a distribution of *BO/T* values ranging 0..4 [24], so that low-end annealing will initially promote the ordering of $BO/T=0$ species. Crystal growth thus stalls when the need for more $BO/T=0$ units can only be met by breaking bonds for units with $BO/T\geq1$. Meanwhile, during the initial stages of annealing, the bulk *BO/T* for this component will have been increased as a direct result of annealing and its average *BO/T* raised towards that of SiO_2 ($BO/T=4$). The stall thus persists until enough energy has been input to the system to allow (the increased number of) bridging bonds to be broken, whereupon the amorphous phase breaks down and further crystal growth becomes possible. This trend is visible in our structural data and is corroborated by the previous observation that the higher the initial annealing temperature, the shorter the spectral stall period [7].

5. CONCLUSIONS

The main conclusions of this work are (1) annealing close to ~1000K can enhance the survivability of an amorphous phase within a silicate grain as it will raise its breakdown temperature by strengthening SiO_2 connectivity; (2) some pre-annealing information may thus survive in the amorphous component of processed grains, even up to relatively high temperatures. This will be relevant to identifying interstellar components among recovered solar system dusts. Analysis of recovered grains that have undergone thermal processing could in principle still reveal information about their pre-processed forms. The next stage of our experiments will be to correlate structural development with spectral development.

REFERENCES

1. M.L. Sitko, C.A. Grady, D.K. Lynch, R.W. Russel and M.S. Hanner, Astrophys. J. 510 (1999) 408.
2. J.P. Bradley, Science 265 (1994) 925.
3. C. Waelkens, L.B.F.M. Waters, M.S. De Graauw, E. Huygen, K. Malfait, H. Plets, B. Vandenbussche, D.A. Beintema, D.R. Boxhoorn, H.J. Habing, A.M. Heras, D.J.M. Kester, F. Lahuis, P.W. Morris, P.R. Roelfsema, A. Salama, R. Siebenmorgen, N.R. Trams, N.R. Van Der Bliek, E.A. Valentijn and P.R. Wesselius, Astron. Astrophys. 315 (1996) L245.
4. M.S. Hanner, T.Y. Brooke and A.T. Tokunaga, Astrophys. J. 502 (1998) 891.
5. A.G.G.M. Tielens, L.B.F.M. Waters, F.J. Molster and K. Justtanont, Astrophys. & Space Sci. 255 (1998) 339.
6. S.L. Hallenbeck, J.A. Nuth and P.L. Daukantas, Icarus 131 (1998) 198.
7. J.R. Brucato, L. Colangeli, V. Mennella, P. Palunbo and E. Bussoletti, Astron. Astrophys. 348 (1999) 1012.
8. F.J.M. Reitmeijer, J.A. Nuth and J.M. Karner, Astrophys. J. 527 (1999) 395.
9. M.S. Hanner, J.A. Hackwell, R.W. Russell and D.K. Lynch, Astrophys J. 425 (1994) 274.
10. S.P. Thompson and C.C. Tang, Astron. Astrophys. 368 (2001) 721.
11. C.C. Tang, G. Bushnell-Wye and R.J. Cernik, J. Synchrotron Rad. 5 (1998) 529.
12. R. Bruckner, J. Non-Cryst. Sol. 5 (1970) 123.
13. T.M. Steel and W.W. Duley, Astrophys. J. 315 (1987) 337.
14. R. Timmermann and H.P. Larson. Astrophys. J. 415 (1993) 820.
15. J.E. Shelby. J. Appl. Phys. 51 (1980) 2589.
16. G.H.A.M. van der Steen and H. van den Boom, J. Non-Cryst. Sol. 23 (1977) 279.
17. J.E. Shelby, J. Appl. Phys. 50 (1979) 3702.
18. J.E. Shelby, J. Non-Cryst. Sol. 179 (1994) 138.
19. Y. Morimoto, T. Igarashi, H. Sugahara and S. Nasu, J. Non-Cryst. Sol. 139 (1992) 35.
20. J.E. Shelby, J. Appl. Phys. 48 (1977) 3387.
23. J.F. Shackleford and J.S. Masaryk, J. Non-Cryst. Sol. 21 (1976) 55.
24. S.P. Thompson, A. Evans and A. Jones, Astron. Astrophys. 308, (1996) 309.
25. B.K. Agarwal, X-ray Spectroscopy. Springer-Verlag, Berlin (1991).
26. I. Farnan, P. Grandinetti, J.H. Baltisberger, J.F. Stebbins, U. Werner, M.A. Eastman and A. Pines, Nature 358 (1992) 31.

Experimental astromineralogy: Circumstellar ferromagnesiosilica dust in analogs and natural samples

Frans J.M. Rietmeijer[a] and Joseph A. Nuth III[b]

[a]Institute of Meteoritics, Department of Earth and Planetary Sciences, University of New Mexico, Albuquerque, NM 87131, USA.

[b]Laboratory for Extraterrestrial Physics, MS 691, NASA Goddard Space Flight Center, Greenbelt, MD 20771, USA.

Hutton's principle of Actualism guides all Earth Sciences research and includes experimental verification of natural processes whereby collected samples are the ground truth to validate experiments. Chondritic aggregate IDPs serve to validate the uniqueness of astromineralogy experiments of kinetically controlled condensation, post-condensation aggregation and thermal alteration of ferromagnesiosilica circumstellar dust. Laboratory experiments constrain the crystallographic and chemical properties of metastable eutectic ferrosilica and magnesiosilica dusts and make very specific predictions about the compositions of ferromagnesiosilica condensates.

1. INTRODUCTION

The first time silicates were postulated to exist in dusty space environments, Astronomy became linked to Earth Sciences. The notion that this dust was initially formed by gas-to-solid condensation of chondritic vapors has no analog in terrestrial environments wherein we can observe this particular formation of silicate minerals. In this paper we will explore the implications to Astronomy of the mineralogical literature. First, when minerals, such as olivine [$(Mg,Fe)_2SiO_4$] and pyroxene [$(Mg,Fe)SiO_3$], are identified in space there is an extensive database on the physical (crystallographic) and chemical properties of these silicates, their occurrences and conditions of formation. In this context terms such as "dirty silicates" or "astronomical silicate" have no meaning although the latter qualifies a non-geological use. This is a matter of nomenclature where two different Sciences will have to meet. Second, the presence of minerals in space can only be interpreted using the founding principle of Geology put forward by James Hutton (1726-1797): "*No powers are to be employed that are not natural to the globe, no action to be admitted except those of which we know the principle*". He called it Actualism but Sir Charles Lyell re-christened this as Uniformitarianism. Earth Scientists have to accept that the processes operating in the past on Earth proceeded exactly as we can observe their operations today on Earth. The presence of minerals in space can constrain the processes in their environments but this formulation of Actualism causes an obvious problem for astronomical application.

The second part of Hutton's original formulation was later interpreted to include experimental verification under laboratory conditions. This change paved the way to attempt experimental quantification of geological processes, such as experimental petrology to determine the pressure and temperature of formation of minerals and mineral assemblages. Earth Scientists literally have the sample in hand that they want to simulate in the laboratory and this provides ground truth to evaluate the uniqueness of the laboratory experiments. This situation does not exist for space dust wherefore there are only infrared spectral data available. By the very nature of the signal such data represent a measurement of 'secondary' dust properties. The primarily properties are determined by the crystallography and chemistry of the solids. Unique mineral determinations based on IR spectra require additional independent information that is generally unavailable to Astronomers. This situation improved when interplanetary dust particles (IDPs) became available. The porous chondritic aggregate IDPs are unique solar system materials from asteroids and periodic comets [1]. Aggregate IDPs offer ground truth to examine the uniqueness of laboratory simulations of space dust whereby the IR spectral information of the analogs can be used to explore dust compositions and physical processes in oxygen-rich astrophysical environments [2].

2. INTERPLANETARY DUST PARTICLES (IDPS)

Classification of unmelted IDPs shows that the early solar system contained a limited number of chondritic and non-chondritic dust types, viz. (1) ferromagnesiosilica principal components (PCs), (2) Fe,Mg- and Ca,Mg,Fe-silicates. (3) Fe,Ni-sulfides (mostly Ni-free and low-Ni pyrrhotite), (4) Fe-metal and -oxides and (5) rare refractory Ca,Ti,Al-rich dust [3,4]. The textures of chondritic aggregate IDPs 10-15 μm in size show that hierarchical dust accretion began with the formation of a fractal matrix of PCs, 90-1,000 nm in diameter [2,3].

The PCs are (1) carbonaceous PCs (refractory hydrocarbons, amorphous, poorly graphitized, pre-graphitic turbostratic carbons, 'buckytubes' and 'buckyballs'[5]; (2) carbon-bearing ferromagnesiosilica PCs of ultrafine (2-50 nm) Fe,Mg-olivines and -pyroxenes, Fe,Ni-sulfides, Fe-oxides in a carbonaceous matrix, and (3) ferromagnesiosilica PCs. The ferromagnesiosilica PCs are (1) Mg-rich, coarse-grained (10 - 410 nm) PCs with Fe/(Fe+Mg) (*fe*) = 0-0.35 [6] and (II) Fe-rich, ultrafine-grained PCs, *fe* = 0.35-0.85 (modal *fe* $\approx$ 0.67) with an amorphous ferromagnesiosilica matrix and embedded Fe,Mg-olivines and -pyroxenes, Fe,Ni-sulfides and Fe,Ni-metal grains (<50 nm) [3]. The PCs are the smallest textural entities in the matrix of aggregate IDPs. They have their analogs in terms of particle size (mass) and composition among the CHON, 'mixed' and 'silicate' particles in comet Halley's dust [6].

This paper emphasizes ferromagnesiosilica (FeMgSiO) PCs that are among the oldest dust in the solar system. We accept that all dust was initially formed by gas-to-solid condensation although there is a discontinuity in size (mass) and composition between the condensates and PCs in aggregate IDPs. The compositions of the FeMgSiO PCs provide ground truth to test the results of our condensation experiments.

3. GAS-TO-SOLID CONDENSATION

Circumstellar dust analogs produced by the kinetically controlled gas-to-solid condensation experiments of Mg-SiO-H_2-O_2 and Fe-SiO-H_2-O_2 vapors yielded amorphous MgSiO and FeSiO grains a few nanometers to 150 nm in diameter. However, pure Fe-, Mg- and Si-oxide condensates were crystalline solids (i.e. minerals). The most significant finding was that the non-stoichiometric MgSiO and FeSiO compositions uniquely matched those of metastable eutectics in the MgO-SiO_2 and FeO/Fe_2O_3-SiO_2 equilibrium phase diagrams [6]. The existence of predictable metastable eutectic compositions is the best proof yet that kinetically controlled gas-to-solid condensation is not a chaotic process but with predictable chemical order. During nucleation in a cooling vapor, the first stable nuclei/molecules that form become seeds for grain growth during continued vapor condensation. Given sufficient activation energy and time, the condensed solids will become arranged into the least-energetic physical and chemical configurations until they equilibrate with the ambient environment. Kinetic factors at high condensation rates might favor metastable high-energy states [6] while surface energy considerations could change the relative stability fields for small grains [7,8]. Condensation in a Mg-Fe-SiO-H_2-O_2 vapor mimicked the metastable eutectics in the separate binary systems. Condensation in the MgO-FeO/Fe_2O_3-SiO_2 system did not lead to the formation of MgO-FeO and MgO-Fe_2O_3 solids because due to complete solid solution in these binary systems there is no metastable eutectic.

Mixed FeMgSiO dust can not form by condensation alone. Although formation of Mg-olivine and Mg-pyroxene is predicted by equilibrium thermodynamics, the condensation of Fe-silicates is not. Instead, iron is predicted to condense as a metal until $T<600K$ when it becomes incorporated into earlier condensed Mg-silicates. Kinetically controlled condensation yields amorphous magnesiosilica and ferrosilica dust with predictable metastable eutectic compositions and pure metal-oxide grains. The experimental data make specific predictions about FeMgSiO dust compositions that are possible as a result of aggregation.

4. AGGREGATION OF MG-RICH FERROMAGNESIOSILICA DUST

The amorphous MgSiO grains are metastable $Mg_3Si_2O_7$, serpentine dehydroxylate, and Si-rich $Mg_6Si_8O_{22}$, or $6MgO{\cdot}8SiO_2$ (smectite dehydroxylate). We note that the chemical composition of an amorphous material is properly presented as a mixture of metal-oxides and not as a mineral structural formula as used here; but using the latter makes it easier to display post-condensation mineralogical evolution]. Amorphous FeSiO solids have two Si-rich metastable eutectic compositions because the FeO/Fe_2O_3 to SiO_2 ratio is sensitive to the Fe oxidation state during condensation. There are also two very-low silica Mg(Si)O and Fe(Si)O metastable eutectics. The condensed dusts that formed in close proximity can mix along tie lines connecting the metastable eutectic MgSiO and FeSiO condensates. The most likely compositions of the resulting mixed FeMgSiO aggregates are uniquely determined by the intersections of these tie lines (Figure 1). The relative positions of the metastable eutectics will restrict spontaneous aggregation to Mg-rich FeMgSiO compositions with *fe* = 0-0.35. These experimentally constrained compositions exactly match those of the coarse-grained FeMgSiO PCs in aggregate IDPs [6]. The amorphous metastable dusts in the aggregates contain a significant amount of internal free energy. As a result, dust aggregates can become compact chemically homogenized amorphous solids. The Mg-rich FeMgSiO PCs (*fe* = 0-0.35) could form spontaneously via aggregation while formation of Fe-rich FeMgSiO dust probably requires an external energy source to initiate fusion of aggregated dust.

5. AGGREGATION OF FE-RICH FERROMAGNESIOSILICA DUST

Metastable MgSiO condensates and Mg-rich FeMgSiO PCs will continue to react with the low-silica metastable Fe(Si)O dust and pure Fe-oxides. This mixing process also occurs along tie lines in the MgO-FeO/Fe_2O_3-SiO_2 ternary system whereby preferred FeMgSiO compositions occur at the intersection of experimentally determined metastable eutectic serpentine dehydroxylate and the $Mg_6Si_8O_{22}$ (Sm-d) - Fe(Si)O mixing-lines (Figure 2). The predicted modal composition of Fe-rich FeMgSiO dust is $fe \approx 0.7$ with a range $fe = 0.35–0.85$.

Condensed dust in the aggregates may still contain internal free energy but efficient fusion into compact amorphous Fe-rich FeMgSiO dust will require an external energy source. The

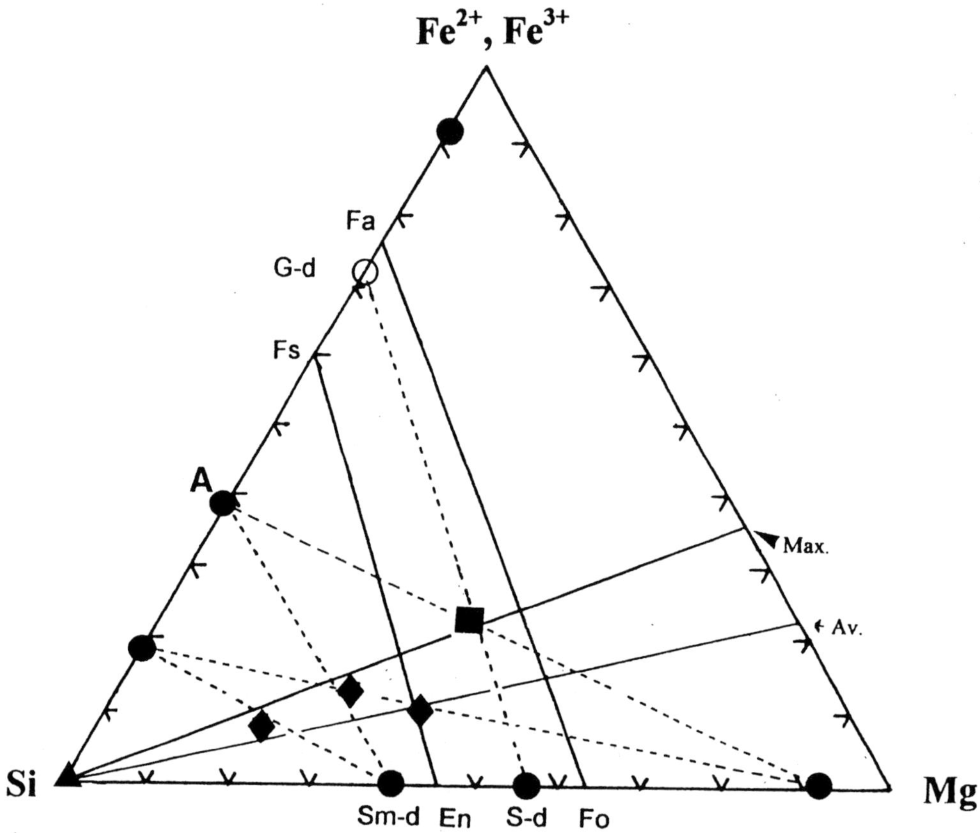

Figure 1. The Mg-Fe-Si (element wt %) diagram showing the compositions of condensed amorphous metastable eutectic solids (dots) [Fe^{2+},Fe^{3+} symbolizes variable Fe^{2+}/Fe^{3+} ratios]. The condensed dusts are (*1*) smectite (Sm-d) and serpentine (S-d) 'MgSiO' dehydroxylates and high-Si FeO dehydoxylates (incl. Si-rich smectite dehydoxylate labeled *A*), (*2*) Fe(Si)O and Mg(Si)O dehydroxylates and (*3*) metal oxide minerals (incl. SiO_2, silica; solid triangle). Mixing lines (dashed) connect the condensed dusts. The compositions of three Mg-rich 'FeMgSiO' PCs (black diamonds) are shown as well as the average *fe* (Av.; 0.23) and maximum (max.) *fe* ratios of these PCs. The solid square is the most Mg-rich composition that possible for Fe-rich *ufg* 'FeMgSiO' PCs. A mixing line (dashed) that connects the S-d and greenalite (G-d; open circle) metastable eutectic dusts is located in between the stoichiometric olivine (Fo-Fa) and pyroxene (En-Fs) lines. Mixed very-low Si "Mg,FeO" dust is not yet seen in aggregate IDPs but it was detected in the coma of comet Halley [9].

predicted compositions exactly match those of the ultrafine-grained (*ufg*) FeMgSiO PCs in chondritic aggregate IDPs (Fig. 2). These particular PCs include units known as GEMS (glass with embedded metal and sulfides) that are believed to result from mixing of glassy Mg-pyroxene and an Fe,Ni-sulfide grain that is a source of Fe,Ni-metal. Mixing is thought to be facilitated by prolonged irradiation and sputtering by energetic H and He nuclei, recoil-mixing, vitrification, and magnesium loss but this mixing mechanism was not verified experimentally. Condensed Fe(Si)O and Fe-oxide could also provide iron for *ufg* FeMgSiO PC formation ([3] for a review).

6. THERMAL ANNEALING

Heat treatment of amorphous condensates and fused FeMgSiO dust causes crystallization in these chemically closed-systems of secondary mineral phases that are unrelated to condensation. In Mg-rich FeMgSiO PCs the results are coexisting coarse-grained Fe,Mg-olivine and Fe,Mg-pyroxene with identical *fe* ratio plus an amorphous aluminosilica residue (Fig. 9 in [6]). The fe-ratios of the silicate minerals reflect this ratio of the parent PC. Heat treatment of serpentine-dehydroxylate condensate yields forsterite and enstatite, viz.

$Mg_3Si_2O_7 = Mg_2SiO_4 + MgSiO_3$ or $Mg_3Si_2O_7 = 1.5Mg_2SiO_4 + 0.5SiO_2$.

The reaction products of the latter, olivine + silica instead of thermodynamically stable pyroxene, reflect the kinetic nature of the thermal reactions [7,6]. Hallenbeck et al. [11] monitored the changes in the IR spectra of MgSiO condensates as represented by the above reactions during thermal annealing. They found that the spectral features of samples in a "stall phase" resembled those of "olivine-rich" comets and that when the "stall phase" had ended the abundances of crystalline Mg-silicates continually increased as a function of heat-treatment time and temperature. Similar reaction rates for MgSiO and Mg-rich FeMgSiO samples [11] suggest that Fe occurs as ferrous iron in stoichiometric forsterite and enstatite.

Thermal annealing of metastable eutectic ferrosilica dusts proceeds differently because this dust such as Si-rich FeSiO smectite-dehydroxylate ('*A*' in Fig. 1) contains ferric iron and the formation pure Fe-olivine is accompanied by formation of Fe^{3+}-bearing Fe-oxides. Trivalent cations will polymerize amorphous silica-rich materials. As a result nucleation and growth in FeSiO condensates will lag behind those in MgSiO and Mg-rich FeMgSiO condensates during heat treatment. This is consistent with the IR spectral development observed during thermal annealing of these different materials by Hallenbeck et al. [11].

7. DISCUSSION

Experimental astromineralogy simulates circumstellar dust formation via vapor phase condensation and post-condensation modification, e.g. thermal alteration [2,5]. Because Si, Fe and Mg are the most abundant rock-forming elements we simulated FeMgSiO dust formation in he MgO-FeO/Fe_2O_3-SiO_2 system. Using Actualism, the guiding principle for the Earth Sciences, and chondritic aggregate IDPs as ground truth samples to validate the uniqueness of the condensation and thermal annealing experiments, we find that

(1) Kinetically controlled gas-to-solid condensation at high delta-T or supersaturation only produces a limited number of amorphous MgSiO and FeSiO dusts with predictable metastable eutectic compositions, and pure metal-oxides,

(2) Condensed solids have considerable "internal crystallographic and chemical free energy",

(3) Formation of FeMgSiO dusts requires aggregation of condensed dusts,
(4) Compact amorphous Mg-rich FeMgSiO dust with *fe* = 0-0.35 forms spontaneously by aggregation of Si-rich metastable eutectic MgSiO and FeSiO dusts,
(5) Compact amorphous Fe-rich FeMgSiO dust with a modal composition of $fe \approx 0.7$ and a range of *fe* = 0.35–0.85 are the result of aggregation and fusion,
(6) Only heat treatment produces stoichiometric crystalline Mg- and Mg,Fe-silicates, and
(7) Pure crystalline iron silicates, Fe_2SiO_4 (olivine) and $FeSiO_3$ (pyroxene) will be rare.

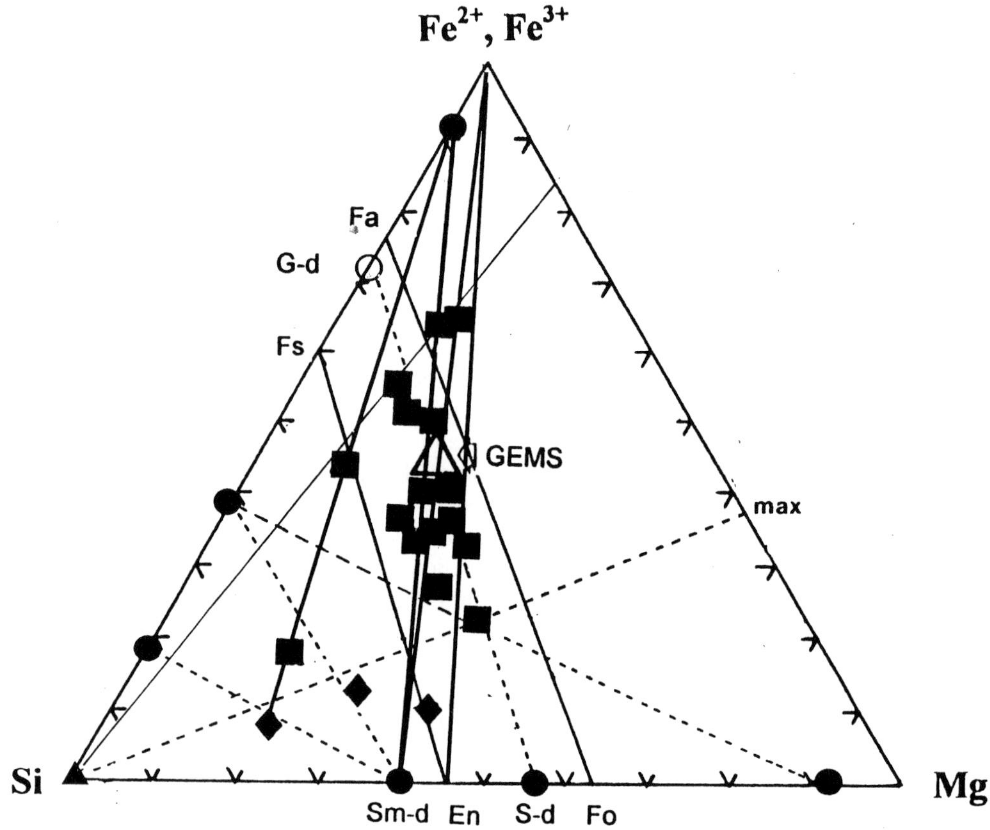

Figure 2. Same Mg-Fe-Si diagram as Fig. 1 showing compositions of Fe-rich *ufg* 'FeMgSiO' PCs (solid squares; Rietmeijer, unpublished data) and GEMS (triangle; [10]). These compositions are defined by mixing lines (heavy solid lines) between the (*1*) smectite dehydroxylate (Sm-d)–Fe(Si)O and Fe-oxides; (*2*) enstatite (En)–Fe(Si)O and Fe-oxides, and (*3*) Mg-rich 'FeMgSiO'–Fe(Si)O and Fe-oxides. Preferred composition occur at intersections of these mixing lines with the serpentine dehydroxylate (S-d)–(G-d) mixing line. The light solid line marks the maximum amount of ferrous iron in the silicates; towards the apex only iron metal exists.

Predictable condensation and aggregation of FeMgSiO dust characterize dust formation in oxygen-rich circumstellar environments. Some implications of our work are:

(1) Circumstellar outflows contain only a limited number of compositionally distinct condensed grain types with relative proportions that may vary among different stars. Our simulations match the observations at far-IR wavelengths by the Infrared Space Observatory (ISO) satellite showing condensation of nearly pure amorphous MgSiO grains that anneal to crystalline Mg-silicates without detectable iron as well as the absence of crystalline Fe-silicates in certain relatively-rare high-mass-loss-rate stellar outflows, and

(2) "Silicate" dust in the jets emanating from the nucleus of comet Halley (Fig. 4a in [9] is silica-rich compared to the CI abundance. It resembles FeMgSiO PCs in chondritic aggregate IDPs.

ACKNOWLEDGMENTS.

The authors thank their collaborators who over the years contributed to this research. This work was supported by NASA grant NAG5-4441 (FJMR).

REFERENCES

1. I.D.R. Mackinnon and F.J.M. Rietmeijer, Reviews in Geophysics 25 (1987) 1527.
2. J.A. Nuth III, S.L. Hallenbeck and F.J.M Rietmeijer, in Laboratory Astrophysics and Space Research (eds. P. Ehrenfreund, K. Krafft, H. Kochan and V. Pirronello) Kluwer Acad. Publ., Dordrecht (1999) 143.
3. F.J.M. Rietmeijer, in Planetary Materials (ed. J.J. Papike) Mineralogical Society of America, Washington, DC. Reviews in Mineralogy 36 (1998) 2-1.
4. F.J.M. Rietmeijer, in Advanced Mineralogy, Vol. 3 (ed A.S. Marfunin) Springer Verlag Berlin-Heidelberg (1998) 22.
5. A. Rotundi, F.J.M. Rietmeijer, J.R. Brucato, L. Colangeli, V. Mennella, P. Palumbo and E. Bussoletti, Planet. Space Sci. 48 (2000) 371.
6. F.J.M. Rietmeijer, J.A. Nuth III and J.M. Karner, Astrophys J. 527 (1999) 395.
7. F.J.M. Rietmeijer, J.A. Nuth and I.D.R. Mackinnon, Icarus 65, (1986) 211.
8. J.A. Nuth III, In The Cosmic Dust Connection (ed. J.M. Greenberg) Kluwer Acad. Publ., Dordrecht (1996) 205.
9. E.K. Jessberger, A. Christoforidis and J. Kissel, Nature 332 (1988) 691.
10. D.J. Joswiak, D.E. Brownlee, J.P. Bradley, D.J. Schlutter and R.O. Pepin, Lunar Planet. Sci. XXVII, (1996) 625.
11. S.L. Hallenbeck, J.A. Nuth III and P.L. Daukantes, Icarus 131 (1998) 198.

VII The Near-Earth Environment

Dust characterisation in the near Earth environment

N. McBride *

Planetary and Space Sciences Research Institute, The Open University
Milton Keynes MK7 6AA, UK

Exposure of impact sensitive surfaces in low Earth orbit (LEO) has led to an improved definition of the near Earth space environment in terms of time-averaged fluxes. In LEO, separation into meteoroids and orbital debris has been possible, and the fluxes to individual faces of a spacecraft can be quite confidently modelled. This paper considers some aspects of dust flux characterisation in the near Earth region, and 'tests' whether the much used Grün et al. dust flux is consistent with recent expansive datasets.

1. INTRODUCTION

In the effort to characterise the interplanetary dust complex in the near Earth region, analyses generally focus on three mains areas. (i) definition of the dust flux at 1 AU (i.e. defining the mass distribution, and the directionality of dust sources), (ii) identification of impactor chemical residues from retrieved space-flown surfaces (i.e. obtaining information on the impactor composition, and the impact processes involved), and (iii) intact capture and retrieval of dust particulates (i.e. obtaining significant amounts of the actual source body). In this paper, I will focus mainly on dust fluxes.

Dust fluxes are generally derived from three main sources: (i) the study of meteors in the atmosphere, (ii) the study of lunar rock sample microcraters, and (iii) data returned from dust impact experiments in space, and the study of retrieved space-flown surfaces. It is the last area that I will be mainly concentrating on in this paper. I will consider some aspects of interpreting flux data from dust instruments (or via inspection of retrieved surfaces), review some of the more important past missions, and then present a consolidation of much of the work on the topic performed by the author and co-workers.

2. MEASURING DUST FLUXES

In characterising the interplanetary dust flux, an important aim is to determine the flux as a function of particle mass. However, when a dust particle impacts a typical dust instrument, a signal is produced (or some damaged-related feature is produced) which is proportional to both the particle mass M and impact velocity v i.e. proportional to $M^a v^b$, where a and b are constants. As the (cumulative) dust flux is generally characterised by a mass distribution, which can be approximated over a given mass region by $F(>M)-kM^{-\alpha}$ (where k is a constant and α is the cumulative mass distribution index;

*Formerly at The Unit for Space Sciences & Astrophysics, University of Kent at Canterbury, UK.

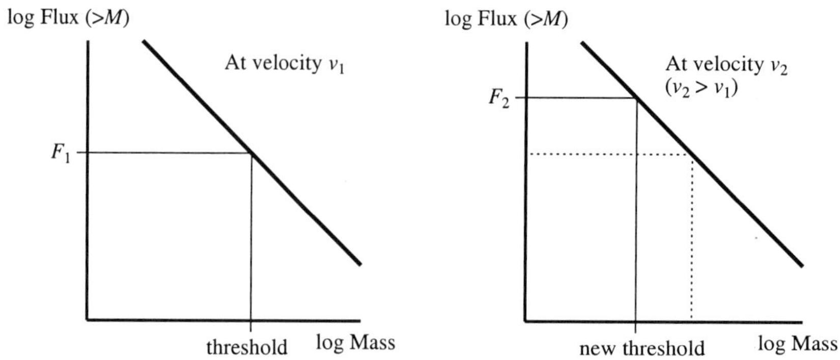

Figure 1. For dust impacting at a given velocity v_1, a detector will have a given mass threshold, giving rise to a flux F_1 being detected (left figure). However, for a velocity v_2 (where $v_2 > v_1$), the mass threshold is lower, so that the detector is sensitive to particles 'further up' the dust mass distribution, and thus detects a flux F_2 (where $F_2 > F_1$).

which is typically of order 1), then it is clear that the number of events giving signals above an instrument detection threshold is dependent on a, b and α. This is demonstrated graphically in Figure 1, where we can visualise a dust flux distribution being 'sampled' by a generic detector. For particles impacting at velocity v_1, the detector has a given mass threshold (left diagram), which gives rise to the detection of a flux F_1. However, for particles with velocity v_2 (where $v_2 > v_1$), the mass threshold is lowered, and thus more (smaller) particles are detected (flux F_2). The ratio of the fluxes at the two velocities is given by $F_2/F_1 = (v_2/v_1)^{\alpha b/a}$. The ratio b/a is often referred to as the factor γ.

For detectors sensitive to impact momentum detection (i.e. $\propto Mv$) then $\gamma = 1$, whereas for detectors sensitive to impact energy ($\propto Mv^2$) then $\gamma = 2$, and for detectors sensitive to impact plasma generation ($\propto Mv^{\sim 3.5}$) then typically γ =3–4. Thus detectors (particularly plasma detectors) have a strong velocity dependence. In interpreting data to deduce flux distributions, it is often not advisable to attempt to directly invert the data, but to use a model incorporating a flux distribution, a velocity distribution, and the detector threshold relationship, and iterate until the fitting is consistent with the data. However to first order, mean velocities can be applied (allowing direct inversion of the data), although the use of weighted means (weighted by $\alpha\gamma$) is generally required.

3. PAST EXPERIMENTS

For the last 40 years or so, efforts have been made to characterise the mass distribution of the dust in near Earth space, and its spatial distribution. Sampling the entire size range of dust (which can broadly be described by the 0.01 μm to cm size regime) is almost impossible for a single detector, and so building a picture of the dust complex at 1 AU has required a combination of data, taking due account of differing thresholds, velocity biases and exposure geometries in space. Without attempting a complete review of *all* past dust detectors, it is worth considering some of the more important missions that have contributed to the understanding of the near Earth dust flux to date (see McDonnell

[1] for an overview of the 'early years').

Some early attempts at measuring fluxes, using piezo-electric microphone detectors on Explorer 8 [2], proved unreliable as the microphones were susceptible to thermal changes associated with low Earth orbit [3], thus giving apparent fluxes that were much too high (by $\sim$3 orders of magnitude). However, more reliable data were obtained by Explorer 16 [4–7] and Explorer 23 [8] which were Earth orbiting satellites (with altitudes $\leq$1200 km) that carried experiments comprising large arrays (1.6 m^2 and 2.1 m^2 respectively) of pressurised canisters, or 'beer cans'. A pressure-sensing switch capable of measuring a 'once only' leak was activated after the (first) perforation of any can (thicknesses 25–55 μm; thus the experiments were sensitive to particles in the $\sim$10 μm size regime). Clearly the array of canisters had a finite lifetime, although calibration of the ballistic limit of the canister wall gives this sort of sensor a fairly reliable mass threshold for a given impact velocity. The Explorer 16 and 23 experiments recorded 55 and 124 impacts respectively.

Extremely large area ($\sim$200 m^2) penetration sensitive capacitance detectors were flown on three Pegasus satellites in low Earth orbit (altitude $\sim$580 km) [9–11]. Penetrations in the charged capacitors were registered by discharges through the dielectric layers (i.e. again the detectors rely on ballistic limit calibrations). Two thicknesses of 2024-T3 aluminium (203 μm and 406 μm) gave reliable data, with sampled particles in the $\sim$10–100 μm size regime (thus overlapping with the radar meteor size regime). Around 2000 impacts were recorded with these detectors.

Pioneers 8 and 9 were spacecraft in heliocentric orbits sampling space between 0.75 AU and 1.08 AU. They carried dust instruments based on impact plasma detection, and thus sampled down to the sub-micron particle size regime. The detectors were mounted on the sides of the spinning spacecraft, which had their spin axes perpendicular to the ecliptic plane. Thus the detectors 'scanned' the ecliptic plane. The data comprised a total of about 800 particle impacts, most of which appeared to come from the solar direction — the small (and possibly very fast) β meteoroids (see [12,13]).

The HEOS-2 spacecraft had a highly eccentric orbit (altitude 350–240,000 km) giving it extended periods away from the increasingly debris-contaminated low Earth orbit region. The dust experiment [14,15] was also based on an impact plasma detection system (which sampled down to the sub-micron particle size regime). As the spacecraft span, with its spin axis along the Earth Sun line, the detector would scan a plane perpendicular to the Earth-Sun line. Thus the solar direction was not well sampled (inhibiting β meteoroid detection), but the Earth-apex direction and antiapex direction fluxes could be resolved. It was found that many more impacts were detected from the Earth apex direction. Several hundred particles were detected in total.

These spacecraft data (i.e. the Pegasus, Pioneer and HEOS data) offer excellent 'tie points' for combination with lunar microcrater data (and indeed meteor data at the larger size regime). This was done by Grün et al. [16] who built on previous work, to define a mean interplanetary dust flux at 1 AU over a wide size regime. One of the key elements of this flux definition was to recognise the 'true' contribution of lunar rock sample microcraters (in the $\sim$ <10 μm size regime). The apparent excess of these small craters [17] was in fact due to secondary impact craters being more prevalent than had first been realised (see [18,19]) and would thus lead to an over-estimation of the deduced flux (by $\sim$ 2 orders of magnitude) if taken at 'face value'. Grün et al. [16] concluded that the interplanetary flux was not represented by the smaller lunar microcraters, and used the spacecraft data instead. The 'Grün flux' is often used, and has become the definitive representation of the *mean* interplanetary flux at 1 AU.

Anisotropies in the near Earth dust environment (such as the Earth apex-antiapex bias, and the β meteoroids) as identified by the Pioneer and HEOS detectors, has been confirmed by more recent experiments such as those on Helios (see [20]), and Hiten. For example, the impact plasma experiment, Munich Dust Counter (MDC) [21] on the Hiten satellite sampled space from a few thousand km from the Earth, to beyond the Moon. The data again identified fast (and small) β meteoroids from the solar direction, superimposed upon an overall bias towards the Earth apex direction.

More recently, results from the SPADUS experiment aboard the Earth orbiting ARGOS satellite (altitude ~850 km), has multilayer penetration sensors to determine high accuracy velocities (and trajectories). Results to date [22], have produced 24 coincident penetrations showing a mixture of meteoroid and debris particles (as might be expected).

Dust measurements in geostationary orbit have also been obtained. Time of flight data obtained via the penetration of thin dielectric films have been made in GEO from the GORIZONT-41 and GORIZONT-43 communication satellites [23]. Velocities were measurable for 76 impacts from particles of sizes 3–100 μm of which 80% were inferred as natural meteoroids (i.e. >12 km s^{-1}). The Geostationary Orbit Impact Detector (GORID), a flight-quality engineering model of the Ulysses impact ionization detector [24] was launched with the Russian Express-2 communications satellite into GEO in 1996, and has since been detecting typically a few impacts per day, many of which are consistent with GEO debris [25] (see also [26] for a consideration of bound and unbound dust particles in GEO in relation to the GORID detections).

As well as remotely operated dust sensors, an extremely powerful technique for understanding the near Earth dust environment, is to retrieve surfaces that have spent significant periods of time in space, and interrogate their surfaces to deduce mean crater fluxes and where possible, the chemical signatures of the impactor.

Retrieval of the SMM (Solar Maximum Mission) spacecraft for repairs in the mid-1980s allowed the study of multi-layer thermal insulation and aluminium louvres. These data [27] covered a wide range of crater diameters from sub-μm to mm dimensions. The precise pointing direction of the louvres is largely undefined, but assumed to be effectively random with respect to Earth. Chemical analysis of crater residues showed that debris particles were important at the smaller size regimes. However, while the geocentric 'random' orientation of the SMM spacecraft gave a good 'snapshot' of *mean* LEO flux environment, is not ideal for resolving the discrete contributions of debris and meteoroids that can impact from different directions in orbit. This information is better served by a spacecraft that maintains its orientation with respect to Earth. Thus the LDEF dataset (see below) essentially supersedes the SSM data in terms of its overall usefulness (or at least, a fuller understanding of the fluxes is obtained by considering SMM *and* LDEF data).

Probably the most expansive and statistically reliable LEO dust impact dataset, has come from the Long Duration Exposure Facility (LDEF) which was deployed in 1984, and stationed at an altitude of ~470 km for 5.8 years. LDEF's large area-time product, the wide range of materials deployed on it, and its gravity gradient stabilised orbit which maintained the geocentric exposure geometry of 14 discrete sides, has meant that the data is of great importance in 'decoding' LEO particulate fluxes.

The small-particle data was best defined by the MAP experiment [28] which comprised foils with thicknesses in the range 2 to 30 μm, located on the north, south, east, west (N, S, E, W) and space faces. Thick target (i.e. non-penetrating) measurements [29] and data from the thermal control surfaces and the longerons and intercostals of the LDEF frame collated by the LDEF Meteoroid and Debris Special Investigator Group (M&D-SIG; e.g. [30,31]) have provided a database of impacts on spacecraft surfaces at a range of sizes. The IDE experiment [32] provided valuable time resolved fluxes of small particles (micron sized and smaller) for the first 10 months of the mission, from capacitor discharge detectors located on the N, S, E, W, space and Earth faces. Intense, but short-lived, 'spikes' exceeding the background by several orders of magnitude and 'multiple orbit event sequences' (MOES) recurring for large numbers of spacecraft orbits were detected [33,34], and interpreted as small debris particulates.

Since LDEF's recovery from orbit in late 1989, other complementary sources of impact data have also been obtained. ESA's EuReCa spacecraft provided impact data on large areas of thermal blanket, solar cell arrays and the science experiment, TiCCE (Timeband Capture Cell Experiment [35]). While the time resolution aspect of TiCCE did not function correctly, it did provide high reliability penetration data from 2.5–9.2 μm foils. EuReCa was oriented with pseudo-fixed Sun and Earth-apex pointing faces, and the data complements the LDEF dataset, and demonstrates a flux bias towards the Earth apex of motion (see McBride et al. [36]).

The recovery of one solar panel after 3.6 years in space from the Hubble Space Telescope (HST) on its servicing mission in December 1993 has provided an additional large area of exposed material to complement the EuReCa solar cell array data (see e.g. Taylor et al. [37]).

The MIR space station has also provided a platform for retrievable experiments such as Echantillons [38] with capture cells, capacitor discharge detectors and aerogel cassettes.

4. LDEF AND THE FLUX MODEL

As stated above, LDEF offers the most useful retrieved surface dataset. Foil penetrations and crater data span the μm to mm size regime, thus the data are self consistent over a very large particle mass range. Furthermore, the geometry of LDEF is also of great benefit for statistically resolving meteoroid fluxes from debris contaminants. LDEF was gravity gradient stabilised, such that it maintained a fixed orientation with respect to Earth. The satellite was an elongated cylinder with 12 side faces (including north, south, east and west faces), and two end faces; Earth and space. The space face is of particular importance, as it had no Earth shielding, and was essentially spinning with respect to interplanetary space (i.e. once per orbit). Additionally, as LDEF's orbit precessed, the exposure of the space face to interplanetary space was to a large extent randomised (see McBride et al. [39] for a more detailed explanation of LDEF's exposure over its entire lifetime) such that the exposure to meteoroids is also essentially random. This is in contrast to the orbital debris exposure which is highly directional. Additionally (and critically), relatively little orbital debris could strike the space face due to the highly oblique impact angles presented to co-orbiting particles (even debris in highly elliptical orbits with low altitude perigee would not contribute significantly due to the highly oblique impact angles). This then makes the space face of LDEF a very good meteoroid detector despite being in LEO.

It is therefore natural, that we would want to investigate in detail whether the LDEF data are consistent with our current understanding of the dust flux at 1 AU, i.e. the Grun et al. [16] flux. Indeed, as the space face of LDEF offers an excellent self-consistent dataset over a wide particle size regime, we would seek to *test* the Grun et al. flux, with the possibility of better defining the accepted interplanetary flux model over the sampled size regime. It is for this reason that considerable effort was put into consolidating the LDEF datasets, and a detailed geometric model of the LDEF exposure was constructed, paying particular attention to the minutiae of the implementation. This is described below.

4.1. The data

For LDEF's 14 faces, all available impact data sources onto aluminium surfaces were collated with well defined selection criteria and specified search areas (see McDonnell et al. [40] for a full description). True crater diameters D_c, defined at the nominal material surface, were derived from crater lip diameters D_r where necessary (using D_c=0.75D_r [29]) and converted to ballistic limit F_{max} values (i.e. maximum thickness of aluminium that could be penetrated) for direct comparison with the high reliability foil MAP penetration data, using the empirical relationship F_{max}=0.87D_c, derived from impact studies [41]. The consolidation of the data involved derivation of 'data-fits' (performed by S.F. Green; see [40]) which passed through the bulk of the flux points, with due regard for statistical significance of each dataset and incomplete sampling. Subjective upper and lower limits were defined to accompany each of the 14 data-fits. Consideration of all 14 data-fits with regard to meteoroids and debris, clearly offers a wealth of information, although it is the space face data-fit which is of the greatest importance for meteoroid investigations due to the relative lack of debris contamination.

4.2. The flux model

To assess the LDEF data, an isotropic meteoroid model was constructed following standard techniques used by a number of workers, but differing somewhat in detail, particularly with the use of a better defined velocity distribution (hence better defining velocity dependent effects). The model includes the following features: the geometry of LDEF's exposure; a meteoroid mass (or flux) distribution; a meteoroid velocity distribution 'at 1 AU'; gravitational flux and velocity enhancement to LEO; velocity dependent Earth shielding to LDEF faces; relative impact direction effects and spacecraft velocity (flux) enhancement; a conversion to impact damage output.

The meteoroid flux distribution can be user-defined, but initially uses the Grün et al. [16] distribution, which gives cumulative mean flux values for a spinning flat plate detector at 1 AU, i.e. outside the gravitational influence of the Earth (but moving in an Earth-like orbit). Note, if we define the isotropic *flux* as F_G then the flux *intensity* is given by F_G/π.

When using a meteoroid flux distribution in a model that will consider the threshold response of various detection techniques (which might be very velocity dependent) one should incorporate a meteoroid velocity distribution rather than just use a mean velocity value. The velocity distribution $n(v_\infty)$ of meteoroids at 1 AU has generally been derived from ground based observations of photographic meteors (corrected for the effect of the

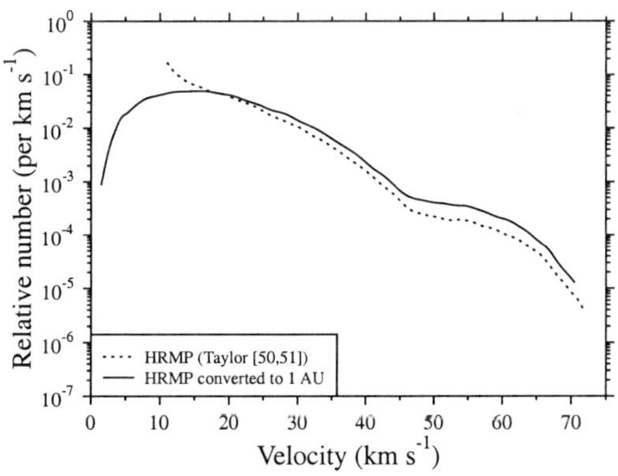

Figure 2. The Harvard Radio Meteor Project (HRMP) meteoroid velocity distribution, following Taylor [50,51], corrected for gravitational enhancement to take the distribution to 1 AU (i.e. as seen from a massless Earth). Also shown for comparison (dotted curve) is the distribution of meteoroids entering the top of the atmosphere.

Earth's gravitational acceleration). Dohnanyi [42] obtained a distribution from 286 observations taken from Hawkins and Southworth [43], and Erickson [44] used the same data but attempted a more rigorous reduction to meteor number with a constant mass threshold. Kessler [45] used 2090 sporadic meteors obtained by McCrosky and Posen [46] to give a more statistically reliable distribution (see [47] for a comparison between these distributions). However, probably the most statistically reliable dataset comes from the Harvard Radio Meteor Project (HRMP) where $\sim$ 20000 meteor observations were obtained [48,49]. This data set is often used in various modelling work. Taylor [50] reappraised the data using an improved analysis of ionisation probability and mass distribution index. Taylor also identified a numerical error in the original code used to reduce the data which resulted in a significant under-estimation of numbers of fast meteors (particularly 50 to 70 km s^{-1} meteors where the under-estimation is by a factor of $\sim$100). We use this Taylor [50,51] corrected velocity distribution of meteoroids encountering the Earth's atmosphere, and convert to the distribution $n(v_\infty)$, which would be seen at 1 AU. This is shown in Figure 2. This gives a useful form which can then be applied to any altitude (e.g. LEO or GEO) within the model accounting for gravitational enhancement.

We consider the flux of particles encountered by a moving flat plate detector (i.e. an LDEF face) by numerically integrating the particle intensity over all viewing angles, particle mass, and particle velocity. The total instantaneous flux contribution to the detector is then given by

$$F = \int_M \int_{v_\infty} \int_\phi \int_\theta \frac{F_G}{\pi} \, G \, n(v_\infty) \, \cos A \, \frac{v_{\rm rel}}{v_E} \, \sin\theta \, d\theta \, d\phi \, dv_\infty \, dM \qquad (1)$$

where θ, ϕ are spherical polar co-ordinates with respect to the spacecraft frame, v_E is the gravitationally enhanced meteoroid velocity given by

$$v_E = \sqrt{v_\infty^2 + v_{\rm esc}^2} \qquad (2)$$

(where v_∞ is the meteoroid velocity at 1 AU, and $v_{\rm esc}$ is the escape velocity at the spacecraft altitude), $v_{\rm rel}$ is the relative velocity of the incoming meteoroid with respect to the spacecraft (i.e. $v_{\rm rel}/v_E$ accounts for the spacecraft moving through the meteoroid environment), and the angle A is the instantaneous impact angle to the face (measured from the face normal). G is the gravitational flux enhancement given by [52]:

$$G = 1 + \frac{v_{\rm esc}^2}{v_\infty^2} \tag{3}$$

(using a realistic 'working range' for v_∞ of $v_\infty > 1$ km s^{-1}).

In numerical evaluation of Equation 1, no instantaneous flux contribution is added if the meteoroid cannot impact the face (i.e. if $A > \pi/2$), or if the spacecraft is shielded by the Earth. A tangent to the Earth's circumference passing through the satellite subtends an angle θ_c with the direction of the Earth's centre. Thus meteoroids cannot strike the spacecraft if approaching from within the cone defined by this angle θ_c (i.e. within the solid angle subtended by the Earth from the point of view of the satellite). If one assumes that meteoroid trajectories can be represented by straight lines, this 'critical' θ_c angle is simply given by $\sin\theta_c = r_e/r$ where r_e is the radius of the Earth, and r is the distance of the satellite from the Earth's centre. In reality the trajectories will be curved due to gravitational influence, and hence the angle θ_c is somewhat velocity dependent (see e.g. Kessler [53]). We are thus dealing with a function, $\theta_c(v_\infty)$, and the angle θ_c is described by

$$\sin\theta_c = \frac{r_e\sqrt{v_\infty^2 + v_{esc}^2(r_e)}}{r\sqrt{v_\infty^2 + v_{esc}^2(r)}} \;. \tag{4}$$

Note that when considering a meteoroid approach direction (and whether it is shielded), one must obviously use the actual meteoroid velocity vector with respect to Earth, and not the relative impact velocity vector.

For an instantaneous $d\theta$, $d\phi$, dv_∞ and dM step, a damage equation may be used to obtain the ballistic limit $\mathrm{F_{max}}$ of a spacecraft surface, i.e. each flux contribution is binned at the appropriate $\mathrm{F_{max}}$ value for the θ, ϕ, v_∞, M element. Clearly, different detection techniques would use an appropriate relationship. The penetration equation used here follows the empirically derived '1992C' equation of McDonnell and Sullivan [54]:

$$\mathrm{F_{max}} = 1.272\ d_p^{1.056}\ v^{0.806} \left(\frac{\rho_p \rho_{\rm Al}}{\rho_{\rm Fe}\rho_t}\right)^{0.476} \left(\frac{\sigma_{\rm Al}}{\sigma_t}\right)^{0.134} \tag{5}$$

where particle diameter d_p and $\mathrm{F_{max}}$ are in cm, impact velocity v is in km s^{-1}, ρ is density, σ is the yield stress, and subscripts p, t, Al and Fe refer to particle, target, aluminium and iron respectively. In the model we use the normal component of impact velocity ($v\cos A$) in Equation 5. As an indication of how $\mathrm{F_{max}}$ depends on particle diameter d_p, $\mathrm{F_{max}} \sim 5d_p$ at 20 km s^{-1}, and $\mathrm{F_{max}} \sim 15d_p$ at 70 km s^{-1} (assuming a particle density of 2500 kg m^{-3} impacting onto aluminium).

It is worth noting that the model, implemented as described here, has been adopted as the European Space Agency's 'reference' isotropic model incorporated into the new release of ESABASE which is ESA's environmental simulation software tool [55].

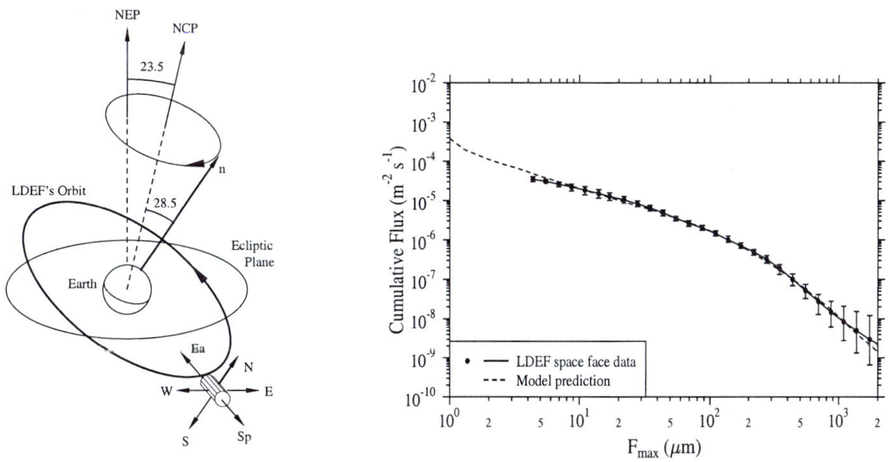

Figure 3. (Left) LDEF's orbit with respect to the Earth and the ecliptic plane. LDEF maintained its geocentric geometry. The main faces are North, South, East, West, Earth and space. Note that the north face normal vector was offset by 8° towards the ram (east) direction and that the space face had a tilt of 1.1° (revolved around the north face normal towards the ram face). This is incorporated into the modelling. (Right) The LDEF data-fit of the space face, and the meteoroid model prediction. Very little orbital debris is expected on the space face, and here we see that the meteoroid model prediction fits very well.

4.3. The results

Figure 3 shows the consideration of the LDEF space face. The diagram (left) reminds us of the LDEF geometry, and it is seen that as the spacecraft orbits (and the orbit precesses) the space face avoids Earth shielding and offers an essentially randomised exposure to space. Figure 3 (right) shows the data-fit for the space face of LDEF compared with the prediction from the meteoroid model described above. The model fits the space face well at all sizes measured (with perhaps a slight underestimation at the smallest sizes). This is a remarkable agreement, and it appears that, considering the uncertainties in the data and modelling procedure (velocity distribution used, impact equation used etc) there is *no need whatsoever* to re-assess the mean flux distribution of Grün et al. [16]. Thus the 'Grün flux', when applied to an appropriate surface (i.e. one that has had an effectively randomised exposure) appears to represent reality very well indeed.

Referring Figure 3 (left), it is seen the east face points to the spacecraft ram direction, whereas the west face is trailing. The east face would thus be particularly susceptible to debris impacts. The west face could also be struck by debris although it is more limited to eccentric orbits near perigee. Figure 4 shows the data-fits for the west and east faces of LDEF with their respective meteoroid model fits. The fit to the west face is reasonable, although underpredicting at larger sizes. The model fits the east face at F_{max}>30 μm, although below this, a striking excess flux is apparent, attributable to 'contamination' by space debris particles (see below).

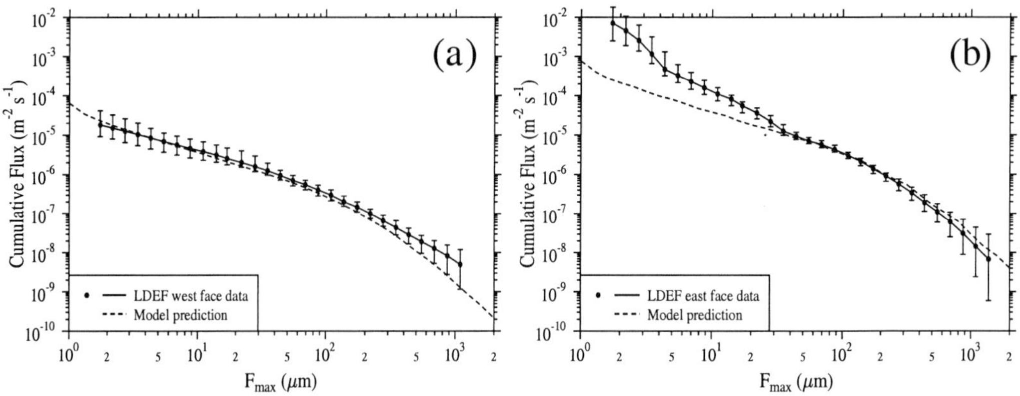

Figure 4. The LDEF data-fits and the meteoroid model predictions, for the LDEF (a) west face, and (b) east face (which is the orbit ram direction). The meteoroid model fits at $F_{max}>30$ μm although below this, the excess is interpreted as being due to small orbital debris particles.

5. SOLAR ARRAY DATA

Although it is undoubtedly an easier task to measure and interpret impacts on ductile materials such as the craters and penetrations on LDEF, the solar arrays of the EuReCa and HST spacecraft offer a valuable complementary source of data from a large area-time product surface. The surfaces have also undergone considerable chemical residue analyses, and so it is useful to compare the data to the meteoroid model, and indeed 'predict' the ratio of the meteoroid to debris flux, for comparison with the chemical analysis data.

The first task is to model the expected exposure to meteoroids for Sun-pointing solar arrays. This is done using the meteoroid model presented above. Although LDEF did not have solar arrays, and clearly did not maintain a Sun-pointing orientation, the data does at least offer a 'solar array analogue', by taking a mean of the space, east, Earth and west faces. This is because Sun-pointing arrays effectively spin with respect to geocentric space, and thus are approximated by the space, east, Earth and west 'LDEF 4-face mean'. The comparison with the meteoroid model and the LDEF 4-face mean solar array analogue is shown in Figure 5a (also shown is the output for an 'apex enhanced' model described in [36,40] which is applicable to non-random exposures; however in this case the isotropic and apex enhanced models are quite similar). As before, for the LDEF east face (Figure 4b) we have good agreement with the model at larger sizes, and an excess attributable to debris at the smaller sizes ($F_{max}<30$ μm).

Figure 5b shows data from the EuReCa solar arrays, and also from the TiCCE experiment penetration foils (which were located between the solar direction and the Earth apex direction, and so approximate quite well to the solar array exposure). The source data for the solar arrays is in the form of conchoidal crater diameter D_{co} (i.e. including the crater shatter zone around the central pit). However this is converted to an equivalent ductile F_{max} value by the expression $F_{max}=(0.3\pm0.1)D_{co}$ defined by Taylor et al. [37]. The plot also includes the meteoroid model fits and the LDEF 4-face mean as in Figure 5a. It is seen that the LDEF 4-face mean solar array analogue agrees with the high reliability

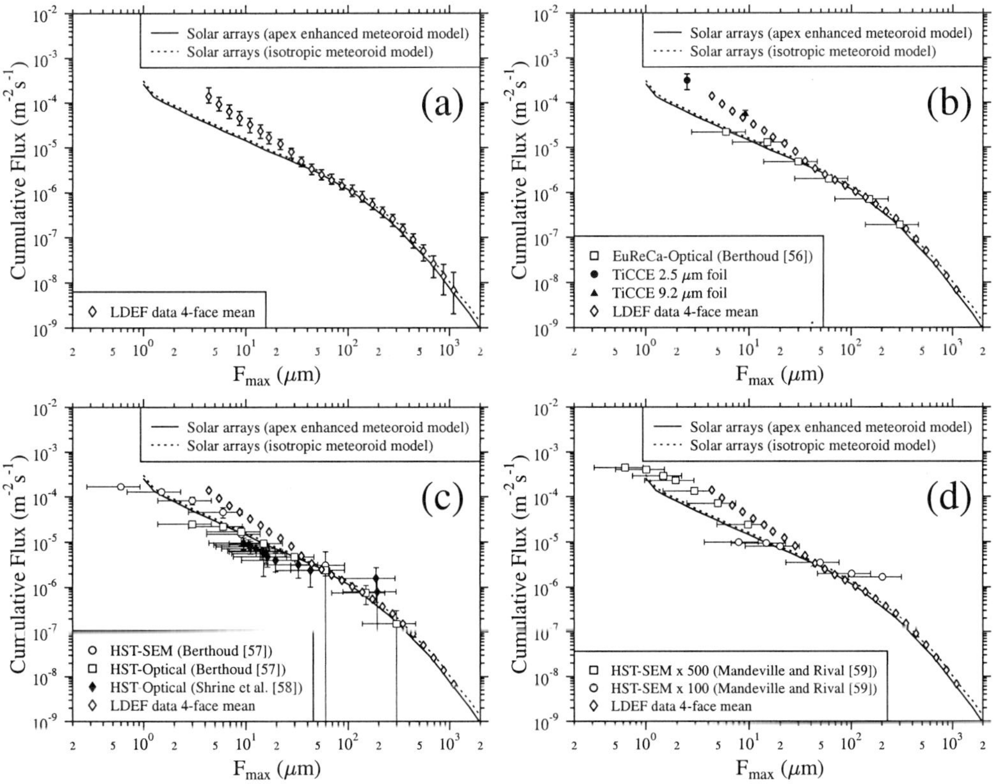

Figure 5. Solar array data. (a) LDEF 4-face mean (i.e. solar array analogue) and model fit; note the debris excess at small sizes. (b) EuReCa solar array data converted to F_{max}, and the high reliability TiCCE foil penetration data. (c) HST solar array data. (d) HST solar array data (as in (c) but with an alternative dataset shown).

TiCCE foil data, as we would expect. The solar array data appears consistent with the model, although we might expect a debris excess to be more apparent at the smaller sizes. It is possible that the data might suffer from incompleteness at the small size regime.

Figures 5c–d show the HST solar array data (converted to F_{max}) again with the model fits and LDEF 4-face mean solar array analogue. The HST array data are broadly consistent with the LDEF analogue (particularly in Figure 5d using the high magnification SEM data) although as before there might be some incompleteness at small sizes.

The summary of this, is that our understanding of the meteoroid flux, as demonstrated by the fits to the LDEF space face data, is consistent with the (admittedly less well defined) solar array data (and the LDEF solar array analogue and TiCCE foils). However in LEO, debris can add a dominant contaminant component at smaller sizes (F_{max}<30 μm). To check and/or confirm this statement, we can turn to chemical analysis. Although detailed analysis of all craters is impractical, recent work by Graham et al. has been particularly successful at obtaining large datasets of impactor residues from solar arrays, using electron backscatter and x-ray energy dispersive spectrometry techniques [60–62].

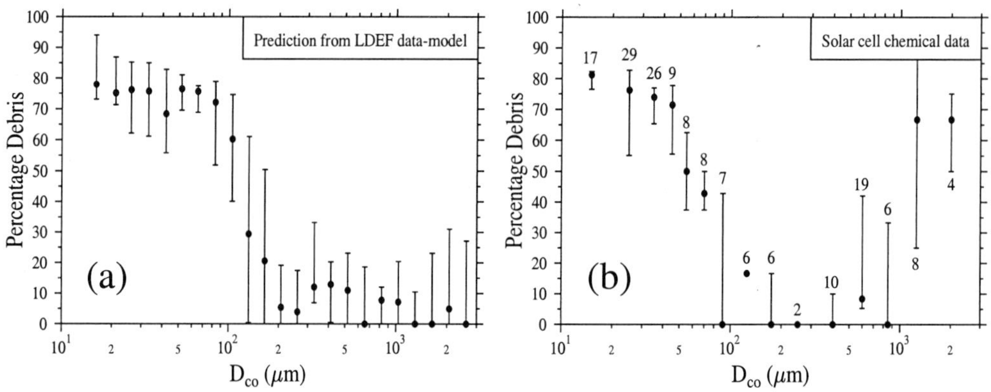

Figure 6. (a) The LDEF 4-face mean data (solar array analogue) has been used with the meteoroid model, to identify the proportion of the impact flux which came from space debris. The data has been converted to D_{co}. (b) The proportion of debris craters identified from the HST solar array chemical analysis. The error bars account only for the uncertainty due to craters where residues could not be identified.

Graham et al. [63] present analyses of 165 craters on the HST solar array, where most residues can be identified as being either from natural meteoroid or man-made debris impactors. We can use the meteoroid modelling, and the LDEF 4-face mean solar array analogue to 'predict' the debris/meteoroid ratio for a LEO solar array, by simply subtracting the meteoroid model flux from the 4-face mean. This is shown in Figure 6a, with F_{max} converted to D_{co} for direct comparison with the HST solar array data shown in Figure 6b (again using the Taylor et al. [37] conversion). In Figure 6a, the upper and lower limits come from the upper and lower limits of the LDEF 4-face mean data-fit points (as was shown in Figure 5a). In Figure 6b, the percentage debris is simply defined as the number of 'natural craters' over the 'debris craters'. The 'error bars' shown are produced by assuming that any unidentified craters were either all debris (upper limit) or all meteoroid (lower limit). The values in parentheses refer to the total number of craters in each logarithmic bin.

It is seen in Figure 6 that in general, the results from the LDEF data and meteoroid modelling agree very well with the HST chemical residue data in the sub-mm regime. At smaller sizes, both results appear to show around 80% debris with a crossover to meteoroid domination at around ~100 μm D_{co} (equivalent to F_{max}=30 μm i.e. as was seen in Figure 4b). This agreement is quite remarkable considering that it is derived in different ways and from completely independent data sources. The HST chemical data (Figure 6b) does show an enhancement of debris at D_{co}>1 mm, not seen in the LDEF data. It should be noted that the data is relatively sparse at this size regime, particularly for LDEF, and perhaps not too strong a conclusion should be drawn at this size. However the HST arrays were in space *after* LDEF was recovered, and so it is also possible that the large particle debris environment suffered a real enhancement at this time.

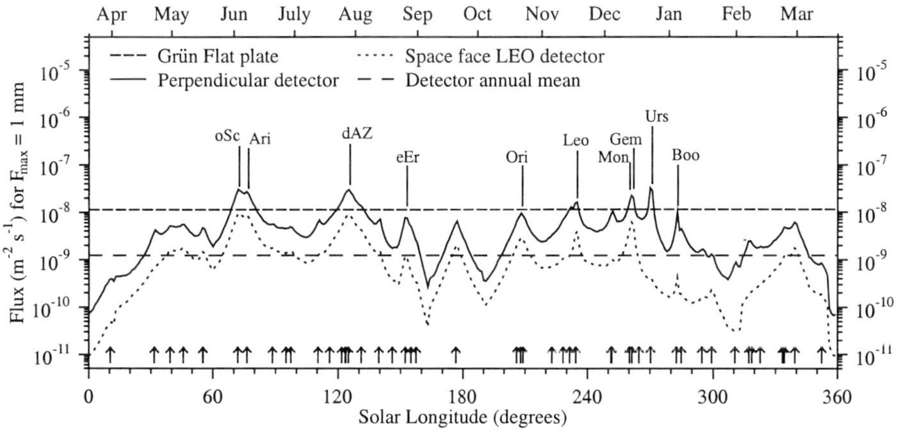

Figure 7. The instantaneous flux throughout the year, of stream meteoroids that penetrate 1 mm of aluminium foil (i.e. F_{max}=1 mm). The arrows indicate the peaks of all 50 streams used, with the 10 most important streams at this F_{max} threshold labelled: Ursids (Urs), δ Aquarids South (dAZ), ω Scorpiids (oSc), Geminids (Gem), daytime Arietids (Ari), Leonids (Leo), Bootids (Boo), Orionids (Ori), Monocerotids (Mon) and ϵ Eridanids (eEr).

6. THE EFFECT OF METEOROID STREAMS

It is clear that the overall annual mean meteoroid influx has a component that comes from recognisable meteoroid streams. As the meteor ionisation efficiency is highly dependent on meteoroid velocity, then a prominent *shower* does not necessarily mean that the meteoroid *flux* at a given mass is higher than the sporadic background. Thus to consider streams fully, one must model them in terms of their meteoroid flux distributions. McBride [64] presents a method for determining the contribution of the annual streams to the total annual mean flux, and calculates the instantaneous flux at large F_{max} throughout the year due to the 50 most prominent meteor showers, using shower parameters from Jenniskens [65] (note that this method of quantitatively accounting for the major streams has been incorporated into the recent upgrade of ESA's ESABASE environmental software package, referred to as the "Jenniskens-McBride model").

Figure 7 shows the instantaneous flux, as a function of solar longitude, for F_{max}=1 mm (i.e. this is the flux of meteoroid impacts that would penetrate 1 mm of aluminium — this being particularly applicable to considerations of significant spacecraft damage). In Figure 7, the upper (solid) curve shows the results obtained using a foil detector mounted perpendicularly to all streams simultaneously, and hence gives a (somewhat unrealistic) upper level for any instantaneous exposure. The lower (dashed) curve is for a foil detector mounted on the space face of a gravity stabilised LEO spacecraft (e.g. like LDEF) with its orbit parallel to the ecliptic plane, and hence represents the most realistic scenario. The lower dashed line represents the mean value of this curve, while the the upper dashed line gives the annual mean level derived for the space face detector using the isotropic meteoroid model discussed above (i.e. using the 'Grün flux' within the model). It is seen that only around 10% of the total mean meteoroid exposure, at this size regime, is obtained purely from the 50 streams. However, the instantaneous contributions from particular streams can exceed the annual mean, and so can be of 'interest' to short exposure missions (e.g. a 2 week Shuttle mission) where the total stream fluence would be highly dependent on the time of year.

7. SUMMARY

Over the last 40 years or so, the near Earth dust flux has been characterised by a combination of data from space-flight dust experiments, lunar rock microcraters, meteor observations, and more recently, retrieved space-flown surfaces. Many new datasets have been obtained since the publication in 1985 of the derived mean annual dust flux at 1 AU by Grün et al. [16]. In particular, the expansive LDEF data that covers a large size regime in a consistent manner. However, the 'Grun flux' appears wholly consistent with the more recent data when applied with care within a geometric meteoroid model. Furthermore, the definition of the level of debris 'contamination' in low Earth orbit, has now been reasonably well identified.

REFERENCES

1. J.A.M. McDonnell, in Cosmic Dust (ed. J.A.M. McDonnell) J. Wiley & Sons, Chichester, (1978) 337.
2. C.W. McCracken, W.M. Alexander, M. Dubin, Nature 192 (1961) 441–442.
3. C. Nilsson Science 153 (1966) 1242.
4. E.C. Hastings Jnr., The Explorer XVI micrometeoroid satellite — Description & Preliminary Results for the period Dec 16, through Jan 13, 1963, NASA TM X-810, 1963.
5. E.C. Hastings Jnr., The Explorer XVI micrometeoroid satellite; supplement I, preliminary results for the period 14 Jan 1963–2 Mar 1963, NASA TM X-824, 1963.
6. E.C. Hastings Jnr., The Explorer XVI micrometeoroid satellite; supplement II, preliminary results for the period 3 Mar 1963 – 26 May 1963, NASA TM X-899, 1963.
7. E.C. Hastings Jnr., The Explorer XVI micrometeoroid satellite; supplement III, preliminary results for the period May 27 through July 22, 1963, NASA TM X-949, 1964.
8. R.L. O'Neal, The Explorer XXIII micrometeoroid satellite — description and results for the period Nov 6 1964 through Nov 5 1965, NASA TN D-4284, 1968.
9. R.J. Naumann, Pegasus measurements of meteoroid penetrations (February 16 — July 20, 1965), NASA TM X-1192, 1965.
10. S. Clifton and R. Naumann, Pegasus satellite measurements of meteoroid penetration (February 16–December 31, 1965), NASA TM X-1316, 1966.
11. J.B. Dozier, in The Micrometeoroid Satellite Project Pegasus, NASA TN D-3505 Chapter V (1996) 65.
12. O.E. Berg and E. Grün, Space Res. 13 (1973) 1047.
13. H.A. Zook and O.E. Berg, Planet. Space Sci. 23 (1975) 183.
14. H. Hoffman, H. Fechtig, E. Grün and J. Kissel, Planet. Space Sci. 23 (1975) 215.
15. H.J. Hoffman, H. Fechtig, E. Grün and J. Kissel, Planet. Space Sci. 23 (1975) 985.
16. E. Grün, H.A. Zook, H. Fechtig and R.H. Giese, Icarus 62 (1985) 244

17. R.J. Allison and J.A.M. McDonnell, Proc. Lunar Sci. Conf. 12(B) (1981) 1703.
18. H.A. Zook, G. Lange, E. Grün and H. Fechtig, Lunar and Planet. Sci. 15 (1984) 965.
19. H.A. Zook, G. Lange, E. Grün and H. Fechtig, in Properties and Interactions of Interplanetary Dust (eds. R.H. Giese and P. Lamy) (1985) 89.
20. E. Grün, N. Pailer, H. Fechtig and J. Kissel, Planet. Space Sci. 28 (1980) 333.
21. E. Igenbergs, A. Hüdepohl, K. Uesugi, T. Hayashi, H. Svedhem, H. Iglseder, G. Koller, A. Glasmachers, E. Grün, G. Schwehm, H. Mizutani, K. Yamamoto and K. Nogami, in Origin and evolution of interplanetary dust, (eds. A.C. Levasseur-Regourd and H. Hasegawa) Kluwer, Tokyo (1991) 45.
22. A.J. Tuzzolino, R.B. McKibben, J.A. Simpson, S. Benzivi, H.D. Voss and H. Gursky, Planet. Space Sci. 49 (2001) 705.
23. L.S. Novikov, K.E. Voronov, N.D. Semkin, V.I. Verhoturov, O.S. Grafodatsky and I.A. Maksimov, in Proc. Second European Conference on Space Debris (eds. B. Kaldeich-Schürmann and B. Harris) ESA SP-393 (1997) 135.
24. E. Grün, H. Fechtig, R.H. Giese, J. Kissel, D. Linkert, D. Maas, J.A.M. McDonnell, G.E. Morfill, G. Schwehma and H.A. Zook, Astron. Astrophys. Suppl. 92 (1992) 411.
25. H. Svedhem, G. Drolshagen, E. Grün, O. Grafodatsky and U. Prokopiev, Adv. Space Res. 25 (1999) 309.
26. N. McBride and D.P. Hamilton, Work Package 5 Technical Note, in Update of Statistical Meteoroid/Debris Models for GEO, final report for ESA contract 13145/98/NL/WK, Unispace Kent CD-ROM, 2000.
27. M.R. Laurance and D.E. Brownlee, Nature (1986) 323.
28. J.A.M. McDonnell, W.C. Carey and D.G. Dixon, Nature 309 (1984) 237.
29. D.H. Humes, in LDEF 69 months in space — first LDEF Post-Retrieval Symposium (ed. A.S. Levine) NASA CP 3134 Part 1 (1991) 399.
30. T.H. See, M.E. Zolensky, R.P. Bernhard, J.L. Warren, J.L., C.A. Sapp and C.B. Dardano, in LDEF — 69 Months in Space: Third Post-Retrieval Symposium (ed. A.S. Levine) NASA CP 3275 part 1 (1995) 257.
31. M.E. Zolensky, T.H. See, R.P. Bernhard, R. Barret, F. Hörz, J.L. Warren, C. Dardano and K.S. Leago, Adv. Space Res. 16 (1995) 53.
32. S.F. Singer, J.E. Stanley, P.C. Kassel and J.J. Wortman, in The Long Duration Exposure Facility (LDEF) mission 1 experiments, NASA SP-473, 1984.
33. J.P. Oliver, S.F. Singer, J.L. Weinberg, C.G. Simon, W.J. Cooke, P.C. Kassel, W.H. Kinard, J.D. Mulholland and J.J. Wortman, in LDEF — 69 Months in Space: Third Post-Retrieval Symposium (ed. A.S. Levine) NASA CP-3275 part 1 (1995) 353.
34. J.D. Mulholland, S.F. Singer, J.P. Oliver, J.L. Weinberg, N.L. Montague, J.J. Wortman, P.C. Kassel and W.H. Kinard, in LDEF 69 months in space — first LDEF Post-Retrieval Symposium (ed. A.S. Levine) NASA CP 3134 Part 1 (1991) 517.
35. T.J. Stevenson, J. Brit. Interplan. Soc. 41 (1988) 429.
36. N. McBride, S.F. Green and J.A.M. McDonnell, Adv. Space Res. 23 (1999) 73.
37. E.A. Taylor, N.R.G. Shrine, N. McBride, S.F. Green, J.A.M. McDonnell and G. Drolshagen, Adv. Space Res 23 (1999) 83.
38. J.C. Mandeville, Adv. Space Res. 10 (1990) 397.
39. N. McBride, A.D. Taylor, S.F Green and J.A.M. McDonnell, Planet. Space Sci. 43 (1995) 757.

40. J.A.M. McDonnell, A.D. Griffiths, S.F. Green and N. McBride, Work Package 7 Technical Note, in Meteoroid and Debris Flux and Ejecta Models, final report for ESA contract 11887/96/NL/JG, Unispace Kent CD-ROM, 1998.
41. J.A.M. McDonnell, S.P. Deshpande, D.H. Niblett, M.J. Neish and P.J. Newman, Adv. Space Res. 13 (1993) 87.
42. J.S. Dohnanyi, Bellcomm. Rep. TR-66-340-1, Washington DC, 1966.
43. G.S. Hawkins and R.B. Southworth, Smithsonian Contrib. Astrophys. 2 (1958) 349.
44. J.E. Erickson, J. Geophys. Res. 73 (1968) 3721.
45. D.J. Kessler, AIAA 7 (1969) 2337.
46. R.E. McCrosky and A. Posen, Smithsonian Contrib. Astrophys. 4 (1961) 15.
47. H.A. Zook, in LDEF — 69 months in space: first post-retrieval symposium (ed. A.S. Levine) NASA CP-3134 Part 1 (1991) 569.
48. R.B. Southworth and Z. Sekanina, Physical and dynamical studies of meteors. NASA CR-2316, Smithsonian Institution, Cambridge MA, 1973.
49. Z. Sekanina and R.B. Southworth, Physical and dynamical studies of meteors: meteor fragmentation and stream distribution studies. NASA contractor report CR-2615, Smithsonian Institution, Cambridge MA, 1975.
50. A.D. Taylor, Icarus 116 (1995) 154.
51. A.D. Taylor, Adv. Space Res. 17 (1995) 205.
52. E.J. Öpik, Proc. R.I.A. 54 (1951) 165.
53. D.J. Kessler, A guide to using meteoroid-environment models for experiment and spacecraft design applications, NASA TN D-6596, 1972.
54. J.A.M. McDonnell and K. Sullivan, in Hypervelocity Impacts in Space (ed. J.A.M. McDonnell) University of Kent at Canterbury (1992) 39.
55. ESTEC: Mathematics and Software Division, ESABASE Reference Manual, ESABASE/GEN-UM-061 Issue 2, April 1995.
56. L. Berthoud, in Proc. Sixth Int. Symp. Materials in a Space Environment, ESA SP-368 (1994) 239.
57. L. Berthoud, in Hubble Space Telescope Solar Array Workshop: Results from Post-Flight Investigations of the Returned HST Solar Array, ESA WPP-77 (1995) 477.
58. N.R.G. Shrine, E.A. Taylor, H. Yano, A.D. Griffiths and J.A.M. McDonnell, in Characteristics and Consequences of Orbital Debris and Natural Space Impactors, SPIE proc. 2813 (1996) 76.
59. J.-C. Mandeville and M. Rival, Impact residue of HST array samples, ESTEC contract report WMA/94-335/GD/HST, 1995.
60. G.A. Graham, A.T. Kearsley, M.M. Grady and I.P. Wright, in Proc. 2nd European Conference on Space Debris, ESA SP-393 (1997) 183.
61. G.A. Graham, A.T. Kearsley, M.M. Grady, I.P. Wright, M.K. Herbert and J.A.M. McDonnell, Int. J. Impact Engng. 23 (1999) 319.
62. G.A. Graham, A.T. Kearsley, M.M. Grady, I.P. Wright, and J.A.M. McDonnell, Adv. Space Res. 25 (2000) 303.
63. G.A. Graham, N. McBride, A.T. Kearsley, G. Drolshagen, S.F. Green, J.A.M. McDonnell, M.M. Grady and I.P. Wright, Int. J. Impact Engng. 26 (2001) 263.
64. N. McBride, Adv. Space Res. 20 (1997) 1513.
65. P. Jenniskens, Astron. Astrophys. 287 (1994) 990.

A new approach to applying interplanetary meteoroid flux models to spacecraft in gravitational fields.

M. J. Matney

Lockheed Martin Space Operations, Houston, TX 77058, USA

Neil Divine in his "Five Populations of Interplanetary Meteoroids" [1] introduced a method of defining the interplanetary meteoroid environment in terms of orbit families. For this work, a new method is introduced to apply orbit populations to compute meteoroid fluxes on spacecraft in interplanetary space and within the gravitational field of a planet or moon. The flux on the target is defined per unit solid angle per unit speed. This differential flux can be related to that outside the gravitational field by use of Liouville's theorem. Integration is performed over bins in solid angle (defining the direction of the meteoroids) and in meteoroid speed. This formulation computes the directional gravitational lensing while avoiding the numerical problems in Divine's method. It is also relatively easy to account for the shadowing of the planet body. This method is even applicable to complex multi-body systems.

1. INTRODUCTION

Much of the use of models to describe the meteoroid environment has been driven by the need to understand the collision risk to spacecraft, especially manned missions. Kessler [2] used the following one-dimensional formula to show the enhancement of the velocity flux within a gravitational field:

$$\left|\vec{v}_G\right|^{-2} \left|\vec{F}_G\right| = \left|\vec{v}_\infty\right|^{-2} \left|\vec{F}_\infty\right| , \tag{1}$$

where the subscripts G and ∞ refer to the values within the gravitational field and at infinity (respectively), F is the flux of meteoroids, and v is their velocity. This equation is only valid for an isotropic flux traveling at a single velocity. Using this equation in conjunction with an isotropic flux distribution such as that of Grün et al. [3], it is straightforward to compute and average flux on a spacecraft or instrument. For many purposes this simple equation is adequate, but for many applications directional and velocity distribution information may be preferred.

Divine [1] introduced a method of defining the interplanetary meteoroid environment in terms of orbit families. As described in [4], the distributions that Divine introduced were not the "classical" distributions. For values of perihelion (r_1), inclination (i), and eccentricity (e), the classical distribution $D(r_1,e,i)\, dr_1\, de\, di$ (note the variables need not be separable) is related to Divine's separable distributions in perihelion (N_1), inclination (p_i), and eccentricity (p_e) by

$$D(r_1,e,i)\, dr_1\, de\, di = 2\pi^2\, r_1^2 \sin i\, (1-e)^{-3/2}\, N_1\, p_e\, p_i\, dr_1\, de\, di\, . \tag{2}$$

The spatial density in interplanetary space from a distribution of meteoroids can be represented as

$$\begin{aligned}\rho(r,\lambda;r_1,e,i)\,dr_1\,de\,di &= \frac{(1-e)^{3/2}\,\frac{1}{4}\,D(r_1,e,i)\,dr_1\,de\,di}{2\pi^3\,r_1\,r\sqrt{r-r_1}\,\sqrt{(1+e)r_1-(1-e)r}\,\sqrt{\sin^2 i-\sin^2\lambda}} \\ &= \frac{\frac{1}{4}\,N_1\,p_i\,p_e\,r_1\sin i\,dr_1\,de\,di}{\pi\,r\sqrt{r-r_1}\,\sqrt{(1+e)r_1-(1-e)r}\,\sqrt{\sin^2 i-\sin^2\lambda}}\end{aligned}, \tag{3}$$

where r and λ are the range and latitude (respectively) relative to the Sun and ecliptic plane (cf. Figure 1). These formulae can be used (in conjunction with the velocity of each orbit distribution) to compute the directional flux in interplanetary space under solar gravity alone for a defined orbital distribution. Such information is useful if a dust-measuring instrument or a susceptible spacecraft part maintains a particular orientation in space and one wishes to compute the expected flux of meteoroids on it.

2. THEORY

Dealing with a directional meteoroid distribution (like that described by Divine) as it encounters a gravitating body such as the Earth is a daunting mathematical task. However, by invoking the often-taught but seldom-used Liouville's theorem, it is possible to greatly simplify the problem. Liouville's theorem states that an ensemble of particles moving under the influence of conservative forces (such as gravity) maintains its density in phase space (see [5]). For an ensemble of N particles traveling from point "A" to point "B" in a conservative field,

$$\frac{N}{\Delta^3 x_A\,\Delta^3 v_A} = \frac{N}{\Delta^3 x_B\,\Delta^3 v_B}, \tag{4}$$

where the Δ terms represent the dimensions of the six-dimensional "box". No matter how the six-dimensional phase space "box" changes in shape due to forces, its volume containing the ensemble remains constant. Defining the functions

$$N = \iiint d^3x\, N(\vec{x}) \quad \text{and} \quad N(\vec{x}) = \iiint d^3v\, N(\vec{x},\vec{v}), \tag{5}$$

where N is the number of particles, $N(\vec{x})$ the spatial density, and $N(\vec{x},\vec{v})$ the phase space density, we can rewrite equation 4 in its differential form where Ω refers to the direction of the velocity,

$$N_A(\overrightarrow{x_A},\overrightarrow{v_A}) = N_B(\overrightarrow{x_B},\overrightarrow{v_B}) \quad \text{or} \quad \frac{N_A(\overrightarrow{x_A},\overrightarrow{v_A})\,d^3v_A}{v_A^{\,2}\,dv_A\,d\Omega_A} = \frac{N_B(\overrightarrow{x_B},\overrightarrow{v_B})\,d^3v_B}{v_B^{\,2}\,dv_B\,d\Omega_B}. \tag{6}$$

The numerators in equation 6 are spatial densities, but defined for a narrow range of velocities. We now define the flux magnitude at a point per unit speed per unit solid angle direction Φ,

$$\Phi(\vec{x},\vec{v}) = \frac{v\, N(\vec{x},\vec{v})\, d^3v}{dv\, d\Omega}. \tag{7}$$

Equation 6 now becomes

$$v_A^{-3}\, \Phi_A(\overrightarrow{x_A},\overrightarrow{v_A}) = v_B^{-3}\, \Phi_B(\overrightarrow{x_B},\overrightarrow{v_B}). \tag{8}$$

Unlike in equation 1, the full directional information is here preserved, and a velocity distribution is used rather than assuming a single velocity.

In order to obtain the meteoroid flux within a gravitational field (point "B") per unit speed and unit solid angle from a particular direction, simply trace the hyperbolic orbit of the meteoroids that came from that particular direction at that speed backwards out of the gravitational field to a point in interplanetary space (point "A") and define the flux per unit speed per unit solid angle from the appropriate direction there. Note that if the orbit projected back out of the gravitational field does not have sufficient energy to leave the gravitational field, then the interplanetary component is zero from that direction. Likewise, if the orbit traced back encounters the body of the gravitating planet or moon, then that flux direction is "shadowed" by the bulk of the body and the flux contribution is zero.

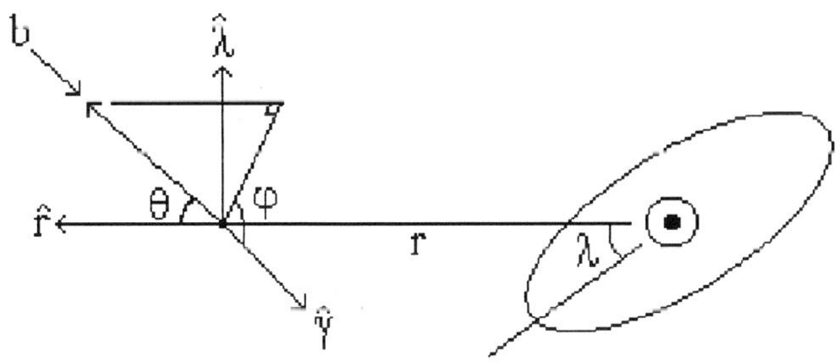

Figure 1. The reference frame for interplanetary flux calculations uses r, the range from the Sun, and λ, the ecliptic latitude. The γ axis represents the direction of increasing ecliptic longitude and the λ-axis of increasing ecliptic latitude. That meteoroids arrive with dimensionless speed b from the direction defined by θ and φ.

This method also works in transforming between reference frames, such as between a spacecraft frame and an Earth-centered frame ("SC" is the spacecraft frame, "E" the Earth-centered inertial frame, and "IS" the heliocentric interplanetary space inertial frame):

$$v_{SC}^{-3}\, \Phi_{SC}(\overrightarrow{x_{SC}},\overrightarrow{v_{SC}}) = v_E^{-3}\, \Phi_E(\overrightarrow{x_E},\overrightarrow{v_E}) = v_{IS}^{-3}\, \Phi_{IS}(\overrightarrow{x_{IS}},\overrightarrow{v_{IS}}). \tag{9}$$

The next step is to compute Φ_{IS} using the reference frame in figure 1 with the definitions

$$b = \sqrt{\frac{r\, v^2}{\mu}}, \quad r_1 = \frac{r\,(1-e)}{2-b^2}, \quad e = \sqrt{b^2 \sin^2\theta\,(b^2-2)+1}, \quad \text{and} \quad \cos i = -\cos\varphi\, \cos\lambda, \tag{10}$$

where μ is the solar gravitational constant. The Jacobian can be constructed for these parameters,

$$dr_1\ de\ di = \frac{2\,b^3\ r\,(1-e)\cos\theta\sin\theta}{(2-b^2)\,e}\ \frac{\sqrt{\sin^2 i - \sin^2\lambda}}{\sin i}\ db\ d\theta\ d\varphi\,, \tag{11}$$

and the spatial density in equation 3 can now be written as

$$\begin{aligned}\rho(r,\lambda;b,\theta,\varphi)\ db\ d\theta\ d\varphi &= \frac{(1-e)^{5/2}\ \{\tfrac{1}{4}\,D(r_1,e,i)\}\ (2\,b^3)\cos\theta\sin\theta}{4\pi^3\ r_1\ e\,(2-b^2)\sqrt{r-r_1}\ \sqrt{(1+e)r_1-(1-e)r}\ (\sin i)}\ db\ d\theta\ d\varphi \\ &= \frac{(1-e)\ \{\tfrac{1}{4}\,N_1\ p_i\ p_e\}\ \ (2\,b^3\ r_1)\cos\theta\sin\theta}{\pi\,e\,(2-b^2)\sqrt{r-r_1}\ \sqrt{(1+e)r_1-(1-e)r}}\ db\ d\theta\ d\varphi\end{aligned} \tag{12}$$

This can be written in terms of the true velocity and solid angle

$$\begin{aligned}\rho(r,\lambda;\vec{v})\ dv\ d\Omega &= \frac{(1-e)^{5/2}\ \{\tfrac{1}{4}\,D(r_1,e,i)\}\ (2\,b^3)\cos\theta}{4\pi^3\ r_1\ e\,(2-b^2)\sqrt{r-r_1}\ \sqrt{(1+e)r_1-(1-e)r}\ (\sin i)}\sqrt{\frac{r}{\mu}}\ dv\ d\Omega \\ &= \frac{(1-e)\ \{\tfrac{1}{4}\,N_1\ p_i\ p_e\}\ \ (2\,b^3\ r_1)\cos\theta}{\pi\,e\,(2-b^2)\sqrt{r-r_1}\ \sqrt{(1+e)r_1-(1-e)r}}\sqrt{\frac{r}{\mu}}\ dv\ d\Omega\end{aligned} \tag{13}$$

Because this density is equivalent to

$$\rho(r,\lambda;\vec{v})\ dv\ d\Omega = N(\vec{x},\vec{v})\ d^3v\,, \tag{14}$$

it is an easy matter to use equation 13 to compute the flux in equation 7 for any point in interplanetary space. Equation 8 or 9 then relates that flux to any point along the trajectory. This method also works for any complex system of gravitating bodies, although computing the meteoroid orbits back in time would be more complex than using simple Keplerian orbits.

REFERENCES

1. N.J. Divine, N. J. Geophys. Res. 98 E9 (1993) 17,029.
2. D. Kessler, NASA TND-6596, March (1972)
3. E.H. Grün, H. A. Zook, H. Fechtig and R. H. Giese, Icarus 62 (1985) 244.
4. M. Matney and D. Kessler, in Physics, Chemistry, and Dynamics of Interplanetary Dust, ASP Conf. Series 104 (1996) 15.
5. R.D. Shelton and A. L. Arbuckle, Final Report, Ballistic Research Labs, Aberdeen Proving Ground, MD, ARBRL-TR-02195 (1979).

The new NASA orbital debris breakup model

J.-C. Liou[a], N. L. Johnson[b], P. H. Krisko[a] and P. D. Anz-Meador[c]

[a]Lockheed Martin Space Operations, Houston, TX77058, USA.

[b]NASA Johnson Space Center, Houston, TX77058, USA.

[c]Viking Science and Technology, Houston, TX77058, USA.

A new breakup model, including on-orbit explosions and collisions, has been developed and implemented in the NASA orbital debris dynamical model EVOLVE 4.0. It is based on more than twenty well-observed on-orbit explosions, one on-orbit collision, and several laboratory hypervelocity impact experiments. The revised size distributions of breakup fragments are quite different from the ones used previously. The new distributions are simpler and well supported by the data. The predicted 10 cm and greater debris populations between 1957 and 1999, based on simulations with the new breakup model, compare well with those derived from radar observations.

1. INTRODUCTION

To model the past, current, and future space debris environment in low Earth orbit (LEO, defined as the region between 200 and 2000 km altitude), the Orbital Debris Program Office at the NASA Johnson Space Center has developed a numerical program, EVOLVE [1, 2]. The first step in EVOLVE is to model historical launches and breakup events, including explosions and collisions. Breakup fragments are then propagated forward in time numerically with other intact objects. The gravitational force of Earth and its J_2 perturbation, solar-lunar perturbations, and atmospheric drag are all included in the propagator. The intact object and debris populations since 1957 are calculated until the end of 1999. The spatial density or number of debris objects of a given size and greater, at a given altitude and at a given time, from EVOLVE can then be compared with radar, optical, or *in situ* debris measurements. Once the current debris environment is modeled properly, EVOLVE takes an assumed launch traffic and an assumed solar activity projection and uses a Monte Carlo approach to model future on-orbit explosions and collisions and predict the future LEO debris environment.

2. THE BREAKUP MODEL

A key module in a dynamical orbital debris model, such as EVOLVE, is the breakup model. A breakup model includes the size distribution of collision or explosion fragments, the area-to-mass ratio of the fragments, and their relative velocity distribution with respect to the parent object. Explosion/collision rate and the fragment size distribution determine the production rate of debris particles while the area-to-mass ratio and the relative velocity distribution determine how they evolve and eventually decay. In this paper, we focus on the size distribution of the breakup fragments. These distributions are different for explosions and collisions, and are discussed separately below.

2.1. Explosion fragment size distribution

The previous EVOLVE breakup model assumes that two fragment size distributions are possible for explosions: high-intensity and low-intensity explosion size distributions [3]. A single event may produce fragments that follow both distributions (e.g., 10% of the fragments follow high-intensity explosion size distribution while the remaining 90% follow low-intensity explosion size distribution).

The high-intensity explosion size distribution has the form:

$$N_{hi,cum} = a \times m_{tot}\left(L^{-b} - c\right) \tag{1}$$

where $N_{hi,cum}$ is the number of fragments greater than diameter L (in meters), m_{tot} is the total mass of fragments (in kg) which follow this size distribution, and a, b, and c are three positive constants. The low-intensity explosion size distribution has the general form of:

$$N_{low,cum} = e^{-aL+b} \tag{2}$$

where $N_{low,cum}$ is the number of fragments greater than diameter L (in meters); a and b are two positive constants that depend on the mass of the parent object.

A recent review of more than twenty well-observed on-orbit explosions, including spacecraft and rocket bodies, indicates that there is not sufficient evidence to support the above-described two types of explosions. Table 1 lists the seven regular rocket body explosions examined. The derived total masses of fragments of these events, based on radar observations, are all within 10% of the published dry masses of their parent objects.

Table 1
Explosion events used to derive the new fragment size distribution

Name	International Designator	Debris Cataloged	Event Date
OPS 7613	1969-082A	259	4-Oct-69
Nimbus 4	1970-025C	372	17-Oct-70
LandSat 1	1972-058B	226	22-May-75
LandSat 2	1975-004B	207	9-Feb-76
LandSat 3	1978-026C	209	27-Jan-81
FengYun 1-2	1990-081D	83	3-Sep-90
Nimbus 6	1975-052B	237	1-May-91

From History of On-orbit Satellite Fragmentations [4]

The cumulative size distributions of the events are shown in Figure 1. They are derived from observations from the US Space Surveillance Network. Fragments smaller than about 10 cm are beyond the sensitivity limit of the radars. Based on the data, it appears that a simple power-law size distribution is adequate to represent these events:

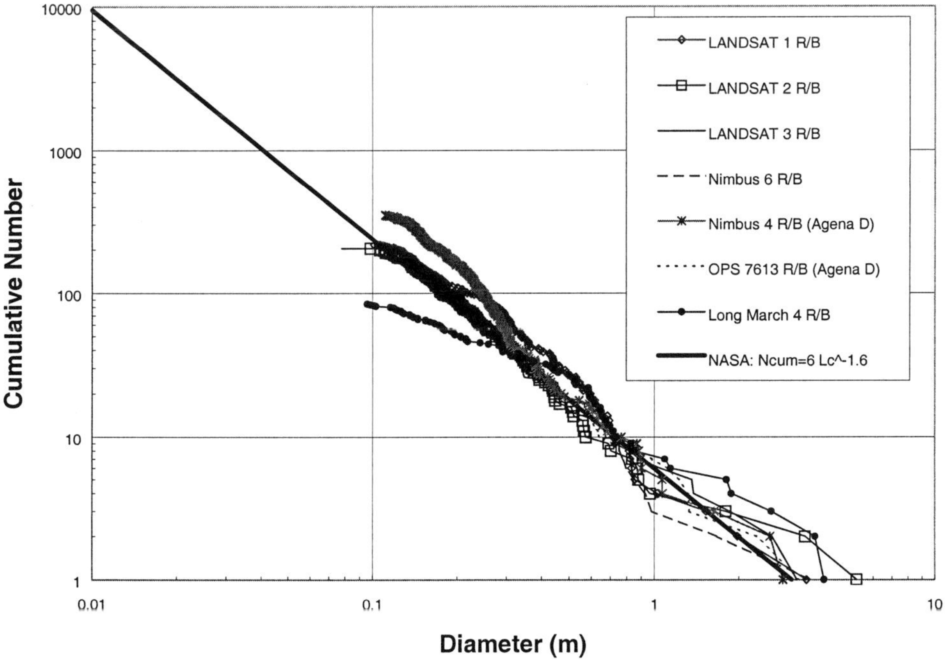

Figure 1. Seven on-orbit explosions and the new NASA explosion model.

$$N_{cum} = sf \times 6 \times L^{-1.6} \tag{3}$$

where N_{cum} is the number of fragments greater than diameter L (in meters), and sf is a scaling factor. Based on the available observations, the scaling factor is 1 for a regular rocket body; 0.1 for a Soviet/Russian Proton ullage motor; 0.1 for a Molniya orbit Soviet Early Warning satellite; 0.25 for an Ukrainian Tsyklon third stage; 0.3 for a Soviet Anti-Satellite (ASAT); 0.5 for a Soviet battery-related explosion; 0.6 for a Soviet Electronic Ocean Reconnaissance Satellite (EORSAT); and 2 for a Titan Transtage.

2.2. Collisional fragment size distribution

The previous EVOLVE breakup model assumed that there were two types of collisions: catastrophic and non-catastrophic. Fragments of a catastrophic collision followed two size distributions, a power-law distribution and a low-intensity explosion distribution. The two components varied with the projectile-to-target mass ratio [3]. The model was complex and again not supported by observational data.

Table 2 lists the hypervelocity impact experiments [5, 6, 7] and one on-orbit collision (P78/SOLWIND) used to derive the new fragment size distribution.

Table 2
Collision experiments/event used to derive the new fragment size distribution

Name	Target Mass (kg)	Projectile Mass (g)	Impact Velocity (km/s)	Catastrophic Collision
Bess 1	---[1]	1.65		No
Bess 2	---[1]	0.37	4.5	No
PSI 1	26	237	5.9	Yes
PSI 2	26	237	3.3	Yes
SOCIT	34.5	150	3.0	Yes
P78/SOLWIND	850	16000	7.6	Yes

[1]The target was a simulated spacecraft wall [5].

Based on the new analysis, fragments from a collision follow a single power-law size distribution:

$$N_{cum} = 0.1 \times M_{tot}^{0.75} \times L^{-1.71} \tag{4}$$

where N_{cum} is the number of fragments greater than diameter L (in meters), and M_{tot} is the total mass (in kg) of the fragments. For a catastrophic collision, M_{tot} is the sum of the target and projectile masses while for a non-catastrophic collision:

$$M_{tot} = m_{projectile} \times V_{impact}^{2} /(km/s)^{2} \tag{5}$$

A catastrophic collision occurs when the ratio of impact kinetic energy to target mass exceeds 40 J/g [7]. The size distributions of the events listed in Table 2 and the predicted size distributions from the new model are shown in Figure 2. The fall-off of P78 fragments is due to the sensitivity limit of the radar. Overall, the new model produces results, extended over 5 orders of magnitude, that agree well with the data. Simulations of EVOLVE with the new breakup model also show good agreement with the NASA Haystack radar observations of the current debris spatial density distribution in the LEO region.

3. CONCLUSIONS

A new breakup size distribution model has been developed to replace the one used previously. The new model has the characteristic of simplicity and is supported by observational data. A better breakup model not only improves our understanding of the nature of the breakup, but also gives us a more accurate description of the current and future debris environment.

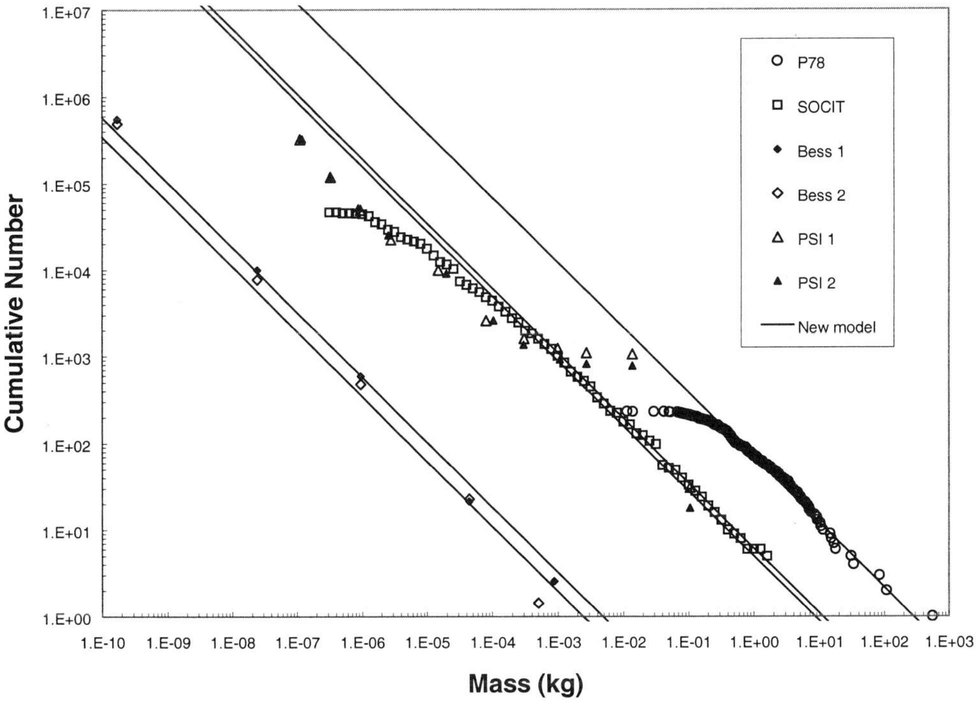

Figure 2. Collisional experiments/event and the new NASA collision model.

REFERENCES

1. N. Johnson, E. Christiansen, R. Reynolds, M. Matney, J.C. Zhang, P. Eichler and A. Jackson, Proc. 2nd European Conference on Space Debris, ESA SP-393 (1997) 225.
2. P. H. Krisko et al., LMSMSS-33020, (2000).
3. R. C. Reynolds et al., LMSMSS-32532, (1998).
4. N. L. Johnson et al., History of On-orbit Satellite Fragmentations (11th Edition), NASA JSC 28383, (1998).
5. T. D. Bess, NASA TN D-8108, (1975).
6. P. E. Nebolisine et al., Physical Science Inc. (PSI) Report TR-399, (1983).
7. D. S. McKnight et al., Kaman Science Corp. Report, NASA contract No. NAS 9-19215 (SOCIT report), (1995).

A CCD search for the Earth-Moon libration clouds around L4

S. Takahashi[a] M. Ishiguro[a], Y. Fujii[a], S. Urakawa[a], C. Yoshishita[a], T. Mukai[a] and R. Nakamura[b]

[a]Graduate School of Science and Technology, Kobe University, Japan

[b]Earth Observing Research Center, National Space Development Agency of Japan

We attempted to detect the Earth-Moon libration clouds around L4 with a cooled CCD camera at Mauna Kea (4200m), Hawaii, on November 17 and 18, 1999. No significant enhancement of brightness larger than $1.5S_{10\odot}$ was found around L_4. It seems that the brightness of the libration clouds in 1999, if it existed, would be fainter than that reported previously by Kordylewski [1] and Roach [2].

1. INTRODUCTION

The Lagrangian points L_4 and L_5 are the triangular solutions of the restricted three-body problem. The L_4 is located in the leading side and the L_5 in the trailing side of an orbit. In the solar system, it is widely known that certain asteroids called the "Trojan family" exist around the Sun-Jupiter Lagrangian points. It is also known that (5261) Eureka is a Mars Trojan [3]. Kuchner et al. [4] tried to find the dust clouds around the Sun-Jupiter L_5 point by analysing the COBE-DIRBE infrared data, but they obtained a negative result.

Concerning the Earth-Moon Lagrangian points, in the 1960s Kordylewski [1] reported the existence of the clouds based on naked eye observations. Roach [2] has shown that the satellite OSO-6 could have caught the enhancement of brightness near the L_4 and L_5, and their size and brightness were deduced. However, their existence could not be confirmed by the observations (e.g. Roosen [5], Roosen and Wolff [6], Bruman [7]). The theoretical approaches have been performed to examine the enhancement of the number density of the dust grains near the Earth-Moon Lagrangian point. Röser [8] has shown that the conditions for supply and loss of dust grains to the libration point regions are not favorable for libration clouds, and he suggested, as a very high upper bound, a surface brightness of $2S_{10\odot}$ at L_4.

Recent CCD Photometry techniques are much superior than those of photo-multipliers used before, and consequently provides a possibility to detect a slight enhancement in the brightness of the zodiacal light (see e.g. Ishiguro et al. [9]). This modern tool would allow us to learn whether these faint clouds exist or not. We will report the results of our search for the libration clouds in the Earth-Moon system, using a cooled CCD camera at Mauna Kea (4200m), Hawaii.

2. OBSERVATIONS AND DATA ANALYSIS

The libration clouds may exist at the triangle points of the Earth and the Moon in the Moon's orbital plane. This restriction prevents us from observing the liberation clouds freely. That is, the detection should be done after the Moon's setting(L_4) and before the Moon's rising (L_5), when the zenith angle is larger than 30°. The sky under such conditions would be hazy and the zodiacal light near the libration points would be brighter compared with at the zenith. Consequently, it is not easy to distinguish the faint light of libration clouds from the background light consisting of airglow and zodiacal light. Furthermore, we have to take care that the Lagrangian point should be far away from the Milky Way and and bright planets/stars during the observation.

We made the observation of the sky region where the liberation clouds are expected, at Mauna Kea (4200m), Hawaii on November 17 and 18, 1999. This high altitude observation site is preferable to reduce the contamination of the scattered light by the atmosphere of the Earth. In addition, the location of Mauna Kea is suitable for the observation of libration clouds because the ecliptic plane in Autumn passes through the zenith, such that the libration clouds could be seen at higher altitude. In addition, the position of the L_4 point was away from the Milky Way in November.

The bright planets, Jupiter and Saturn, were located about 15° away from the Lagrangian point on November 17 and 18, 1999. It was expected that their scattering light would somewhat pollute the frame. However, since a good chance of detection is so rare (i.e. only two or three times per year) we performed the observation in November at Mauna Kea. In order to avoid the contamination of bright planets, the L_4 point could not be sited at the center of the frame. A cooled CCD camera (Mutoh CV-16), at the temperature of −30°C during the observation, was used with a wide-field lens (Sigma 24mm, F=2.8). Its angular resolution using 2×2 binning, is 2.50' pixel^{-1} and the field of view is 32°×21°. A broad band filter designed to transmit light between 439–524nm, where no bright emission lines of the airglow exist, was used. The exposure time was set to be 3 minutes.

Flat frames, taken by using an integrating sphere of the National Institute of Polar Research, were adopted in the data reduction process. We removed stars from the frames and tried to estimate the sky brightness which consists of zodiacal light, airglow from upper atmosphere, integrated starlight of unresolved stars and scattered light from the Earth's atmosphere, by using similar methods to those in Ishiguro et al. [9]. We applied the Fourier filtering technique to enhance the structures of sky brightness with a scale smaller than 10°. In order to reduce the noise arising from the fluctuation of detected photons and dark current, the brightness in the resulting frame was averaged over 24×24 pixels (1°×1°).

Our data reduction process is suitable for enhancing clouds with a scale of about 1°–10°. If a cloud had a scale of about 6° as reported by Roach [2], and a brightness above the detection limit of about $1S_{10\odot}$, we would be sure to find the clouds in the resulting frame.

3. RESULTS

Figure 1 is one of the Fourier filterd frames taken on the 17th November 1999, where the L_4 position is included. The X and Y axes correspond to the frame of the CCD array,

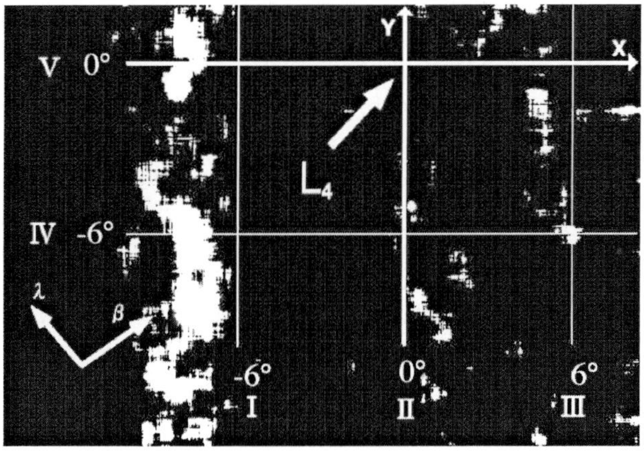

Figure 1. The area, including L_4, taken at 11^h51^m UT on 17th Nov. 1999. The L_4 point with R.A. 02^h27^m and Dec. 37°25' is at the origin of the arbitrary coordinate system. The bright regions indicate the areas with brightnesses larger than $1.0S_{10\odot}$. The bright region on the left side along $X = -7° \sim -8°$ is not a real cloud, but the ghost caused by the Fourier filtering process. A cut of the brightness along the lines marked I–V is shown in Figure 2.

and the L_4 point is at the origin of the co-ordinates. The directions of ecliptic latitude (β) and longitude (λ) are also indicated in Figure 1. The Fourier filtering was done along the X axis in the frame. The L_4 point is not located at the center of the frame because we had to purge the bright areas contaminated by Jupiter and Saturn from the image. Both planets were placed at about 15° away from the L_4 point. The contamination from bright planets is removed by the Fourier filtering.

The bright area in Figure 1 denotes the sky region with a brightness greater than $1.0S_{10\odot}$. The fluctuations of the brightness along the lines at $X = 0°, \pm 6°$ and $Y = O°, -6°$ are illustrated in Figure 2. The 1σ noise level of brightness in Figure 2 is about $0.5S_{10\odot}$ for each case. The remaining cloud structure on the left side of the image in Figure 1 along the $X = -7° \sim -8°$ is not a real cloud because it is produced at the edge of the frame in the Fourier filtering process.

No significant enhancement of brightness appears to show the existence of a real cloud near the L_4 point. Roach [2] reported that the libration cloud had the size of about 6° and drifts around the libration point within the oval region with the size of about 6° along the ecliptic and of about 2° perpendicular to the ecliptic. The area imaged in Figure 1 does not contain all of the oval region proposed by Roach [2]. Therefore, we examined other frames taken in the neighboring sky regions by applying the same procedure for analysis and found no significant cloud structure near the L_4 point.

We can estimate the upper limit of the brightness of the expected cloud. The noise fluctuation in Figure 1 is $3\sigma = 1.5S_{10\odot}$ so the cloud brightness might be below $1.5S_{10\odot}$. Our results suggest that the libration clouds might be the transient objects, if they really

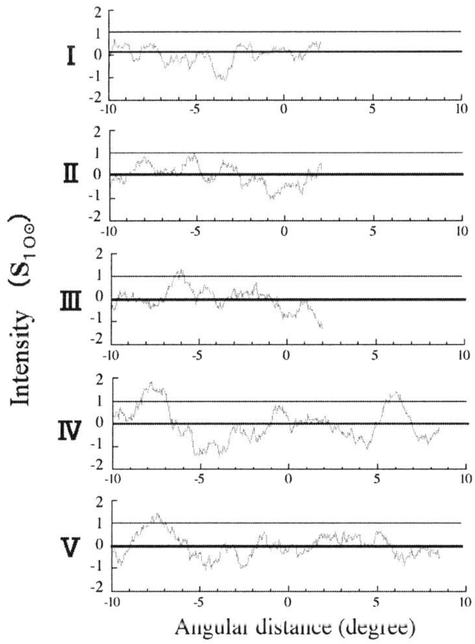

Figure 2. The brightness distribution along the line defined in Figure 1. Each number on the line corresponds to that in Figure 1. The L_4 point is at angular distance = 0° in the cases of II and V. Noise fluctuation is under $1.0S_{10\odot}$. Enhancements at $X = -7° \sim -8°$ in the cases of numbers IV and V are not real libration clouds (see Figure 1). No significant cloud with the size of about 6° appears.

existed in the previous observations. Further trials to detect them would be needed to study their existence in more detail.

REFERENCES

1. K. Kordylewski, Acta Astron. 11 (1961) 165.
2. J.R. Roach, Planet. Space Sci. 23 (1975) 173.
3. K.A. Innanen, J. R. Astron. Soc. Can. 85 (1991) 151.
4. M.J. Kuchner, W.T. Reach and M.E. Brown, Icarus 145 (2000) 44.
5. R.G. Roosen, Icarus 9 (1968) 429.
6. R.G. Roosen and C.L. Wolff, Nature 224 (1969) 571.
7. J.R. Bruman, Icarus 10 (1969) 197.
8. S. Roser, Interplanetary Dust and Zodiacal Light 48 (1975) 124.
9. M. Ishiguro, R. Nahmura, Y. Fujii, K. Morishige, H. Yano, H. Yasuda, S. Yokogawa and T. Mukai, Astrophys. J. 511 (1999) 432.

The chemistry and origin of micrometeoroid and space debris impacts on spacecraft surfaces.

G.A.Graham[a], A.T.Kearsley[b], G.Drolshagen[c], M.M.Grady[d], I.P.Wright[a] and H.Yano[e]

[a]Planetary and Space Science Research Institute, The Open University, Walton Hall, Milton Keynes MK7 6AA, UK.

[b]Space Science Research, School of BMS, Oxford Brookes University, Headington, Oxford OX3 0BP, U.K.

[c]TOS-EMA, ESTEC, European Space Agency, Keplerlaan 1, NL-2201 AZ Noordwijk, NL.

[d]Mineralogy Department, The Natural History Museum, Cromwell Road, London SW7 5BD, U.K

[e]Planetary Science Division, The Institute of Space and Astronautical Science, Japan

Laboratory investigations of impact residues captured on the solar cells from the Hubble Space Telescope and on insulation foils from the Space Flyer Unit demonstrate preservation of abundant and diverse micrometeoroid and space debris remnants. Micrometeoroid residues often appear as complex melts of poly-mineralic origin derived from silicates, carbonates, metals and metal sulfides. The space debris includes paint-flakes, metal alloys and possible reactor coolant, but the most abundant components are aluminium and aluminium oxide remnants from solid rocket motor operation. The impactor origins have now been compared with the theoretical flux models for Low Earth Orbit.

1. INTRODUCTION

Our understanding of small particle populations has been aided by laboratory investigations of cosmic dust particles (approximately 1-400μm diameter). Such studies have focused mainly on material collected from 'terrestrial' locations, e.g. the ocean floor, polar ices and the stratosphere, e.g. [1-3]. However, these particles may have undergone selection and alteration during atmospheric entry, e.g. [4], and it is desirable to achieve some sampling outside of the Earth's atmosphere. The Giotto spacecraft investigated particles from Comet Halley using a dust-impact detection system [5], but did not return samples for further examination. Particle analysers have flown on several interplanetary and earth orbital missions. They have yielded important information, yet have not returned samples to Earth. Spacecraft deployed in near Earth orbits, e.g. low Earth orbit (LEO), do offer opportunities to retrieve material that can be analysed in the laboratory. LEO is an interesting environment to sample, it not only enjoys the passage of cosmic dust particles, but also contains an orbital population of artificial particles, space debris generated by human activity. Space debris is diverse in nature, from paint fragments and human waste on a sub-millimetre scale, to spent rocket bodies on the metre scale [6]. Dedicated *in-situ* sampling techniques have been developed and successfully deployed in LEO, e.g. the microabrasion foil experiment flown on the STS-3 Space Shuttle mission [7] and the COMET-1 experiment flown on the Salyut 7 spacecraft [8]. However, perhaps the most extensive sampling of particles in LEO was carried out by NASA's Long Duration Exposure Facility (LDEF), in orbit for 69 months [9]. As part of LDEF's scientific payload, there were numerous experiments dedicated to the collection of micrometeoroids and space debris, analysed upon return to Earth, e.g. [10]. Detailed analysis was also carried out on non-dedicated surfaces that had experienced impact damage from micro-particle hypervelocity collisions, e.g. [11]. Subsequently, the utility of non-dedicated collector surfaces returned from LEO has been demonstrated by the success of post-flight investigation of thermal blankets and aluminium thermal control covers from the Solar Maximum satellite [12] and [13].

Until recently [16], the full potential of residue analysis has not been apparent. Retrieval of one of the two solar array panels from the Hubble Space Telescope (HST) during the first service mission in 1993, and return of the Space Flyer Unit (SFU) have provided contrasting substrates for analysis of particle residues by analytical electron microscopy. Herein we discuss the chemistry of preserved micrometeoroid and space debris remnants on solar cells of the HST and on aluminised Kapton multi-layer insulation (MLI) foils from SFU.

2. EXPERIMENTS

2.1. Flight Details of the HST and the SFU

In 1993, during the first service mission of the Hubble Space Telescope, the "V2" solar panel was successfully replaced and returned to Earth. Prior to the retrieval, the array had been in LEO for 1320 days at an operation orbit of ~600km. Individual solar cells from the returned array were investigated extensively in a detailed post-flight investigation program, e.g. [14]. The SFU was retrieved from LEO after 301 days of exposure in an operational orbit of ~480km. As with the solar cells from the HST, the surfaces from the SFU were extensively examined and were subject to detailed post-flight investigations, [15].

2.2. The Structure of Solar Cells and the MLI-Foils

The HST solar cells are multi-layered structures. The top layer of borosilicate glass rests upon silicone resin, which is underlain by the silicon layer. On the back of the silicon is another resin layer that also contains metallic silver connectors. The complex assembly is supported by a backing-tape of resin-bonded glass-fibre mesh. A full, detailed description of the solar cell composition is given in [16]. The SFU MLI-foils consist of 12 layers of aluminised Kapton film interspersed with Dacron nets, a full description is given in [17].

2.3. Laboratory Methodology

The analytical work was carried out on a Jeol 840 scanning electron microscope (SEM) with 2nA beam current and at 20kV accelerating voltage. The samples were carbon-coated to reduce the effects of electrical charging during SEM investigations. Most of the electron imaging was carried out using a solid-state back-scattered electron detector. The X-ray elemental maps and X-ray spectra were acquired using an Oxford Instruments eXL energy-dispersive spectrometer (EDS) microanalyser with a Pentafet detector, fitted with an ultra-thin window (this allows the detection of X-rays from light elements such as carbon). A full description of the analytical protocol is given in [17]. The interpretation of samples of material from LEO, whether micrometeoroids or space debris, is a complex task, the particles will have had encounter velocities of ≥ 5km/s relative to the collector surface. The particles experience extensive shock deformation, melting, and often even complete vapourisation during the impact process. Hypervelocity impact events may modify the original chemical composition of an impactor, fractionating volatile from refractory elements. Micrometeoroid residues may therefore not necessarily retain the stoichiometric chemical signature of a parent mineral, and quantitative analysis of residue is usually impractical. The classification of micrometeoroid and space debris residues employed was based on similar chemical and mineralogical criteria to those used for LDEF, e.g. [19].

3. DISCUSSION

3.1. Impact Morphology

The impact damage suffered by solar cells and the MLI-foils is very different. The glass and silicon layers of solar cells are brittle, and therefore generate complex crater structures that contain radial and conchoidal fractures (figure 1). Extensive breakage results in spallation, and most of the central melt pit may be lost. The MLI-foils, in contrast, show a simple perforation, sometimes with overturned collar, on the top or bottom of the foil layer (figure 2). Dispersed impactor remnants may be preserved on one or more layers of the MLI-foils.

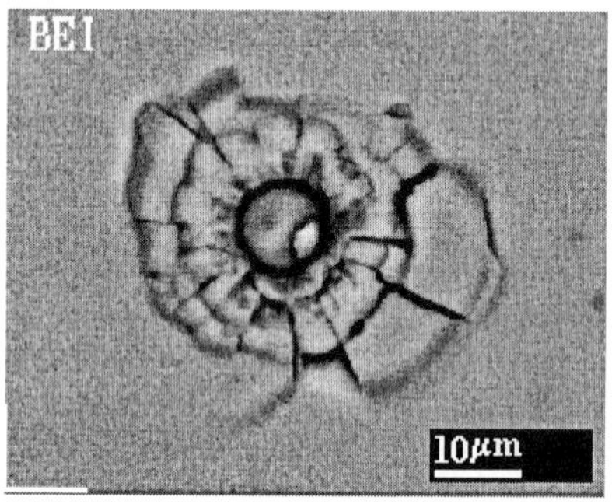

Figure 1. BEI of a typical small impact crater on the HST solar cells.

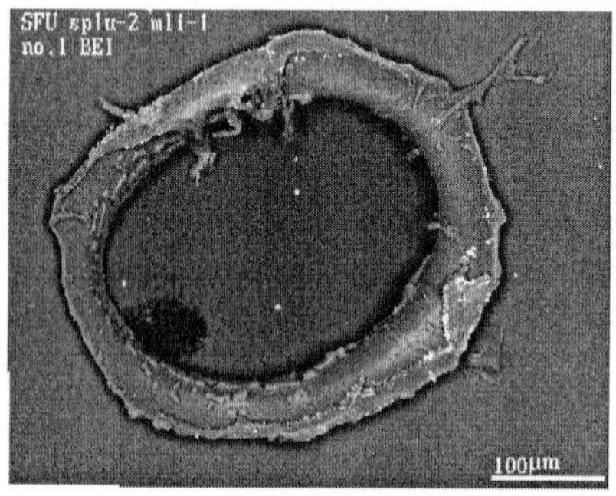

Figure 2. BEI of the top layer of an MLI-foil containing an impact feature.

3.2. Micrometeoroid and Space Debris Chemistries on HST Solar Cells

Our initial survey of HST craters with conchoidal diameter (D_{co}) between 100 and 1000μm identified that micrometeoroid remnants were dominant [19]. Residues were composed of remnants from silicate minerals, calcite, metal sulfides and metals. Residues often appeared as complex poly-mineralic melts within the melt pit. The second survey, of 10-100μm D_{co} craters, identified the most common impactor as space debris. Aluminium and Aluminium Oxide residues (from solid rocket motor operation) were dominant, particularly in craters below 30μm diameter. The micrometeoroid residues identified were again remnants of silicates and metal sulfides. The most interesting discovery was a residue composed of Si and C, this may represent interstellar material. Rarer space debris remnants included steels and paint flakes. One small impact crater appeared enriched in K, perhaps from alkali-metal reactor coolant. During the 1320 days that the HST solar array was in orbit, a Russian 'RORSAT' nuclear-powered satellite in a higher orbit was reported to have had leakage from the liquid Na and K coolant system. The attributions as to impactor origin from these chemical studies have now been compared with predictions derived from Long Duration Exposure Facility flux data, and a recent meteoroid flux model [19]. Interestingly, there is close agreement between predictions and observations.

3.3. Micrometeoroid and Space Debris Chemistries on SFU MLI-Foils

The micrometeoroid material captured on MLI foils shows residue to be in greater abundance and larger in size (figure 3) than is preserved on HST solar cells (with discrete micron-size rather than probable nanometer-sized grains). As on HST solar cells, micrometeoroid remnants are dominated by Mg-Fe residues (figure 4). It could be assumed that these are melted or condensed remnants of stoichiometric silicate minerals, but it is possible that they are relatively unaltered remnants of Glass-Embedded-Metal-Sulfide particles (GEMS), as have been identified in IDPs [20]. Other identified micrometeoroid residues include Fe-Ni sulfide and plagioclase feldspar. The space debris chemistries have included components of steels and other metallic alloys. The investigation of the MLI-foils is still on-going.

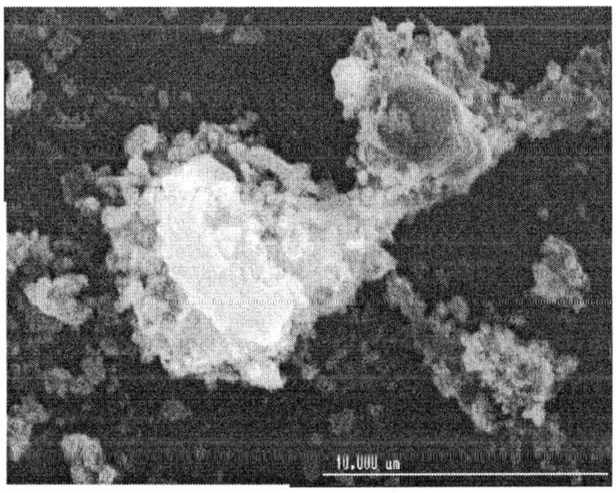

Figure 3. Secondary electron image of a residue fragment preserved on an MLI foil layer.

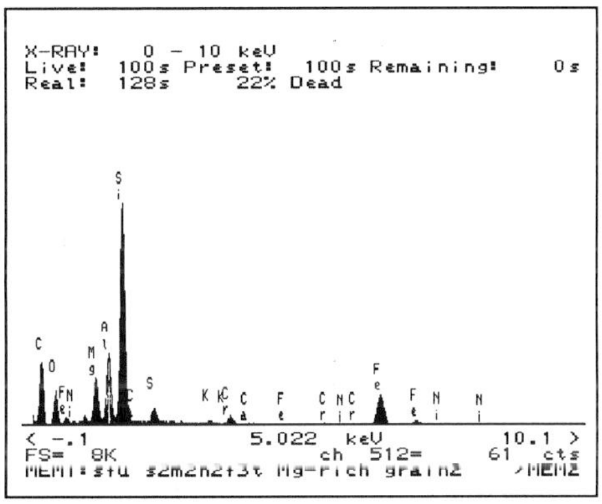

Figure 4. EDS of the fragment, enriched in Mg, Fe and Si, suggesting a remnant of a mafic silicate mineral, or of glass.

4. CONCLUSIONS

Important information about the chemistry of the micro-particles present in LEO can be obtained from analysis of non-dedicated surfaces. Returned spacecraft surfaces should continue to be examined for large numbers of residues, and will yield a valuable resource for interpretation of fluxes. Such studies complement micro-particle sampling by low-density capture-cell technologies such as aerogel [21]. Together, they offer the opportunity to sample large quantities of material, including some that may be relatively intact.

REFERENCES

1. D.E. Brownlee, B.A. Bates and M.M. Wheelock, Nature 309 (1984) 693.
2. M.M. Maurette, C. Hammer, D.E. Brownlee, N. Reeh and H.H. Thomsen, Science 233 (1986) 869.
3. J.P. Bradley, M.S. Germani and D.E. Brownlee, Earth Planet. Sci. Lett. 93 (1989) 1.
4. M.J. Genge, M.M. Grady and R. Hutchison, Meteoritics Planet Sci. 31 (1996) 627.
5. J.C. Zarnecki, J. Brit. Interplan. Soc. 41 (1988) 403.
6. National Research Council, Orbital debris: A technical assessment, National Academy of Science, (1995).
7. J.A.M. McDonnell, W.C. Carey and D.G. Dixon, Nature 309 (1984) 237.
8. J. Borg, J.-P. Bibring, Y. Langevin, P.H. Salvetat and B. Vassent, Meteoritics 28 (1993) 641.
9. M.E. Zolensky. T.H. See, R.P. Bernard, R. Barrett, F. Horz, J.L. Warren, C. Dardano, K.S. Leago, D. Kessler and T.R. Foster, Adv. Space Res. 16 (1995) (11)53.
10. R.P. Bernhard, T.H. See and F. Hörz, In LDEF 69 Months in Space - 2nd Post-Retrieval Symp., NASA CP-3194 (1993) 551.
11. R.P. Bernhard, C. Durin and M.E. Zolensky, In LDEF 69 Months in Space - 2nd Post-Retrieval Symp., NASA CP-3194 (1993) 541.
12. J.L. Warren, H.A. Zook, J.H. Allton, U.S. Clanton, C.B. Dardano, J.A. Holder, J.A., et al. Proc. 19th Lunar and Planet. Sci. Conf. (1989) 641.
13. F.J.M. Rietmeijer and G.E. Blandford, J. Geophys. Res. 93 (1988) 11943.
14. G. Drolshagen, Proc. Hubble Space Telescope Array Workshop, ESA WPP-77 (1995) 295.
15. H. Yano, S. Kibe, S.P. Deshpande and M.J. Neish, Adv. Space Res. 20 (1997) 1489.
16. G.A. Graham, A.T. Kearsley, M.M. Grady and I.P. Wright, Proc. 2nd European Conf. On Space Debris, ESA SP-393 (1997) 183.
17. H. Yano, K. Morishige, S.P. Deshpande, Y. Maekawa, S. Kibe, M.J. Neish and E.A. Taylor, Adv. Space Res. 25 (2000) 293.
18. M.E. Zolensky, H.A. Zook, F. Hörz, D.R. Atkinson, C.R. Coombs, A.J. Watts et al. LDEF 69 Months in Space - 2nd Post-Retrieval Symposium, NASA CP-3194 (1993) 277.
19. G.A. Graham, N. McBride, A.T. Kearsley, G. Drolshagen, S.F. Green, J.A.M. McDonnell, M.M. Grady and I.P. Wright, Int. J. Impact Engng. 26 (2001) 263.
20. J.P. Bradley, Science 256 (1994) 925.
21. P. Tsou, J. Non-Crystalline Solids 186 (1995) 415.

VIII Evidence from Meteorites

The nature and significance of meteoritic matter.

Monica M. Grady

The Natural History Museum, Cromwell Road, London SW7 5BD. UK.

Meteorites are fragments broken from asteroids, with a compositional variation that spans a whole range of planetary materials, from completely unmelted and unfractionated stony chondrites to highly fractionated and differentiated iron meteorites. Meteorites, and components within them, carry records of all stages of Solar System history.

1. INTRODUCTION

Meteorites are pieces of rock and metal that fall to the Earth. Almost all are fragments broken from asteroids during collisions, taking between around 0.2 – 100 million years to journey from the asteroid belt to the Earth. There are also currently about 20 meteorites that come from the Moon, and a similar number from Mars. Meteorites are the only physical materials available on the Earth that allow direct study of the original dust from which the Solar System formed. The relationship between meteorites and the primordial material from which they aggregated as asteroids is shown in Figure 1.

Different components within different types of meteorite can place a reasonably accurate relative chronology on many of the processes that have affected Solar System bodies; study of rare, exotic grains within meteorites can also assist in tracing the evolution of the Sun's galactic neighbourhood.

2. METEORITE CLASSIFICATION

Before discussing what can be learnt from meteorites, I will give an overview of the different types of meteorites and their classification. Meteorites exhibit a compositional variation that spans a whole range of planetary materials, from completely unmelted and unfractionated stony chondrites to highly fractionated and differentiated iron meteorites. Classification of meteorites into groups is one way of identifying materials that might be associated in space and time, e.g., through accretion in closely neighbouring regions of the solar nebula, or having suffered similar processes of heating, melting, differentiation and/or hydrothermal alteration. However, the classification scheme is incomplete, and there are many meteorites that do not fit comfortably into the framework. Clasts and inclusions within meteorites also frequently defy ready assignation to recognised meteorite groups.

Asteroidal meteorites are classified according to their gross composition, into stones (dominantly silicate rocks), irons (almost completely metal) and stony-irons (as the name suggests, an approximate mixture of silicate and metal). Sub-classification is then on the basis of the processing (melting and differentiation) that the meteorites experienced on their parent bodies (Figure 2). Unmelted stones are chondrites. Melted meteorites are subdivided into the stony achondrites, stony-irons and iron meteorites. There is then a further complicated

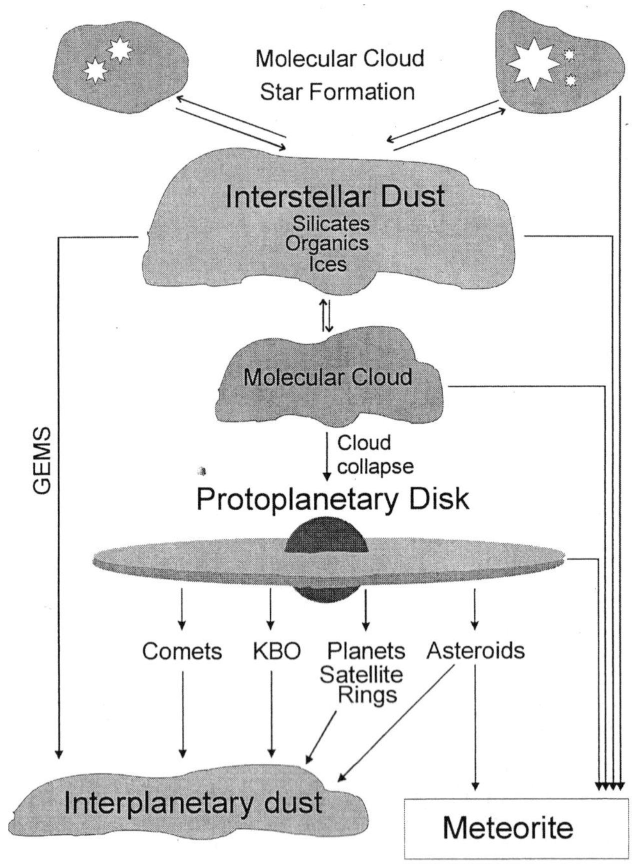

Figure 1. Schematic relationship between interstellar dust and meteorites

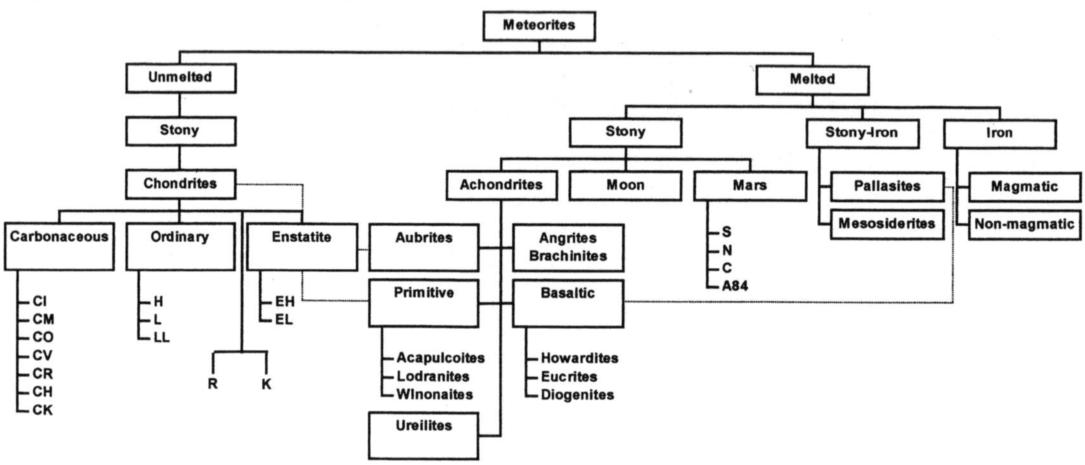

Figure 2. Classification of meteorites

hierarchy of sub-divisions into groups and classes. There is not always a clear cut distinction between types: e.g., there are many iron meteorites that contain silicate inclusions related to chondritic meteorites. Both melted and unmelted stony meteorites are made from the same elements as terrestrial rocks: Si, O, Fe, Mg, Ca and Al. Like terrestrial rocks, these meteorites are assemblages of silicate minerals: pyroxene, olivine and plagioclase, but unlike terrestrial rocks, they also contain iron-nickel metal and sulphides.

2.1. Chondrites

Chondrites are stony meteorites that have not melted since their aggregation early in the history of the Solar System. They have unfractionated elemental compositions that (apart from the most volatile of elements, such as hydrogen and helium) are close to the composition of the Sun, and thus of the original material from which the Solar System formed. Chondrites, therefore are the most significant meteorites for understanding early Solar System chronology, since they are the most primitive of all meteorites, having experienced only mild thermal or hydrothermal metamorphism since accretion into parent-bodies. Chondrites are composed of high temperature components (CAIs, chondrules) set in a matrix of fragmented chondrules mixed with sulphides, metal and minerals formed at lower temperatures (clay minerals, carbonates, sulphates, organic matter). The CAIs (for Calcium, Aluminium-rich Inclusions) are irregular-shaped refractory inclusions (up to ~ 1 cm in size) of the oxide and silicate minerals spinel, hibonite, melilite, etc. (Figure 3). CAIs frequently exhibit complex mineralogical zoning, both in their rims and in their cores.

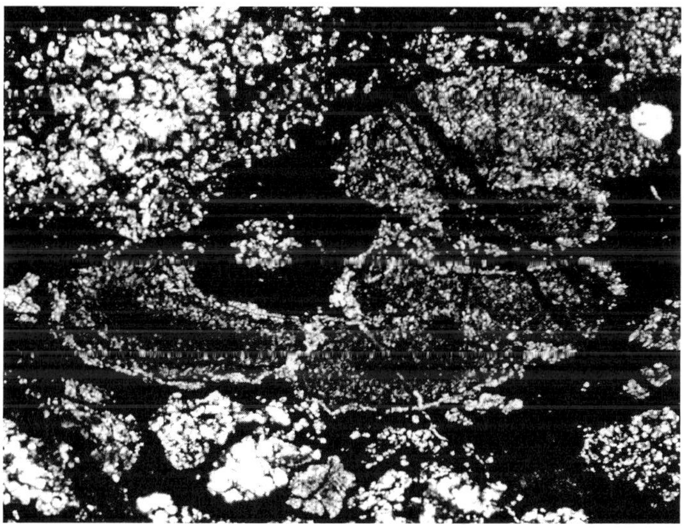

Figure 3: CAI in the Leoville carbonaceous chondrite (Thin section, plane polarised light. Magnification x 20; field of view = 4mm).

Chondrules are spherical to sub-spherical assemblages of olivine, pyroxene and feldspar, up to 1 mm in diameter, that have been partially or totally melted prior to parent-body accretion (Figure 4). Chondrules are a study in themselves; the most comprehensive recent reference is the workshop proceedings edited by Hewins et al. [1]. CAIs and chondrules are primary solids, generally presumed to have formed in the nebula prior to aggregation into parent-bodies, although their precise formation processes are not completely understood. The lower-temperature components, phyllosilicates, carbonates, sulphates, etc, are secondary products of fluid activity within parent-bodies. Also preserved within chondritic matrix are unaltered interstellar dust grains: sub-micron-sized diamonds, together with micron-sized silicon carbide, graphite and aluminium oxide grains. These materials were introduced into the pre-solar nebula from neighbouring stars, prior to parent-body aggregation, and thus pre-date the major chondritic components.

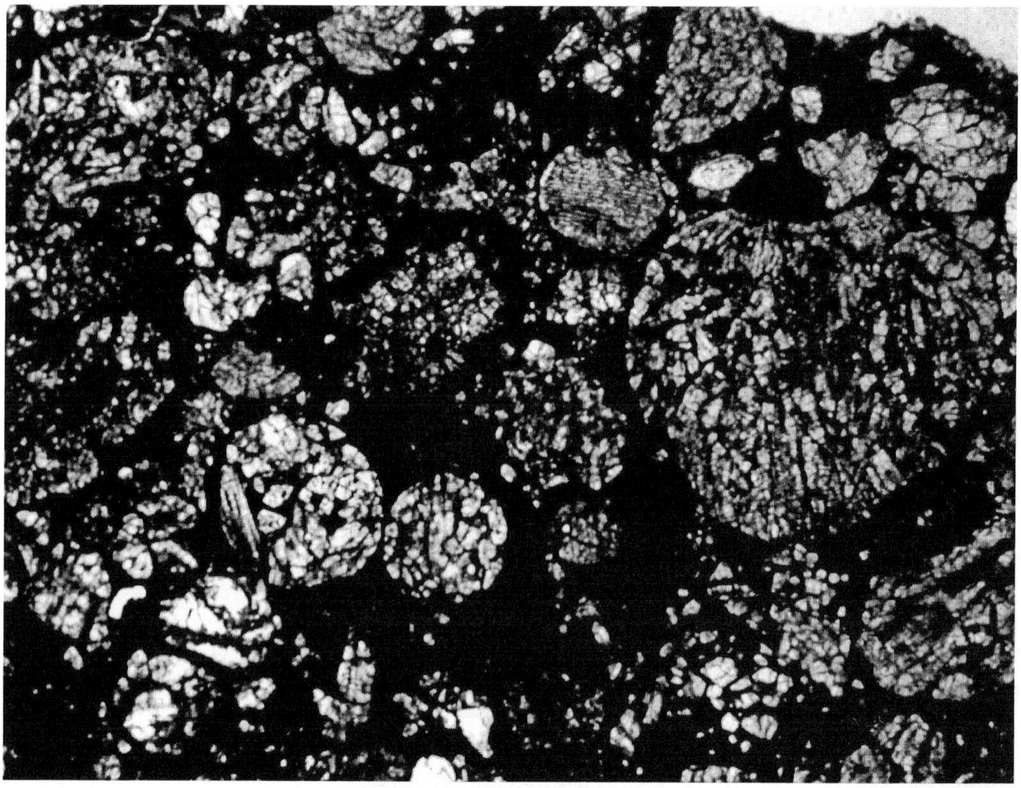

Figure 4: Chondrules in the Krymka LL3 ordinary chondrite (Thin section, plane polarised light. Magnification x 20; field of view = 4mm).

Chondrites sub-groups are distinguished on the basis of chemistry, matrix, metal and chondrule contents and chondrule properties (size, type, etc.). The differences between the classes are primary, i.e. were established as the parent bodies accreted in different regions of the solar nebula. The sub-groups are also distinguished in terms of the oxygen isotopic composition of their major silicate minerals (Figure 5), again generally taken to be a reflection of primordial nebula heterogeneity, but possibly modified as a result of widespread fluid-solid exchange processes.

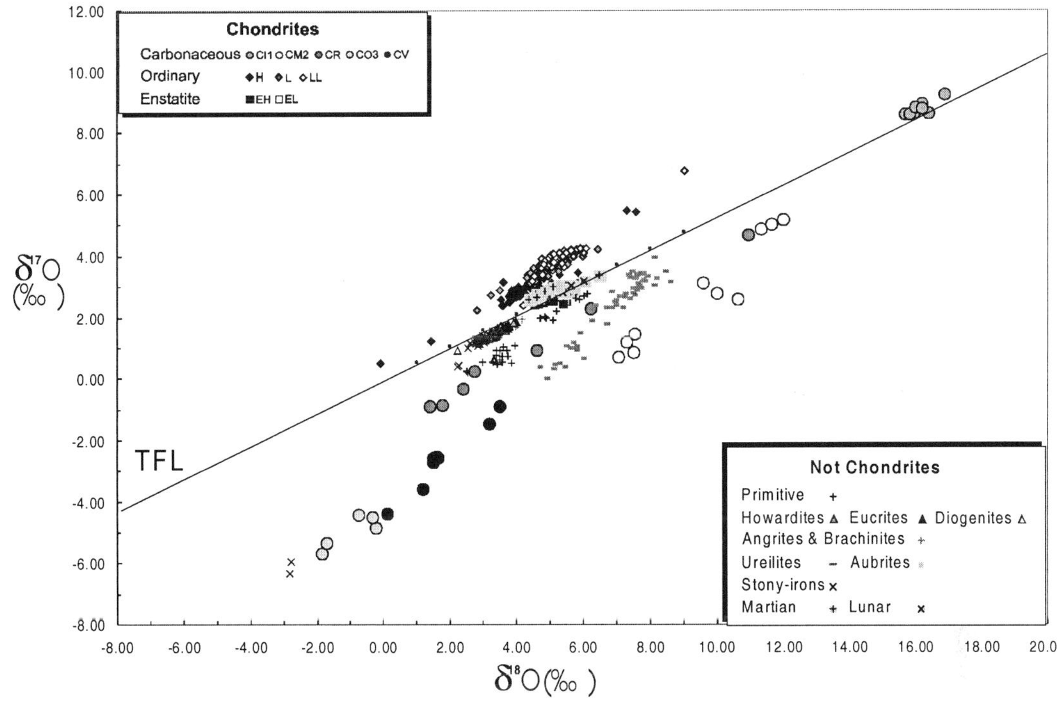

Figure 5. Oxygen isotopic composition of whole-rock meteorites. TFL – Terrestrial Fractionation Line: the array on which data from terrestrial and lunar materials fall. Data from Clayton et al. (1991); Clayton and Mayeda (1996, 1999), [2-4].

Subsequent to accretion, most meteorites have experienced varying extents of thermal metamorphism or aqueous alteration. These secondary processes occurred on the meteorites' parent bodies, and did not affect the overall composition of the chondrites. Mobilisation and re-distribution of elements such as Fe, Mg and Ca occurred as primary silicates either became homogeneous in composition through heating, or through formation of secondary mineral phases (e.g., clay minerals, carbonates) during aqueous alteration. Tertiary processes of brecciation and shock melting are also recognised, as are features resulting from weathering during the terrestrial lifetime of a meteorite. A thorough review of the properties and classification of chondritic meteorites is given by Brearley and Jones [5].

2.2. Achondrites

Broadly speaking, an achondrite is a stony meteorite that formed from a melt on its parent body. Thus achondrites have differentiated compositions, having lost a large fraction of their primordial metal content, and generally do not contain chondrules. There are many different groups of achondrites, some of which can be linked together to form associations allied with specific parents (Figure 1). The separate associations have little, if any, genetic relationship to each other and, like the chondrites, can be distinguished on the basis of oxygen isotope composition (Figure 5). One sub-set of achondrites, the primitive achondrites, comprises several groups of meteorites that have only experienced limited partial melting of their chondritic precursors. It has been suggested that the primitive achondrites might be a bridge between chondrites and achondrites. In contrast, the Howardite-Eucrite-Diogenite association (HED) is a suite of generally brecciated igneous rocks ranging from coarse-grained orthopyroxenites (diogenites) to cumulates and fine-grained basalts (eucrites). The howardites are regolith breccias, rich in both solar wind gases and clasts of carbonaceous material. The HEDs all have similar oxygen isotopic compositions; one candidate for the HED parent body is the asteroid 4 Vesta. For a much more thorough review of the mineralogy and classification of achondrites, see Mittlefehldt et al. [6].

2.3. Stony-irons

These are meteorites with very approximately equal proportions of silicate minerals and iron-nickel metal. Stony-irons are sub-divided into two big groups, mesosiderites and pallasites, which have very different origins and histories. Pallasites have a most unusual appearance, and are perhaps the most strikingly beautiful of all meteorites. They are an approximately equal mixture of iron-nickel metal and silicates, predominantly olivine, and are presumed to represent material from the core-mantle boundary of their parent bodies. The mesosiderites are a much more heterogeneous class of meteorites than the pallasites. They are a mixture of varying amounts of iron-nickel metal with differentiated silicates, the whole assemblage of which seems to have been brecciated.

2.4. Irons

Iron meteorites are composed of iron metal, generally with between 5 - 20 wt. % nickel, and account for approx. 5% of all observed meteorites falls. The mineralogy of iron meteorites is dominantly an intergrowth of two phases, the iron-nickel alloys kamacite and taenite. Kamacite, or α-Fe,Ni, has a body-centred cubic structure and a Ni content < 7wt.%, whilst taenite, or γ-Fe,Ni, is face-centred cubic and approx. 20-50 wt.% Ni. Iron meteorites are highly differentiated materials, products of extensive melting processes on their parents. One major division of iron meteorites is into 'magmatic' irons (those that have solidified by fractional crystallisation from a melt) and 'non-magmatic' irons (those that seem not have completely melted, and possibly formed during impact processes). The irons are sub-divided into 13 different groups on the basis of nickel and trace element chemistries (Ga, Ge and Ir contents). The non-magmatic irons are groups IAB and IIICD, and show a wide range in nickel contents. Many irons defy chemical classification, and simply remain "Ungrouped". Prior to classification on the basis of trace element chemistry, iron meteorites were classified in terms of their metallographic structure. Laths of kamacite intergrown with Ni-rich phases form the 'Widmanstätten pattern' revealed in polished and etched iron meteorites. The width of the kamacite lamellae is related to the cooling history of the parent bodies. The structural classification of iron meteorites was redefined and made systematic by V.F. Buchwald [7].

2.5. Non-asteroidal meteorites

<u>Lunar meteorites</u>: there are currently 18 known lunar meteorites, 15 of which have been collected in Antarctica. Several are gabbroic or basaltic in nature, but the majority are anorthositic regolith breccias. Material returned by the Apollo and Luna missions sampled a fairly restricted area of the lunar surface. The discovery of meteorites from the Moon is a fortunate opportunity that has allowed study of a more diverse range of lunar materials. Papike et al. [8] reviewed the mineralogy of lunar samples, which allows the significance of lunar meteorites to be placed in context with Apollo and Luna rocks.

Martian meteorites: there are currently 16 meteorites (20 separate stones) that almost certainly originate on Mars. The martian origin rests on the age, composition and noble gas inventory of the meteorites (see http://www.nhm.ac.uk/mineralogy/staff/grady/mars.html for a brief review). The meteorites are all igneous in nature, and can be sub-divided into 4 groups: the Shergottites (after the type specimen Shergotty), further sub-divided into basaltic and lherzolitic types. Nakhlites (after Nakhla) are shallow cumulates that have been exposed to the martian hydrosphere, and thus contain rich assemblages of carbonates, sulphates and halite. Chassigny, the only member of its group, is an olivine-rich dunite. ALH 84001 is the sole orthopyroxenite, and is rich in carbonates. For a much more thorough review of the mineralogy and classification of martian meteorites, see McSween and Treiman [9].

3. WHAT CAN BE LEARNT FROM METEORITES?

Solar System history starts with the collapse of an interstellar molecular cloud to a protoplanetary disk (the solar nebula) and continues through a complex process of accretion, coagulation, agglomeration, melting, differentiation and solidification, followed by the secondary influences of bombardment, collision, break-up, brecciation and re-formation. The rocks which are accessible for study at the Earth's crust are not representative of the original material that accreted from the solar nebula. Study of meteorites allows a more complete understanding of the processes undergone by the material that resulted in the Earth of today. The last decade has seen a greater understanding of the processes that led to the formation of the Sun and Solar System. Advances have resulted from astronomical observations of star formation regions in molecular clouds, the recognition and observation of protoplanetary disks and planetary systems around other stars, and also from refinement of chronologies based on short-lived radionuclides. The main events that led to the formation of the Earth and the Solar System can be followed by several different short-lived radiometric age-dating chronometers, utilising components from meteorites that formed at various stages in Solar System evolution. The subsequent processes of differentiation, segregation and core-formation in planetary-sized bodies can also be traced by reference to more long-lived nuclear chronometers.

3.1. Timescales of accretion

Gravitational instability within an interstellar molecular cloud results in collapse of a fragment of the cloud to form a protoplanetary disk [10-12]; the mechanism that triggers cloud collapse is not clear: several possibilities have been suggested, e.g. a shock wave from a nearby supernova or ejection of a planetary nebula from an AGB star [13]. The oldest components in meteorites are the CAIs, with ages (determined by U-Pb dating) of 4566^{+2}_{-1} [14]. This age can be taken as a baseline from which the date of formation of other meteoritic components can be measured, and has led to the establishment of a relative timescale for the production of CAIs and chondrules, based on the presence of the decay products from short-lived radionuclides. The 'canonical' model for accretion rationalised that the short-lived radionuclides were produced externally to the presolar nebula (probably in a supernova), injected into the collapsing dust cloud and incorporated into the CAIs, etc, on timescales shorter than that of the radionuclide half life. For example, the presence of ^{26}Mg (from the decay of ^{26}Al; $T_{1/2}$ ~ 0.73 Myr) within CAIs shows that the CAIs formed whilst ^{26}Al was still "live" in the solar nebula, i.e., agglomeration took place over a very short timescale, < 1 Myr

[15]. The ^{41}Ca-^{41}K chronometer, with $T_{1/2} \sim 0.15$ Myr, implies even more rapid formation of CAIs, with an interval between nucleosynthesis and agglomeration of < 0.3 Myr [16]. Measurements on aluminium-rich chondrules using the Al-Mg chronometer also imply that chondrules formed several million years after CAI production, and that chondrule-formation took place over an extended period of time, commencing very shortly (perhaps 1-5 Myr) after CAI formation [17, 18].

However, the use as a chronometer of isotope abundance differences between chondritic components is dependent on the models of short-lived isotope formation. Over the past few years, an alternative mechanism to production of the radionuclides in a supernova has been proposed, whereby the species are formed by nuclear reactions close to the energetic early sun, followed by expulsion to planet-forming regions by the X-wind [19]. In this scenario, no chronological significance can be attached to the lack of daughter products from short-lived radionuclides – any effects are spatial, rather than chronological. Currently, the debate between proponents of the exotic (i.e., supernova) *versus* local (i.e., young sun) origin of short-live radionuclides is vigorous. The detection, in meteoritic components, of the decay products of short-lived isotopes thought not to be produced in supernovae, (e.g., ^{10}B, from the decay of ^{10}Be, $T_{½} \sim 1.5$ Myr) were at first taken to support the local production of such species by spallation [20], however, refinement of models of supernovae have suggested that ^{10}Be might be formed by spallation in the outer envelope of a supernova [21]. Most recently, evidence for ^{7}Li (from the decay of ^{7}Be, $T_{½} \sim 52$ days) in CAIs has again turned the tide of the argument towards local production of short-lived isotopes [22]. Efforts are also being directed towards detection in meteoritic components of products from radionuclides that can only be formed by spallation, e.g. ^{50}V [23].

No matter what the mechanism for production of short-lived radionuclides, it is probable that the Solar System formed by the heterogeneous accretion of components. The growth of planet-sized bodies from micron-sized dust grains is controlled by several factors, such as the nature of the initial grains (fluffy or compact) and the degree of turbulence within the nebula, and has been modelled by many authors [24]. End-member models for planetesimal formation are coagulation of material by gravitational instability in a quiescent nebula [25] or by coagulation during descent to the midplane of a turbulent nebula [26-27]. The aggregation of interstellar dust ($\leq 0.1\mu m$ in diameter) into increasingly large bodies, eventually forming kilometre-sized planetesimals and culminating in the asteroids and planets, took place over a time interval of some 8 Myr following formation of the CAIs [14].

3.2. Timescales of differentiation

In addition to illuminating the earliest history of the Solar System, and providing examples of the types of material from which the Solar System formed, meteorites also enable studies of the planetary differentiation processes that occurred subsequent to aggregation. Once the proto-parents had aggregated, internal heat from radioactive decay, combined with gravitational energy and collisional energy from planetesimal bombardment kept the bodies molten. As cooling progressed, reduction reactions within the convecting system resulted in production of metal-rich cores and silicate-rich crust-mantle structures. Achondrites record episodes of incipient melting and differentiation, and iron and stony-iron meteorites represent the highly-differentiated end-member compositions that result from core formation and metal-silicate segregation on parent bodies. The timescale over which planetary melting and core formation occurred can be deduced using several radiometric decay schemes.

Igneous activity on differentiated stony asteroids is traced by data from ^{53}Mn-^{53}Cr chronometry ($T_{1/2}$ ~ 3.7 Myr). For the HED assemblage, whose parent-body is believed to asteroid 4 Vesta, core formation apparently occurred < 4 Myr after CAI formation; subsequent volcanic activity on the asteroid continued for a further 4 Myr or so [28]. This compressed timescale indicates the relatively short period that elapsed between accretion of primordial dust into small proto-planets, and the onset of volcanism, melting, and differentiation.

The recently-established ^{182}Hf-^{182}W chronometer [29] has been used to infer the timescale over which core formation occurred on both asteroidal and planetary bodies. The strongly lithophile ^{182}Hf is partitioned into silicates, relative to the more siderophile W during differentiation, and subsequent variations in Hf/W are caused by decay of ^{182}Hf to ^{182}W ($T_{1/2}$ ~ 9 Myr). Models based on the ^{182}Hf-^{182}W chronometer indicate that core formation on iron meteorite parent bodies occurred fairly rapidly, within 5-10 Myr of the onset of CAI formation, whilst formation of the Earth's core took place more gradually, some 50 Myr or so after the differentiation of iron meteorite parent-bodies [29]. Based on measurements from martian meteorites, it appears that core formation on Mars also occurred across an extended timescale, within ~ 30 Myr of CAI formation [30].

3.3. Mars from martian meteorites

Satellite imagery has shown that Mars has had ice and liquid water flowing across or just under its surface during past epochs. The first detailed topographic maps of Mars were produced with data from NASA's *Mariner 9* orbiter mission of 1971-1972, in which channel and valley networks, volcanoes, canyons and craters were imaged at reasonable resolution for the first time. *Mariner 9* was also the mission that returned pictures of layered (possibly sedimentary) terrain in Mars' polar regions. In 1976, NASA's two *Viking* landers sent back many images of Mars' landscape, showing panoramic scenes of broken boulders distributed over flat dusty plains. *Viking* also measured the elemental composition of both Mars' atmosphere [31] and surface soils [32]. In combination with the *Mariner* data, the *Viking* results have allowed a picture of Mars as a rocky planet with a significant geological history to be built up. The recent Pathfinder mission of 1997 recorded spectacular images of a rock-strewn plain, with tantalising glimpses of rounded pebbles and possible layered structures and hollows within some of the rocks. Inferences drawn from the images are that some of the rocks might be sedimentary, possibly even conglomerates, thus implying a significant fluvial history. Paler, almost cream-coloured patches in the soil might be areas leached by fluid action, hard-grounds or evaporite deposits. The poorly-sorted landscape of part-rounded pebbles and boulders has been interpreted as the type of landscape remaining after catastrophic flooding, further evidence of the stability of liquid water at some time in Mars' past. Recent data from the laser altimeter aboard the *Mars Global Surveyor* have been interpreted as evidence for a very large ocean across Mars' northern hemisphere [33]. For water to have been present at Mars' surface, the martian atmosphere must have been much thicker, and surface temperatures much warmer than they are today [34]. A thicker atmosphere engenders greater protection from solar radiation; in a previous epoch, Mars would have had a warmer and wetter climate and provided all the conditions suitable for the emergence of life [35]. Given this framework, it is not surprising that so much interest has been focussed on a potential martian biosphere. But many questions about Mars remained unanswered, particularly about the fluvial and seismic history of the planet. Vital to interpretation of much of the chemical data returned by spacecraft missions have been data obtained from martian meteorites [36].

Measurements made on martian meteorites complement data obtained from spacecraft exploration of Mars. So, for instance, the recognition of complex assemblages of salt (halite) with carbonates and clay minerals in nakhlites [37] has allowed interpretation of the scale and mode of fluid flow on the surface of Mars: results from the nakhlites imply that when water was present on the surface, it was warm and briny, and restricted in flow [38]. In other words, it might have been locked in enclosed basins that occasionally overflowed in episodes of flash flooding. Petrogenetic analysis of a second group of martian meteorites, the shergottites, has allowed development of theories of martian magma genesis [39, 40]. Analyses of martian meteorites have also added impetus to the resurgence of interest in exobiology. The observation that at least one of the SNCs, EETA79001 (a specimen found in the Elephant Moraine region of Antarctica in 1979) contained indigenous (i.e., martian) organic material associated with carbonate minerals [41] sparked a debate on the possibility of martian meteorites containing evidence for extraterrestrial life. This work was followed, in 1994, by the discovery of similarly enhanced levels of organic carbon in close association with carbonates in a second Antarctic martian meteorite, ALH 84001 [42, 43]. ALH 84001 became a further focus of interest in 1996 when a team of scientists led by David McKay of NASA's Johnson Space Centre in Houston reported the description of nanometre-sized features within carbonate patches in ALH 84001, and claimed to have found evidence for a primitive "fossilised martian biota" [44]. Identification of the "nanofossils" remains controversial, since much of the evidence is circumstantial and relies on the coincidence between a number of otherwise unrelated characteristics of the meteorite (carbonate minerals in ALH 84001; organic compounds and magnetite associated with the carbonates, and the external morphology of the carbonates, see ref. 38, for a review of the main arguments for and against the interpretation of McKay et al.). Indeed, the formation conditions of both the carbonate and magnetite, the relevance of the organic compounds and interpretation of the morphology of the "nanofossils" have all been subject to detailed investigation by several groups of scientists. Although many members of the scientific community are sceptical about interpretation of the structures in ALH 84001 as martian nanofossils, discussion of the features has stimulated an enormous amount of interest in the possibilities of life on Mars.

3.4. Interstellar grains

A volumetrically insignificant, but scientifically critical component within chondrites is their complement of interstellar and circumstellar grains. The grains comprise several populations of nanometre-sized diamond and micron-sized silicon carbide, graphite and aluminium oxide [45]. The presence of the grains was first inferred in the late 1970s to early 1980s, on the basis of the isotopic composition of noble gases released by pyrolysis of density separates and size-fractions of residues produced from demineralisation of bulk chondrites [46]. The unusual isotopic signatures of the noble gases implied the existence of several different hosts; analyses of acid-resistant residues suggested that the hosts might be carbon-rich. In 1983, combustion of a set of residues yielded the first carbon and nitrogen isotopic compositions of the grains [47, 48]. Together with the noble gas results, these measurements pointed to a variety of extra-solar sources for the grains including supernovae and red giant stars. Over the subsequent 20 or so years, increasingly sophisticated instrumentation, such as the ion microprobe and the analytical transmission electron microscope, have revealed different generations of interstellar grains, including grains within grains [49]. Astrophysicists have been able to construct, then fine-tune, models of stellar evolution on the basis of the combined isotopic data resulting from analysis of the grains [50, 51]. Currently, grains from at least 15 different extra-solar sources have been isolated from chondritic meteorites. These grains were presumably introduced into the pre-solar nebula prior to its collapse and the onset of proto-planet formation. The presence of such a variety of grains indicates that the Sun did not form in isolation, but was part of a busy neighbourhood in which AGB stars, supernovae, novae and planetary nebula co-existed. When coupled with astronomical observations (both ground- and space-based) of interstellar and circumstellar dust clouds, the meteorite data are enabling astronomers to model star and protoplanetary disk formation processes [52]. It is hoped that continued study of pre-solar grains with a new generation of instrumentation will allow both relative and absolute chronologies to be constructed for the different grain populations, leading to models for the history and evolution of this region of the Galaxy.

4. SUMMARY

The cartoon outline in Figure 1 encapsulates the fundamental relationship that meteorites have with the other members of the Solar System family. Populations of interstellar grains in primitive asteroidal meteorites reveal the constituency of the neighbourhood in which the Sun initially formed, helping to constrain astrophysical models of stellar evolution. Chondrites mark the change that took place between simple aggregates of interstellar dust and their accumulation into parent bodies as a protoplanetary disk formed and then evolved into the Solar System. The presence of the decay products of short-lived radionuclides within components in chondritic meteorites provides a chronology for the formation of the earliest solid materials within the nebula. Primary mineral assemblages in chondrites trace the heterogeneity of the solar nebula, whilst secondary minerals reveal the extent of hydrous and thermal alteration on the parents. Differentiated meteorites record the extent, timing and duration of melting and subsequent magmatic activity on their parents, indicating the short timescale on which parent bodies aggregated, then differentiated. Iron and stony-iron meteorites act as markers for iron-silicate segregation and core formation. Turning from asteroidal meteorites to those from other sources: lunar meteorites extend the range and diversity of materials available for study from Earth's only natural satellite, whilst the families of martian meteorites assist in interpretation of magmatic and fluvial processes on Mars, and have stimulated studies into the potential existence (extinct or extant) of life on Mars.

Meteorites are the only physical objects that can be analysed directly in the laboratory, hence they are an unparalleled resource for Solar System, and Galactic, exploration. Past, and future space missions to comets, asteroids and Mars rely on measurements from meteorites to aid in interpretation of returned data. Astronomers studying interstellar and circumstellar dust by both ground- and space-based techniques are turning more frequently to results from meteorites, and the grains separated from them, to help with understanding the size, shape and composition of the dust. Direct measurement of the isotopic composition of the several populations of interstellar grains isolated from meteorites, and the gases trapped within them, are enabling astrophysicists to model more accurately stellar evolutionary sequences, and to confirm (or refute) predictions made from these models. All in all, and again with reference to Figure 1, meteorites are not at the bottom of a hierarchy of stellar and planetary objects, rather they are the foundation on which all theoretical, observational and experimental data rest, the root that underpins and firmly anchors our comprehension and appreciation of nebular history.

ACKNOWLEDGEMENTS

This work is part of the *Earth Materials, History and Processes* Theme of the Natural History Museum; financial support by the Particle Physics and Astronomy Research Council is gratefully acknowledged. I thank Simon Green and Tony Mcdonnell for the invitation to present this review, and Ian Wright and Sara Russell for commenting on early drafts of the manuscript.

REFERENCES

1. R. Hewins, R.H. Jones and E. R. D. Scott (Editors) Chondrules and the Protoplanetary Disk, Cambridge Univ. Press. 346pp. (1996).
2.. R. N. Clayton, T. K. Mayeda, J. N. Goswami and E. J. Olsen, Geochim. Cosmochim. Acta 55 (1991) 2317.
3. R. N. Clayton and T. K. Mayeda, Geochim. Cosmochim. Acta 60 (1996) 1999.
4. R. N. Clayton and T. K. Mayeda, Geochim. Cosmochim. Acta 63 (1999) 2089.
5. A. J. Brearley and R. H. Jones, in: Planetary Materials (ed. J. J. Papike). Min. Soc. America, Reviews in Mineralogy No. 36 (1998).
6. D. W. Mittlefehldt, T. J. McCoy, C. A. Goodrich and A. Kracher, in: Planetary Materials (ed. J. J. Papike). Min. Soc. America, Reviews in Mineralogy No. 36 (1998).
7. V. F. Buchwald, Handbook of Iron Meteorites, Univ. of Calif. Press. 3 Volumes (1975).
8. J. J. Papike, G. Ryder and C. K. Shearer, in: Planetary Materials (ed. J. J. Papike). Min. Soc. America, Reviews in Mineralogy No. 36 (1998).
9. H. Y. MacSween, Jr. and A. H. Treiman, in: Planetary Materials (ed. J. J. Papike). Min. Soc. America, Reviews in Mineralogy No. 36 (1998).
10. A. P. Boss, Astrophys. J. 345 (1989) 554.
11. A. P. Boss, Astrophys. J. 417 (1993) 351.
12. G. W. Wetherill, Ann. Rev. Earth Planet. Sci. 18 (1990) 205.
13. A. G. W. Cameron, Ann. Rev. Astron. Astrophys. 26 (1988) 441.
14. J. Allègre, G. Manhès and C. Göpel, Geochim. Cosmochim. Acta 59 (1995) 1445.
15. G. J. MacPherson, A. M. Davis and E. K. Zinner, Meteoritics 30 (1995) 365.
16. G. Srinivasan, S. Sahijpal, A. A. Ulyanov and J. N. Goswami, Geochim. Cosmochim. Acta 60 (1996) 1823.
17. S. S. Russell, G. Srinivasan, G. R. Huss, G. J. Wasserburg and G. J. MacPherson, Science 273 (1996) 757.
18. M. Wadhwa and S. S. Russell, in: Protostars and Planets IV (eds. V. Mannings, A. P. Boss and S. S. Russell) Univ. Arizona Press, Tucson (2000) 995.
19. F.H. Shu, H. Shang, A.E. Glassgold and T. Lee, Science 277 (1997) 1475.
20. K.D. McKeegan, M. Chaussidon and F. Robert, Science 289 (2000) 1334.
21. A.G.W. Cameron, Astrophys. J. 562 (2001) 456.
22 M. Chaussidon, F. Robert and K.D. McKeegan, Lunar Planet. Sci. XXXIII, (2002) No. 1563.
23. M. Gounelle, S.S. Russell, D. Weiss, E. Mullane, T.F.D. Mason and B.J. Coles, Meteoritics Planet. Sci. 36 (2001) A71.
24. J. L. Lissauer, Ann. Rev. Astron. Astrophys. 31 (1993) 129.
25. V. S. Safronov, Moscow: Nauka. Transl. by IPST in 1972 as NASA TT F-677 (1969).

26. S. J. Weidenschilling, B. Donn and P. Meakin, in: The Formation and Evolution of Planetary Systems (eds. H.A. Weaver and L. Danly). Cambridge Univ. Press (1989) 131.
27. S. J. Weidenschilling and J. N. Cuzzi, in: Protostars and Planets III (ed. L.H. Levy & J.L. Lunine). Univ. Arizona Press. (1993).
28. G. W. Lugmair and A. Shukolyukov, Geochim. Cosmochim. Acta 62 (1998) 2863.
29. A. N. Halliday, D.-C. Lee, J. N. Christensen, M. Rehkämper, W. Yi, X. Luo, C. M. Hall, C. J. Ballentine, T. Pettke and C. Stirling, Geochim. Cosmochim. Acta 62 (1998) 919.
30. D.-C. Lee and A. N. Halliday, Nature 388 (1997) 854.
31. A. O. Nier, M. B. McElroy and Y. L. Yung, Science 194 (1976) 68.
32. B. C. Clark, A. K. Baird, R. J. Weldon, D. M. Tsusaki, L. Schnabel and M. P. Candelaria, J. Geophys. Res. 87 (1982) 10059.
33. J. W. Head III, H. Hiesinger, M. A. Ivanov, M. A. Kreslavsky, S. Pratt and B. J. Thomson, Science 286 (1999) 2134.
34. S. W. Squyres and J. F. Kasting, Science 265 (1994) 747.
35. C. P. McKay, R. L. Mancinelli, C. R. Stoker and R. A. Wharton Jr., in: Mars (eds. H. H. Kieffer, B. M. Jakosky, C. W. Snyder and M. S. Matthews) Univ. Arizona Press, Tucson (1992) 1234.
36. H. Y. McSween Jr., Meteoritics 29 (1994) 757.
37. J. C. Bridges and M. M. Grady, Meteoritics and Planetary Sciences 34 (1999) 407.
38. J. C. Bridges and M. M. Grady, Earth Planet. Sci. Lett. 176 (2000) 267.
39. T. J. McCoy, G. J. Taylor and K. Keil, Geochim. Cosmochim. Acta 56 (1992) 3571.
40. M. Wadhwa, H. Y. McSween Jr. and G. Crozaz, Geochim. Cosmochim. Acta 58 (1994) 4213.
41. I. P. Wright, M. M. Grady and C. T. Pillinger, Nature 340 (1989) 220.
42. M. M. Grady, I. P. Wright, C. Douglas and C. T. Pillinger. Meteoritics 29 (1994) 469.
43. M. M. Grady, I. P. Wright and C. T. Pillinger, in: The Search for Life on Mars (ed. J. A. Hiscox) British Interplanetary Society (1999) 78.
44. D. S. McKay, E. K. Gibson Jr., K. L. Thomas-Keprta, H. Vali, C. S. Romanek, S. J. Clemett, X. D. F. Chillier, C. M. Maechling and R. N. Zare, Science 273 (1996) 924.
45. E. Anders and E. K. Zinner, Meteoritics 28 (1993) 490.
46. L. Alaerts, R. S. Lewis, J. Matsuda and E. Anders, Geochim. Cosmochim. Acta 44 (1980) 189.
47. R. S. Lewis, E. Anders, I. P. Wright, S. J. Norris and C. T. Pillinger, Nature 305 (1983) 767.
48. P. K. Swart, M. M. Grady, C. T. Pillinger, R. S. Lewis and E. Anders, Science 220 (1983) 406.
49. T. J. Bernatowicz, S. Amari, E. K. Zinner and R. S. Lewis, Astrophys. J. 373 (1991) L73.
50. D. D. Clayton, in: Meteorites and the Early Solar System (eds. J. F. Kerridge and M. S. Matthews). Univ. Arizona Press, Tucson (1989) 1021.
51. R. Gallino, M. Busso, G. Picchio and C. M. Raiteri, Nature 348 (1990) 298.
52. A. K. Speck, A. M. Hofmeister, M. J. Barlow and R. J. Sylvester, Meteoritics Planet. Sci. 35 (2000) A150.

Antarctic micrometeorites collected by the Japanese Antarctic Research Expedition teams during 1996 – 1999

T. Noguchi [a], H. Yano [b], K. Terada[c], N. Imae [d], T. Yada [e], T. Nakamura [e] and H. Kojima [f]

[a] Department of Materials and Biological Science, Ibaraki University, 2-1-1, Bunkyo, Mito, Ibaraki, 310-8512, Japan

[b] Planetary Science Division, Institute of Space and Astronautical Science, 3-1-1 Yoshinodai, Sagamihara, Kanagawa, 229-8510, Japan

[c] Department of Earth and Planetary System Science, Hiroshima University, Higashi-Hiroshima, Hiroshima, 739-8526, Japan

[d] National Institute of Polar Research, 1-9-10 Kaga, Itabashi, Tokyo, 173-8515, Japan.

[e] Department of Earth and Planetary Sciences, Kyushu University, 33, Hakozaki, Fukuoka, 812-8581 Japan

The Japanese Antarctic Research Expedition (JARE) teams have started collection of unmelted and melted micrometeorites (MMs) in Antarctica since 1996. Some results of the consortium studies are: (1) relatively common occurrence of magnesiowüstite (MW) in unmelted MMs, (2) coexistence of MW with low-Ca pyroxene in moderately heated MMs, and (3) evidence that MMs were formed as small particles, rather than fragments of larger bodies, within < a few Ma.

1. RECOVERY OF MICROMETEORITES (MMs) IN THIS STUDY

1.1. Recovery of MM candidates from the dust samples collected at the Dome Fuji Station in 1996 and 1997

In 1996 and 1997, the JARE teams collected MMs at the Dome Fuji Station, located at Queen Maud Land [1]. Fine-grained deposits in a water tank at the station were processed to enhance the abundance of MMs by filtration, separation by differences of sedimentation rate, and in some cases, magnetic separation [1,2]. The MM candidates were handpicked and investigated by SEM. More than 230 MMs were identified from the deposits collected in 1996 [2]. EDS data show that they have mostly undifferentiated chondritic elemental abundance. As heated during atmospheric entry to variable extents, they represent a wide range of mineralogy from phyllosilicate-dominated to barred olivine-dominated types [1, 2].

1.2. MM collection in 1998-1999 at Yamato Mountains

From November 1998 to January 1999, MMs were collected in bare ice areas of the Yamato Mountains. Thirty-six tons of ice were melted and filtered by a system to melt ice and filter the melted water [3]. After transferring the samples to Japan, captured particles on the sieves were removed to small petri dishes carefully by distilled water or acetone in a clean room. To understand the basic characteristicss of the samples, 5 among 24 sampling sites were selected to pick up all extraterrestrial dust particles. For example, a size fraction of 40-100 μm contained about 200 MMs. The number of MMs of this size fraction in the 24 sites is estimated as ~4800. Allocation of these samples to qualified researchers in the world began in the middle of November 2000.

2. A BRIEF SUMMARY OF A COMBINED STUDY ON MMs COLLECTED IN 1996

230 MMs collected in 1996 were investigated by the consortium [e.g., 1, 2, 6-10]. Here we will introduce some of the interesting results and also refer to some preliminary results of MMs collected in 1999.

2.1 Bulk mineralogy

Bulk mineralogy of 28 individual unmelted to partially melted MMs was determined by X-ray diffraction using Gandolfi camera. Four mineralogical types of MMs were identified. They are characterized as follows: (1) dehydrated-smectite dominant, olivine and low-Ca pyroxene being absent, while primary Fe-sulfide and magnetite present, (2) phyllosilicate absent while olivine in low crystallinity, and Fe-sulfide and magnetite present, (3) presence of variable amounts of olivine, low-Ca pyroxene, Fe-sulfide, and magnetite, and (4) abundant olivine and magnetite with small amounts of low-Ca pyroxene and Fe-sulfide.

It has already been suggested that anhydrous mineral assemblages are often seen in unmelted MMs due to dehydration of phyllosilicates by heating [4, 5]. However, our study revealed that the species and relative abundance of anhydrous minerals formed by dehydration of phyllosilicates are governed by the extent of heating. Olivine and magnetite increase as heating proceeds, while once low-Ca pyroxene first increases and then decreases as heating proceeds. This study also revealed that type 3 (that is, low Ca pyroxene abundant) MMs often

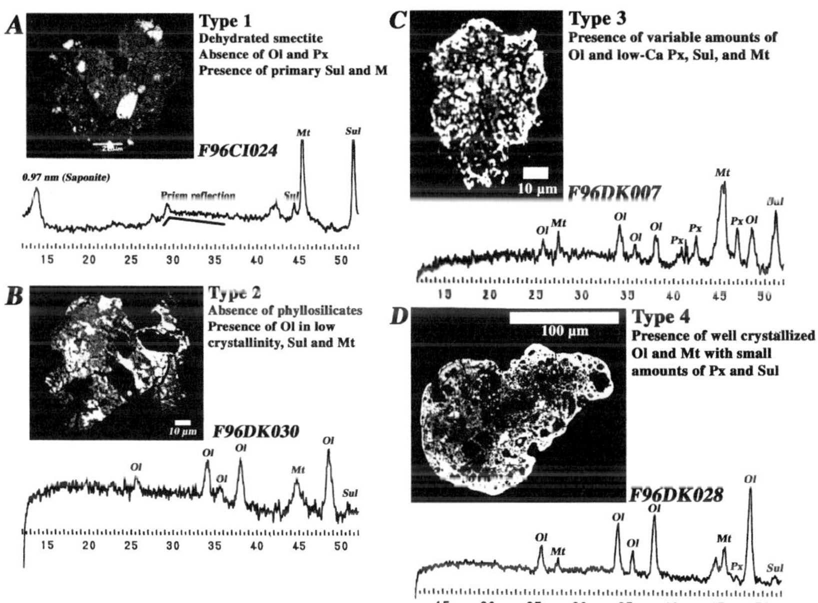

Figure 1. Backscattered electron images of MMs with XRD charts of them. These MMs show four types of typical mineral assemblages in MMs. Abbreviations: Ol: olivine, Px: pyroxene, Mt: magnetite, Sul: Fe-sulfide.

contain Mg- and Fe-oxide (magnesiowüstite); the first discovery of magnesiowüstite in MMs.

2.2 TEM observation of MMs

Only one phyllosilicate-bearing MM (type 1) was found among 83 MMs collected in 1996 [7]. It contains abundant saponite. Its mineralogy is similar to the phyllosilicate-bearing MM previously reported [4]. However, this MM contains abundant aggregates of magnesiowüstite (hereafter, MW) and Fe sulfide (Figure 2). The aggregates are surrounded by bundles of saponite. Most of the MW grains in such aggregates are < 20 nm across. During the preliminary study of MMs collected in 1998-1999, another saponite-rich MM was found. It also contains MW-bearing aggregates. As well as the absence of serpentine, MW-bearing aggregates may be also an unique feature of saponite-rich MMs. Although whether such aggregates are primary constituents or not, it is more plausible that they were formed from minerals that decomposed during weak atmospheric entry heating. MW in the recently found MM contains Mn as well as Mg and Fe. Its composition suggests that MW in this MM was formed from Mg, Fe, Mn-bearing minerals such as Mg, Fe, Mn-bearing carbonates found in CI chondrites [10].

As described in 2.1, MW is often found in MMs that experienced relatively weak heating (type 3). TEM observation revealed that MW coexists with low-Ca pyroxene intimately (Figure 3). These minerals have typical morphology of recrystallization. Because these minerals are absent in the peripheries of MMs, where abundant anhydrous minerals are olivine and magnetite, mixture of MW and low-Ca pyroxene may have been formed from phyllosilicates under slightly reducing conditions during atmospheric entry.

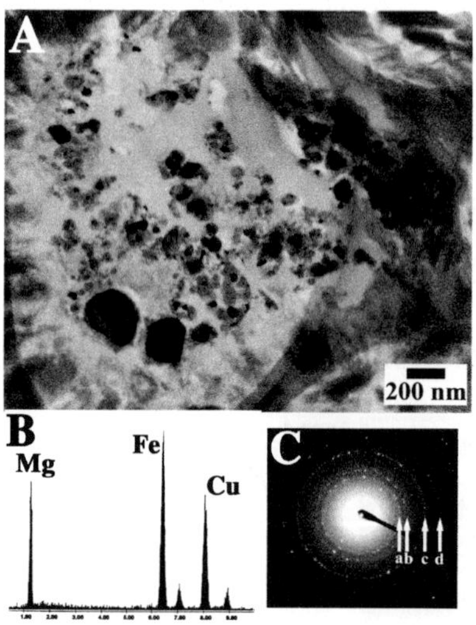

Figure 2. (A) A TEM photograph of a MW and Fe-sulfide aggregate in a smectite-rich type 1 MM. (B) MW contains only Mg and Fe peaks. (Cu peaks from the supporting grid) (C) SAED pattern showing diffraction rings from MW.

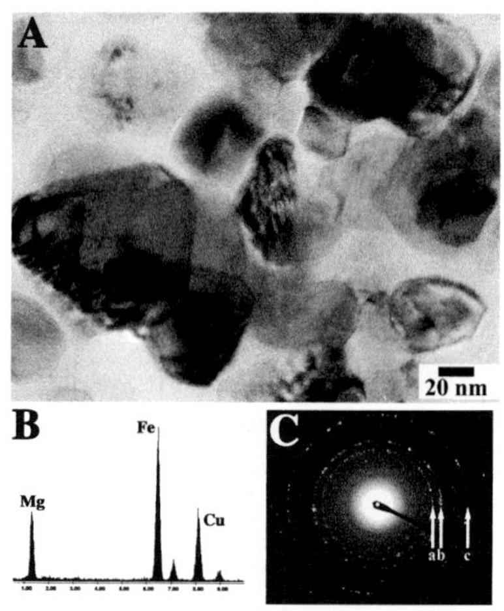

Figure 3. A TEM photomicrograph of MW in a moderately heated type 3 MM.. MW in them coexists with low-Ca pyroxene. Grain boundaries of these minerals show typical texture of recrystallization. Their EDS spectrum and SAED pattern indicate that the crystals are MW.

2.3 Noble gas analysis

Noble gas isotopic compositions in fine particles including individual MMs were analyzed by stepped pyrolysis [10]. Figure 4 shows a three-isotope diagram of Ne in MMs collected in 1996. The high ratio of solar/cosmogenic gas in MMs suggests that they had been small particles in space to have wide surface areas exposed to solar winds, being consistent with the results of previous works [9-14]. These MMs are not particles produced by breaking-up of hydrous carbonaceous chondrites during the atmospheric entry; otherwise a set of such particles must have Ne isotopic ratio close to an average value of hydrous carbonaceous chondrites enriched in cosmogenic gases. Based on these noble-gas compositions, they emerged into interplanetary space less than a few million years ago as individual particles and came to the earth in 1950-1980.

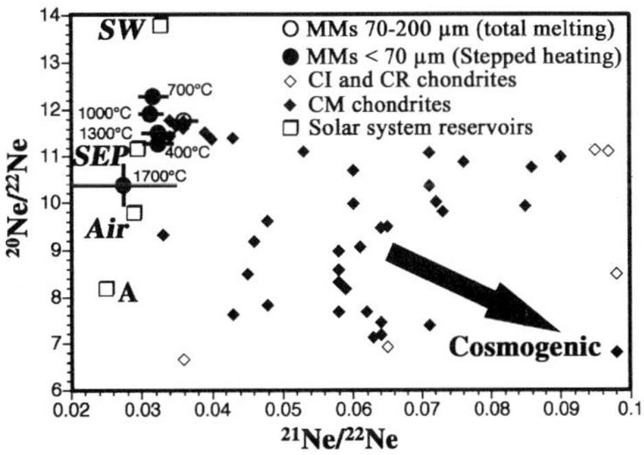

Figure 4. Three isotope diagram of Ne.

3. CONCLUSIONS

The JARE teams have been collecting MMs in Antarctica since 1996. The consortium studies of these MMs revealed that MW is common among unmelted MMs and that mixtures of MW and low-Ca pyroxene is characteristic of MMs that have experienced moderate atmospheric entry heating. In addition, noble-gas compositions revealed that MMs were in interplanetary space for < a few Ma as individual particles and came to the Earth in 1950-80.

REFERENCES

1. T. Nakamura et al., Antarct. Meteorite Res. 12 (1999) 183.
2. T. Noguchi et al., Antarct. Meteorite Res. 13 (2000) 270.
3. T. Yada and H. Kojima, Antarct. Meteorite Res. 13 (2000) 9.
4. W. Klöck and F.J. Stadermann, in Analysis of Interplanetary Dust (eds. M.E. Zolensky, T.L. Wilson, F.J.M. Rietmeijer and G.J. Flynn, AIP Press (1994) 51. .
5. A. Greshcke et al., Meteoritics Planet. Sci. 33 (1998), 267.
6. T. Nakamura et al., Papers presented to the 24th Symp. Antarct. Meteorites (1999) 125.
7. T. Noguchi and T. Nakamura, Antarct. Meteorite Res. 13 (2000) 285.
8. I. Nakai et al., Antarct. Meteorite Res. 13 (2000) 285.
9. T. Osawa et al., Antarct. Meteorite Res. 13 (2000) 322.
10. T. Nakamura and N. Takaoka, Antarct. Meteorite Res. 13 (2000) 311.
11. M. Zolensky and H.Y. McSween, in Meteorites and the Early Solar System (1987) 114.
12. C.T. Olinger, M. Maurette, R.M. Walker and C.M. Hohenberg, Earth Planet Sci. Lett. 100 (1990) 77.
13. A. Nier and D.J. Schlutter, Meteoritics 25 (1990) 263.
14. M. Maurette, C. Olinger, M.C. Michellevy, G. Kurat, M. Pourchet, F. Brandstatter and M. Bourotdenise, Nature 351 (1991) 44

The possibility of abiogenic synthesis of complex biochemical compounds on surfaces of dust particles.

Michael B. Simakov and Eugenia A. Kuzicheva

Institute of Cytology, Russian Academy of Sciences, St.Petersburg 194064, RUSSIA

The small bodies in the Solar system are rich in organic compounds. Different classes of organic molecules, many of which have common biochemical significance, such as amino acids, their precursors and nucleic acids bases have been identified in comets, meteorites and interplanetary dust particles. It is important to test how far the processes of the chemical evolution could develop on the surface of these bodies under the action of different open space energy sources. The abiogenic synthesis of nucleotides and dipeptides in solid films has been investigated under the action of VUV (λ=145 nm) and high energy protons. In order to test the processes of chemical evolution in open space, the investigated mixtures are exposed to space conditions on board satellite "BION-11". The abiogenic synthesis of nucleotides is also observed. Our results indicate that such complex compounds as nucleotides and dipeptides could be synthesized in the early stages of Solar system evolution and safely transported to Earth by comets, meteorites and IDP, where they could take a place in prebiological evolution.

1. INTRODUCTION

The problem of life origins is a very intriguing question in contemporary science. Geological data indicate that complex photosynthetic organisms, possibly related to cyanobacteria, existed 3.5 billion years (Gyr) ago [1]. Moreover, there is evidence for biological mediated carbon isotope fractionation in the 3.85 Gyr old Isua metasediments, which indicates that the photoautotrophy biochemical processes must have existed at that time [2].

On the other hand, it is stated that impacts by meteorites and comets at an early stage of the Earth's history until 3.8-4.2 Gyr ago would have made our planet uninhabitable [3,4]. So we can see, that life arose rapidly on Earth once conditions became clement and this process could be very rapid, almost instantaneous on the geological time scale. Only 300 million years appear to be left for terrestrial chemical evolution to lead to primitive, yet fully developed organisms. There is a very short time interval for production of the biochemical compounds in the atmosphere and on the surface of planet, but this kind of matter's evolution had to occur before the biological one. Besides, geological data and theoretical models point to the Earth's early atmosphere to be composed mainly of carbon dioxide, carbon monoxide, nitrogen and water vapor [5,6]. It is not easy to form biologically important compounds such as components of peptides and nucleic acids from these gases by any energy sources.

We can suggest that the chemical evolution which had proceeded in the early stages of the Solar system origin, would lead to the appearance of very complex organic compounds. These compounds have been delivered to early Earth at the stage of the "late heavy bombardment" and later [7]. So, the highly complex organic molecules have been proposed as a starting material for early biological activity on Earth, or on other planets in the Universe [8,9].

The main attention in experimental work is now devoted to energetic chemistry in "simulated space ice conditions", such as ice mantles of interstellar dust particles (IDP) [10,11] and cometary ices [12]. These laboratory and theoretical studies were stimulated by related astronomical observations showing the signature of complex carbonaceous material in circumstellar and interstellar environments [13]. The laboratory experiments yielded some interesting results. Kobayashi with co-workers [14] performed a proton irradiation experiment using a gas mixture of CO, N_2 and H_2O and found a wide variety of amino acids with high yield in irradiating products after acid-hydrolysis. A mixture of a source of carbon (CO, CH_4 or C_3H_8), H_2O and NH_3 after irradiation with high energy protons also yielded several amino acids [15]. The another "simulated ISD ices", ice mixtures of methanol, ammonia and water also yielded amino acid precursors after irradiation with high energy protons [16]. One of the RNA bases, uracil, was identified in the product after irradiation of a mixture CO, N_2 and H_2O by high energy protons [17]. However, the reactions of complification of organic compounds could also take place on the solid surfaces of small bodies in the Solar system, such as asteroids, meteorites, IDP and satellites of other planets (Mars, Jupiter, Saturn). These bodies have tremendous total surface area and are subjected to irradiation by ultraviolet photons, the solar wind, solar flare ions and galactic or extra-galactic cosmic rays.

In space most of the chemical evolution toward complex molecules takes place in the solid phase. The aim of our work is to understand better, using laboratory simulation and spacecraft opportunities, the chemical evolution of biologically important compounds on the surfaces of small bodies in the Solar system. We present laboratory data relevant to ultraviolet irradiation and cosmic ray bombardment of the solid organic films in such environments. In our experiment in laboratory and on board of the space satellite "BION-11" we have planned to establish how far the chemical evolution could develop on the surfaces of small bodies under the action of open space energy sources.

2. EXPERIMENTAL RESULTS

We investigated two types of reactions:

(1) Abiogenic synthesis of nucleotides from mixtures of nucleoside + inorganic phosphate;

(2) Abiogenic synthesis of dipeptides from mixtures of simple amino acids.

The reaction mixture in the form of a solid film contains nucleoside and dihydrogen phosphate or two different amino acids. Seven different nucleosides (thymidine, cytidine, uracil, adenosine or deoxyadenosine, guanosine or deoxyguanosine) and inorganic phosphate and four mixtures of aromatic (tyrosine or triptophan) and aliphatic (glycin or alanine) amino acids are investigated. Mixtures were prepared by air drying of aqueous solutions and irradiated by different sources of energy:

(1) VUV-light of 145 nm;

(2) high energy protons (2-6 MeV); and

(3) were installed on the surface of biosputnik in outstanding container (OC) when they were exposed to the action of all spectra of the open space energy sources during the entire time of flight - 327 hours (24.12.1996 - 07.01.1997).

As a result of VUV irradiation of the solid mixture of nucleoside and inorganic phosphate the natural monophosphates of corresponding nucleosides were found [18,19]. The main products were nucleoside-5'-monophosphates and some amount of by-products (2'- and 3'-monophosphates, 2'3'- and 3'5'-cyclomonophosphates). The yields of products were small, less than a few per cent, however the effectiveness of the abiogenic synthesis on VUV irradiation is higher than on UV-radiation and heat. When investigated films were irradiated with protons, the full mixture of nucleotides also was found. The maximal summary yield of nucleotides was 9.43% (under action of 2 MeV protons). The main product was 5'AMP (3.19%) as under action of VUV. Preferable 5'-monophosphate formation is indicative of more advantageous spatial position of 5'hydroxyl group in carbohydrate residue in comparison with 2' and 3' hydroxyl groups. The 2'3'-cyclomonophosphate has been formed in solid films as well. It is interesting to note the yield of 2'3'cAMP has values more than those of 2'- and 3'-monophosphates. In our previous papers we had proposed that interaction between all three nucleotides exists under VUV radiation [18,19] and g-radiation [20]. Kinetic analyses carried out in a recent paper [21] suggested that 2'3'-cyclomonophosphate formation arose from 3'- or 2'-nucleoside derivatives. In this work it was also shown that the 5'-OH site is inherently favored over the 3'- or 2'-OH in phosphorylation. We can see that 5'-nucleotides are predominantly formed under action of all investigated sources of energy as well as in open space energy. Other nucleotides (2'and 3') and cyclic nucleotides, such as 2'3'cAMP and 2'3'cCMP also have been found in the all investigated films.

The films containing a mixture of amino acids yielded various dipeptides after they were exposed to VUV-radiation and protons [22,23]. Polymerization is an essential step in prebiological evolution and we have shown that this process probably could take place even at early stages of Solar System formation, before planet accretion, on surface of small bodies.

It was important to test how far the process of chemical evolution could take place on the surface of space bodies under the action of all energy sources of open space. In space flight experiment on board the "BION-11" satellite the solid films from mixtures of different nucleosides and inorganic phosphate are exposed to space conditions. The abiogenic synthesis of such nucleotides as 5'-, 2'- and 3'- monophosphates of thymidine, cytidine, adenosine and deoxyadenosine is observed. The cyclomonophosphates are synthesized also [24].

The main products of the reactions in all experiments were 5'monophosphates of corresponding nucleosides as in the laboratory investigations.

3. PREBIOLOGICAL SIGNIFICANCE

The results of our laboratory experiments on the action of UV radiation with different wavelengths (145-260 nm), high-energy particles, g-radiation and heat show that many kinds of open space energy could be effective in the chemical evolution of nucleic acid precursors and oligopeptides [18-20,22,23]. The abiogenic synthesis of nucleotides and dipeptides can proceed in open space conditions in the early stages of Solar System evolution. The results of experiments on board "BION-11" enable us to conclude, that 5'-nucleotides are predominantly formed by the action of the full spectrum of space energy sources on dry thin films of nucleoside plus inorganic phosphate. Other nucleotides also have been found in the investigated films.

Of course, an irradiation of a few hours in a laboratory or a week of spaceflight cannot reproduce billions of years of slow evolution on the surfaces of small bodies that will never be reproduced in the laboratory. However, they might shed light on the basic tendency of the course of the chemical evolution.

So, the organic compounds, which had been delivered on the primordial Earth would have had a very complex structure and reached the second stage of the chemical evolution — polymerization. The small bodies could have imported prebiotic, complex molecules such as nucleotides and olygopeptides, to early Earth, which have allowed the fast synthesis of all molecules thought to be necessary for the origin of life. As shown by Basiuk and Douda [25] the safe delivery of these compounds depends on such factors as size of the space body, nature and altitude of the heating and it's chemical composition, and may be very successful in some cases. The comets, carbonaceous chondrites and other primitive bodies of the Solar System including interplanetary and interstellar dust particles are responsible for essentially all carbon compounds and water currently present in our planet [25].

REFERENCES

1. J.W. Schopf, Science 260 (1993) 640.
2. S.J. Mojzsis, G. Arrhenius, K.D. McKeegan, T.M. Harrison, A.P. Nutman and C.R.L. Friend, Nature 384 (1996) 55.
3. K.A. Maher and D.J. Stevenson, Nature 331 (1988) 612.
4. C.F. Chyba, Geochim.Cosmochim.Acta 57 (1993) 3351.
5. H.D. Holland, The chemical evolution of the atmosphere and oceans. Princeton Univ. Press. (1984)
6. J.F. Kasting, Science 259 (1993) 920.
7. E. Anders, Nature 342 (1989) 255.
8. J. Oro, T. Mills and A. Lazcano. Orig. Life Evol. Biosph. 21 (1992) 267
9. G.J. Flynn, Earth, Moon and Planets 72 (1996) 469.
10. L.J. Allamandola, S.A. Sandford and G.J. Valero, Icarus 76 (1988) 225.
11. J.M. Greenberg, W.A. Schutte and A. Li, Adv. Space Res. 23 (1999) 289.
12. R.L. Hudson and M.H. Moore. Icarus 140 (1999) 451.
13. W.M. Irvine, Orig. Life Evol. Biosph. 28 (1998) 365.
14. K. Kobayashi, T. Kaneko and T. Saito, Adv. Space Res. 24 (1999) 461.
15. K. Kobayashi, T. Kasamatsu T. Kaneko, J.Koike, T. Oshima, T. Saito, T. Yamamoto and H. Yanagawa, Adv.Space Res., 16 (1995) 21.
16. T. Kasamatsu, T. Kaneko, T. Saito and K. Kobayashi, Bull.Chem.Soc.Jpn. 70 (1997) 1021.
17. K. Kobayashi and T. Tsuji, Chemistry Letters (1997) 903.
18. E.A. Kuzicheva, M.B. Simakov, I.L. Mal'ko, N.Y. Dodonova and N.B. Gontareva, Adv.Space Res. 18 (1996) 65.
19. M.B. Simakov, E.A. Kuzicheva, and I.L. Malko, Adv.Space Res. 19 (1997) 1059.
20. E.A. Kuzicheva, V.F. Zhukova M.B. Simakov, Zhurn. Evol. Biokh. i Fisiol. 29 (1993) 339.
21. R. Reimann and G. Zubay, Orig. Life Evol. Biosph. 29 (1999) 229.
22. M.B. Simakov, E.A. Kuzicheva, I.L. Mal'ko and N.Y. Dodonova, Adv.Space Res. 18 (1996) 61.
23. M.B. Simakov, E.A. Kuzicheva, N.Y. Dodonova and A.E. Antropov, Adv.Space Res. 19 (1997) 1063.
24. E.A. Kuzicheva and M.B. Simakov, Adv.Space Res. 23 (1999) 387.
25. V.A. Basiuk and J. Douda, Planet.Space Sci. 47 (1999) 577.
26. A. Delsemme, Icarus 146 (2000) 313.

Microanalysis of cosmic dust – Prospects and challenges.

G.A. Graham[a], A.T. Kearsley[b], M.J. Burchell[c], J.A. Creighton[d] and I.P. Wright[a].

[a]Planetary and Space Science Research Institute, The Open University, Walton Hall, Milton Keynes MK7 6AA, UK.

[b]Space Science Research, School of BMS, Oxford Brookes University, Headington, Oxford OX3 0BP, UK.

[c]Unit for Space Sciences & Astrophysics, School of Physical Sciences, University of Kent, Canterbury CT2 7NR, UK.

[d]Chemistry, School of Physical Sciences, University of Kent, Canterbury CT2 7NR, UK.

Our laboratory experiments have used a light gas gun to accelerate olivine mineral grains to 5.1 km/s prior to their impact upon, and capture within, aerogel (density 96 kg m^{-3}). The composition and mineralogy of the trapped particles were investigated using the non-destructive techniques of raman spectroscopy and analytical scanning electron microscopy. The raman spectra obtained for an impacted olivine show wider peak width than those obtained from the analysis of a pristine grain, this may indicate that the particles do undergo a degree of alteration at the crystallographic scale during the impact event.

1. INTRODUCTION

Of the 15,000-60,000 tons of extraterrestrial material that bombards the Earth per annum, 95% is in the sub-micron size range, e.g. [1]. This cosmic dust may be a more representative sample of small body material in the solar system than are meteorites. The possible origins of the dust may include comets, asteroids and interstellar dust etc., e.g. [2]. Many of the laboratory investigations looking at dust particles have been based on material collected from a number of terrestrial locations: i.) deep-sea spherules collected from the sea-floor [3], ii.) micrometeorites collected from the Antarctic Ice Sheets [4] and iii.) interplanetary dust particles (IDPs) collected from the stratosphere [5]. These collections have been extensively investigated over the past 30 years using a wide range of analytical techniques [6]. Yet, whilst substantial information has been obtained in terms of mineralogical and chemical compositions and processes within the particles, the transit through the Earth's atmosphere does make them susceptible to selection and modification process, e.g. [7]. Therefore, it is almost impossible to unambiguously state the parent body origin of the particles. However, in this first decade of the twenty first century, we should see samples returned from missions that have been purpose-built to collect particles from examples of the two most likely sources of dust: i.) NASA STARDUST mission to Comet Wild 2 [8] and ii.) The Japanese MUSES-C mission to an asteroid [9]. This paper will discuss techniques that might be applicable to the particles that we hope will be returned in 2006 by the STARDUST mission.

Capturing intact particles from a comet will not be an easy task as the impact encounter velocities will be ≥5 km/s. Studies of particles sampled in low Earth orbit (LEO) at similar velocities using a variety of dedicated collector surfaces, e.g. [10] and non-dedicated surfaces, e.g. [11] have shown that the particles undergo extreme shock and melting and are generally highly altered, if not destroyed completely. To overcome these problems and to collect near-intact and pristine particles, STARDUST is using silica aerogel [12], a low-density material. Previous experimental work, e.g. [13] and [14] has shown the suitability of the aerogel as a capture cell and the technology has already been used to collect particles in low Earth orbit, e.g. [14] and [15].

Previous studies, e.g. [15], have focused on the interpretations of the physical characteristics of the impacted particle and the associated features in the aerogel generated by the hypervelocity impact event, e.g. track length and morphology [14]. In this paper we present chemical analyses and proxy crystallographic information from olivine particles captured in aerogel in the laboratory.

2. EXPERIMENTS

The laboratory experiments were carried out using a two stage light-gas-gun at the University of Kent. The aerogel targets for the study had a density of 96 kg m^{-3}. The olivine projectiles (125-250µm diameter grains from the coarse Admire pallasite, approximate homogeneous composition:- forsterite 88 : fayalite 12) were selected on the basis that mafic silicates are one of the major mineral components of cosmic dust particles. The projectiles were sabot mounted and accelerated in the aerogel targets at approximately 5.1km/s using a buck-shot technique.

The raman spectroscopy was carried out on an Instruments SA model HR640 spectrograph with a liquid nitrogen cooled CCD detector, linked to an Instruments SA Raman scattering module based on an Olympus BX40 microscope. The analytical conditions for the work were: a spot size of approximately 25µm, and a He-Ne laser (wavelength 632.8 nm) was used at 20 mW power.

The electron microscopy was carried out using a Jeol 840 scanning electron microscope at 2nA beam current and at 20kV accelerating voltage. The samples were carbon-coated to reduce the effects of electrical charging during SEM investigations. The X-ray elemental maps and energy dispersive X-ray spectra (EDS) were acquired using an Oxford Instruments eXL energy-dispersive spectrometer microanalyser (EDS) with a Pentafet detector fitted with an ultra-thin window (this allows the detection of X-rays from light elements – e.g. carbon). A full description of the analytical protocol for this type of work is given in [16].

3. DISCUSSION

When the STARDUST samples are returned to Earth, one of the first fundamental tasks will be to identify the bulk chemical composition of the particles. This can be carried out using a number of analytical techniques – see [6] for a comprehensive review. The choice of which technique to apply is clearly important, for example whilst analytical SEM and electron microprobe analysis are the conventional techniques in many meteoritic investigations, the samples require a conductive coating (typically carbon or gold). Coating the samples with carbon would potentially contaminate them to the extent that highly specialised analytical techniques such as carbon stable isotopic mass spectrometry [17] can not be carried out reliably. The chemical information derived from X-ray microanalysis may give strong clues as to the likely composition of the particle, but is unlikely to unambiguously reveal the full mineralogical identity, as this also requires crystallographic parameters to be determined. X-ray microanalysis is also unable to differentiate materials of similar composition but widely differing states of shock deformation. Therefore, it is desirable to use a technique that is non-destructive, does not contaminate the sample, can provide some crystallographic information by proxy and has a small spatial sampling area, for example: raman spectroscopy. This technique obtains a vibrational spectrum of the crystal lattice of the sample by illuminating it with a laser and collecting the spectrum of inelastically scattered photons. The detected spectrum contains a suite of peaks whose identity and morphology are characteristic of the chemical structure of the sample. In other words raman spectroscopy is a chemical analysis technical rather than an elemental one, it asks "*what is it?*" rather than "*what is it made of?*". As each component of the sample that is illuminated by the laser may contribute to the spectrum seen, it is fortunate that the aerogel does not significantly absorb or scatter the laser light. Consequently, only the selected impacted particle material is analysed.

The olivine grains trapped in the aerogel following a light-gas-gun shot were subjected to this technique and dispersive raman spectra were obtained (figure 1). The raman spectrum obtained for the impacted olivine showed a wider peak width than that obtained from the analysis of a pristine grain. This may indicate that the particle has undergone a degree of alteration at the crystallographic scale. A previous investigation [18] studied the thermal and mechanical alteration that particles impacted into aerogel targets may have undergone.

Whilst raman spectroscopy does appear to be a very good technique to carry out *in-situ* mineralogical analysis of the trapped particles, a separate study [19] has suggested that during analysis the there can be small-scale localised thermal alterations to the examined particle. This observation clearly requires further work to determine the extent of damage.

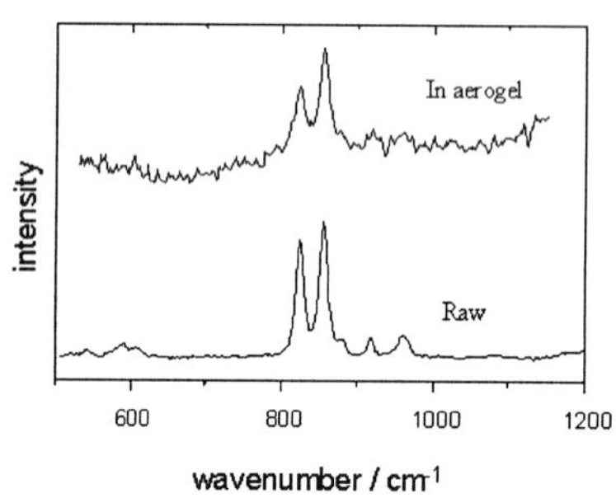

Figure 1. The raman spectra obtained for an olivine grain captured in aerogel and a unaltered grain.

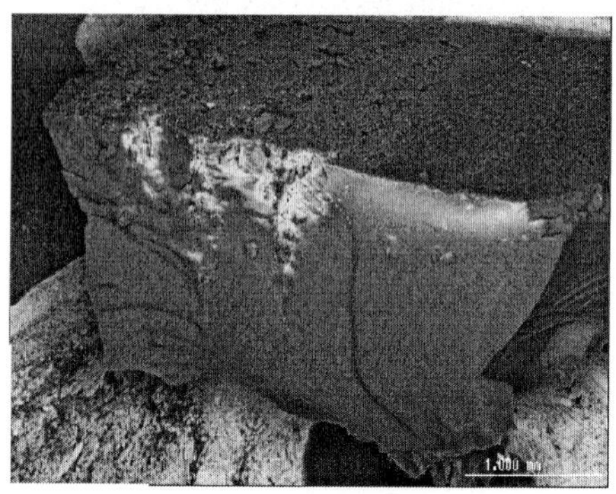

Figure 2. Secondary electron image of 3 tracks generated in a block of aerogel by olivine projectiles.

After the raman spectroscopic analysis, trapped grains were examined in a SEM, using a combination of secondary electron imaging (SEI), back-scattered electron imaging (BEI) and X-ray elemental microanalysis. The key goal of the SEM study was to locate the position of the particles within tracks, and if possible examine their surface morphology. To enable the detailed examination of the trapped grains the aerogel block was cut so that the penetrating track could be seen at, or very close to, the surface (figure 2). It is possible to see that there were at least three major tracks generated in this small block. It appears that in one of the tracks the impacting particle had exploded on penetration and fine debris was dispersed along the length of the track – obviously this particle was not 'intact' following capture.

As the aerogel substrate has a very simple bulk composition, i.e. Si K-alpha dominates EDS spectra of pure aerogel, the best technique to locate extraneous impactor debris is X-ray elemental mapping for Mg and Fe K-line emission. This can rapidly define the position of the olivine grains. In our study several individual grains were located by this method (figure 3) and, interestingly, it was observed that the grains typically had a layer of aerogel adhering to their surface. EDS spot analysis of the individual grains produced a spectrum typical of an olivine (figure 4). From this very preliminary investigation it is not possible to determine whether the olivine grains have undergone any shock or crystallographic alteration as a result of the impact event. It is planned that the analytical work will be extended to include the use of transmission electron microscopy, in which it will be possible to examine the crystallographic structure of grains extracted from the aerogel.

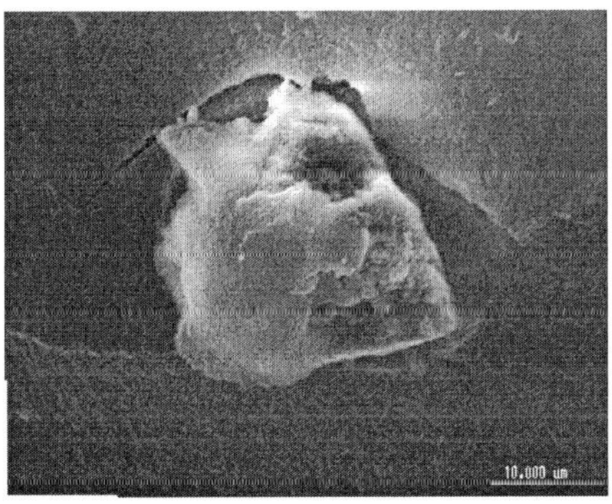

Figure 3. Secondary electron image of an individual olivine grain captured in the aerogel.

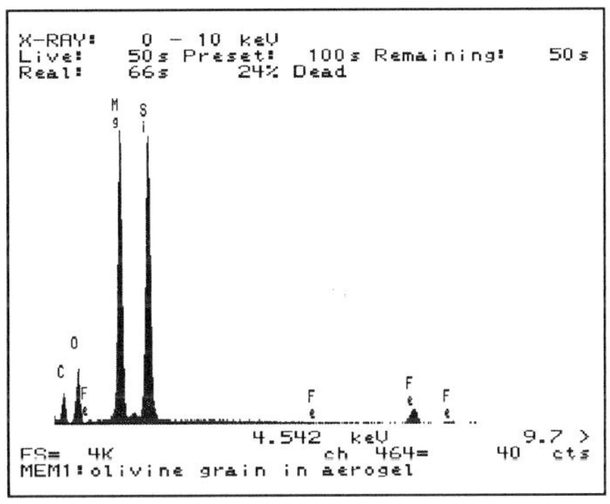

Figure 4. The energy-dispersive spectrum obtained for the olivine grain.

4. CONCLUSIONS

The study of olivine grains impacted into aerogel has offered the opportunity to assess the suitability of raman spectroscopy as a basic non-destructive technique that allows the identification of mineral composition. The raman spectra obtained for an impacted olivine show a wider peak width than that obtained from the analysis of a pristine grain. This may indicate that the particles do undergo a degree of alteration at the crystallographic scale. The study has also demonstrated that detailed X-ray microanalysis of the trapped grains is feasible using an analytical SEM. X-ray elemental mapping has been particularly useful in locating grains within the tracks of the impacted aerogel block. The study will now be extended to include more minerals and complex meteorite powders. STARDUST will make available some of the most scientifically significant material to be investigated in the laboratory since the return of lunar samples from the Apollo missions. It is therefore clearly important for the cosmic dust community to prepare for the prospects and challenges that await.

REFERENCES

1. S.G. Love and D.E. Brownlee, Science, 265 (1993) 550.
2. D.E. Brownlee, in Analysis of Interplanetary Dust Particles (eds. M.E. Zolensky, T.L. Wilson, F.J.M. Rietmeijer and G. Flynn) AIP Conf. Proc. 310, AIP Press, Woodbury New York (1994) 5.
3. D.E. Brownlee, L.B. Pilachowski and P.W. Hodge, LPSC X, (1979) 157.
4. M. Maurette, C. Olinger, M. Christophe, G. Kurat, M. Pourchet, F. Brandstätter and M. Buorot-Denise, Nature, 351 (1991) 44.
5. J.P.Bradley, M.S. Germani and D.E. Brownlee, Earth Planet. Sci. Lett., 93 (1989) 1.
6. M.E. Zolensky, C. Pieters, B. Clark and J.J. Papike, Planet. Sci. 35 (2000) 9.
7. S.G. Love and D.E. Brownlee, Icarus 89 (1991) 26.
8. D.E. Brownlee, P. Tsou, D. Burnett, B. Clark, M.S. Hanner, F. Horz, J. Kissel, J.A.M. McDonnell, R.L. Newburn, S. Sandford, Z. Sekanina, A.J. Tuzzolino and M. Zolensky, Meteorit. Planet. Sci., 32, (abstract) (1997) A22.
9. H. Yano, Earth Planets Space 51 (1999) 1233.
10. R.P. Bernhard, T.H. See and F. Hörz, in LDEF 69 Months in Space - 2nd Post-Retrieval Symp., NASA CP-3194 (1993) 551.
11. G.A. Graham A.T. Kearsley, M.M. Grady, I.P. Wright and J.A.M. McDonnell, Adv. Space Res. 25 (2000) 303.
12. P. Tsou, J. Non-Crystalline Solids 186 (1995) 415.
13. P. Tsou, D.E. Brownlee, M.R. Lawrence, L. Hrubesh and A.L.Albee, LPSC XIX (1988) 1205.
14. F. Hörz, G. Cress, M.E. Zolensky, T.H. See, R.P. Bernhard and J.L. Warren, Optical analysis of impact features in aerogel from the Orbital Debris Collection Experiment on the Mir station. NASA/TM-1999-209372 (1999).
15. M.J. Burchell, R. Thomson and H. Yano, Planet. Space Sci. 47 (1999) 186.
16. G.A. Graham, A.T. Kearsley, M.M. Grady and I.P. Wright, Proc. 2nd European Conf. On Space Debris, ESA SP-393 (1997) 183.
17. G.A. Graham, I.P. Wright, M.M. Grady, M Perreau, M. Maurette and C.T. Pillinger, Meteoritics & Planet. Sci. 31 (abstract) (1996) A53.
18. T.E. Bunch, P. Schulz, P. Cassen, D. Brownlee, M. Podolak, J. Lissaur, R. Reynolds and S. Chang, Icarus 91 (1991) 76.
19. D.E. Brownlee, personal Communication (2000).

Index

AUTHOR INDEX

KEYWORD INDEX

Page numbers refer to start of appropriate paper(s)